Current Progress in Atmospheric Science

Current Progress in Atmospheric Science

Editor: Evie Hughes

www.statesacademicpress.com

States Academic Press,
109 South 5th Street,
Brooklyn, NY 11249, USA

Visit us on the World Wide Web at:
www.statesacademicpress.com

ISBN: 978-1-63989-132-0 (Hardback)

Cataloging-in-Publication Data

Current progress in atmospheric science / edited by Evie Hughes.
    p. cm.
Includes bibliographical references and index.
ISBN 978-1-63989-132-0
1. Atmospheric science. 2. Atmosphere. 3. Earth sciences. I. Hughes, Evie.
QC861.3 .C87 2022
551.5--dc23

Table of Contents

Permissions

List of Contributors

Index

Preface

The study of physics, chemistry, and dynamics of the Earth's atmosphere is known as atmospheric science. Atmospheric chemistry refers to the study of the chemistry of Earth and other planets' atmosphere. Atmospheric physics is the branch of atmospheric science in which physics is applied to the earth's atmosphere. Atmospheric dynamics refers to the study of motion systems of meteorological importance, integrating observations at multiple locations, times and theories. The atmospheric sciences are conventionally divided into three topical areas of meteorology, climatology and aeronomy. This book is a valuable compilation of topics, ranging from the basic to the most complex advancements in the field of atmospheric science. It aims to shed light on some of the unexplored aspects of atmospheric science and the recent researches in this field. As this field is emerging at a rapid pace, the contents of this book will help the readers understand the modern concepts and applications of the subject.

This book unites the global concepts and researches in an organized manner for a comprehensive understanding of the subject. It is a ripe text for all researchers, students, scientists or anyone else who is interested in acquiring a better knowledge of this dynamic field.

I extend my sincere thanks to the contributors for such eloquent research chapters. Finally, I thank my family for being a source of support and help.

Editor

An observation campaign of precipitable water vapor with multiple GPS receivers in western Java, Indonesia

Eugenio Realini[1,4]*, Kazutoshi Sato[1,5], Toshitaka Tsuda[1], Susilo[2] and Timbul Manik[3]

Abstract

A campaign was conducted from 23 July to 5 August 2010 to measure atmospheric precipitable water vapor (PWV) using five Global Positioning System (GPS) receivers, stationed at four different locations in Jakarta and Bogor, western Java, Indonesia. Radiosondes were launched at an interval of 6 h to validate the GPS-derived PWV data. The validation resulted in a root mean square error of 2 to 3 mm in PWV. The influence of atmospheric pressure and temperature on GPS-derived PWV was evaluated. A regular semi-diurnal pressure oscillation was observed, showing an amplitude ranging from 3 to 5 hPa, which corresponds to 1.1 to 1.8 mm in PWV. A nocturnal temperature inversion layer was observed in the radiosonde profiles, which resulted in an error of about 0.5 mm in PWV. From 26 to 29 July, there was a passage of distributed rain clouds over western Java, moving southwestward from the equator toward the Indian Ocean. A second precipitation event, with scattered rain clouds forming locally near Bogor, occurred on 2 August. Both events were observed also by a C-band Doppler Radar operated near Jakarta. The highest peak of GPS-derived PWV (about 67 mm) registered during the campaign occurred on 27 July, coinciding with the distributed rainfall event. Spatial variations in the estimated PWV between the four sites were enhanced before both the analyzed rainfall events, on 27 July and 2 August. Peaks in the temporal variability of PWV were also observed in conjunction with the two events. The results indicated a relation between the space-time inhomogeneity of GPS-PWV and rainfall events in the tropics.

Keywords: GPS; Radiosonde; PWV; Indonesia

Background

Accurate prediction of local heavy rainfall in Indonesia is difficult with current mesoscale numerical prediction models, partly because of the lack of a meteorological observation network. Global Navigation Satellite Systems (GNSS), which include the U.S. Global Positioning System (GPS), are widely used for civilian applications, including scientific studies and atmospheric monitoring. Bevis et al. (1992) initiated the interdisciplinary scientific field of GNSS meteorology, which derives the vertically integrated amount of water vapor in a column of atmosphere, i.e., the precipitable water vapor (PWV), from the analysis of tropospheric delays in GNSS signals.

The number of continuously operated GNSS stations has rapidly increased in Indonesia recently, up to the current 100 (IUGG 2011). These stations, operated by the Indonesian Geospatial Information Agency (Badan Informasi Geospatial (BIG)), provide the means for continuous monitoring of PWV in the Indonesian region and constitute a distributed GNSS observation network uniquely and usefully located in the tropical region. GPS-derived PWV measurements were earlier used for the study of the meteorological disturbances that occur in some areas of the tropics. For example, Wu et al. (2003) found a distinct diurnal variation in PWV from GPS measurements at a station located in Sumatra Island, Indonesia. PWV estimated by a GPS network installed along the west coast of Sumatra Island was also used to study diurnal variations in tropical convection in relation to the different Madden-Julian oscillation (MJO) phases (Fujita et al. 2011).

*Correspondence: eugenio.realini@g-red.eu
[1] Research Institute for Sustainable Humanosphere (RISH), Kyoto University, Gokasho, Uji 611-0011, Japan
[4] Present address: Geomatics Research & Development (GReD) srl, Como, Italy
Full list of author information is available at the end of the article

The GNSS meteorology technique consists in retrieving the amount of PWV in the column of air above a GNSS station antenna. The presence of atmospheric water vapor along the slant path from each satellite to the station antenna causes a delay in the signal propagation: these delays are used to estimate the delay in the vertical direction (zenith total delay (ZTD)). The ZTD is then separated into its hydrostatic ('dry') and non-hydrostatic ('wet') components, respectively called zenith hydrostatic delay (ZHD) and zenith wet delay (ZWD). Details about this procedure are given in the section 'PWV retrieval method'. The ZWD is converted to PWV by means of the weighted mean temperature of the atmosphere (T_m), which can be calculated from the profiles of temperature and partial pressure of water vapor. However, a simplified relation between the measured surface temperature (T_s) and T_m is often employed, since these profiles are not always available. Bevis et al. (1992) performed a regression analysis between T_s and T_m using radiosonde data collected in North America. Wang et al. (2005) compared and analyzed global estimates of T_m from three different data sets from the year 1997 to 2002. They found that by following the Bevis model, problems in estimating the value of T_m arise because of the erroneously large diurnal atmospheric temperature cycle it produces, owing to the diurnally invariant $T_s - T_m$ relationship and large diurnal variation in surface temperature. They also evidenced a cold bias of T_m derived from the Bevis model in the tropics and subtropics and a warm bias in the middle and high latitudes.

The present study aims to investigate the applicability of the GPS meteorology technique for estimating PWV in and around Jakarta, Indonesia. To this aim, we examined whether the $T_s - T_m$ relation holds true in the case of the Indonesian tropical environment. We should stress here that the main purposes of this study are to analyze the horizontal inhomogeneity of PWV with multiple GPS receivers and to identify its relation to tropical convective rain events. Earlier studies, although not using GPS, analyzed the effects of the trans-equatorial flow and the MJO on torrential rains, respectively on Java Island (Wu et al. 2007) and on western Java Island (Wu et al. 2013). However, to our knowledge, the study presented here is the first attempt to quantitatively investigate the relation between space-time variations of GPS-derived PWV and severe rain events over western Java.

We conducted an observation campaign to monitor the PWV around Jakarta for a period of over 10 days during July to August 2010. A total of 21 radiosondes were launched, and GPS observations from four different sites, located in the west part of the Java Island, were analyzed. During this campaign, active storm clouds passed over the GPS sites, and the consequent time variations in the GPS-derived PWV values were successfully measured. An effort was made to discuss a possible relationship between the time variations in GPS-derived PWV and the rainfall events.

Methods

Experimental setup

The GPS experiment was carried out from 23 July to 5 August 2010, in collaboration with the Indonesian Geospatial Information Agency (BIG), the National Institute of Aeronautics and Space (LAPAN), and the Bandung Institute of Technology (ITB). Figure 1 shows the location of the GPS receiver sites, the radiosonde launch site, the routine weather station that was used in this work, and the location of the C-band Doppler Radar (CDR) that was used to evaluate the distribution of rain clouds over the area of interest. This CDR is installed in Serpong (Yamanaka et al. 2008) and, during the campaign described in this paper, was operated by the Japan Agency for Marine-Earth Science and Technology (JAMSTEC) in collaboration with the Indonesian Agency for the Assessment and Application of Technology (BPPT). In addition to the continuously operated GPS stations of BIG, indicated as CJKT (Jakarta) and CTGR (Tangerang) in Figure 1, we installed two more receivers, one at the LAPAN observatory in Pekayon (PKYN) and one close to the IGS (International GNSS Service) station of BAKO, named BAK2. Note that these last two GPS stations (i.e., BAKO and BAK2) were operated at about 80 m of distance one from the other, within the main office

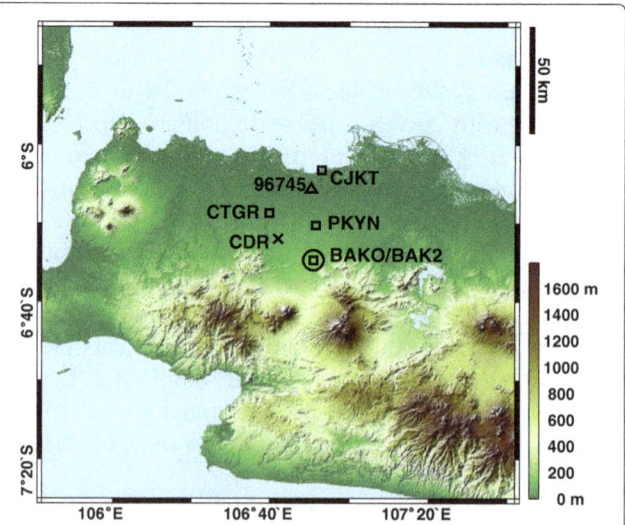

Figure 1 Experimental setup. Spatial distribution of the GPS stations (squares), with marker names CJKT (Jakarta), CTGR (Tangerang), PKYN (Pekayon), BAKO and BAK2 (Bakosurtanal); the BMKG weather monitoring station (triangle), identified by its WMO code; the radiosonde launch site (circle); the C-band Doppler Radar (CDR) at Serpong (cross).

campus of BIG. The extent of the experimental domain was about 40 km (north-south) × 25 km (east-west). BAKO station was chosen as a reference for the comparison described in the section 'Comparison between GPS- and radiosonde-derived PWV' both because of its proximity to the radiosonde launch site (about 100 m) and because it is an official IGS station, therefore subject to routine checks and validation. Observations from all the GPS stations were sampled at 1 s interval. All the GPS sites were located to the north of a mountainous region of Java Island, facing the ocean. The topography shows a gradually decreasing altitude from south to north, with the largest height difference among stations being about 130 m, between BAKO and CJKT. Twenty-one radiosondes (Vaisala RS92-SGPD) were launched from the BIG campus (near BAKO and BAK2 stations), with a time interval of 6 h, at 0030, 0630, 1230 (at times postponed to 1630), and 1830 local time (LT), to obtain detailed vertical profiles up to about 30 km altitude. The average vertical resolution of radiosonde observations was about 10 m. Surface pressure and temperature were continuously monitored only at CTGR, using a Vaisala PTU-300, with a rate of 30 s. Hence, we have also referred to the 3-hourly meteorological records collected by the weather stations of the Meteorological, Climatological and Geophysical Agency of Indonesia (BMKG). Atmospheric surface pressure was also measured at the radiosonde launch point, by means of a calibrated aneroid pressure gauge, used for the ground-check of the radiosonde pressure sensor. Pressure and temperature observations from the BAKO station weather sensor were also made available starting from 2 August.

The GPS observations were processed by using RTNet ver.3.3.0 (Rocken et al. 2001), which estimates the ZTD as a Kalman filter parameter by the precise point positioning (PPP) method (Zumberge et al. 1997). The satellite orbits and clock offsets used for the processing are those provided in the IGS final products. Other settings include a processing rate of 30 s, satellite elevation cutoff of 10°, the dry and wet formulations of the Global Mapping Function (Böhm et al. 2006) for mapping slant delays to the zenith direction, and the GOT4.8 model (Ray 2013) to account for the ocean loading. RTNet was earlier applied for retrieving PWV (or more precisely, for estimating the ZTD that was then converted to PWV), and the results were compared with radiosondes and radiometers (Fujita et al. 2008, Sato et al. 2013), yielding a root mean square error (RMSE) of about 2 mm.

It should be noted that the IGS final products are generally available with a latency of 2 to 3 weeks; therefore, they can be used only for non-real-time analyses like those described in this paper. Setting up a nowcasting system based on GPS/GNSS would require using near real-time

satellite orbits and clocks, which are generally less accurate. However, it must be taken into account that the geodetic community is already tackling the problem of providing near real-time products with higher accuracy, as for example by the IGS Real-Time Service (RTS) (IGS 2012), which is currently being tested also for water vapor retrieval (Li et al. 2014).

PWV retrieval method

The ZHD for each GPS station was calculated by using the Saastamoinen model (Saastamoinen 1973), as

$$\mathrm{ZHD} = 0.0022768\,P\,(1 + 0.00266\,\cos 2\phi + 0.00028\,H)$$
(1)

where P is the surface pressure, ϕ is the latitude of the GPS station antenna, and H is its orthometric height in kilometers. The ground pressure at each station was retrieved as explained in the section 'Pressure analysis and adjustment'. The EGM2008 geoid model (Pavlis et al. 2012) was interpolated at each station location for converting the ellipsoidal height to orthometric height. The zenith wet delay (ZWD) was obtained by subtracting the ZHD from the ZTD, and the amount of precipitable water vapor was then retrieved from the wet delay as

$$\mathrm{PWV} = Q \cdot \mathrm{ZWD} .$$
(2)

The conversion coefficient Q is a function of the weighted mean temperature of the atmosphere T_m, which can be computed on the basis of radiosonde observations (details are provided in the section 'Regression relation between T_s and T_m'); however, since radiosonde launch sites sufficiently close to GPS stations are often not available, T_m is generally modeled as a function of the surface temperature T_s, as described in the next subsection. Q is calculated by the formula proposed by Askne and Nordius (1987):

$$Q = \frac{10^8}{\rho\,R_\mathrm{w}((k_3/T_\mathrm{m}) + k_2')}$$
(3)

where ρ is the density of liquid water (1,000 kg m^{-3}), R_w is the specific gas constant of water vapor (461.5 J kg^{-1} K^{-1}), T_m is a modeled weighted mean temperature of the atmosphere (in K) and $k_2' = k_2 - mk_1$, where m is the ratio of molar masses of water vapor and dry air ($M_\mathrm{w}/M_\mathrm{d} = 0.622$). The atmospheric refractivity constants k_1, k_2, and k_3 used in this work are those proposed by Bevis et al. (1994), yielding $k_2' = 22.1$ K mbar^{-1} and $k_3 = 3.739\,10^5$ K^2 mbar^{-1}.

The amount of PWV was estimated from radiosonde observations by integrating the water vapor density between the ground altitude (h_1) and the altitude at which each radiosonde measured the 99% of the total accumulated water vapor (h_2, ranging from 7 to 9 km), discretized according to the radiosonde altitude steps, i.e.,

$$\text{PWV} = \int_{h_1}^{h_2} \rho_w \, dh = \sum_{i=1}^{n} \rho_{d,i} \, r_i \, \Delta h_i \qquad (4)$$

where ρ_w is the water vapor density, n is the total number of layers between h_1 and h_2, and $\rho_{d,i}$, r_i, Δh_i are respectively the dry air density, the mixing ratio, and the altitude step for layer i. Dropping the index i for the sake of simplicity, the dry air density ρ_d is expressed as

$$\rho_d = \frac{M_d}{R} \frac{P}{T} \qquad (5)$$

with the molar mass of dry air $M_d = 0.0289644$ kg mol^{-1} and the universal gas constant for air $R = 8.31432$ J mol^{-1} K^{-1}. T is the observed air temperature in Kelvin. The mixing ratio r is defined as the dimensionless ratio of the mass of water vapor to the mass of dry air. Based on the ideal gas law, the following relations can be derived:

$$r = 0.622 \frac{e}{P - e} \approx 0.622 \frac{e}{P} \qquad (6)$$

where e is the partial pressure of water vapor and P is the air pressure. The product between Equations 6 and 5 required to apply Equation 4 thus yields a quantity independent of P.

The partial pressure of water vapor e is computed from radiosonde observations as follows:

$$e = \frac{U}{100} e_s \qquad (7)$$

where U is the observed relative humidity and e_s is the saturation vapor pressure, calculated by the following equation (WMO 2008):

$$e_s = 6.112 \exp\left(\frac{17.62 \, T}{243.12 + T}\right) \qquad (8)$$

with T the observed temperature in degrees Celsius.

Regression relation between T_s and T_m

Since water vapor is mostly concentrated in the lower atmosphere, T_m is expected to be closely correlated to the surface temperature T_s. Bevis et al. (1992) derived the following equation from radiosonde observations over 2 years, under various conditions, taken from 27° N to 65° N from 13 stations in North America, from 0 to 1.6 km of altitude

$$T_m = 70.2 + 0.72 \, T_s . \qquad (9)$$

Shoji (2010) also derived a similar relation using radiosonde observations collected during 1 year from 20 stations in Japan, which was reported to be more appropriate for the Japanese regional meteorological conditions. Equations 10 and 11 were derived at 0900 and 2100 LT, respectively, to separately represent day and night conditions

$$T_m = 27.008 + 0.8688 \, T_s \qquad (10)$$

$$T_m = 20.072 + 0.8956 \, T_s . \qquad (11)$$

After comparing our experimental results from radiosondes with the models described in Equations 9 to 11 (see the next two paragraphs), we decided to employ Equation 10 as a reasonable approximation for both our daytime and nighttime results.

Figure 2 shows the temperature profiles obtained in our radiosonde experiments. It is evident that a temperature inversion layer appeared near the ground at night (launches at 0030 LT) and early morning (launches at 0630 LT) every day. The observed relation between T_s and T_m during this campaign was evaluated in order to check which of the existing T_s-T_m models was more suitable to be applied in our experiment. For each radiosonde launch, the weighted mean temperature of the atmosphere T_m was computed as (Bevis et al. 1992; Davis et al. 1985)

$$T_m = \frac{\int e/T \, dh}{\int e/T^2 \, dh} \qquad (12)$$

where e is the partial pressure of water vapor (computed from the radiosonde humidity and temperature readings, see Equation 7) and T is the radiosonde temperature reading. The integration was performed from the ground level up to the maximum altitude reached by the radiosonde (it is worth pointing out that, in this equation, the partial pressure of water vapor e acts as a weight for the temperature observations, which will therefore have a decreasing influence on the result as the altitude increases).

Figure 3 shows scatter diagrams between T_s and T_m obtained from the radiosonde observations collected in this campaign; on the left, the results when considering the full temperature profiles, while on the right by assuming ground level at 300 m height, in order to remove the effect of the inversion layer. The daytime results show a reasonable agreement with the Bevis model (Equation 9) in both cases. However, the nighttime results indicate a considerable bias from the Bevis model, showing a positive deviation of 2 to 4 K. While the agreement improves when removing the inversion layer, a significant bias is still present. T_m, therefore, tends to be overestimated by the Bevis model at night due to the inversion layer that occurs near the ground. The nighttime results tend to

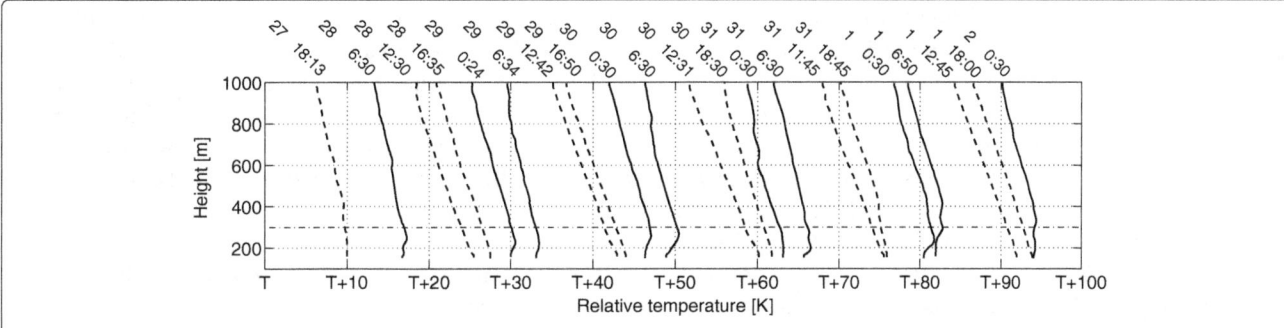

Figure 2 Radiosonde temperature profiles. Temperature profiles observed during daytime (dashed red line) and nighttime (solid blue line) radiosonde launches. Profile plots are shifted along the x-axis, in order to improve the figure readability; therefore, the x-axis only shows temperature increments of 10 K. The horizontal dash-dotted line indicates the minimum height threshold (300 m) used to exclude the inversion layer for the regression analysis.

agree better with the Shoji nighttime model (Equation 11) compared to the Bevis model, in both cases. However, the daytime results show a negative bias with respect to the Shoji daytime model (Equation 10). The effects of the T_m deviation on the GPS-PWV estimates were investigated. If the model T_m introduces a 3 K error with respect to the actual weighted average temperature, the corresponding error in PWV is about 0.5 mm. It is worthwhile noting that even if a mismodeling of this order of magnitude is introduced, it would not have a significant impact on the results described in this paper: in fact, as described later on in the section 'Spatial variations of GPS-derived PWV around Jakarta', the expected GPS-derived PWV error standard deviation in this campaign is about 1 mm, while the observed horizontal PWV fluctuations have a standard deviation ranging from 1 to 4 mm. It should also be pointed out that, in the particular case of this campaign, the choice between the three previously described models

is not critical: since all stations would be similarly affected, the impact on the inter-station PWV differences would be very limited. Therefore, we decided to use the Shoji daytime model for the whole campaign, as it is a reasonable compromise to approximate both daytime and nighttime results.

In general, it is difficult to reach a firm conclusion on the resulting regression relations in our campaign, as the number of available radiosonde observations is too small. An appropriate relationship between T_s and T_m for the Indonesian area should be constructed, also by considering regional weather conditions, such as the inversion layer occurrence. To reach this goal, one could investigate more radiosonde profiles from routine soundings over Indonesia, e.g., by accessing radiosonde observations by BMKG, which are available through the Global Telecommunications Service (GTS) datasets. However, since the observable temperature ranges in the tropics are generally

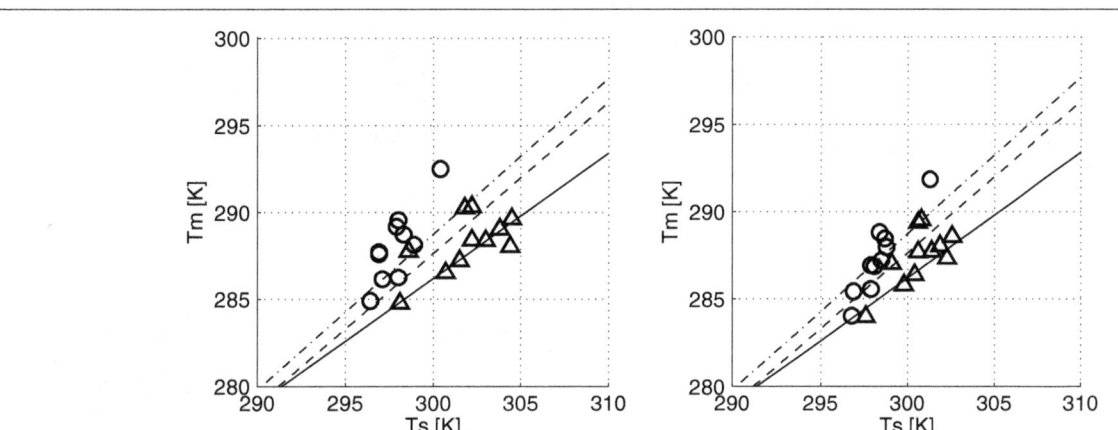

Figure 3 Scatter diagrams of T_s and T_m. Relationship between the surface temperature T_s and the weighted mean temperature of the atmosphere T_m. On the left, by using profiles above the ground level; on the right, by using profiles above 300 m of height (red triangles, daytime observations; blue circles, nighttime observations; black solid line, Bevis model; red dashed line, Shoji daytime model; blue dash-dotted line, Shoji nighttime model).

limited to about 10 K, it is difficult to carry out a reliable regression analysis as it was done for Bevis and Shoji models.

Pressure analysis and adjustment

Observations from five GPS receivers were used during this experiment, with marker names CJKT (orthometric height 13 m), CTGR (48 m), PKYN (69 m), BAKO (141 m), and BAK2 (150 m); the last two receivers were located closely within the campus of BIG. However, surface pressure was measured continuously at CTGR only. As pressure data is very important in estimating GPS-derived PWV, to separate the hydrostatic and wet components of the signal delay, we investigated the accuracy of the pressure measurements at CTGR and the feasibility of adjusting CTGR pressure measurements to different heights in order to have sufficiently accurate pressure measurements for each GPS station. The original surface pressure measured at CTGR (P_{CTGR}, black solid line in Figure 4) was converted to the pressure at the height of station 96745 (P_{96745}, orthometric height 8 m) by using the following relationship, derived from the barometric formula in Berberan-Santos et al. (1997)

$$P_{96745} = P_{CTGR} \exp\left(-\frac{g\, M_d\, (H_{96745} - H_{CTGR})}{R^*\, T_{ISA}}\right)$$

(13)

where g is the gravitational acceleration constant (9.80665 m s^{-2}), M_d is the molar mass of dry air (0.0289644 kg mol^{-1}), R^* is the gas constant for air (8.31432 J mol^{-1} K^{-1}), and T_{ISA} is the international standard temperature of the atmosphere at sea level (288.15 K). H_{CTGR} and H_{96745} are the orthometric heigths of CTGR and 96745 stations, respectively. Figure 4 shows the computed pressure P_{96745} time series as a dashed line, while the weather station measurements are represented by white circles. Their agreement is very good, with a bias (computed-measured) of -0.03 hPa and a standard deviation of 0.55 hPa. Next, we applied Equation 13 to compare the height-corrected CTGR pressure data with the aneroid barometer readings (orthometric height 141 m, black circles in Figure 4), obtaining a bias of -0.72 hPa and a standard deviation of 1.24 hPa. The larger random error is likely due to the lower precision of the aneroid barometer and to possible human error in reading the result. Lastly, the comparison with the observations of the weather sensor of BAKO station (dark grey line in Figure 4) yielded a bias of -0.19 hPa and a standard deviation of 0.23 hPa. Given that a pressure error of 1 hPa roughly corresponds to 0.4 mm of error in PWV, we assume that the height-compensated pressure data at CTGR can be used for PWV estimation. This dataset is thus used to infer the pressure values at CJKT, PKYN, and BAKO/BAK2 stations.

It is well known that a 12-h oscillation in the pressure is evident at low latitudes because of atmospheric tides

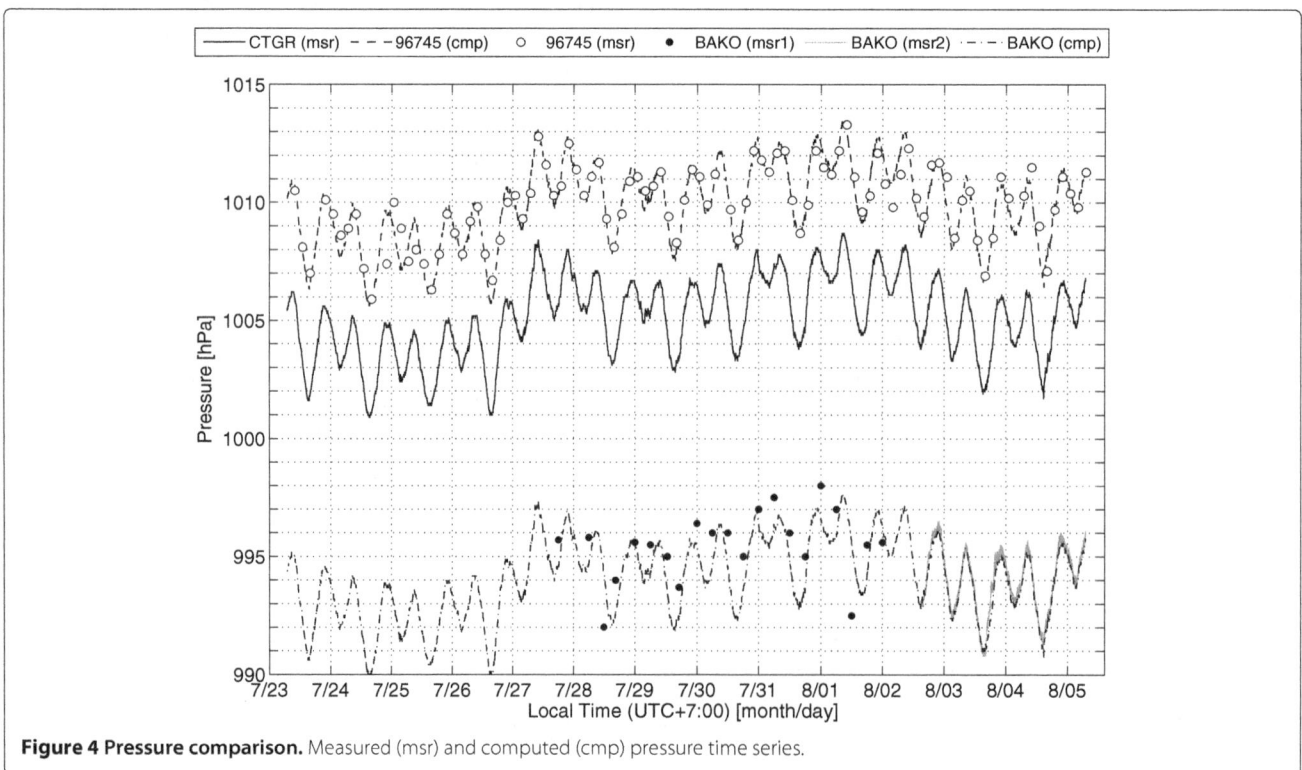

Figure 4 Pressure comparison. Measured (msr) and computed (cmp) pressure time series.

(Covey et al. 2011; Lindzen and Chapman 1969). The magnitude and local time of the maximum value for the semi-diurnal variations in Figure 4 are generally consistent with results reported in the literature. It is important to note that the influence of the observed semi-diurnal pressure variations (3 to 5 hPa) on the estimated PWV is about 1.1 to 1.8 mm; therefore, at low latitudes, it is important to consider the effects of semi-diurnal tides for GPS-derived PWV retrieval. Pressure measurements with a time resolution higher than 1 h is recommended. The effects of space-time variations of pressure and temperature may not be significant to estimate GPS-derived PWV for meteorological phenomena with horizontal scales larger than several tens of kilometers, with relatively small horizontal gradients in pressure and temperature, as it is the case for this work. However, if a precise local-scale spatial distribution of PWV is investigated by means of a denser GPS network, it becomes important to accurately retrieve the PWV by referring to the simultaneous measurements of pressure and temperature at each GPS site.

Temperature adjustment

While pressure can be adjusted with good accuracy on the basis of the altitude difference between two stations, the surface temperature is generally more variable spatially; thus, it cannot be reliably inferred from one station to another. In this work, the only correction applied to CTGR surface temperature in order to use it with the other stations is the following height adjustment (Bai and Feng 2003):

$$T_{station} = T_{CTGR} + 0.0065 \times (H_{CTGR} - H_{station}) \quad (14)$$

where T_{CTGR} and $T_{station}$ are respectively the temperature measured at CTGR and that inferred at the other station,

and H_{CTGR} and $H_{station}$ are the orthometric heights of the two stations. However, in the specific case of the campaign described in this paper, this adjustment is hardly significant, being at most 0.9 K when considering the maximum altitude difference between stations, that is between CJKT (13 m) and BAK2 (150 m), which corresponds to about 0.16 mm in PWV.

Results and discussion

Comparison between GPS- and radiosonde-derived PWV

Figure 5 shows the 21 radiosonde-derived PWV values and the time series of GPS-derived PWV using BAKO station. The GPS-derived PWV for the comparison was computed both by using the ground temperature measured at CTGR station, after having applied the height correction shown in Equation 14 (line in Figure 5), and by using the temperature readings of the radiosonde sensor at launch (circles in Figure 5). Each radiosonde PWV value was compared to the averaged GPS PWV values over a time span of 30 minutes from the radiosonde launch, in order to take into account the radiosonde measurement time. Two significantly larger values (i.e., exceeding twice the standard deviation of the differences) were identified at 1845 LT on 31 July and 0030 LT on 1 August. A possible explanation for these two anomalies could be the presence of strong disturbances localized in proximity of BAKO station, causing strong spatial and time variation of water vapor: sharp increments in the PWV time series of BAKO are evident in Figure 5, and increased PWV variability in both time and space for the whole network was observed (see Figure 6). The comparison results are shown in Table 1, which reports mean, standard deviation and RMSE of the PWV difference between radiosonde and GPS, either by using the temperature measured at CTGR (T_{CTGR}) or that measured by the radiosonde sensor at launch (T_{sonde}), including and

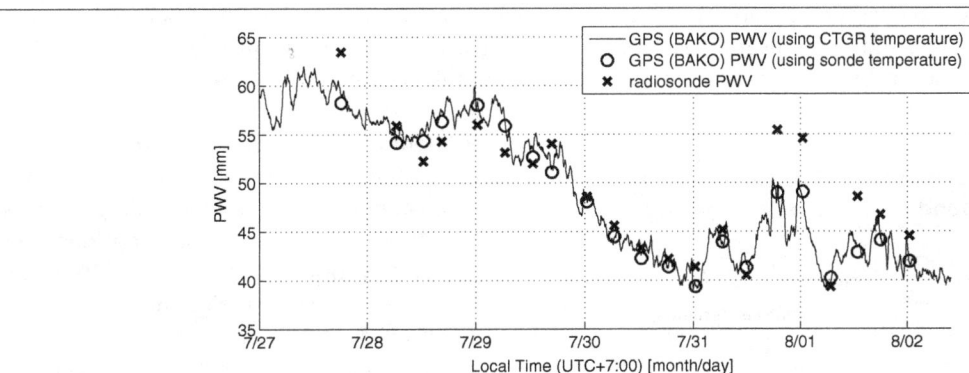

Figure 5 Radiosonde-GPS comparison. Comparison between the PWV retrieved from radiosondes (crosses) and by the GPS station BAKO. The GPS-derived PWV was computed by using the temperature measured at CTGR station (line) and that measured by the radiosonde sensor at launch (circles).

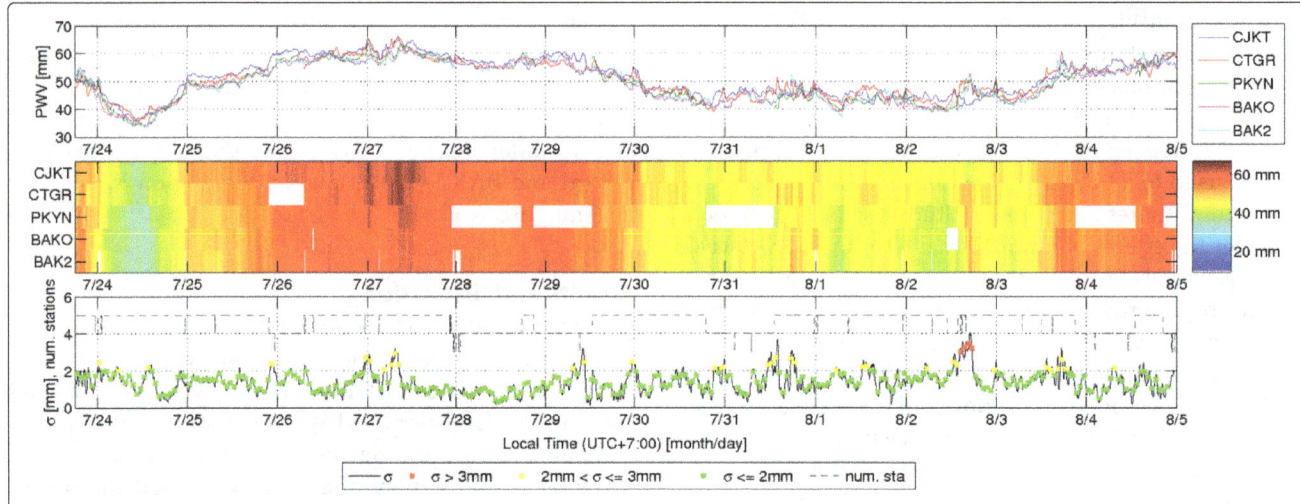

Figure 6 Spatial variations of GPS-derived PWV. Line (top) and matrix (middle) plots of GPS-derived PWV time series; inter-station standard deviation (bottom), every 30 s (solid line) and hourly (crosses). The number of available stations at each epoch is overlaid on the bottom plot as a dashed line.

excluding the two anomalous values. The magnitude of the deviation is consistent with that reported in previous studies (Fujita et al. 2008; Sato et al. 2013). As regards the bias, previous comparisons of GPS-derived PWV with respect to Vaisala radiosondes, both in mid-latitude (by using the RS80H model (Deblonde et al. (2005) and the RS90 model (Van Baelen et al. 2005)) and tropical zones (RS80-15G model (Sapucci et al. 2007), RS80A, RS80H, and RS90 models (Wang et al. 2007), and RS92 model (Yoneyama et al. 2008)), evidenced a negative (dry) bias of the radiosonde measurements. This systematic error, especially significant during daytime, is generally attributed to the solar heating of the radiosonde humidity sensor (Vömel et al. 2007). However, examples exist in the literature that report also positive (moist) biases of Vaisala radiosonde measurements with respect to GPS. For example, Fernández et al. (2010) reported both moist and dry biases of Vaisala RS80 radiosonde observations compared to GPS in Argentina, varying according to different radiosonde launch sites and different models used for estimating the weighted mean temperature of the

atmosphere T_m. The European network of GPS stations (EUREF Permanent Network (EPN)) routinely compares GPS- and radiosonde-derived zenith tropospheric delays and publishes up-to-date time series on its website (EPN 2001): several of the published time series of radiosonde-GPS differences exhibit moist biases[a], generally of less than 1 mm (PWV) in magnitude. It must be noted that these examples of moist biases were reported in mid-latitudes. Wang and Zhang (2008) demonstrated that the subsequent improvements to Vaisala radiosonde models from the RS80A to the RS92 have significantly mitigated the impact of the dry bias with respect to GPS-derived PWV, particularly for PWV values ranging from 40 to 60 mm. However, it should also be pointed out that the timespan of our observation campaign is quite limited compared to those that would be needed to reliably quantify the difference between GPS-derived and radiosonde-measured PWV from a statistical point of view; more extensive radiosonde and GPS observation campaigns would be needed in order to quantify the nature of the bias (if any) in this region.

Spatial variations of GPS-derived PWV around Jakarta

Figure 6 shows the time series of GPS-derived PWV during the whole campaign (top and middle panels); all the five stations showed consistent time variations over long timespans (i.e., they showed consistent low-frequency signals) with maximum spatial variations (i.e., inter-station PWV differences) ranging from 5 to 10 mm. The matrix representation in the middle panel was chosen in order to try to identify the movement of large water vapor fluctuations as they are detected by different stations; the stations are ordered from the northernmost (top row, CJKT) to

Table 1 Mean, standard deviation, and RMSE of the difference between radiosonde and GPS results for PWV retrieval (in mm)

	T_{CTGR}	T_{sonde}	T_{CTGR}	T_{sonde}
			(excluding anomalies)	
Mean	1.1	1.3	0.5	0.7
st.dev.	3.0	3.0	2.4	2.3
RMSE	3.2	3.3	2.4	2.4

st.dev., standard deviation.

the southernmost (bottom row, BAK2) to improve the readability of the plot. The PWV showed a local minimum around midday on 24 July, after which it started to increase rapidly in the late afternoon and continued to increase until the early morning of the 27 of July, with some intermittent fluctuations, reaching a peak of about 67 mm. Then it gradually decreased on 29 to 30 July, stabilizing within a range of 40 to 50 mm until 3 August, when it started to increase again. Overall, Figure 6 indicates that a moist air block passed over the campaign site,

causing the detection of a large amount of water vapor on 26 to 29 July. It is also worth noting that from 31 July to 3 August a diurnal variation in PWV is regularly repeated, with an enhancement in the afternoon (further details are given in the section 'Temporal variations of GPS-derived PWV around Jakarta').

The measurement error in PWV retrieval by GPS was roughly evaluated by comparing the PWV time series of the two co-located stations, i.e., BAKO and BAK2: the mean and standard deviation of their difference were 0.02

Table 2 Precipitation events detected by the CDR during the campaign

Date	Time (LT)	Event	max. reflectivity (dBZ)	max. st.dev. (mm)	Notes
24 July	1332 to 1922	No data	-	2.3	
25 July	1120 to 2200	Scattered weak echoes	20	2.1	
25 July	2200 to 2359	Localized echoes	30	2.4	Echoes approaching the observation area from northeast
26 July	0000 to 0200	Localized echoes over CJKT	30	1.5	CJKT PWV higher than other stations
26 July	0300 to 0700	Scattered echoes	30	2.1	Echoes homogeneously covering the observation area
26 July	2330 to 2359	Scattered strong echoes	50	2.6	Echoes approaching the observation area from northeast
27 July	0000 to 2359	Scattered strong echoes	60	3.1	Details in the section 'Scattered rain events on 27 July 2010'
28 July	0000 to 1100	Scattered echoes	30	1.1	Echoes scattered over the whole observation area
28 July	1720 to 2359	No data	-	1.4	
29 July	0000 to 1820	No data	-	3.1	
31 July	1120 to 1430	Localized strong echoes	50	3.6	Strong winds with spiral motion over the observation area
31 July	1430 to 2359	No data	-	3.0	
1 August	0000 to 1220	No data	-	2.1	
1 August	1240 to 1430	Localized strong echoes	50	2.5	Strong winds with spiral motion over the observation area
1 August	1430 to 1920	No data	-	2.3	
1 August	1920 to 2300	Localized echoes	40	2.0	Echoes near BAKO/BAK2 stations and moving westward
2 August	0000 to 2359	Localized strong echoes	60	4.0	Details in the section 'Localized heavy rain on 2 August 2010'
3 August	0300 to 0910	Scattered weak echoes	20	1.8	Moving through the observation area from north to south
3 August	1220 to 1500	Localized strong echoes	50	2.6	Strong wind bursts with spiral motion over the observation area
3 August	1500 to 1605	No data	-	1.8	
3 August	1605	Localized strong echoes	50	1.8	Echoes near BAKO/BAK2 stations
3 August	1605 to 2210	No data	-	3.2	
3 August	2210 to 2359	Localized strong echoes	50	1.8	Echoes near CJKT station
4 August	0000 to 0930	Scattered weak echoes	20	2.2	Echoes east and north of the observation area
4 August	1245 to 1500	Localized strong echoes	50	2.4	Echoes near BAKO/BAK2, PKYN, and CTGR stations

Timespans not reported in the table are those during which the CDR was functioning, but no echoes over the observation area were measured.
max., maximum; st.dev., standard deviation.

and 0.94 mm, respectively. Any inter-station PWV variation exceeding 2 mm (i.e., twice the standard deviation) was thus considered as containing significant geophysical information, i.e., spatial fluctuations of water vapor. In the bottom panel of Figure 6, we have plotted the time variation of the standard deviation of the five PWV results. The standard deviation showed sharp peaks reaching up to 3 to 4 mm in the morning of 27 and 29 July and in the afternoon of 31 July and 2 to 3 August. In particular, it exceeded 3 mm for few hours in the afternoon of 2 August. It should be noted that the absolute PWV values estimated for each station depend on the station altitude; hence, a fraction of the inter-station standard deviation is due to the altitude difference. However, this fraction is expected to be constant over time; therefore, it does not affect the time variability of the standard deviation. In any case, it is advisable to deploy new GNSS stations (or to select stations in existing networks) at approximately the same altitude.

Table 2 shows an overview of the rain clouds detected by the CDR in Serpong. The highest values of standard deviation appear to be associated to the strongest echoes detected by the CDR over the observation area. A thorough statistical study on the relation between the characteristics of rain events and the horizontal PWV inhomogeneity in western Java is planned for our future works.

The next two subsections will focus on the precipitation events associated with two of these spatial variations: a relatively wide precipitation encompassing the Jakarta area (CJKT, CTGR, PKYN stations) on 27 July, coinciding with the local maximum of PWV in this campaign, and a localized heavy rainfall near Bogor (BAKO/BAK2 stations) on 2nd August, coinciding with the highest peak of standard deviation.

Characteristics of two precipitation phenomena observed during the observation campaign
Scattered rain events on 27 July 2010

The meteorological conditions during the rainfall events on 27 July were investigated by referring to satellite images on infrared and water vapor channels provided by the Japan Meteorological Agency, as well as the CDR in Serpong. The infrared satellite images indicated that convective clouds were intermittently generated in the western Pacific region around the Borneo Island; at times, these cloud systems moved southwestward from the equatorial region toward the Indian Ocean passing over the Java Island. On 26 July, the water vapor satellite images indicated moist air associated with a cloud area that approached from the Borneo Island toward the Java Island. Considerable rainfall occurred on 27 July in Jakarta and surrounding areas; the BMKG rain gauges registered 20 to 30 mm of 3-hourly precipitation,

mostly during the morning. Convective clouds were sporadically generated around the north coast of the Java Island on that day, and the cloud system moved southwestward encountering the mountainous region south of Jakarta. Then, the convective clouds rapidly developed in a wide area that included the campaign sites. As the cloud system moved southwestward, three distinct rain events were detected by the CDR, from 0000 to 0200 LT (Figure 7, top), from 0400 to 0600 LT (Figure 7, middle), and from 0700 to 0930 LT (Figure 7, bottom)[b]. The

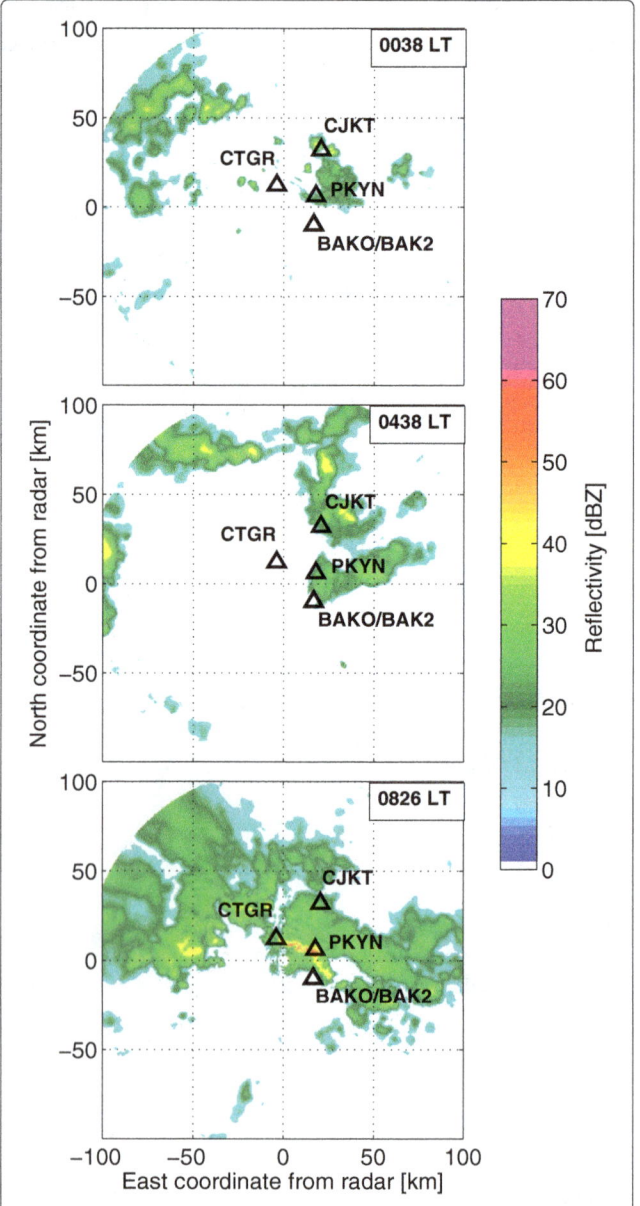

Figure 7 Radar observations on 27 July. Serpong C-band Doppler Radar reflectivity observations at 5 km height on 27 July 2010 at 0038 LT (top), 0438 LT (middle), and 0826 LT (bottom).

spatial inhomogeneity of PWV detected by the GPS stations, as indicated by the inter-station standard deviation in Figure 8, also shows three distinct peaks, which correspond in time to the three rain events. During the first event (from 0000 to 0200 LT), the GPS-derived PWV was larger at CJKT, CTGR, and PKYN than at BAKO/BAK2, suggesting a horizontal gradient of the water vapor distribution from the north (coast) to the south (mountains) and causing the first peak in standard deviation among the GPS stations. The second event (from 0400 to 0600 LT) was similarly characterized by a north-south (coast to mountains) gradient, with significantly higher water vapor values at CJKT compared to the other stations; this station may have sensed increased amounts of water vapor coming from the ocean, which could be associated to the precipitation clouds that developed north of CJKT, as shown in the middle panel of Figure 7. The third event (from 0700 to 0930 LT) saw a distinct offset between the PWV estimated by CJKT and CTGR stations and that estimated by the other three stations, leading to an increment of the standard deviation lasting about 3 h, from 0600 to 0900 LT. Toward the end of this timespan, also PKYN, BAKO, and BAK2 started registering a PWV increment, coinciding with the rain cloud expansion to cover the whole GPS network, with the consequent decrease of inter-station inhomogeneity (bottom panel of Figure 7).

Localized heavy rain on 2 August 2010

The rain event on 2 August was identified thanks to the highest peak in standard deviation shown in Figure 6. The rain gauges of the three BMKG weather stations, in fact, did not detect it since its extents were limited to a relatively small area (about 15×15 km^2) close to Bogor, where no weather station data were available. The rain event was also relatively short in time, about 2 h, with CDR echoes appearing from 1500 to 1700 LT. Figure 9 shows the CDR reflectivity observations at three instants related to significant features of the PWV time series monitored by the GPS network, shown in Figure 10, or to significant observational evidence from the CDR: at 1228 LT, the standard deviation surpassed the threshold of 2 mm, which in the previous example was found to anticipate the formation of rain clouds, while the CDR still did not detect strong reflectivity (Figure 9, top panel); at 1528 LT, the first radar echoes, which appeared west of BAKO/BAK2 stations at 1500 LT, reached their maximum values, about 55 dBZ (Figure 9, middle panel); shortly after their disappearance, a second set of echoes appeared east of BAKO/BAK2 stations, with even smaller spatial extents (about 10×15 km^2), again with maximum reflectivity values of about 55 dBZ (Figure 9, bottom panel). Unfortunately, the CDR data are missing from 1540 to 1658 LT; thus, it is not possible to know exactly when the second set of echoes appeared. The first event seems to be associated to a

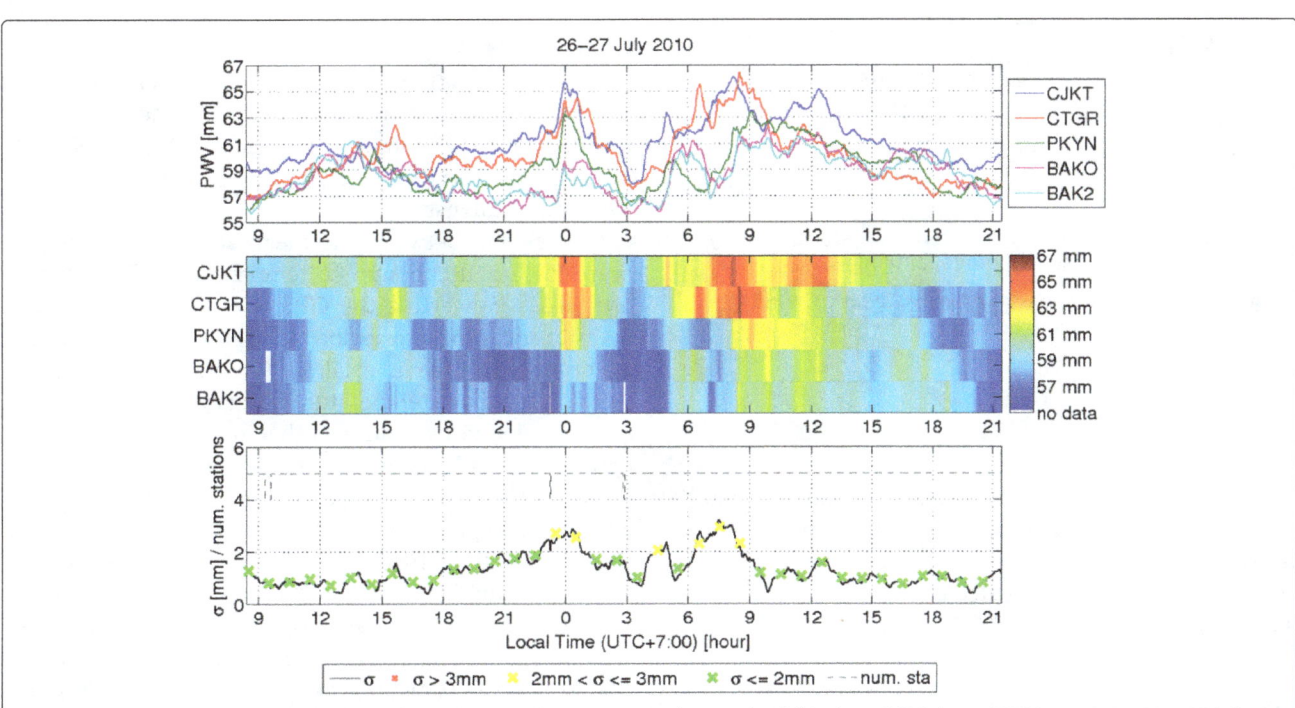

Figure 8 Spatial variations of GPS-derived PWV on 27 July. Line (top) and matrix (middle) plots of GPS-derived PWV time series 26 to 27 July 2010 (in LT); inter-station standard deviation (bottom), every 30 s (solid line) and hourly (crosses). The number of available stations at each epoch is overlaid on the bottom plot as a dashed line.

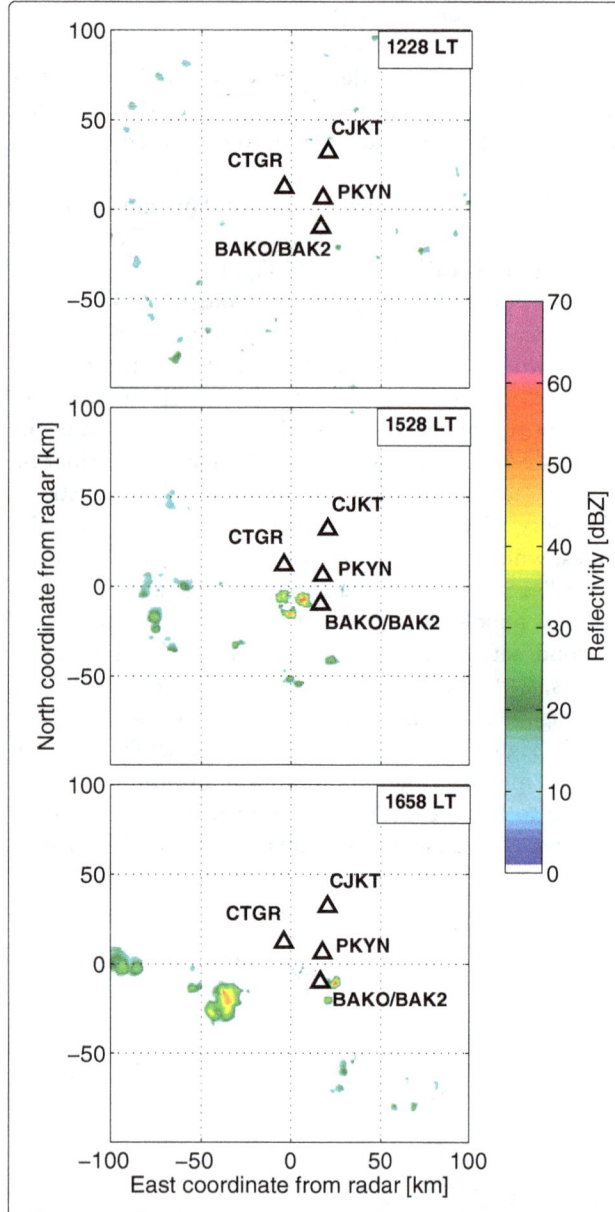

Figure 9 Radar observations on 2 August. Serpong C-band Doppler Radar reflectivity observations at 5 km height on 2 August 2010 at 1228 LT (top), 1528 LT (middle), and 1658 LT (bottom).

gradual increase of PWV from the coast, moving south-westward (first detected by CJKT station, then CTGR, and lastly PKYN). The second event seems to be related to a steep increase in PWV detected by both BAKO and BAK2 stations (see Figure 10).

Temporal variations of GPS-derived PWV around Jakarta
The time variability of PWV for each GPS station during the entire campaign was evaluated by means of time

derivatives of PWV estimates, with a time step of 10 min. The time step was chosen empirically, based on the magnitude of the temporal fluctuations observed in the PWV time series. The results are plotted in Figure 11, revealing the presence of several peaks of PWV time variations. The time variations were stronger during the second part of the campaign, in August, as it can partially be seen also in the top and middle panels of Figure 6. The strongest fluctuations were detected by the inland stations, while CJKT, on the coast, had generally smoother time series. The main fluctuations, in August, appeared more often in the late afternoon (1600 to 1700 LT), and preliminary comparisons with CDR observations suggest that they were consistently associated to the formation of rain clouds. Minor fluctuations appeared almost every day also during the night, within a timespan of about ± 1 h from midnight, usually for all stations. These nocturnal fluctuations are generally not associated to clouds detected by the CDR, with some exceptions such as that between 26 and 27 July, which can be seen in the top panel of Figure 8: while in this case there was also a significant spatial variation more clearly associated to the rain clouds seen in the top panel of Figure 7, in other cases all stations exhibited temporal fluctuations of similar magnitude, the spatial variation was not significant, and rain clouds were not detected by the CDR. Further investigations on these anomalous time variations are currently ongoing.

Conclusions
A PWV observation campaign with five GPS receivers was carried out in Jakarta and Bogor, Indonesia for about ten days from 23 July to 5 August 2010. A total of 21 radiosondes was launched, with a time interval of 6 h from 27 July to 2 August.

The relation between the mean atmospheric temperature (T_m) and the surface temperature (T_s), which is crucial for converting the GPS wet delay to PWV, was investigated. The regression relations between T_m and T_s proposed by Bevis et al. (1992) and by Shoji (2010) were compared to temperature observations by radiosondes in order to evaluate whether they could be reliably applied for PWV retrieval in tropical regions. A temperature inversion layer frequently appeared below 300 m altitude at night, causing an overestimation of T_m by 2 to 4 K, which, in turn, introduced an error in PWV of about 0.5 mm. While this error was deemed acceptable for the purpose of our campaign, future work is required to establish an appropriate relation for the Indonesian region, by statistical analysis of routine radiosonde soundings.

Atmospheric pressure and temperature were continuously monitored only at one of the GPS stations. However,

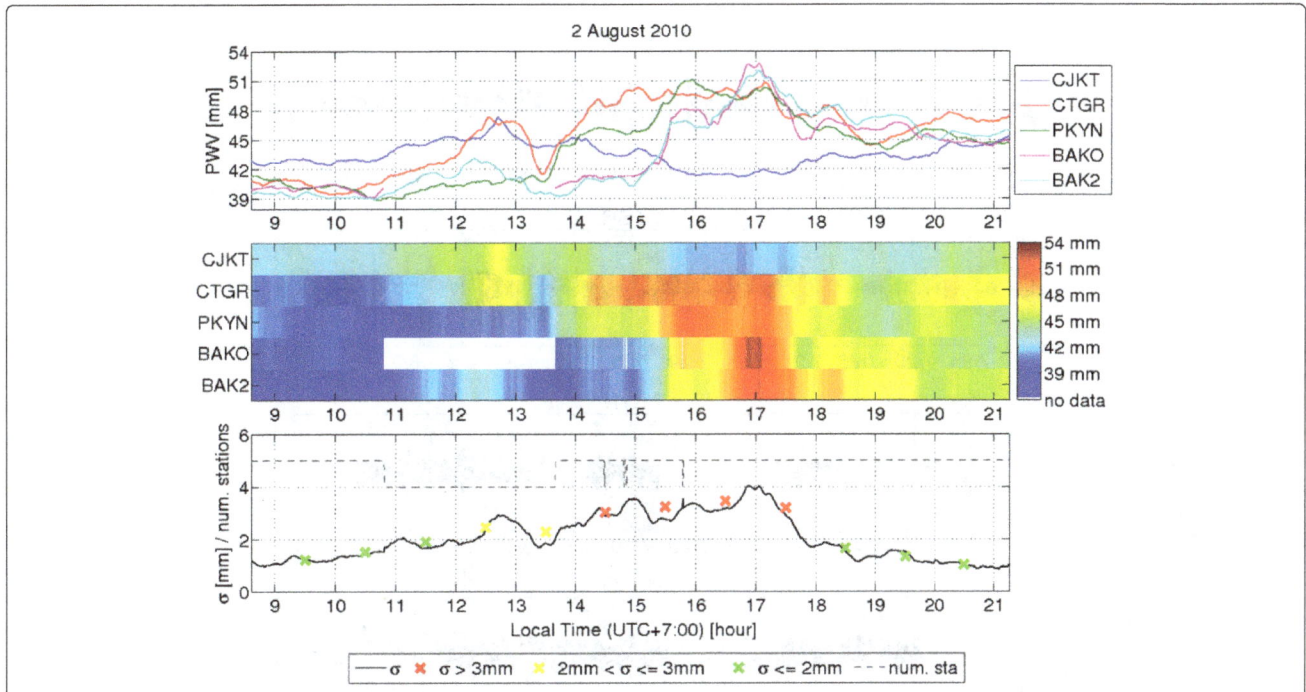

Figure 10 Spatial variations of GPS-derived PWV on 2 August. Line (top) and matrix (middle) plots of GPS-derived PWV time series on 2 August 2010 (in LT); inter-station standard deviation (bottom), every 30 s (solid line) and hourly (crosses). The number of available stations at each epoch is overlaid on the bottom plot as a dashed line.

by compensating its pressure readings considering the height difference, it was possible to infer pressure time series for all the GPS stations. These inferred values were validated by comparison with 3-hourly pressure records taken at a routine weather station located in the study area, as well as an aneroid barometer and an additional pressure sensor, yielding errors lower than 1 hPa, i.e., about 0.4 mm in PWV. Regular semi-diurnal pressure tides with amplitudes of 3 to 5 hPa were confirmed, corresponding to 1.1 to 1.8 mm in PWV. It is therefore advisable to employ ground pressure readings for PWV retrieval from GPS observations instead of relying on modeled pressure values.

C-band Doppler Radar observations and infrared satellite images indicated a passage of clouds moving southwestward from the equator toward the Indian Ocean through the Java Island during 26 to 29 July, and 3-hourly precipitation of 20 to 30 mm was recorded on the 27 of July by rain gauges within the observation area. An increase in GPS-derived PWV for all stations, up to about 67 mm, coincided with this precipitation event. A second heavy rain event, characterized by significantly smaller spatial scales, was observed in the southern area of the GPS network. Spatial variations in the retrieved PWV were observed to have a strong correlation with both these events. In general, the spatial inhomogeneity of PWV started increasing before

the observation of strong radar echoes within the GPS observation area, suggesting the possibility to use it as an index to predict precipitation events or at least as a parameter within a more complex index including other meteorological observations. The time variation of PWV observed by each station exhibits a correlation with the diurnal cycle of water vapor, with stronger fluctuations appearing during the late afternoon (1600 to 1700 LT), usually in conjunction with precipitation. Anomalous time fluctuations were observed during the night, within ±1 h from midnight, which require further investigations.

The GPS meteorology technique was proved to be useful to investigate severe weather conditions over Indonesia, complementing existing meteorological observation systems. The currently existing Indonesian network of GPS stations (IUGG 2011) could thus be useful not only for geodesy or seismology but also for meteorological research and applications. The analysis of space-time variations of GPS-derived PWV has potential to allow the identification of precursors for nowcasting local convective rain, which might trigger hazardous events such as floods and landslides. A fundamental requirement to achieve this goal is the near real-time processing of GPS observations (i.e., by using near real-time satellite orbits and clocks), which is the target of our future works.

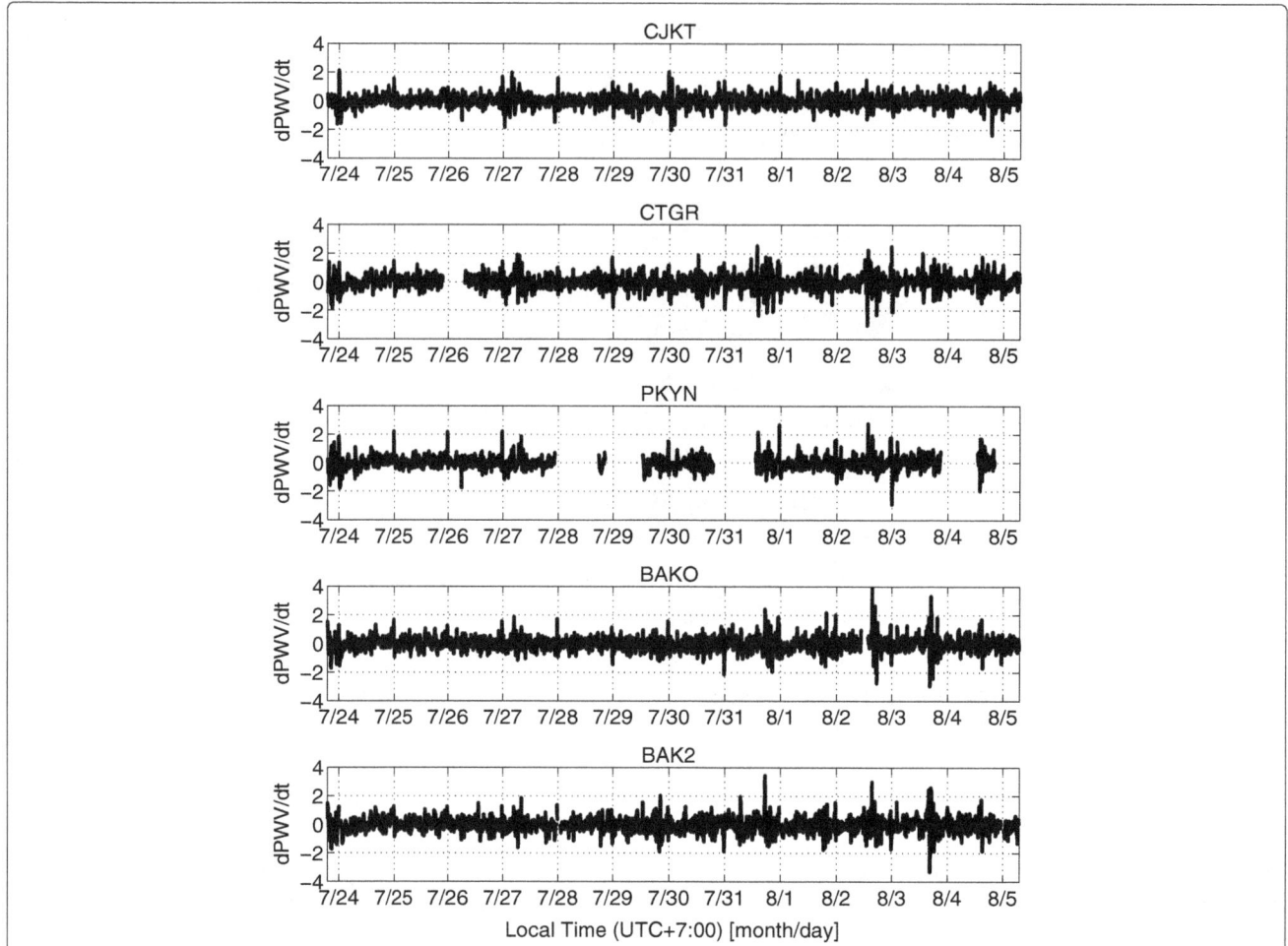

Figure 11 Time variations of GPS-derived PWV. Time derivative of PWV (dt = 10 min); the y-axis represents the time variation of PWV in mm over 10 min. The different colors represent time variations exceeding ±1 mm/10 min (the threshold was arbitrarily chosen to highlight the peaks).

Endnotes

[a] e.g., GPS stations BELF, BUCU, GARI, GRAZ, HERS, HERT, and TERS, all compared to RS92SGP soundings.

[b] All the CDR reflectivity maps in this paper are constant altitude plan position indicator (CAPPI) plots, at 5 km altitude, produced by means of the open source software library Python ARM Radar Toolkit (Py-ART 2013), developed at the Atmospheric Radiation Measurement (ARM) Climate Research Facility (USA).

Competing interests

The authors declare that they have no competing interests.

Authors' contributions

ER processed the GPS observations for PWV retrieval, carried out the pressure analysis and adjustment, and participated in the the PWV space-time analyses, CDR comparisons, and manuscript draft. KS conducted the field experiments, processed the radiosonde data, analyzed the temperature profiles, studied the regression relation between T_s and T_m and participated in the manuscript

draft. S provided GPS observations of the stations operated by BIG, and TM participated in the radiosonde field experiments. TT proposed and guided the research direction of this work, provided insights and interpretations related to the meteorological aspects and atmospheric physics, and reviewed the results. All authors read and approved the final manuscript.

Acknowledgements

This research was supported by the Sustainability/Survivability Science for a Resilient Society Adaptable to Extreme Weather Conditions (GCOE-ARS), which is one of the Global Center of Excellence programs at Kyoto University, under the Ministry of Education, Culture, Sports, Science and Technology of Japan (MEXT). It was also supported by the coordination funds for promoting space utilization provided by MEXT. We appreciate the devoted efforts by the staff of BIG and LAPAN for conducting the GPS and radiosonde experiments. We extend our gratitude to Dr. Hiromu Seko and Dr. Yoshinori Shoji of the Meteorological Research Institute (Japan Meteorological Agency), who gave us constructive comments and warm encouragement. We are grateful to Dr. Shuichi Mori of JAMSTEC and Dr. Fadli Syamsudin of BPPT for the support regarding the CDR raw data, which was obtained under the financial support by the Science and Technology Research Partnership for Sustainable Development (SATREPS) of the Japan Science and Technology Agency (JST) and the Japan International Cooperation Agency (JICA). Lastly, we would like to thank the members of the Py-ART mailing list for their help in improving the reflectivity plots obtained with the excellent Py-ART open source software library.

Author details

[1]Research Institute for Sustainable Humanosphere (RISH), Kyoto University, Gokasho, Uji 611-0011, Japan. [2]Indonesian Geospatial Information Agency (BIG), Cibinong 16911, Indonesia. [3]National Institute of Aeronautics and Space (LAPAN), Bandung 40173, Indonesia. [4]Present address: Geomatics Research & Development (GReD) srl, Como, Italy. [5]Present address: Satellite Navigation Office, Satellite Applications Mission Directorate I, Japan Aerospace Exploration Agency (JAXA), Tsukuba, Japan.

References

Askne J, Nordius H (1987) Estimation of tropospheric delay for microwaves from surface weather data. Radio Sci 22(3):379–386

Bai Z, Feng Y (2003) GPS water vapor estimation using interpolated surface meteorological data from Australian automatic weather stations. J Global Positioning Syst 2(2):83–89

Berberan-Santos MN, Bodunov EN, Pogliani L (1997) On the barometric formula. Am J Phys 65:404

Bevis M, Businger S, Herring TA, Rocken C, Anthes RA, Ware RH (1992) GPS meteorology: remote sensing of atmospheric water vapor using the Global Positioning System. J Geophys Res Atmos (1984–2012) 97(D14):15787–15801

Böhm J, Niell A, Tregoning P, Schuh H (2006) Global Mapping Function (GMF): a new empirical mapping function based on numerical weather model data. Geophys Res Lett 33(7):07304

Covey C, Dai A, Marsh D, Lindzen RS (2011) The surface-pressure signature of atmospheric tides in modern climate models. J Atmos Sci 68(3):495–514

Davis J, Herring T, Shapiro I, Rogers A, Elgered G (1985) Geodesy by radio interferometry: effects of atmospheric modeling errors on estimates of baseline length. Radio Sci 20(6):1593–1607

Deblonde G, Macpherson S, Mireault Y, Héroux P (2005) Evaluation of GPS precipitable water over Canada and the IGS, network. J Appl Meteorol 44(1):153–166

EPN (2001) EUREF Permanent GNSS Network. http://www.epncb.oma.be/. Accessed 5 Aug 2014

Fernández L, Salio P, Natali M, Meza A (2010) Estimation of precipitable water vapour from GPS measurements in Argentina: validation and qualitative analysis of results. Adv Space Res 46(7):879–894

Fujita M, Kimura F, Yoneyama K, Yoshizaki M (2008) Verification of precipitable water vapor estimated from shipborne GPS, measurements. Geophys Res Lett 35(13):1–5

Fujita M, Yoneyama K, Mori S, Nasuno T, Satoh M (2011) Diurnal convection peaks over the eastern Indian Ocean off Sumatra during different MJO phases. J Meteorol Soc Jpn 89(0):317–330

IGS (2012) Real-Time Service. http://rts.igs.org/. Accessed 5 Aug 2014

IUGG (2011) Indonesian National Committee country report. Technical report, International Union of Geodesy and Geophysics

Li X, Dick G, Ge M, Heise S, Wickert J, Bender M (2014) Real-time GPS sensing of atmospheric water vapor: precise point positioning with orbit, clock and phase delay corrections. Geophys Res Lett 41(10):3615–3621

Lindzen RS, Chapman S (1969) Atmospheric tides. Space Sci Rev 10(1):3–188

Pavlis NK, Holmes SA, Kenyon SC, Factor JK (2012) The development and evaluation of the Earth Gravitational Model 2008, (EGM2008). J Geophys Res Solid Earth (1978–2012) 117(B4):1–38

Py-ART (2013) The Python ART Radar Toolkit. http://arm-doe.github.io/pyart/. Accessed 5 Aug 2014

Ray R (2013) Precise comparisons of bottom-pressure and altimetric ocean tides. J Geophys Res C Oceans 118(9):4570–4584

Rocken C, Mervart L, Lukes Z, Johnson J, Kanzaki M, Kakimoto H, Iotake Y (2001) Testing a new network RTK software system In: Proceedings of the 17th international technical meeting of the satellite division of the institute of navigation (ION GNSS 2004), California, 21–24 Sept 2004, pp 2831–2839

Saastamoinen J (1973) Contributions to the theory of atmospheric refraction. Bull Geod (1946–1975) 107(1):13–34

Sapucci LF, Machado LA, Monico JF, Plana-Fattori A (2007) Intercomparison of integrated water vapor estimates from multisensors in the Amazonian region. J Atmos Ocean Tech 24(11):1880–1894

Sato K, Realini E, Tsuda T, Oigawa M, Iwaki Y, Shoji Y, Seko H (2013) A high-resolution, precipitable water vapor monitoring system using a dense network of GNSS receivers. J Disast Res 8(1):37–47

Shoji Y (2010) Accurate estimation of precipitable water vapor using ground-based GPS, observation network and its data assimilation into a mesoscale numerical weather prediction model. PhD thesis, Kyoto University

Vömel H, Selkirk H, Miloshevich L, Valverde-Canossa J, Valdés J, Kyrö E, Kivi R, Stolz W, Peng G, Diaz J (2007) Radiation dry bias of the Vaisala RS92 humidity sensor. J Atmos Ocean Tech 24(6):953–963

Van Baelen J, Aubagnac J-P, Dabas A (2005) Comparison of near-real time estimates of integrated water vapor derived with GPS, radiosondes, and microwave radiometer. J Atmos Ocean Tech 22(2):201–210

Wang J, Zhang L (2008) Systematic errors in global radiosonde precipitable water data from comparisons with ground-based GPS measurements. J Clim 21(10):2218–2238

Wang J, Zhang L, Dai A (2005) Global estimates of water-vapor-weighted mean temperature of the atmosphere for GPS applications. J Geophys Res Atmos (1984–2012) 110(D21):1–17

Wang J, Zhang L, Dai A, Van Hove T, Van Baelen J (2007) A near-global, 2-hourly data set of atmospheric precipitable water from ground-based GPS measurements. J Geophys Res Atmos (1984–2012) 112(D11):1–17

WMO (2008) Guide to meteorological instruments and methods of observation (WMO-No. 8) 7th edn.. World Meteorological Organization, Geneva

Wu P, Hamada J-I, Mori S, Tauhid YI, Yamanaka MD, Kimura F (2003) Diurnal variation of precipitable water over a mountainous area of Sumatra Island. J Appl Meteorol 42(8):1107–1115

Wu P, Hara M, Fudeyasu H, Yamanaka MD, Matsumoto J, Syamsudin F, Sulistyowati R, Djajadihardja YS (2007) The impact of trans-equatorial monsoon flow on the formation of repeated torrential rains over Java Island. Sci Online Lett Atmosphere 3:93–96

Wu P, Arbain AA, Mori S, Hamada J-i, Hattori M, Syamsudin F, Yamanaka MD (2013) The effects of an active phase of the Madden-Julian oscillation on the extreme precipitation event over western Java Island in January 02013. Sci Online Lett Atmosphere 9:79–83

Yamanaka MD, Hashiguchi H, Mori S, Wu P-M, Syamsudin F, Manik T, Hamada J, Yamamoto MK, Kawashima M, Fujiyoshi Y, et al. (2008) HARIMAU radar-profiler network over the Indonesian maritime continent: a GEOSS early achievement for hydrological cycle and disaster prevention. J Disaster Res 3:78–88

Yoneyama K, Fujita M, Sato N, Fujiwara M, Inai Y, Hasebe F (2008) Correction for radiation dry bias found in RS92 radiosonde data during the MISMO field experiment. Sci Online Lett Atmosphere 4:13–16

Zumberge J, Heflin M, Jefferson D, Watkins M, Webb F (1997) Precise point positioning for the efficient and robust analysis of GPS data from large networks. B3 102:5005–5017

2

Outcomes and challenges of global high-resolution non-hydrostatic atmospheric simulations using the K computer

Masaki Satoh[1]* [ID], Hirofumi Tomita[2], Hisashi Yashiro[2], Yoshiyuki Kajikawa[2,3], Yoshiaki Miyamoto[4,2], Tsuyoshi Yamaura[2], Tomoki Miyakawa[1], Masuo Nakano[5], Chihiro Kodama[5], Akira T. Noda[5], Tomoe Nasuno[5], Yohei Yamada[5] and Yoshiki Fukutomi[6]

Abstract

This article reviews the major outcomes of a 5-year (2011–2016) project using the K computer to perform global numerical atmospheric simulations based on the non-hydrostatic icosahedral atmospheric model (NICAM). The K computer was made available to the public in September 2012 and was used as a primary resource for Japan's Strategic Programs for Innovative Research (SPIRE), an initiative to investigate five strategic research areas; the NICAM project fell under the research area of climate and weather simulation sciences. Combining NICAM with high-performance computing has created new opportunities in three areas of research: (1) higher resolution global simulations that produce more realistic representations of convective systems, (2) multi-member ensemble simulations that are able to perform extended-range forecasts 10–30 days in advance, and (3) multi-decadal simulations for climatology and variability. Before the K computer era, NICAM was used to demonstrate realistic simulations of intra-seasonal oscillations including the Madden-Julian oscillation (MJO), merely as a case study approach. Thanks to the big leap in computational performance of the K computer, we could greatly increase the number of cases of MJO events for numerical simulations, in addition to integrating time and horizontal resolution. We conclude that the high-resolution global non-hydrostatic model, as used in this five-year project, improves the ability to forecast intra-seasonal oscillations and associated tropical cyclogenesis compared with that of the relatively coarser operational models currently in use. The impacts of the sub-kilometer resolution simulation and the multi-decadal simulations using NICAM are also reviewed.

Keywords: K computer, NICAM, Intra-seasonal oscillations, Madden-Julian oscillation, Tropical cyclone, Global non-hydrostatic model

Introduction

The K computer began operation in September 2012 at RIKEN Advanced Institute for Computational Science, Japan. The Strategic Programs for Innovative Research (SPIRE), an initiative launched in October 2010 in Japan, aims to leverage the unparalleled power of the K computer to produce the world's most cutting-edge simulation studies in five key strategic fields. SPIRE also aims more generally to boost the creation of promotional frameworks for computer science and technology and to yield significant social breakthroughs. Field 3 of SPIRE, "Advanced prediction research for natural disaster prevention and reduction," consists of weather and climate simulations, including earthquake and tsunami simulations (Oishi et al. 2015; Tsuboi et al. 2016; Ando et al. 2016; Hyodo et al. 2016). Studies using meso-scale atmospheric modeling (Saito et al. 2013; Kunii 2014a, 2014b; Chen et al. 2015a, 2015b; Duc et al. 2015; Ito et al. 2015; Kobayashi et al. 2015a; Seko et al. 2015; Yokota et al. 2016) and global modeling are also included under weather simulations. Here, we review the outcome of the global high-resolution atmospheric modeling studies conducted from 2012 to 2016 under

* Correspondence: satoh@aori.u-tokyo.ac.jp
[1]The University of Tokyo, 5-1-5 Kashiwanoha, Kashiwa, Chiba 277-8568, Japan
Full list of author information is available at the end of the article

Field 3 of SPIRE and preview the challenges for the next K computer project, called the FLAGSHIP2020 project (Future LAtency core-based General-purpose Supercomputer with HIgh Productivity). Running from 2014 to 2019, FLAGSHIP2020 is in preparation for the Post-K supercomputer, which was originally planned to start operation in 2020. (Recently, it was announced that the operation of the Post-K supercomputer has been delayed by at least 2 years.)

For global atmospheric modeling using the K computer, new research using the non-hydrostatic icosahedral atmospheric model (NICAM; Tomita and Satoh 2004; Satoh et al. 2008, 2014) has been pursued by enhancing horizontal resolution, ensemble sizes, or duration of integration time. In particular, we established the following research targets at the beginning of the project: (1) enable extended-range forecasts for a timeframe of 10–30 days, for better simulations of intraseasonal oscillations (ISO) and (2) enable projection of tropical cyclones. Before the K computer became available, NICAM was known to more accurately reproduce individual ISOs in case studies at a horizontal grid spacing between 3.5 and 14 km (Miura et al. 2007a; Oouchi et al. 2009a, 2009b; Yanase et al. 2010). At that time, simulations such as these were performed mainly on the Earth Simulator at the Japan Agency for Marine-Earth Science and Technology (JAMSTEC). Multi-year simulations were also conducted with the 7 km grid spacing NICAM using the Athena computer (operated by the University of Tennessee's National Institute for Computational Science and hosted by Oak Ridge National Laboratory in the USA) and showed better simulations of boreal summer ISO (Satoh et al. 2012b; Kinter et al. 2013). The development of the K computer offered the opportunity to obtain more robust performance from NICAM by drastically increasing ensemble size and integration time together with resolution.

Review

This article reviews the outcomes of the NICAM studies using the K computer under Field 3 of SPIRE and previews challenges for the forthcoming Post-K supercomputer under the FLAGSHIP2020 project. In the "Computational aspects" section, we describe how NICAM was adapted for use on the K computer. Following this, we list major outcomes using NICAM with the K computer in the sections "Sub-kilometer global simulation," "Asian summer monsoon," "Madden-Julian oscillation," "Boreal summer intra-seasonal oscillation and tropical cyclogenesis," "Atmospheric Model Intercomparison Project-type experiments," and "Projection of tropical cyclones." The challenges to overcome in the near future are reviewed in "Future perspectives," and we conclude our review in the final section, "Conclusions."

Computational aspects
Background

The K computer was the first 10 petaflop supercomputer (10^{15} floating-point operations per second; Yokokawa et al. 2011). This massively parallel scalar machine was twice ranked as the world's top supercomputer in the TOP500 List in 2011 (Top 500 2011a, 2011b). Furthermore, it showed the top performance in new benchmarks such as the high-performance conjugate gradient (Kumahata et al. 2016) and Graph500 (Graph 500 2014). Two recipients of the Gordon Bell Prize used the K computer to perform their award-winning work (Hasegawa et al. 2011; Ishiyama et al. 2012). In the research area of earth science, which Field 3 of SPIRE also covers, an urban-scale earthquake simulation study using the K computer was selected as a finalist for the Gordon Bell Prize (Ichimura et al. 2014, 2015). These results reflect the K computer's high performance not only in benchmarks, but also in applications.

The K computer system possesses characteristics advantageous for weather and climate research, which require data-intensive simulation models. The performance of these models is limited not only by the number of calculations processed in the computer's central processing unit but also by data transfer processes, such as the memory-cache transfer, inter-node communication, and file input/output (I/O). The K computer has relatively high memory throughput (0.5 byte per floating-point operation) compared to the today's massively parallel scalar supercomputers. The asynchronous, distributed file I/O system of the K computer minimizes waiting time for frequent output of the atmospheric variables during the simulation. The K computer is also highly resilient and has a lower machine failure rate (Yamamoto et al. 2014) compared to other peta-scale supercomputers, which is very important for long-term climate simulations with a large number of computational nodes.

At an early stage of the development of the K computer, NICAM was selected as one of the target applications and was used in many of the computer's performance optimizations. In experiments with a small number of nodes, we achieved about 10% of the performance efficiency (Terai et al. 2014; Yashiro et al. 2016b), which is almost competitive with that of the state-of-the-art weather and climate models. NICAM also shows good scalability, up to almost the full-node of the K computer in the weak scaling test.

Single-node performance

For general model applications, a small number of major calculation parts occupy most of the floating-point operation and elapsed time. These parts are called hot spots, and we usually focus on these hot spots for the optimization of computational performance. However, most weather and climate models do not have a clear hot spot. Every part of the source code contains a small computation that consumes a little time, a characteristic called a flat profile. Weather and climate models generally have a flat profile, where every part of the source code has the potential to deteriorate the total performance. The effort to optimize NICAM on the K computer aimed to overcome the flat profile (Yashiro et al. 2016b). Optimizations were applied mainly to the calculation parts (kernels) such as divergence, gradient, and Laplacian operations. These kernels are a type of stencil calculation, which updates the grid point value using the value of surrounding grids. Because the finite-volume method is used for numerical discretization in NICAM, these kernels are the most important part for solving the fluid dynamics equations of the atmosphere. After optimization, each kernel showed improved performance (Terai et al. 2014). However, we achieved only a 1% gain in sustained performance over the total simulation, because the ratio of elapsed time of the kernels to total time decreased. This result was related to the effect of Amdahl's law (Amdahl 1967). From this experience, we realized that we needed to find the time-consuming, less computationally intensive parts of the source code and remove them. As a result of these efforts, we achieved about 10% of the sustained performance, which was appropriate to the memory performance of the K computer. We also succeeded in reducing the model execution time by 30%.

Scalability

Terai et al. (2014) performed a weak scaling test using NICAM on the K computer and showed good scalability of elapsed time and sustained performance of the NICAM (Fig. 1). The test was conducted by changing the number of nodes and the horizontal grid spacing. Horizontal grid spacing was decreased from 56 km to 440 m, which corresponded to an increase in the number of nodes from 5 to 81,920. The total time was divided into three parts: (1) the dynamical step, which covered the time taken to solve the fluid dynamics; (2) the physical step, which covered the physics components of the calculations, such as cloud microphysics, atmospheric radiation, and boundary layer turbulence; and (3) the communication step, which comprised the inter-node data transfer using message passing interface (MPI) (referred to as comm_data_transfer in Fig. 1). The computations and communications were completely separated using barrier statements of MPI (MPI_barrier), in which all tasks must synchronize. The result shows good scalability of total elapsed time. The timing of the dynamical step does not change as the number of nodes is increased, because the number of operations remains almost the same. The increase in the communication time as the number of nodes increases is not significant because most of the communications are conducted node-to-node using halo exchanges. The increase in the total elapsed time is mainly caused by increases in the time of the physical step due to the load imbalance between the processes. Terai et al. (2014) found that the imbalances occurred mainly in the cloud microphysics scheme. They found that the global map of the number of floating-point operations in the cloud microphysics showed inhomogeneous distribution, similar to that of ice hydrometeors. Overall, we achieved a maximum performance of 0.87 petaflops with 81,920 nodes (655,360 cores), as shown by Terai et al. (2014).

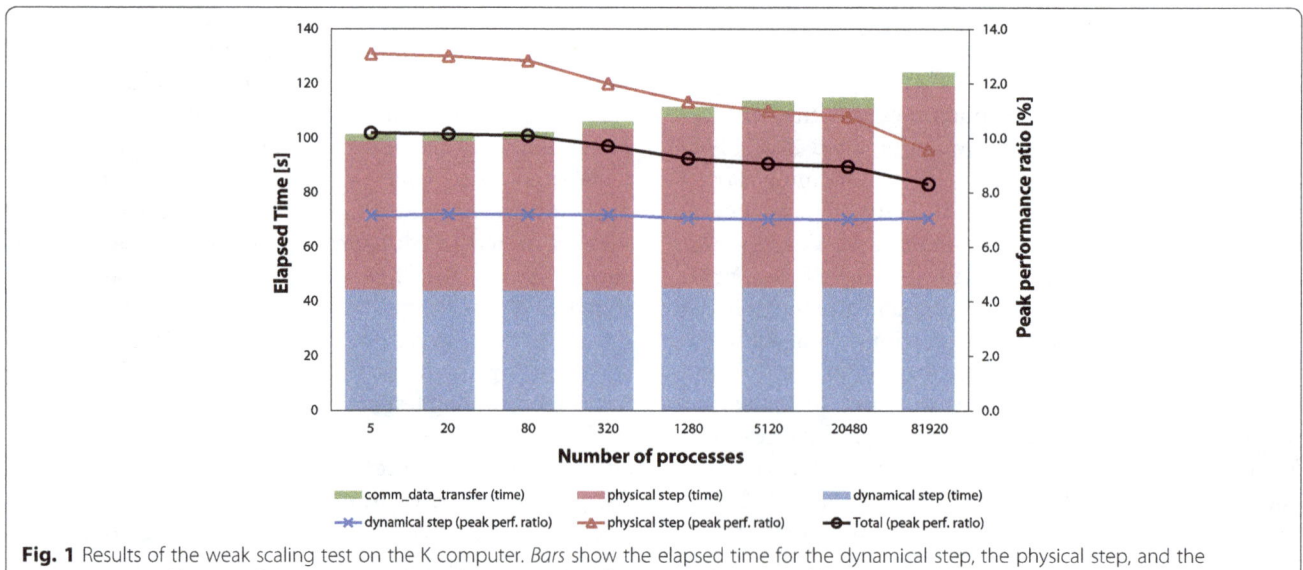

Fig. 1 Results of the weak scaling test on the K computer. *Bars* show the elapsed time for the dynamical step, the physical step, and the communication (data transfer) step. *Lines* indicate the sustained performance. After Fig. 2 of Terai et al. (2014)

The results of the strong scaling test are different from those of the weak scaling test (Fig. 2). In the strong scaling test, we used NICAM with horizontal grid spacing of 14 km and 38 vertical layers. The number of grid points in each process decreased from 33,800 to 100 as the number of nodes increased from 80 to 40,960. The model showed good scalability up to 2560 nodes. The number of simulation days per wall-clock day increased in proportion to the number of nodes. However, the performance saturated beyond 2560 nodes, due to the relative increase in communication time; at 40,960 nodes, the communication time occupied over 50% of total elapsed time. This result led to production runs being conducted with 640 nodes in the studies described in the following sections. For long-term simulations in future studies, strong scalability will need to be improved. Reducing the number of subroutine calls within the MPI communication is one possible solution.

Pre- and post-processing

Owing to the comprehensive optimizations of NICAM, we achieved high efficiency and good scalability. However, the elapsed pre- and post-processing times were relatively increased in the total simulation time for the following reasons. The remapping process has one of the heaviest workloads in post-processing. To use common analysis and visualization tools, we usually remap the meteorological variables from an icosahedral grid to a geodesic (latitude-longitude) grid. For example, the data size of the global 0.87 km horizontal grid spacing simulations was 6 TB for one snapshot of all 3-D and 2-D variables (Miyamoto et al. 2013, 2015; Kajikawa et al. 2016). The required time for remapping and analyzing all data was estimated to be more than 2 months using typical computer clusters, which is much longer than the time required for simulation on the K computer. To solve this problem, we developed programs for parallel analysis on the K computer without remapping, thereby reducing the post-processing time from several months to less than 1 h. This improvement was due to massively parallel I/O rather than parallel computation. In the future, simulation sizes are expected to increase even more drastically. The most important consideration is the total elapsed time of the workflow in simulation studies, which includes pre- and post-processing, analysis, and visualization. All computational work will need to be carried out on the supercomputer to keep this time to a minimum.

Future directions

NICAM has been selected as the target application for the development of Japan's next-generation flagship computer (the Post-K supercomputer), together with the local ensemble transform Kalman filter (LETKF). NICAM and LETKF play an important role in the process of application-architecture co-design. Generally, computations of weather and climate models are not expected to be carried out faster than improvement in microprocessors, because the rate of improvement in memory throughput is slower than the floating-point operation speed. Thus, the computational performance of NICAM will be improved by modifications of the algorithm and its optimization (e.g., changes of loop order, data layout, and call structure).

Active use of mixed-precision floating-point operation is one possible approach to further enhance computational performance. The bit length of floating-point value affects the data size of memory transfer and the compression ratio of the output data. The effect of less precise values on the simulation results should be investigated. In recent years, the development of components and schemes such as the atmosphere-ocean coupling system with coupling library Jcup (Arakawa et al. 2014) and the ensemble-based data assimilation system (Terasaki et al. 2015) have expanded the capabilities of the simulations using NICAM. While taking these developments into account, we must increase the speed of total workflow. Maximizing the power of future supercomputers will enable simulations with even more sophisticated physics and/or higher resolution and will ensure that it becomes a powerful tool for further understanding weather and climate phenomena.

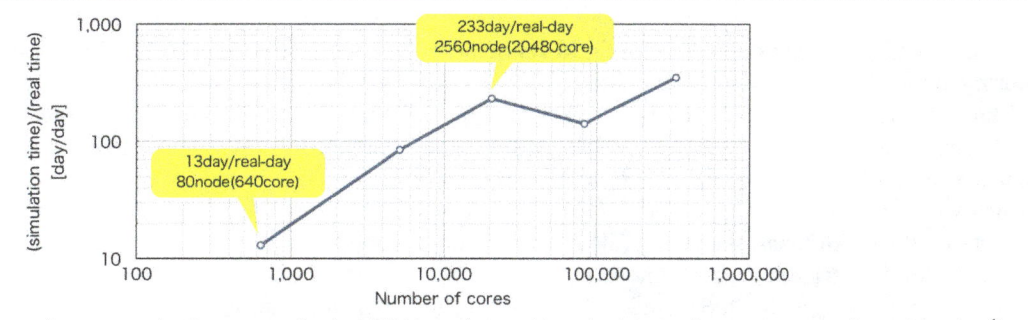

Fig. 2 Results of the strong scaling test on the K computer for the NICAM simulations. *Vertical axis* is simulation time per real time [day day^{-1}]. *Horizontal axis* is the number of cores

Sub-kilometer global simulation

Background

One of the greatest challenges in climate modeling is achieving more accurate simulations of clouds (Stevens and Bony 2013; Palmer 2014). Among various types of naturally occurring clouds, deep moist convective clouds have important roles in determining the structure of atmospheric circulation through the release of latent heat. This deep convection is also part of the cloudy atmospheric disturbances that sometimes cause natural disasters. Since it has been impractical to simultaneously simulate all-scale phenomena in global atmospheric models, from deep convection on a scale of $O(1)$ km to global circulation on a scale of $O(10^4)$ km, deep convection has been expressed as a parameterization in conventional general circulation models (GCMs) (e.g., Arakawa and Schubert 1974; Kain and Fritsch 1990). NICAM was developed to simulate the global atmospheric circulation without cumulus parameterization (Satoh et al. 2008, 2014). Previous studies using NICAM have shown advanced performance in simulating large-scale convective systems and disturbances mainly using the Earth Simulator (e.g., Miura et al. 2007a; Sato et al. 2009; Nasuno and Satoh 2011; and references in Satoh et al. 2014). However, their grid spacing (several kilometers) was coarser than or similar to that of the observed convection scale. The highest resolution simulations using the Earth Simulator were the 3.5 km horizontal grid spacing simulations by Tomita et al. (2005) and Miura et al. (2007a, 2007b). Meanwhile, studies using cloud-resolving models in a confined area, such as large eddy simulation (LES) models, have indicated a change in deep convection characteristics at grid spacing less than 1 km (Petch et al. 2002; Bryan et al. 2003). Under Field 3 of SPIRE, we successfully conducted the first-ever global atmospheric simulation with sub-kilometer grid spacing using NICAM, which opened the door for simulations resolving deep convective cores in global atmospheric models. This work also allowed us to highlight the resolution dependency of simulated convection in the global model as well as its statistical features.

Results

Miyamoto et al. (2013) conducted a set of grid-refinement experiments with grid spacing that varied from 14 to 0.87 km. Further analysis of these experiments is shown by Miyamoto et al. (2015), Kajikawa et al. (2016), and Yashiro et al. (2016a), and the details of the experimental settings are described in Miyamoto et al. (2013) and Kajikawa et al. (2016). Although the large-scale characteristics of global cloud distributions are almost unchanged between the simulations with 14, 3.5, and 0.87 km grid spacing, the simulation with sub-kilometer resolution provided a more detailed description of the smaller scale cloud structure (Fig. 3). The resolution dependency showed that the properties of simulated convection were qualitatively changed between 3.5 and 1.7 km grid spacing. We defined convective cores in the simulations and made a composite structure of convective cores averaged over the globe. In simulations at finer resolution than the 3.5 km grid spacing, the convective core is expressed by multiple grid points instead of a single grid point for the 3.5 km grid spacing; the dependency of the number of convective cores on resolution changes, with smaller dependency for finer resolutions (Fig. 4); and the minimum distance between convective cores is larger than four grid points for the finer resolutions than the 3.5 km grid spacing (Fig. 4). From these results, the horizontal grid spacing around 2 km can be viewed as the minimum required resolution for a deep convective resolving model over the entire Earth, although the required grid spacing might depend on numerical discretization methods.

Miyamoto et al. (2015) conducted detailed analysis of deep convection in the simulation with the finest 0.87 km grid spacing. They defined cloud disturbances such as the Madden-Julian oscillation (MJO), tropical cyclones (TCs), mid-latitude lows, and fronts and analyzed the statistical features of deep convection associated with these disturbances. The cloud-top height, strength of upward motion, and the environmental field of the convection were different in each type of disturbance. For example, the MJO convection was generated under high convective available potential energy (CAPE) and relatively weak lower convergence, whereas the TC convection was accompanied by low CAPE with strong convergence.

Kajikawa et al. (2016) analyzed the grid-refinement experiments to address the difference in the resolution dependence of the simulated convection by location and environment over the globe. This study extends the work of Miyamoto et al. (2013), which mainly discussed globally averaged convection properties. Kajikawa et al. (2016) showed that convective clouds over the tropics are more dependent on resolution than those in mid-latitudes; on the other hand, no significant difference is found in the resolution dependence of convection properties between land and ocean. Convection within cloud disturbances such as the MJO and TCs also shows large resolution dependency. Consequently, deep convection that is not categorized as cloud disturbances makes a large contribution to the global mean convection properties in Miyamoto et al. (2013). It should be stressed that most deep convection is expressed by multiple grid points in the 0.87 km grid spacing simulation even in tropical cloud disturbances.

Yashiro et al. (2016a) investigated the diurnal cycle of precipitation in this series of grid-refinement experiments and also found that the resolution dependency of

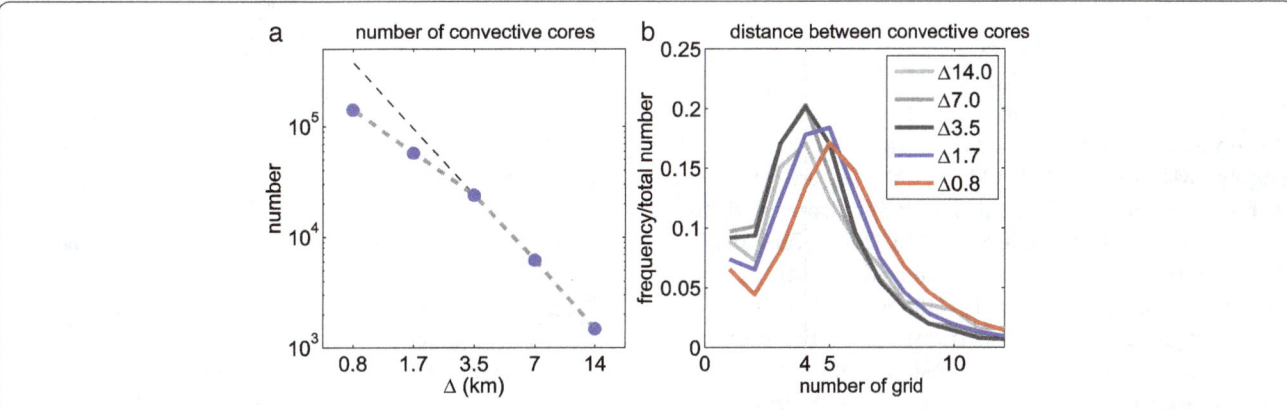

Fig. 3 Resolution dependencies of simulated cloud distribution. Global view of simulations with **a** 14 km, **b** 3.5 km, and **c** 0.87 km grid spacing. **d–f** Focus over the western North Pacific region of (**a–c**), respectively. After Fig. 3 of Kajikawa et al. (2016)

Fig. 4 Resolution dependencies of convective features. **a** Number of convective cores, and **b** grid distance to convective cores. The *dashed line* in (**a**) indicates a log Δ^4 crossing at the point of simulation results with 14 km grid spacing as a reference. After Fig. 4 of Miyamoto et al. (2013)

its characteristics changed at a grid spacing around 2–3 km. The simulations using 1.7 and 0.87 km grid spacing better reproduced the observed peak time of the diurnal cycle compared to the experiments with coarser grid spacing. This could be a result of the better representation of deep convective cores, as shown by Miyamoto et al. (2013) and Kajikawa et al. (2016).

Future directions

We showed the resolution dependency of the simulated convection properties and their differences in various environments through a set of grid-refinement experiments. The simulation with sub-kilometer grid spacing revealed new results that differed from previous studies; for example, deep convective cores were expressed by multiple grid points. Higher spatial resolution and more realistic convection simulations are required to better understand cloud disturbances from the meso-scale to the synoptic scale (Prein et al. 2015). Moreover, longer integration at sub-kilometer grid spacing would enable us to investigate the interaction between convection and larger-scale cloud disturbances. Using the Post-K super-computer, we estimate that we can perform experiments at further finer resolution, such as 220 m, to resolve more detailed structure of clouds.

Asian summer monsoon
Background

The Asian monsoon system is characterized by a seasonal cycle of convective activity and atmospheric circulations (e.g., Murakami and Matsumoto 1994). The variability of the monsoon system's effects on weather and climate over Asia is largely achieved through local convective activities. This variability is also affected by synoptic scale tropical disturbances. For example, the monsoon onset is often triggered by tropical disturbances or ISO. Recently, global climate modeling has made advances in many aspects of Asian summer monsoon simulations (Sperber et al. 2013; Ogata et al. 2014; Lee and Wang 2014). However, numerical representations of the seasonal transition of the monsoon, such as the monsoon onset, with scale interactions remain challenging. Meanwhile, NICAM has shown an advantage over conventional GCMs in simulating tropical disturbances, such as TCs (Fudeyasu et al. 2008; Taniguchi et al. 2010; Nakano et al. 2015) and MJO (Miura et al. 2007a; Miyakawa et al. 2014). Under Field 3 of SPIRE, we investigated the seasonal transition of the monsoon and anticipated a potential extension of its predictability using a large number of ensemble NICAM simulations, in contrast to previous studies that used simulations of only a single or a few members.

Experimental setup

Kajikawa et al. (2015) investigated the process of the Indian summer monsoon onset in 2012 based on ensemble simulations using NICAM with a grid spacing of about 14 km and 38 vertical levels. Moist processes were calculated using the NICAM single-moment water 6-cloud microphysics scheme (NSW6; Tomita, 2008) without any convective parameterization. Sea surface temperature (SST) was forecasted using a slab ocean model to allow diurnal variation and nudged to the persistent SST anomaly at the initial time with an e-folding time (that is, the time in which SST is nudged to specified values) of 7 days. The ensemble simulations were started at 00 UTC each day during the period from 10 May to 10 June 2012, and integrated for 30 days. Using the same set of NICAM ensemble simulations, Yamaura et al. (2013) clarified the role of TCs on the migration of the Baiu front.

Results

The onset of the 2012 Indian summer monsoon was announced on 5 June that year by the Indian Institute of Tropical Meteorology. The NICAM simulations using initial conditions 2 weeks before the onset (15 May) reproduced the onset very well (Fig. 5). Both the observations and simulations included the abrupt onset with northward and eastward migrating tropical disturbances over the Bay of Bengal and the Arabian Sea in late May. Based on a comparison with the Japan Meteorological Agency's operational Ensemble Prediction System, Kajikawa et al. (2015) pointed out that the better reproducibility of the tropical disturbances seen in the NICAM results extended the predictability of the Asian summer monsoon transition phase.

Yamaura et al. (2013) specifically used the NICAM simulations initialized at 29 and 30 May 2012 to clarify

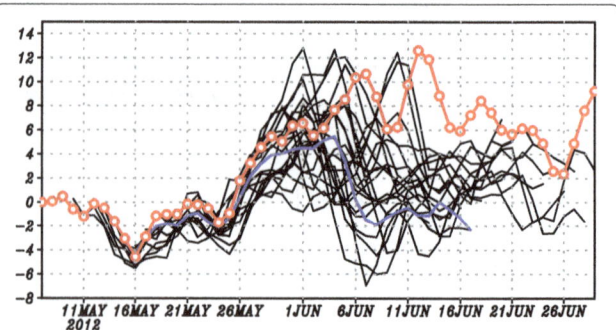

Fig. 5 Time series of the Indian summer monsoon indices for the ensemble simulations. The *black curves* are for each run, the *blue curve* is for averaging five members with initial conditions covering May 15–19, and the *red curve* is for the actual observations. The Indian summer monsoon index is defined as the meridional shear of zonal winds at 850 hPa over the Indian continent (40°E–80°E, 5°N–15°N minus 70°E–90°E, 20°N–30°N). After Fig. 1 of Kajikawa et al. (2015)

the role of a TC in the migration of the Baiu front. Observations revealed that the Baiu front shifted northward with the TC. A simulation integrated from 29 May reproduced both the TC and the northward migration of the Baiu front, whereas the Baiu front stayed stagnant in another simulation integrated from 30 May. Since the TC in the latter simulation developed farther eastward than the observed TC, Yamaura et al. (2013) suggested that the potential extension of the predictability of the Baiu front migration was due to the realistic reproducibility of TCs.

Seasonal simulations of boreal summer using the global 7 km grid spacing NICAM have captured the Asian summer monsoon as a case study (Oouchi et al. 2009a, 2009b; Satoh et al. 2012b; Kinter et al. 2013). However, these simulations have large biases in precipitation and lower tropospheric wind fields, particularly over the western North Pacific. These biases were partly due to insufficient spin-up of the land surface model, which was upgraded in the experiments conducted during Field 3 of SPIRE.

Another achievement of this project was the first-ever 30-year simulations (see "Atmospheric Model Intercomparison Project-type experiments" section; Kodama et al. 2015). The 30-year simulation successfully simulated the climatology of the annual cycle of the western North Pacific monsoon and its inter-annual variability (see Figures 14 and 15 of Kodama et al. 2015). In Murakami and Matsumoto (1994), the climatological Asian monsoon was defined by differences between the annual maximum and minimum outgoing longwave radiation (OLR) (Fig. 6). The NICAM simulation adequately represents each monsoon sub-system in the observation (e.g., northern Australia-Indonesia monsoon, Southeast Asian monsoon, western North Pacific monsoon); although, the magnitude is stronger by approximately 20–40 W m^{-2}. The NICAM simulation also included a northward extension of the Southeast Asian monsoon

and the western North Pacific monsoon which are related to the northward displacement of a low-level westerly and a subtropical high; this in turn led to earlier termination of the Baiu season and to a weaker monsoon trough (Kodama et al. 2015).

Future directions

Both NICAM ensemble simulations and climate simulation showed realistic behavior of the Asian summer monsoon system including the relationship between TCs and the Baiu front. However, similar to other climate models, NICAM suffers from model biases. In particular, the Asian summer monsoon simulations are severely affected by the warm temperature bias over continents. In addition, the mid-latitude jets were stronger in the simulation than in observations partly because a gravity drag parameterization scheme was turned off in the 14 km grid spacing NICAM simulations. In future, simulations of Asian summer monsoons will need to reduce such biases in order to improve their performance; this is currently being tested through development of the land surface model and the introduction of the gravity wave drag scheme.

NICAM simulations with atmosphere and ocean coupling are also being investigated, because realistic responses of convection to basin-scale ocean surface conditions are strongly related to ocean-atmosphere coupling (Wang et al. 2005). In addition, process-oriented diagnosis in comparison with other climate models (Sperber et al. 2013; Ogata et al. 2014; Kusunoki and Arakawa 2015; Zou and Zhou 2015) is planned to understand the effects of basic-state biases on monsoon systems and to make the necessary improvements. The multi-scale processes of monsoons and the Baiu front under global warming conditions in cloud-system resolving simulations are another ongoing research topic (e.g,. Krishnamurthy et al. 2014).

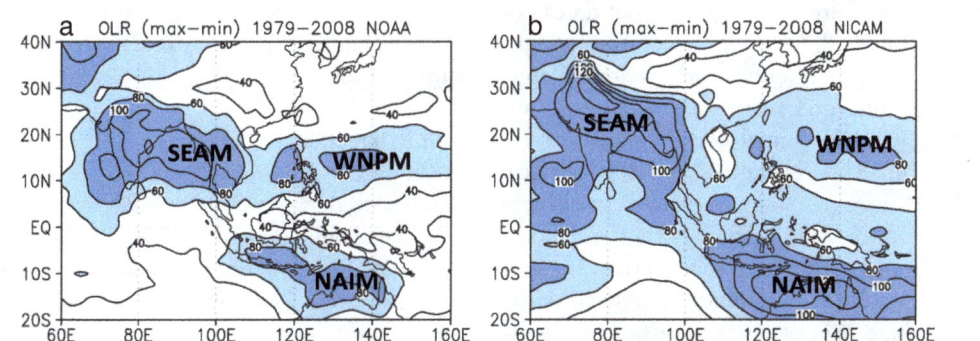

Fig. 6 Horizontal distribution of OLR comparing Asian monsoon between simulations and observation. **a** National Oceanic and Atmospheric Administration (NOAA) OLR and **b** OLR of the NICAM AMIP-type simulation defined by the difference between the annual maximum and minimum OLR in the pentad mean climatology (1979–2008). The northern Australia-Indonesia monsoon (NAIM), Southeast Asian monsoon (SEAM), and western North Pacific monsoon (WNPM) are identified after Murakami and Matsumoto (1994)

Madden-Julian oscillation

Background

The MJO, which is often described as an eastward propagating atmospheric pulse with a typical lifespan of 30–60 days, is one of the dominant disturbances in the equatorial atmosphere at the intra-seasonal timescale (Madden and Julian 1972). It is viewed as a key source of predictability for the extended-range forecasts that target the time range of 10–30 days. More than 40 years since its foundation, the majority of GCMs still struggle to produce a satisfactory MJO (Hung et al. 2013). NICAM had enjoyed success in producing a particular MJO event, maintaining a highly accurate eastward propagation speed along with realistic precipitation and zonal wind structures (Miura et al. 2007a). In Field 3 of SPIRE, we utilized the increased computational resource of the K computer to conduct ensemble simulations of MJOs using NICAM to statistically assess the model's ability to predict MJOs (Miyakawa et al. 2014). Applying a widely used evaluation method proposed by Gottschalck et al. (2010), NICAM was diagnosed to have a prediction limit of 26–28 days. The precipitation patterns associated with MJO phases also compared well with observations.

Experimental setup

The following procedure based on the real-time multivariate MJO (RMM) indices (Wheeler and Hendon 2004) was applied to the boreal winter (October–March) MJOs during an evaluation period (2003–2012) to assign initial dates for the experiment. We identified target MJO cases when a succession of observational RMM plots move counterclockwise through phase 2 to phase 5 (corresponding to the MJO convective envelope traveling from the western Indian Ocean to the eastern Indonesian Archipelago) without retreating more than one phase and had an average amplitude greater than 1 over the span between phases 2 and 5. For each of the MJO cases extracted, the first dates on which the RMM plot falls in phases 8, 1, and 2 are assigned as initial dates. For the 19 MJO cases identified, 54 initial dates were assigned. We used the European Centre for Medium-Range Weather Forecasts interim reanalysis (ERA-Interim) dataset (Dee et al. 2011) for initial conditions of the atmosphere and ocean. The resolution and the physical schemes of NICAM were the same as those described in the "Asian summer monsoon" section except for modified cloud microphysics parameters that control the formation and fall speeds of precipitation particles and minor updates of land-process/boundary layer scheme parameters. A mixed-layer ocean model was applied, and its SST was nudged to externally provided SST at an e-folding time of 7 days. The external SST was defined as the sum of the time-varying mean annual cycle and constant anomalous component. The difference from the mean annual cycle was averaged over the week before the initial date to derive the anomalous component. Thus, no information observed after the initial date was used in the simulations. Further details are found in Miyakawa et al. (2014).

Results

NICAM maintained a valid (>0.6) MJO skill score for 26–28 days depending on the initial MJO phase, and for 27 days when all 54 cases were included. Composites of observed and simulated precipitation anomalies for MJO phases 3, 5, and 7 are shown in Fig. 7. The horizontal structure of the simulated precipitation anomalies closely resembles the observations, even for phase 7, for which the average lead time was 28 days. However, precipitation anomalies were overestimated over the central Pacific in phase 5. Such a bias is occasionally found in 14 km grid spacing NICAM, perhaps due to under-resolved local vertical moisture transport, which could be a common problem for models that explicitly calculate cloud systems without cumulus parameterization. The simulation series includes an MJO case that occurred in November–December 2011 during the Cooperative Indian Ocean Experiment on Intraseasonal Variability in Year 2011 and Dynamics of the MJO field campaign (CINDY2011/DYNAMO; Yoneyama et al. 2013). Figure 8 shows the observed and simulated convective signals that clearly propagate at a very similar eastward speed over the tropical Indian to the western Pacific Ocean (10°S–10°N, 40°E–160°W). Time-height sections of observed and simulated zonal wind over Gan Island (not shown) further reveal that the model captures the depth and timing of the intrusion of the westerly winds as well as the deepening of moisture with time, both of which are key aspects of MJO propagation. Despite a lead time of nearly 4 weeks, the model also appears to capture the occurrence of the next MJO-like signal in mid-December. The successful prediction of the event provides the opportunity to compare a global non-hydrostatic model simulation with a major in situ observation for the same MJO event for the first time.

Future directions

Although NICAM appears to be one of the unique models that can produce MJO events with high accuracy, we believe that the model still does not live up to its potential in terms of MJO prediction. The limit in available resources has prevented us from applying ensemble simulations of significant size for each event. However, a next-generation supercomputer designed to be roughly 20–100 times more powerful than the K computer should be able to accommodate such simulations. The initialization process of this experiment was a simple

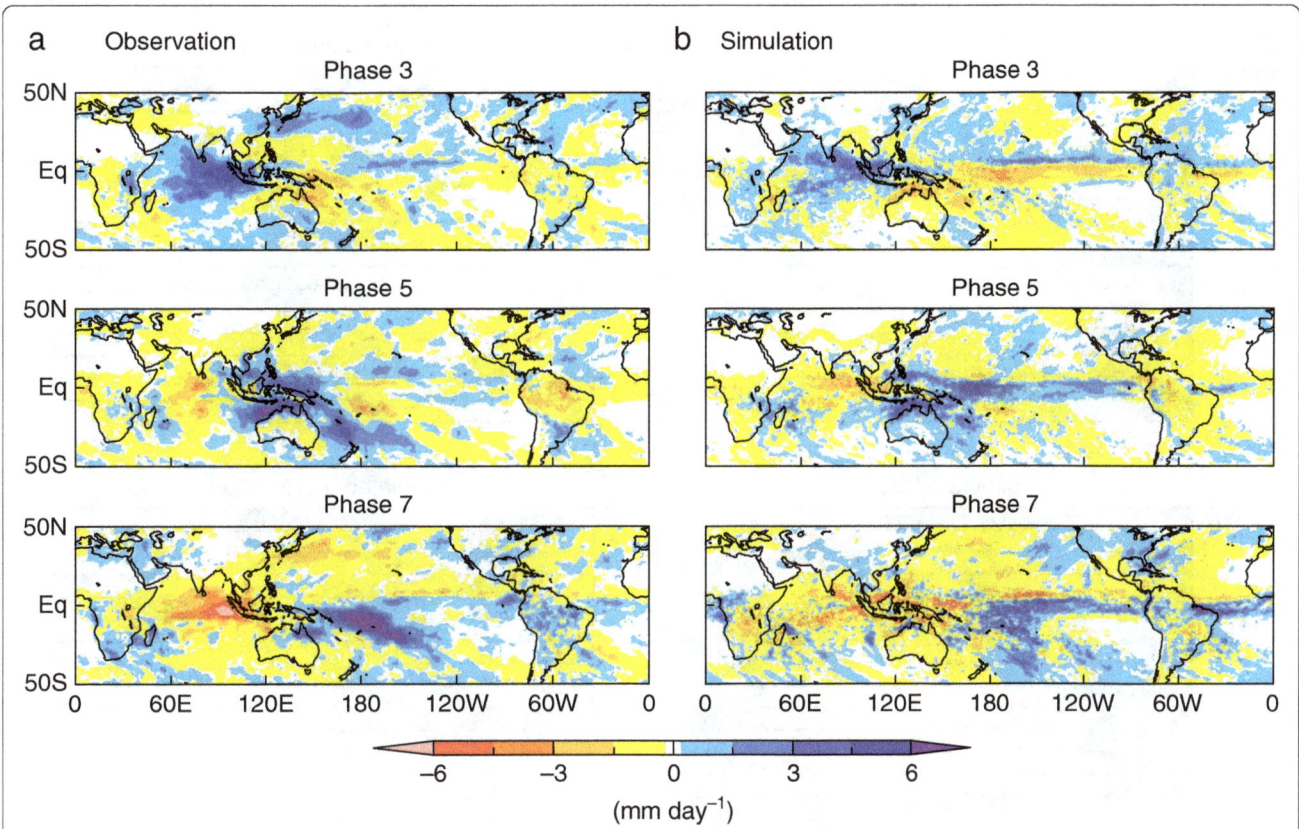

Fig. 7 Precipitation anomaly composites for different MJO phases. **a** Observed 1° daily product version 1.2 by the Global Precipitation Climatology Project (Huffman et al. 2001), and **b** simulation using NICAM. The average number of days from the initial dates of the simulations are 16 days (phase 3), 25 days (phase 5), and 28 days (phase 7). Resolution of the simulation output is lowered to 1° mesh to equal the resolution of the observational data set. Anomalies are calculated as deviations from 40-day mean values in each case. After Fig. 2 of Miyakawa et al. (2014)

interpolation from the ERA-Interim dataset, which allows significant initial shock, but a new data assimilation system specifically designed for NICAM is being developed. The use of a simple mixed-layer ocean model limits the model skill in situations in which dynamical features of the ocean drastically affect the SST conditions. A fully coupled version of NICAM with a three-dimensional dynamical ocean is now available and currently being tested (NICAM-COCO coupled model; Satoh et al. 2014).

Boreal summer intra-seasonal oscillation and tropical cyclogenesis

Background

Tropical cyclones cause huge socio-economic impacts, so more accurate TC forecasts are demanded. Because forecasting tropical cyclogenesis (TC genesis) is still a challenge in numerical weather forecasting, the predictability of TCs at sub-seasonal to seasonal time scales (2 weeks to 2 months) is promoted worldwide (Vitart et al. 2012, 2017). TC genesis is affected by large-scale environmental conditions, such as low-level cyclonic vorticity, small vertical wind shear, the convective

instability, and mid-level humidity (Gray 1975, 1979). The boreal summer intra-seasonal oscillation (BSISO; Wang and Rui 1990; Wang and Xie 1997) modulates the large-scale environment and TC activity. For example, Nakazawa (2006) pointed out that the active phase of BSISO in the 2004 summer season successively generated TCs, and persistent steering flow led to a record-breaking number of TC landfalls in Japan. Therefore, the accurate prediction of BSISO and the modulation of large-scale environmental conditions associated with BSISO are believed to lead to accurate forecasting of TC genesis.

Studies using NICAM show that the model is a promising tool for accurate BSISO and TC genesis forecasting. For example, Oouchi et al. (2009a) demonstrated that NICAM successfully simulated modulations of BSISO and that TC genesis was captured 3 weeks in advance. However, the authors examined only a limited number of TC genesis cases using a single run because of the limitations in computational resources. The K computer enables us to examine the predictability of the BSISO and TC genesis using the ensemble approach. Here, we examine the predictability

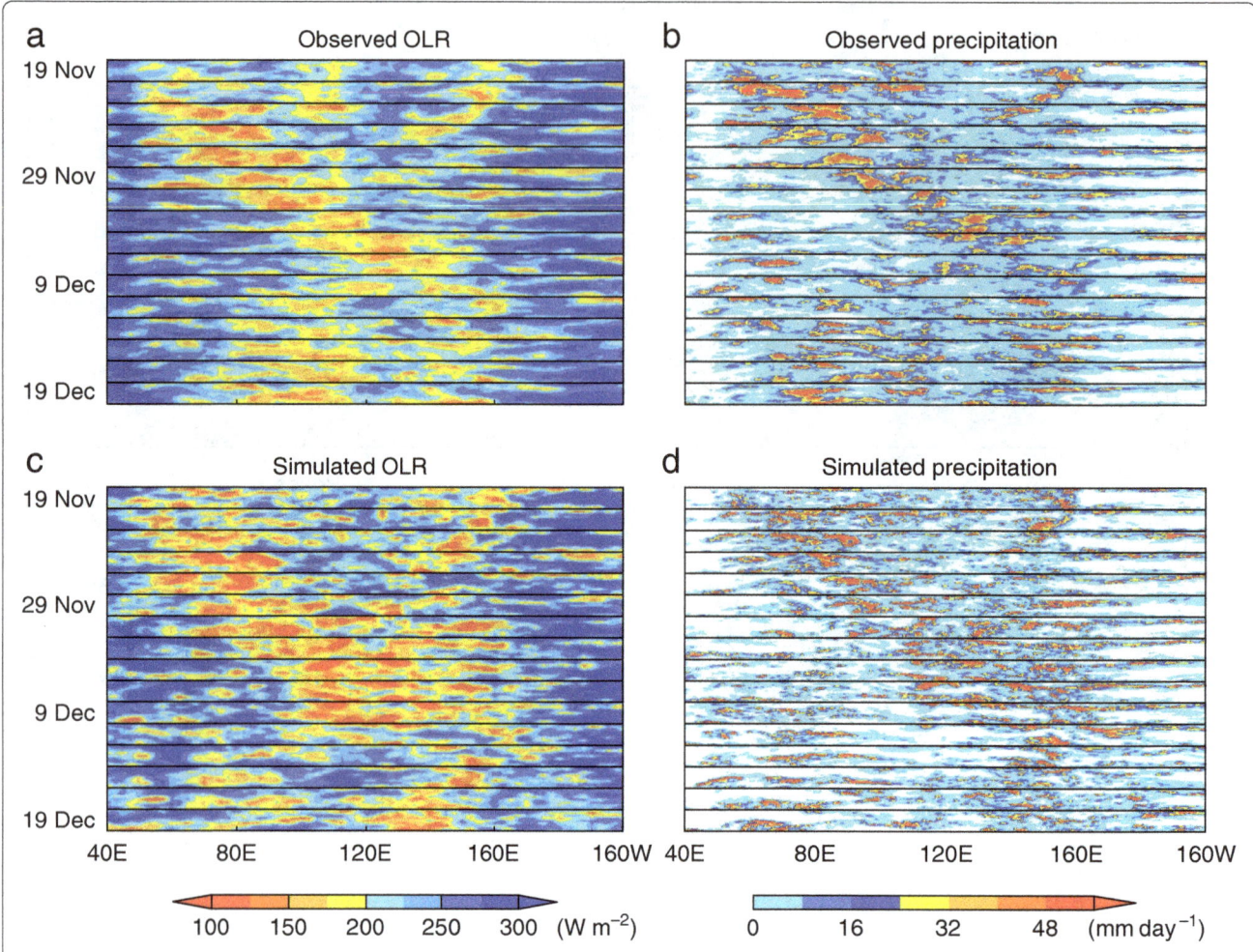

Fig. 8 Precipitation anomaly composites for different Madden-Julian oscillation (MJO) phases. **a** Interpolated OLR by NOAA polar-orbiting series of satellites (NOAA-OLR; Liebmann and Smith, 1996) and **c** OLR by a single simulation initialized at 00 UTC 17 November 2011. **b** Precipitation by the 3B42 product of the Tropical Rainfall Measuring Mission (TRMM) satellite (Huffman et al. 2007) and **d** precipitation by the simulation. The figures consist of slices that show horizontal snapshots of the tropical Indian Ocean to the western Pacific Ocean (10°S–10°N, 40°E–160°W). The resolution of the simulated OLR is lowered to 2.5° mesh to equal the resolution of the NOAA-OLR data set. The resolutions of the TRMM 3B42 data set and the simulated precipitation are lowered to 1° mesh. After Fig. 3 of Miyakawa et al. (2014)

of eight cases of TC genesis that occurred successively in August 2004.

Experimental setup

Nakano et al. (2015) performed 30-day simulations for each day of August 2004; that is, a total of 31 ensemble 30-day simulations were performed. The initial conditions of the atmosphere and SST were derived from the ERA-Interim dataset (Dee et al. 2011). The resolution and the physical schemes of NICAM were the same as those described in the "Asian summer monsoon" section. This setting is "forecast-mode" because the model never knows the observed SST during the time integration.

A candidate vortex that corresponds to an observed TC should pass within 10° of the location of the observed TC genesis within 1 day of the observed TC genesis time. The candidate vortex should meet TC criteria (e.g., surface wind speed >17.5 m s^{-1}) within 5 days of the observed TC genesis time. Detailed descriptions of the experimental setup and method of TC detection are provided in Nakano et al. (2015).

Results

Table 1 summarizes the results of the TC genesis forecast. The model captured six out of the eight cases of TC genesis 1 week in advance with a low missing rate (all were above 70%). Moreover, four cases of TC genesis in late August were captured 2 weeks in advance with high probability (all were above 40% and 3 of 4 were above 50%). The genesis of TC Songda was well captured 2 weeks in advance, though not 3 weeks in advance. Note that two TC genesis events (Malou and Malakas) were not captured, probably because they were

Table 1 Simulation performance for forecasting TC genesis

	Malou	Meranti	Rananim	Malakas	Megi	Chaba	Aere	Songda
Central pressure (hPa)	996	960	950	990	970	910	955	925
Lifetime (days)	0.875	4.75	4.5	2.75	4.125	11.75	6.25	11
Week 1	0	100	75	43	86	71	71	86
Week 2	N/A	N/A	N/A	N/A	80	43	57	57
Week 3	N/A	N/A	N/A	N/A	N/A	N/A	N/A	29

Each of the eight TCs occurring in August 2004 is listed with their observed central pressure and lifetime, and the simulation's hitting ratio (%) in weeks 1, 2, and 3. Adapted from Table 1 of Nakano et al. (2015).

very weak (minimum central pressure >990 hPa) and had short lifespan (less than 3 days).

Figure 9 shows the zonal wind at 850 hPa and the OLR, which is related to convective activities. In the analyzed field, the shear line between easterly wind in the tropics and monsoonal westerly wind extended eastward, accompanied by active convection in mid-August. This eastward extension was maintained throughout late August. TC Songda was generated on the shear line. The experiments initialized between August 5 and 11 (3 weeks before Songda's genesis) captured the eastward extension of the shear line in mid-August but it retreated in late August. The experiments initialized between August 12 and 18 (2 weeks before Songda's

genesis) captured the eastward extension of the shear line in late August. Therefore, we conclude that the eastward extension of the shear line is closely related to Songda's genesis, and if models reproduce the longitudinal evolution of the shear line, the associated TC genesis will be simulated effectively.

Future directions

Once we have achieved more reliable predictability of TC genesis with a longer lead time, our next targets are the probabilities of TC tracks, intensities, and landfall. Such information could be calculated from the large samples of ensemble simulations. Since the NICAM-LETKF system (Terasaki et al. 2015) is an ensemble-

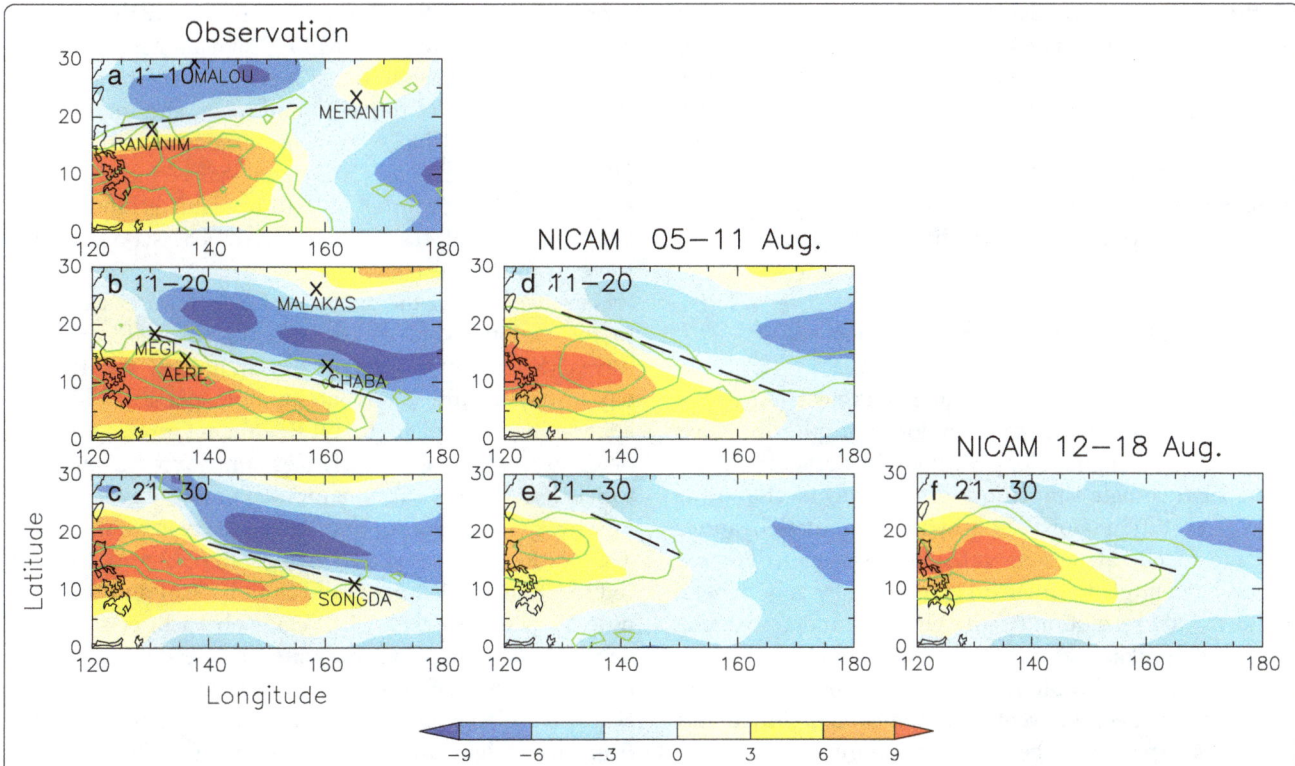

Fig. 9 Comparison of 10-day mean OLR and zonal wind at 850 hPa between observation and simulation. Observed and ensemble mean of 10-day mean OLR (*green contour*) and zonal wind at 850 hPa (*color shading*) averaged during **a** 1–10 August, **b, d** 11–20 August, and **c, e, f** 21–30 August 2004. **d, e** show NICAM simulation initialized from 5 to 11 August; **f** shows 12–18 August. **a** *Black crosses* indicate the locations of tropical cyclogenesis. OLR values less than 200 W m⁻² are shown with contour intervals of 20 W m⁻². *Broken lines* depict the shear line. After Fig. 1 of Nakano et al. (2015)

based data assimilation system, it can be used to make the large number of initial atmospheric fields fit to the model's dynamics and physics. It is expected that the probability forecast of TC tracks and landfall probability would be affected by the mid-latitude atmospheric conditions including the structure and evolution of the jet stream and the subtropical high. Given that the mid-latitude jet stream and the subtropical high are affected by the weather systems in the tropics such as MJO (Kawamura et al. 1996; Moon et al. 2013; Molinari and Vollaro 2012), we should further examine the relationship between mid-latitude model biases and those in the tropics to improve probability forecasting of TC tracks. For forecasts of TC intensity, current atmospheric models have typical biases: they are not able to reproduce rapid intensification, and TCs in the models continuously intensify even in the mid-latitude, where TCs begin to weaken, then resulting in overintensification. A higher resolution model is needed to improve the representation of the structure of the TC's inner core, which results in accurate simulation of rapid intensification (Rogers et al. 2013; Wang and Wang, 2014). Interestingly, a recent study shows that high-resolution global models reduce forecast error in TC intensity and tracking forecasts (Nakano et al. 2017). The atmosphere-ocean interacting processes should also be incorporated in the numerical model to avoid overintensification (Wada et al. 2014; Zarzycki 2016). Simulations of TC intensities are sensitive to details of physical schemes such as the boundary layer scheme and the cloud microphysics scheme (Kanada et al. 2012). More improvement and development of NICAM together with evaluations using available observational data will be necessary to pursue the aforementioned future directions in TC predictability studies.

Atmospheric Model Intercomparison Project-type experiments

Atmospheric Model Intercomparison Project (AMIP) is a standard experimental protocol for atmospheric GCMs to test models' climatology with multi-decadal simulations. Such climate simulations are now achievable using a high-resolution global non-hydrostatic model, and a new era of climate-system research has begun. For example, the simulated intensities of TCs strongly depend on horizontal resolution (Camargo 2013) and convection scheme (Murakami et al. 2012). Thus, by taking advantage of high-resolution global non-hydrostatic simulations without deep convection schemes, the climatology of TCs is expected to be simulated more informatively, thereby facilitating further investigations of the probability and geographical distributions of TC intensities and structures and their projected changes under a warmer climate (Oouchi et al. 2006; Murakami et al. 2012;

Knutson et al. 2015). The global non-hydrostatic climate model is also a powerful tool for simultaneously investigating phenomena at various spatiotemporal scales of atmospheric disturbances, such as planetary-scale atmospheric general circulations, MJO, tropical waves, diurnal meso-scale convective events, severe rainfall, and atmospheric gravity waves and their interactions (Holt et al. 2016). In addition, better representation of clouds, convection, and circulations is key to reducing uncertainties of climate sensitivity (Bony et al. 2015; Stevens and Bony 2013).

Motivated by the aforementioned demands for AMIP-type experiments with high-resolution global models, we performed present (Kodama et al. 2015) and future (Satoh et al. 2015) climate simulations using the 14 km grid spacing NICAM. The model settings were the same as those described in the "Asian summer monsoon" section. Specifically, we should note that instead of following the strict AMIP protocol, the slab ocean model and nudging technique were employed instead of imposing a fixed SST condition to obtain better performance of the distribution of tropical precipitation, as shown in Kodama et al. (2015). The simulations were performed to reproduce the climates of 1979–2008 (Kodama et al. 2015) and to project the climates of 2074–2100 (Satoh et al. 2015).

Kodama et al. (2015) reported the simulated climatology of basic state, TCs, MJO, Asian monsoon, diurnal precipitation cycle, and quasi-biennial oscillation (QBO). In-depth analysis of the MJO and mean and intense precipitation around Japan were presented in Kikuchi et al. (2016) and Fujita et al. (2017, personal communication), respectively. Fukutomi et al. (2015) evaluated the activity and structure of the simulated tropical synoptic scale waves over the western Pacific.

Here, we extend the analysis of Fukutomi et al. (2015) to the various types of convectively coupled equatorial waves (Kiladis et al. 2009), which govern the tropical day-to-day weather. Spatiotemporal spectral and filter analyses were applied to OLR, precipitation, and wind fields to diagnose OLR variances contributed from equatorial Kelvin waves, equatorial Rossby waves, mixed Rossby gravity waves, and tropical depression waves. The National Oceanic and Atmospheric Administration (NOAA) OLR (Liebmann and Smith 1996) and the Japanese 55-year Reanalysis (JRA-55; Kobayashi et al. 2015b) datasets were used to evaluate the results. Here, the equatorial wave-filtered OLR variances are plotted (Fig. 10). Geographical distributions of tropical waves over the Indian and Pacific Oceans were qualitatively simulated, though their overall activities tended to be weaker than the observed ones (Fig. 10). Significant biases were found over Central and South America, the Atlantic, and Africa in all the wave components. Kelvin

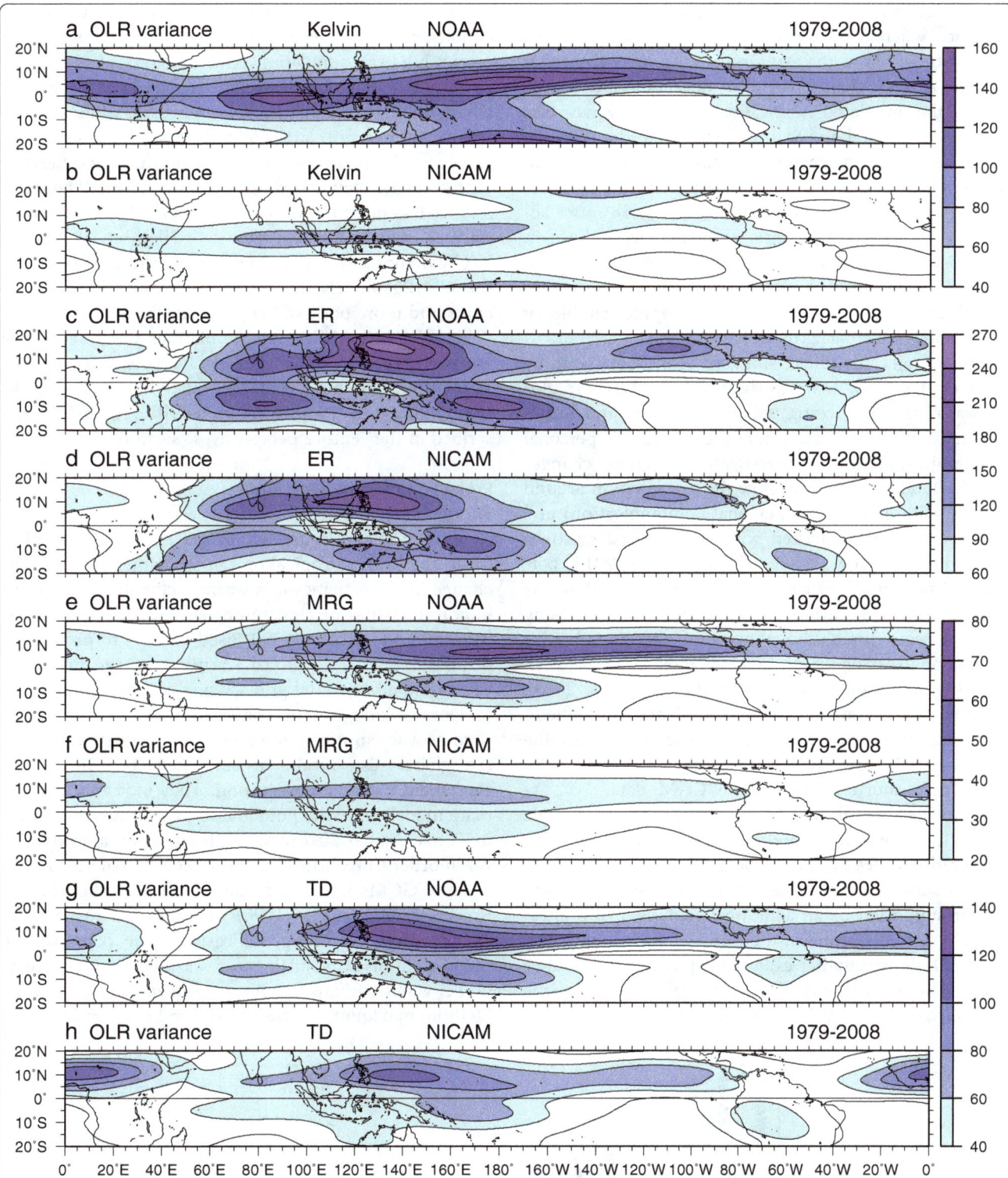

Fig. 10 Comparison of climatology of the OLR variances of equatorial waves between observation and simulation. Annual mean climatology (1979–2008) of the OLR variances in (W m⁻²)² for Kelvin wave (Kelvin) (**a**, **b**), equatorial Rossby waves (ER) (**c**, **d**), mixed Rossby gravity waves (MRG) (**e**, **f**), and tropical depression waves (TD). The observed NOAA OLR variances are shown by (**a**, **c**, **e**, **g**), whereas the NICAM simulated OLR variances are shown by (**b**, **d**, **f**, **h**)

waves were very weakly simulated, particularly over the Atlantic, whereas simulated equatorial Rossby waves were stronger over the Amazonian region and weaker over the Atlantic than those of the observation. Distribution of the simulated mixed Rossby gravity waves was biased over Central and South America and the Atlantic. Tropical depression waves were weaker over Central and South America and the western Atlantic, whereas the African easterly waves in the simulation were more active over the eastern Atlantic compared with those in the observation. The reasoning behind the model biases in convectively coupled equatorial waves will be further investigated from the perspective of reproducibility of the seasonal scale background fields.

Future projected climate simulations were analyzed by Satoh et al. (2015), in which the response of TCs to global warming was investigated using present and future climate simulation data. Yamada et al. (2017, personal communication) further investigated structural changes in TCs. In addition to TCs, how the precipitation around Japan (Fujita et al., 2017, personal communication) and the precipitation associated with extratropical cyclones respond to global warming (Kodama et al. 2017, personal communication) has been investigated. Before the K computer was available, projection studies using NICAM were limited to short time-slice simulations such as one year or season (Yamada et al. 2010; Satoh et al. 2012a; Tsushima et al. 2014; Noda et al. 2014, 2015), and the insufficient integration period made it difficult to achieve statistical significance of the global warming response. It is now possible to confirm the results derived from shorter-integration NICAM data. For example, Chen et al. (2016) performed seasonal scale sensitivity experiments and showed a large longwave cloud radiative feedback, irrespective of the cloud microphysics schemes that were employed. This result was further supported by an additional analysis of the interannual variability found in the NICAM AMIP-type data. Figure 11 shows another example depicting the distribution of high-cloud size that was evaluated using the method described by Noda et al. (2014). Qualitatively,

the size distributions under the present and the future conditions show a result similar to those evaluated in a 1-year NICAM simulation data in Noda et al. (2014). Responding to global warming, the number of smaller clouds (<40 km radius) notably increases, while changes to the number of larger clouds (≥40 km) are much smaller and not significant. We found that the difference in the smaller clouds is statistically significant. The result implies less organization of tropical cloud systems in the warmer world. As analyzed by Chen et al. (2016), NICAM shows larger positive high cloud feedback than the CMIP5 models due to the increase of high cloud cover and more positive longwave feedback.

The NICAM AMIP-type simulations described above are the first step toward global cloud-resolving climate simulation. Our next plan, the High Resolution Model Intercomparison Project (HighResMIP), is briefly described in the "Future perspectives" section.

Projection of tropical cyclones
Background
A primary energy source of TCs is latent heat supplied from the warm tropical ocean (Emanuel 2003). The change in TC activities in a warmer climate is one of the important issues of climate change (Walsh et al. 2016). Broccoli and Manabe (1990) used a GCM to project the responses of TCs to global warming for the first time. They used a horizontal grid spacing of a few hundred kilometers, which was too coarse to resolve TC structures such as a warm core. However, their simulation showed the global distribution of simulated TCs in general agreement with the observation. They indicated that climate models have the potential to predict future changes in TCs and advocated reducing the horizontal grid interval in order to realistically reproduce TC structures.

Many GCMs have used finer resolutions than that of Broccoli and Manabe (1990) to assess the response of TCs to global warming, indicating some consensus in the change of TC statistics in a warmer world: decrease in TC genesis frequency, increase in intense TCs, mean lifetime maximum intensity, and rainfall around TCs

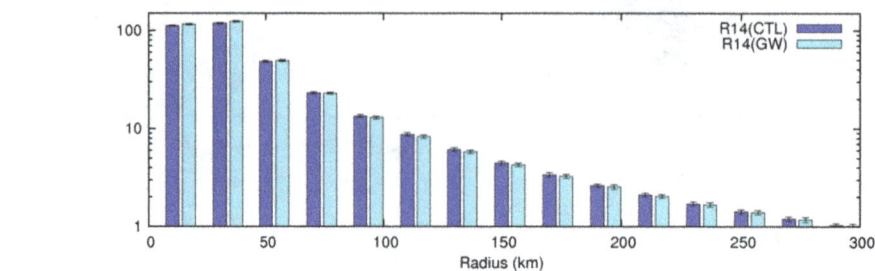

Fig. 11 Size distribution of high clouds showing its changes between present and future climate simulations. Size distribution of high clouds in present (*dark blue*) and future (*light blue*) climate simulations using the NICAM AMIP-type simulation are analyzed by following the method described by Noda et al. (2014). Standard variations are designated at each bin

(Knutson et al. 2010; Christensen et al. 2013; Walsh et al. 2016). However, Emanuel (2013) showed an increase in global annual frequency of TCs under warmer climate conditions using a downscaling method in which incipient vortices are embedded into large-scale climate conditions projected by CMIP5 models. To achieve a better understanding of the relationships between TC genesis and climate change, the Hurricane Working Group (HWG) established by the US Climate and Ocean: Variability, Predictability and Change (CLIVAR) group coordinated common experiments (Walsh et al. 2015). Although the downscaling methodology of Emanuel (2013) projected an increase in TC genesis under the warmer climate condition projected by the HWG models, there were no models that generated a substantial increase in global TC frequency. The reduction of global TC numbers is associated with a decrease in upward mass flux (Sugi et al. 2002, 2012), and an increase in saturation deficit of the mid-troposphere (Emanuel et al. 2008). The mechanism underpinning the decrease in TC numbers, however, remains controversial.

Due to recent advances in supercomputing and GCMs, fine-resolution models with grid spacing less than 25 km have been used to investigate future change in TC activities (Murakami et al. 2012; Manganello et al. 2014; Knutson et al. 2015; Roberts et al. 2015; Wehner et al. 2015). These models reproduced TC structure more realistically. The advent of the K computer enabled researchers to perform long-term simulations using a high-resolution global non-hydrostatic model (Kodama et al. 2015; Satoh et al. 2015).

In this section, future changes in TC activities are investigated using the outputs of the aforementioned NICAM AMIP-type simulation (referred to as the present day (PD) simulation; Kodama et al. 2015) and a warmer climate simulation (referred to as the global warming (GW) simulation; Satoh et al. 2015). Kodama et al. (2015) documented the model settings and experimental design of the PD simulation. The model settings of the GW simulation were identical to those of the PD simulation, and the experimental design is described in Satoh et al. (2015).

Results

Figure 12 shows geographical distributions of TC genesis density for the PD and GW simulations and the future change (GW − PD). The PD simulation reproduces the observed distribution of TC genesis fairly well; although, the model has clear bias particularly over the north Atlantic for less TC genesis (See Fig. 10 of Kodama et al. 2015). The GW simulation showed a salient decrease in TC genesis over the eastern North Pacific (Fig. 12c). Satoh et al. (2015) documented the annual global TC number of 104.0 for the PD simulation and 81.8 for the

GW simulation. The decrement in the TC number due to global warming is 21.3%, which is within the range of previous studies (Knutson et al. 2010; Christensen et al. 2013). Figure 12c shows that TCs are mainly decreased over the north eastern Pacific.

As the mechanism of the decrement was not completely understood, Satoh et al. (2015) used outputs of the PD and GW simulations to propose a new concept based on the relationship between convective mass flux and TCs. They extracted a convective mass flux associated with TCs and defined its contribution rate to convective mass flux in the tropics (R) as follows:

$$R = \frac{M_T}{T_{int}M},$$

where M_T is convective mass flux associated with TCs (defined as integration over the area with a radius smaller than 500 km from the center of TCs), T_{int} is the integration period, and M is the total convective mass flux over the tropical domain (30°N–30°S). Since $M_T = NTm_T$, where N is the number of global TCs, T is the average lifetime of TCs, and m_T is the average convective mass flux per single TC, we can rewrite this as (Eq. 15 of Satoh et al. 2015)

$$N = R\frac{T_{int}M}{Tm_T}.$$

Satoh et al. (2015) found that the contribution of TC (R) in GW is almost unchanged or less than that in PD. The convective mass flux in the tropics (M) generally decreases in GW (Vecchi and Soden 2007), and our results confirm this change. In the analysis of the TC convective mass flux, Satoh et al. (2015) also found that the convective mass flux associated with TCs and its fractional area become greater as TCs become more intense (Fig. 13a and b, respectively). They also found that the TC convective mass flux and its frictional area in GW are greater than those in PD for the same intensity of TC (Fig. 13a and b, respectively). These results are contrasted by Fig. 13c, where the area-averaged convective mass flux, or vertical velocity, does not show a robust change with warming. Intense TCs increase under a warmer climate, leading to greater convective mass flux associated with TCs. In these simulations, the incidence of intense TCs that develop to less than 945 hPa increases by 38.6% due to global warming (Fig. 14). Thus, Satoh et al. (2015) showed that m_T increases as warming. Assuming that R is unchanged between PD and GW, this constraint implies that TC number decreases in order to intensify TCs under a warmer climate.

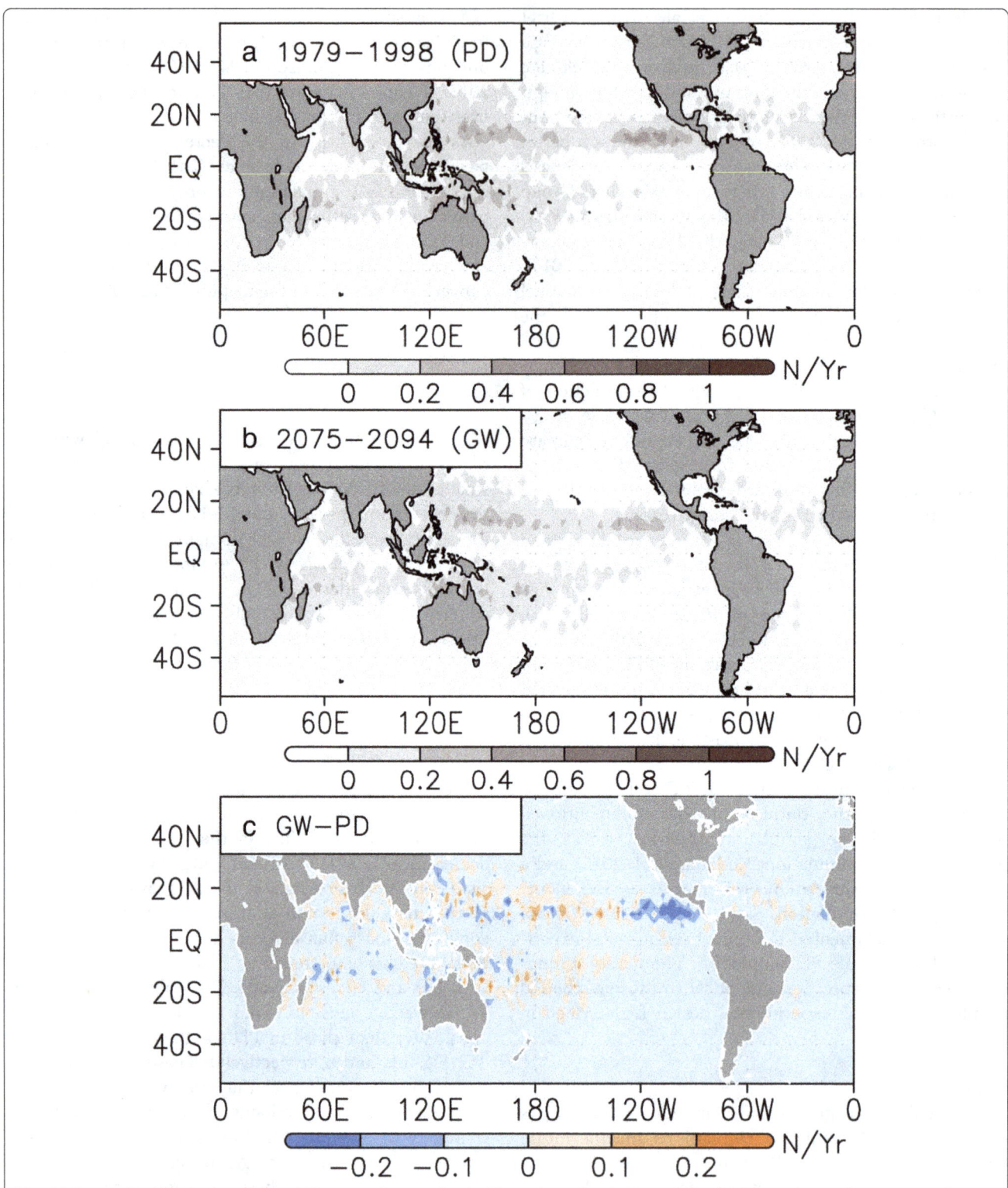

Fig. 12 Geographical distributions of annual TC genesis density. **a** for the PD simulation for 1979–1998; **b** for the GW simulation of 2074–2093; **c** shows the difference between the GW and PD simulations. TC genesis density is defined as the number of TCs generated in each 2.5° × 2.5° grid box per year

Future directions

Walsh et al. (2016) highlighted the importance of the following topics for future research on TCs and climate change: detection methods as a source of uncertainty, lack of climate theory of TC formation, confidence in detection and attribution of observed changes in TCs to

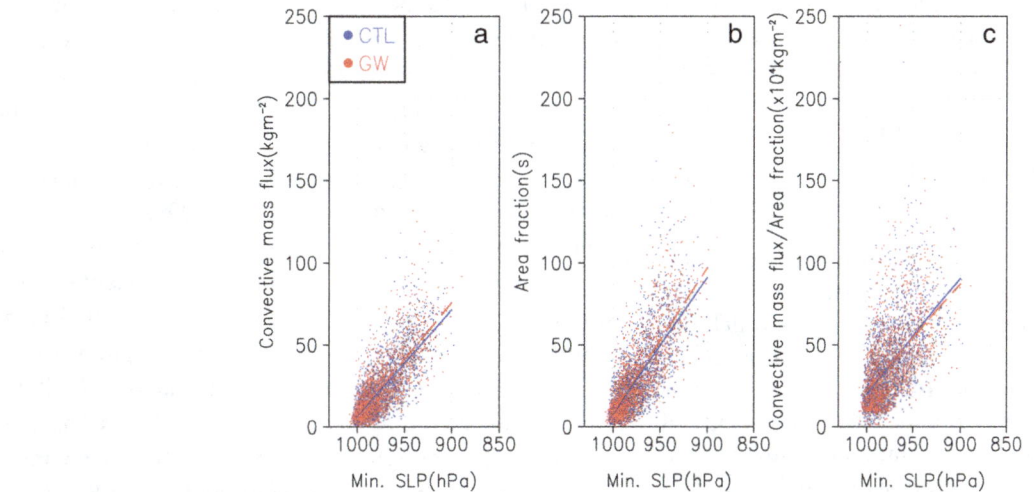

Fig. 13 Scatter plots of the TC properties versus the minimum sea-level pressure. Scatter plot represents **a** the relationship between the minimum sea-level pressure (Min. SLP) and the convective mass flux over the TC domain [*A*], **b** the relationship between the min. SLP and the fractional area of the strong updrafts [*B*], and (**c**) the relationship between the min. SLP and the area-averaged convective mass flux [*A/B*] for all the simulated TCs. The PD simulation is shown by *blue dots* and the GW simulation by *red dots*; *blue* and *red lines* show regressions for the PD and GW simulations, respectively. The convective mass flux and the fractional area of the strong updrafts are accumulated for the lifetime of each TC, so that these have units of kg m^{-2} and s, respectively. Only the vertical velocities exceeding the threshold value of 0.5 m s^{-1} were used. The values at an altitude of 4 km are shown. After Fig. 5 of Satoh et al. (2015)

date, impact of uncertainty in projections of large-scale climate fields, and correct simulations of TC frequency response in high-resolution climate models. It is expected that high-resolution models will be useful tools for investigating these topics. Thanks to the K computer, we completed the first-ever climate simulation with a non-hydrostatic GCM using 14 km grid spacing

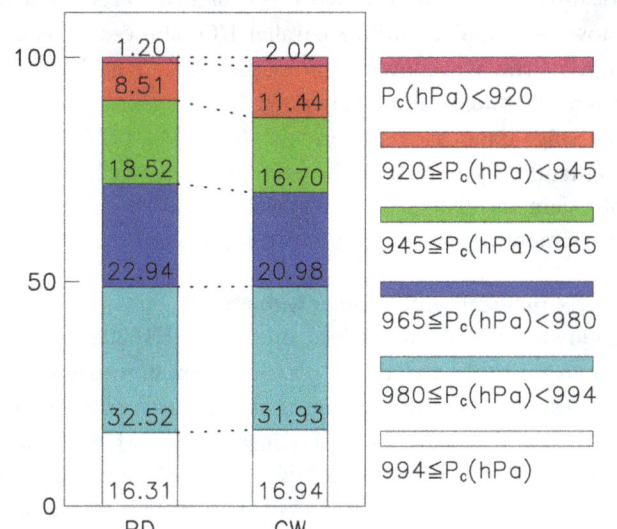

Fig. 14 Normalized frequencies (%) for each TC intensity from the PD and GW simulations. The *two boxes* represent the number of TCs categorized into six intensities using the lifetime minimum central sea-level pressure (Pc). The number in each colored box indicates the number of TCs categorized in each intensity level

(Kodama et al. 2015; Satoh et al. 2015). The outputs from this work will contribute important new findings on TC climatology.

Higher-resolution models have simulated more realistic TC climatology (Walsh et al. 2016). However, TC distributions simulated in high-resolution models still show inconsistency with observed TC activities such as fewer TCs over the north Atlantic (Kodama et al. 2015; Murakami et al. 2015; Roberts et al. 2015), more TCs over the western North Pacific (Manganello et al. 2012; Murakami et al. 2015) and the eastern North Pacific (Kodama et al. 2015; Roberts et al. 2015), weak maximum wind speed compared with low central sea-level pressure (Roberts et al. 2015), and an overestimation of mean TC intensities at higher latitudes (Murakami et al. 2012; Kodama et al. 2015). Reducing these biases will enhance confidence in GCM simulations of future change. For example, Lloyd and Vecchi (2011) investigated the observation that the passage of TCs causes SST cooling, indicating that this oceanic feedback suppresses intensification of TCs. Atmosphere-ocean coupled models will improve the overestimation of TC intensities at higher latitudes.

In addition, high-resolution models simulate TC structures more realistically. An increase in the convective mass flux associated with TCs due to global warming is not caused by an enhancement of updraft velocities (Fig. 13c), but by an increase in the area of the updraft region around TCs (Fig. 13b). This suggests that global warming influences the dynamical structure of TCs. The scale of damage caused by a disaster associated with TCs is measured by TC gale force wind radii as well as its frequency, intensity, and track. Recent studies have focused on future changes in the size of TCs (Kanada et al. 2013; Manganello et al. 2014; Knutson et al. 2015). High-resolution global non-hydrostatic models should also make it possible to statistically evaluate future changes of TC structure.

Ohno and Satoh (2015) revealed that the flow from the lower stratosphere near the TC eyewall plays a key role in constituting the upper-level warm core which develops TC intensity prior to reaching the mature stage. To increase confidence in this theory, it must be compared with an authentic observation with high temporal and spatial resolution. Today, geostationary satellites can observe atmospheric wind vectors associated with TCs in the upper-troposphere with high temporal and spatial resolution (Bessho et al. 2016). Outputs of high-resolution models should be evaluated with such new satellite data, promoting our understanding of TCs.

Future perspectives
Post-K supercomputer project

Among the outcomes of Field 3 of SPIRE summarized in this review, the series of studies using the sub-kilometer global atmospheric simulation (Miyamoto et al. 2013, 2015; Kajikawa et al. 2016) are typical examples of "capability computing" using the K computer; that is, a single complex problem solved in the shortest time possible using the maximum computing power available. Although this kind of simulation does not always lead to the maximum output or outcome, the knowledge gained has great potential to bolster next-generation research. In much larger computer system such as the Post-K supercomputer, problems of this size will be easily handled with various configurations. We can expect to get much more information from these supercomputers in terms of "capacity computing," or the solution of smaller or less complex problems using more efficient computing power.

A research strategy similar to that mentioned above has already been adopted. In the era of the Earth Simulator, Miura et al. (2007a) take a capability computing approach to demonstrate the ability to represent the MJO with 3.5 and 7 km grid spacing using NICAM. Using a large number of ensemble simulations, Miyakawa et al. (2014) statistically confirmed the MJO predictability suggested by Miura et al. (2007a) and suggested that high-resolution global non-hydrostatic models like NICAM generally have better skill in MJO predictability than GCMs. Thus, combining the two strategies of capability computing and capacity computing may be an effective approach for future research using the Post-K supercomputer.

The project to develop the next-generation flagship supercomputer is now underway. The Post-K supercomputer is designed by using homogeneous CPU architecture. The CPU will be a many-core processor, and accelerators are not included in the system. The system will consist of a six-dimensional mesh/torus interconnected network (Tofu; Ajima et al. 2009) and a three-level hierarchical storage system. The co-design is based on collaboration between computational scientists and computer scientists, because a system constructed using commodity computers along current computer trends cannot satisfy the socio-scientific demands of all research areas, including earth sciences. NICAM has been selected as a co-design application from the viewpoint of a structured stencil calculation. In addition, LETKF has also been selected in this co-design project as an application requiring the dense-matrix calculation. NICAM-LETKF (Terasaki et al. 2015) is one of the most important target applications for the development of the new computer system. The current target problem for the NICAM-LETKF is as follows: in the data assimilation cycle, 1000 ensembles by NICAM with 3.5 km grid spacing and 100 vertical levels are conducted every 6 h. The prediction using 440 m horizontal grid spacing is integrated for several months. In terms of the computational performance, the workflow between the simulation code and the data assimilation system, including I/O and communication in a massively parallel supercomputer becomes a serious problem. A framework proposed by Yashiro et al. (2016c), which was designed to reduce data movement and to utilize parallel I/O, showed a reduction in the workflow's total elapse time. Based on the scientific outcomes summarized in the present review, the big data assimilation by NICAM-LETKF using tremendous volumes of observational data, such as satellite data, is expected to further contribute to the improvement of mid-range predictions around the globe.

Post-K supercomputer studies with NICAM

Following the SPIRE project, the FLAGSHIP2020 project began in 2014 to examine possible maximum use of the Post-K supercomputer. Among the research subjects defined for the weather and climate simulation studies, seamless simulations using NICAM have been proposed. Seamless means beyond a week to seasonal prediction, especially for the predictability of TC statistics such as TC genesis, tracks, and intensities. Furthermore, seamless studies cover a year to decadal scales, and even to the centennial scale, as aimed for by the HighResMIP study, which is introduced below. Successful extended-range forecasts of MJO and BSISO (Miyakawa et al. 2014; Nakano et al. 2015) suggest greater predictability of TCs beyond a week to a few weeks. Through ensemble simulations, many aspects of TCs such as TC genesis, tracks, and intensities will be obtained. In addition, together with the development of the high-resolution atmosphere and ocean coupled model (NICAM-COCO), the seasonal scale predictability of ISO and associated TC characteristics will be explored. For further multi-year simulation, we will examine the potential predictability of the near future from a year to a decadal scale, particularly focusing on TC statistics (as described in the "Projection of tropical cyclones" section). These longer time-scale simulations have also been examined in the "Atmospheric Model Intercomparison Project-type experiments" section and are further pursued in the next subsection, "High Resolution Model Intercomparison Project."

High Resolution Model Intercomparison Project

As a part of Coupled Model Intercomparison Project phase 6 (CMIP6), the High Resolution Model Intercomparison Project (HighResMIP; Haarsma et al. 2016) is now under way. In this project, climate modeling with a horizontal resolution of 25–50 km will be performed for more than half a century. The aim is that such fine-resolution climate model data will be available for everyone within a few years as CMIP6 continues. Following the HighResMIP protocol, we plan to perform 56 and 28 km grid spacing NICAM for 1950–2014 and 14 km grid spacing NICAM for ten years. Currently, a series of sensitivity experiments are being performed to improve climatology in the simulation. Our aim in participating in the HighResMIP is to investigate multi-scale phenomena including tropical and extratropical cyclones, MJO, tropical waves, convection, and clouds in a seamless manner (see Section 7.2 of Haarsma et al. 2016). Process-based assessment using satellite datasets will also be key to evaluating and constraining the high-resolution climate model.

Conclusions

NICAM-related studies using the K computer under Field 3 of SPIRE have produced a wealth of valuable weather and climate modeling research and present clear challenges that may be solved with the Post-K super-computer. Key achievements of these studies include high-resolution sub-kilometer global simulations, ensemble simulations for MJO and BSISO, and simulations of TC genesis. The longer multi-decadal simulation by NICAM was conducted for the first time with the global non-hydrostatic model. The climatology of the 14 km

grid spacing NICAM shows comparable results with other climate models; although, biases similar to those of other models exist. The results suggest that, even though it has been arguable whether deep convective parameterization is required for the 14 km grid spacing model, the simulation without the parameterization reproduces climatology consistent with observation. Thanks to the K computer, we obtained considerable evidence of NICAMs usability for weather and climate simulations. Further research using the Post-K supercomputer in the next decade is expected to contribute even more crucial knowledge to this exciting research field.

Abbreviations

BSISO: Boreal summer intra-seasonal oscillation; CMIP6: Coupled Model Intercomparison Project phase 6; ERA-Interim: European Centre for Medium-Range Weather Forecasts interim reanalysis; FLAGSHIP2020: Future LAtency core-based General-purpose Supercomputer with HIgh Productivity; GCM: General circulation model; GW: The global warming simulation; HighResMIP: High Resolution Model Intercomparison Project; I/O: Input/output; ISO: Intra-seasonal oscillations; JAMSTEC: Japan Agency for Marine-Earth Science and Technology; LETKF: Local ensemble transform Kalman filter; MJO: Madden-Julian oscillation; NAIM: Northern Australia-Indonesia monsoon; NICAM: Non-hydrostatic icosahedral amospheric model; NOAA: National Oceanic and Atmospheric Administration; NSW6: NICAM single-moment water 6-cloud microphysics scheme; OLR: Outgoing longwave radiation; PD: The present day simulation; QBO: Quasi-biennial oscillation; RMM: Real-time multivariate MJO; SEAM: Southeast Asian monsoon; SPIRE: Strategic Programs for Innovative Research; SST: Sea surface temperature; TC: Tropical cyclone; TC-genesis: Tropical cyclogenesis; TRMM: Tropical Rainfall Measuring Mission; WNPM: Western North Pacific monsoon

Acknowledgements

All the simulations analyzed in this study were performed on the K computer at the RIKEN Advanced Institute for Computational Science (Proposal number hp120279, hp120313, hp130010, hp140219, and hp150213). This study was supported by Strategic Programs for Innovative Research (SPIRE) Field 3 (Projection of Planet Earth Variations for Mitigating Natural Disasters), which is promoted by the Ministry of Education, Culture, Sports, Science and Technology (MEXT), Japan. All the authors acknowledge Ms. Hisae Tamura for formatting the manuscript and INLEXIO for the English editing.

Funding

This study was supported by the Strategic Programs for Innovative Research (SPIRE) Field 3 (Projection of Planet Earth Variations for Mitigating Natural Disasters) and Future LAtency core-based General-purpose Supercomputer with HIgh Productivity (FLAGSHIP2020 project), which are promoted by the Ministry of Education, Culture, Sports, Science and Technology (MEXT), Japan.

Availability of data and materials

The NICAM data simulated by the K computer will be available. The following three papers are to be submitted or in review process and referred to as personal communication in the text: Fujita et al. (2017), Kodama et al. (2017), and Yamada et al. (2017). Please contact the authors for these data and unpublished manuscripts:
Fujita M, Kodama C, Yamada Y, Noda AT, Nakano M, Nasuno T, Satoh M (2017) Change in precipitation around Japan induced by tropical cyclones as projected by the NICAM AMIP-type 25-year simulation.
Kodama C, Stevens B, Mauritsen T, Seiki T, Satoh M (2017) Evidence for larger-than average increases in rainfall from intense cyclones with warming.
Yamada Y, Satoh M, Sugi M, Kodama C, Noda AT, Nakano M, Nasuno T (2017) Response of tropical cyclone activity and structure to a global warming using a high-resolution global non-hydrostatic model.

Authors' contributions

MS led the studies using NICAM in SPIRE and FLAGSHIP2020 and coordinated this review. HY wrote the "Computational aspects" section. YK and YM wrote the "Sub-kilometer global simulation" section, and the experiments described in this section were carried out by the team at the Advanced Institute for Computational Science: YM, YK, HY, TY, and HT. YK and HY wrote the "Asian summer monsoon" section, and HY conducted the experiments described in this section. TM wrote the "Madden-Julian oscillation" section and conducted the experiments described in this section. MN wrote the "Boreal summer intra-seasonal oscillation and tropical cyclogenesis" section and conducted the experiments described in this section. CK, ATN, TN, YY, and YF wrote the "AMIP-type experiments" section, and YY and MS wrote the "Future projection of tropical cyclones" section. CK, ATN, and YY conducted the experiments described in these two sections. HT, MS, and CK wrote the "Future perspectives" section. MS is responsible for the whole manuscript and wrote the "Introduction" and "Conclusions" sections. All authors read and approved the final manuscript.

Authors' information

MS and HT started the development of NICAM around the year 2000, and the NICAM studies have been conducted mainly in these three institutes since then: Atmosphere and Ocean Research Institute of the University of Tokyo, JAMSTEC, and the Advanced Institute for Computational Science.

Competing interests

The authors declare that they have no competing interests. For more information, visit the websites, http://nicam.jp/ and http://www.jamstec.go.jp/hpci-sp/en/.

Author details

[1]The University of Tokyo, 5-1-5 Kashiwanoha, Kashiwa, Chiba 277-8568, Japan. [2]RIKEN Advanced Institute for Computational Science, 7-1-26, Minatojima-minami-machi, Chuo-ku, Kobe, Hyogo 650-0047, Japan. [3]Research Center for Urban Safety and Security, Kobe University, 1-1, Rokko-dai, Nada-ku, Kobe 657-8501, Japan. [4]Rosenstiel School of Marine and Atmospheric Science, University of Miami, 4600 Rickenbacker Causeway, Miami, FL 33149, USA. [5]Japan Agency for Marine-Earth Science and Technology, 3173-15, Showa-machi, Kanazawa-ku, Yokohama, Kanagawa 236-0001, Japan. [6]Institute for Space-Earth Environmental Research, Nagoya University, Furo-cho, Chikusa-ku, Nagoya 464-8601, Japan.

References

Ajima Y, Sumimoto S, Shimizu T (2009) Tofu: a 6D mesh/torus interconnect for exascale computers. Computer 42:36–40. doi:10.1109/MC.2009.370

Amdahl GM (1967) Validity of the single processor approach to achieving large scale computing capabilities. In: AFIPS '67 (Spring) Proceedings of the April 18-20, 1967, spring joint computer conference. ACM, New York, pp 483–485. doi:10.1145/1465482.1465560

Ando K, Hyodo M, Baba T, Hori T, Kato T, Watanabe M, Ichikawa S, Kitahara H, Uehara H, Inoue H (2016) Parallel-algorithm extension for tsunami and earthquake-cycle simulators for massively parallel execution on the K computer. Int J High Perform Comput Appl 30:454–468. doi:10.1177/1094342016636670

Arakawa A, Schubert WH (1974) Interaction of a cumulus cloud ensemble with large-scale environment, Part 1. J Atmos Sci 31:674–701

Arakawa T, Inoue T, Sato M (2014) Performance evaluation and case study of a coupling software ppOpen-MATH/MP. Procedia Comp Sci 29:924–935. doi:10.1016/j.procs.2014.05.083

Bessho K, Date K, Hayashi M, Ikeda A, Imai T, Inoue H, Kumagai Y, Miyakawa T, Murata H, Ohno T, Okuyama A, Oyama R, Sasaki Y, Shimazu Y, Shimoji K, Sumida Y, Suzuki M, Taniguchi H, Tsuchiyama H, Uesawa D, Yokota H, Yoshida R (2016) An introduction to Himawari-8/9—Japan's new-generation geostationary meteorological satellites. J Meteor Soc Jpn 94:151–183. doi:10.2151/jmsj.2016-009

Bony S, Stevens B, Frierson DMW, Jakob C, Kageyama M, Pincus R, Shepherd TG, Sherwood SC, Siebesma AP, Sobel AH, Watanabe M, Webb MJ (2015) Clouds, circulation and climate sensitivity. Nat Geosci 8:261–268. doi:10.1038/ngeo2398

Broccoli AJ, Manabe S (1990) Can existing climate models be used to study anthropogenic changes in tropical cyclone climate? Geophys Res Lett 17:1917–1920. doi:10.1029/GL017i011p01917

Bryan GH, Wyngaard JC, Fritsch JM (2003) Resolution requirements for the simulation of deep moist convection. Mon Weather Rev 131:2394–2416. doi:10.1175/1520-0493(2003)131<2394:RRFTSO>2.0.CO;2

Camargo SJ (2013) Global and regional aspects of tropical cyclone activity in the CMIP5 models. J Clim 26:9880–9902. doi:10.1175/JCLI-D-12-00549.1

Chen G, Zhu X, Sha W, Iwasaki T, Seko H, Saito K, Iwai H, Ishii S (2015a) Toward improved forecasts of sea-breeze horizontal convective rolls at super high resolutions. Part I: configuration and verification of a down-scaling simulation system (DS3). Mon Weather Rev 143:1849–1872. doi:10.1175/MWR-D-14-00212.1

Chen G, Zhu X, Sha W, Iwasaki T, Seko H, Saito K, Iwai H, Ishii S (2015b) Toward improved forecasts of sea-breeze horizontal convective rolls at super high resolutions. Part II: the impacts of land use and buildings. Mon Weather Rev 143:1873–1894. doi:10.1175/MWR-D-14-00230.1

Chen Y-W, Seiki T, Kodama C, Satoh M, Noda AT, Yamada Y (2016) High cloud responses to global warming simulated by two different cloud microphysics schemes implemented in the Nonhydrostatic Icosahedral Atmospheric Model (NICAM). J Clim 29:5949–5964. doi:10.1175/JCLI-D-15-0668.1

Christensen JH, Krishna Kumar K, Aldrian E, An S-I, Cavalcanti IFA, de Castro M, Dong W, Goswami P, Hall A, Kanyanga JK, Kitoh A, Kossin J, Lau N-C, Renwick J, Stephenson DB, Xie S-P, Zhou T (2013) Climate phenomena and their relevance for future regional climate change. In: Stocker TF, Qin D, Plattner G-K, Tignor M, Allen SK, Boschung J, Nauels A, Xia Y, Bex V, Midgley PM (eds) Climate Change 2013: The Physical Science Basis. Contribution of Working Group I to the Fifth Assessment Report of the Intergovernmental Panel on Climate Change. Cambridge University Press, Cambridge

Dee DP, Uppala SM, Simmons AJ, Berrisford P, Poli P, Kobayashi S, Andrae U, Balmaseda MA, Balsamo G, Bauer P, Bechtold P, Beljaars ACM, van de Berg L, Bidlot J, Bormann N, Delsol C, Dragani R, Fuentes M, Geer AJ, Haimberger L, Healy SB, Hersbach H, Hólm EV, Isaksen L, Kållberg PK, Köhler M, Matricardi M, McNally AP, Monge-Sanz BM, Morcrette J-J et al (2011) The ERA-Interim reanalysis: configuration and performance of the data assimilation system. Q J R Meteorol Soc 137:553–597. doi:10.1002/qj.828

Duc L, Kuroda T, Saito K, Fujita T (2015) Ensemble Kalman Filter data assimilation and storm surge experiments of tropical cyclone Nargis. Tellus A 67:25941. doi:10.3402/tellusa.v67.25941

Emanuel KA (2003) Tropical cyclones. Annu Rev Earth Planet Sci 31:75–104. doi:10.1146/annurev.earth.31.100901.141259

Emanuel KA (2013) Downscaling CMIP5 climate models shows increased tropical cyclone activity over the 21st century. Proc Natl Acad Sci U S A 110:12219–12224. doi:10.1073/pnas.1301293110

Emanuel KA, Sundararajan R, Williams H (2008) Hurricanes and global warming: results from downscaling IPCC AR4 simulations. Bull Am Meteorol Soc 89:347–367. doi:10.1175/BAMS-89-3-347

Fudeyasu H, Wang YQ, Satoh M, Nasuno T, Miura H, Yanase W (2008) Global cloud-system-resolving model NICAM successfully simulated the lifecycles of two real tropical cyclones. Geophys Res Lett 35:L22808. doi:10.1029/2008gl036003

Fukutomi Y, Kodama C, Yamada Y, Noda AT, Satoh M (2015) Tropical synoptic scale wave disturbances over the western Pacific simulated by a global cloud-system resolving model. Theor Appl Climatol 124:737–755. doi:10.1007/s00704-015-1456-4

Gottschalck J, Wheeler M, Weickmann K, Vitart F, Savage N, Lin H, Hendon H, Waliser D, Sperber K, Nakagawa M, Prestrelo C, Flatau M, Higgins W (2010) A framework for assessing operational model MJO forecasts: a project of the CLIVAR Madden-Julian Oscillation working group. Bull Am Meteorol Soc 91:1247–1258

Graph 500 (2014) The Graph 500 List., http://www.graph500.org. Accessed 10 Mar 2017

Gray WM (1975) Tropical cyclone genesis. Atmospheric Science Paper, No. 234. Colorado State University, Fort Collins

Gray WM (1979) Hurricanes: their formation, structure and likely role in the tropical circulation. In: Shaw DB (ed) Meteorology over the tropical oceans. Royal Meteorological Society, Reading, pp 155–218

Haarsma RH, Roberts MJ, Vidale PL, Senior CA, Bellucci A, Bao Q, Chang P, Corti S, Fučkar NS, Guemas V, von Hardenberg J, Hazeleger W, Kodama C, Koenigk T, Leung LR, Lu J, Luo J-J, Mao J, Mizielinski MS, Mizuta R, Nobre P, Satoh M, Scoccimarro E, Semmler T, Small J, von Storch J-S (2016) High Resolution Model Intercomparison Project (HighResMIP v1.0) for CMIP6. Geosci Model Dev 9:4185–4208. doi:10.5194/gmd-9-4185-2016

Hasegawa Y, Iwata JI, Tsuji M, Takahashi D, Oshiyama A, Minami K, Boku T, Shoji F, Uno A, Kurokawa M, Inoue H, Miyoshi I, Yokokawa M (2011) First-principles calculations of electron states of a silicon nanowire with 100,000 atoms on the K computer. In: Proceedings of 2011 International Conference for High Performance Computing, Networking, Storage and Analysis. ACM Press, New York. doi:10.1145/2063384.2063386

Holt LA, Alexander MJ, Coy L, Molod A, Putman W, Pawson S (2016) Tropical waves and the quasi-biennial oscillation in a 7-km global climate simulation. J Atmos Sci 73:3771–3783. doi:10.1175/JAS-D-15-0350.1

Huffman GJ, Adler RF, Morrissey MM, Bolvin DT, Curtis S, Joyce R, McGavock B, Susskind J (2001) Global precipitation at one-degree daily resolution from multi-satellite observations. J Hydrometeorol 2:36–50. doi:10.1175/1525-7541(2001)002<0036:GPAODD>2.0.CO;2

Huffman GJ, Bolvin DT, Nelkin EJ, Wolff DB, Adler RF, Gu G, Hong Y, Bowman KP, Stocker EF (2007) The TRMM multi-satellite precipitation analysis (TMPA): quasi-global, multiyear, combined-sensor precipitation estimates at fine scale. J Hydrometeorol 8:38–55. doi:10.1175/JHM560.1

Hung M-P, Lin J-L, Wang W, Kim D, Shinoda T, Weaver SJ (2013) MJO and convectively coupled equatorial waves simulated by 20 CMIP5 models. J Clim 26:6185–6214. doi:10.1175/JCLI-D-12-00541.1

Hyodo M, Hori T, Kaneda Y (2016) A possible scenario for earlier occurrence of the nest Nankai earthquake due to triggering by an earthquake at Hyuga-nada, off southwest Japan. Earth Planets Space 68:6. doi:10.1186/s40623-016-0384-6

Ichimura T, Fujita K, Tanaka S, Hori M, Lalith M, Shizawa Y, Kobayashi H (2014) Physics-based urban earthquake simulation enhanced by 10.7 BlnDOF x 30 K time-step unstructured FE non-linear seismic wave simulation. In: SC14: International Conference for High Performance Computing, Networking, Storage and Analysis (2014). IEEE Press, Piscataway, pp 15–26. doi:10.1109/SC.2014.7

Ichimura T, Fujita K, Quinay PEB, Maddegedara L, Hori M, Tanaka S, Shizawa Y, Kobayashi H, Minami K (2015) Implicit nonlinear wave simulation with 1.08T DOF and 0.270T unstructured finite elements to enhance comprehensive earthquake simulation. In: Proceedings of the International Conference for High Performance Computing, Networking, Storage and Analysis. ACM Press, New York. doi:10.1145/2807591.2807674

Ishiyama T, Nitadori K, Makino J (2012) 4.45 Pflops astrophysical N-body simulation on K computer: the gravitational trillion-body problem. In: SC '12 Proceedings of the International Conference on High Performance Computing, Networking. Storage and Analysis. IEEE CS Press, Los Alamitos, pp 1–10

Ito K, Kuroda T, Saito K, Wada A (2015) A large number of tropical cyclone intensity forecasts around Japan using a coupled high-resolution model. Weather Forecast 30:793–808

Kain JS, Fritsch JM (1990) A one-dimensional entraining detraining plume model and its application in convective parameterization. J Atmos Sci 47:2784–2802

Kajikawa Y, Yamaura T, Tomita H, Satoh M (2015) Impact of tropical disturbance on the Indian summer monsoon onset simulated by a global cloud-system-resolving model. SOLA 11:80–84. doi: 10.2151/sola.2015-020

Kajikawa Y, Miyamoto Y, Yoshida R, Yamaura T, Yashiro H, Tomita H (2016) Resolution dependence of deep convections in a global simulation from over 10-kilometer to sub-kilometer grid spacing. Prog Earth Planet Sci 3:16. doi:10.1186/s40645-016-0094-5

Kanada S, Wada A, Nakano M, Kato T (2012) Effect of planetary boundary layer schemes on the development of intense tropical cyclones using a cloud-resolving model. J Geophys Res 117:D03107. doi:10.1029/2011JD016582

Kanada S, Wada A, Sugi M (2013) Future changes in structures of extremely intense tropical cyclones using a 2-km mesh nonhydrostatic model. J Clim 26:9986–10005. doi:10.1175/JCLI-D-12-00477.1

Kawamura R, Murakami T, Wang B (1996) Tropical and midlatitude 45-day perturbations over the western Pacific during the northern summer. J Meteorol Soc Jpn 74:867–890

Kikuchi K, Kodama C, Nasuno T, Miura H, Satoh M, Noda AT, Yamada Y (2016) Tropical intraseasonal oscillation simulated in an AMIP-type experiment by NICAM. Climate Dynam 48:2507–2528. doi: 10.1007/s00382-016-3219-z

Kiladis GN, Wheeler MC, Haertel PT, Straub KH, Roundy PE (2009) Convectively coupled equatorial waves. Rev Geophys 47. doi:10.1029/2008RG000266

Kinter JL, Cash B, Achuthavarier D, Adams J, Altshuler E, Dirmeyer P, Doty B, Huang B, Marx L, Manganello J, Stan C, Wakefield T, Jin E, Palmer T, Hamrud M, Jung T, Miller M, Towers P, Wedi N, Satoh M, Tomita H, Kodama C, Nasuno T, Oouchi K, Yamada Y, Taniguchi H, Andrews P, Baer T, Ezell M, Halloy C et al (2013) Revolutionizing climate modeling with Project Athena: a multi-institutional, international collaboration. Bull Am Meteorol Soc 94:231–245. doi:10.1175/Bams-D-11-00043.1

Knutson TR, Mcbride JL, Chan J, Emanuel K, Holland G, Landsea C, Held I, Kossin JP, Srivastava AK, Sugi M (2010) Tropical cyclones and climate change. Nat Geosci 3:157–163. doi:10.1038/NGEO779

Knutson TR, Sirutis JJ, Zhao M, Tuleya RE, Bender M, Vecchi GA, Villarini G, Chavas D (2015) Global projections of intense tropical cyclone activity for the late twenty-first century from dynamical downscaling of CMIP5/RCP4.5 scenarios. J Clim 28:7203–7224. doi:10.1175/JCLI-D-15-0129.1

Kobayashi K, Kitamura D, Ando K, Ohi N (2015a) Parallel computing for high-resolution/large-scale flood simulation using the K supercomputer. Hydrol Res Lett 9:61–68

Kobayashi S, Ota Y, Harada Y, Ebita A, Moriya M, Onoda H, Onogi K, Kamahori H, Kobayashi C, Endo H, Miyaoka K, Takahashi K (2015b) The JRA-55 reanalysis: general specifications and basic characteristics. J Meteorol Soc Jpn 93:5–48. doi:10.2151/jmsj.2015-001

Kodama C, Yamada Y, Noda AT, Kikuchi K, Kajikawa Y, Nasuno T, Tomita T, Yamaura T, Takahashi HG, Hara M, Kawatani Y, Satoh M, Sugi M (2015) A 20-year climatology of a NICAM AMIP-type simulation. J Meteorol Soc Jpn 93:393–424. doi:10.2151/jmsj.2015-024

Krishnamurthy V, Stan C, Randall DA, Shukla RP, Kinter JL (2014) Simulation of the south Asian monsoon in a coupled model with an embedded cloud-resolving model. J Clim 27:1121–1142

Kumahata K, Minami K, Maruyama N (2016) High-performance conjugate gradient performance improvement on the K computer. Int J High Perform Comput Appl 30(1):55–70. doi:10.1177/1094342015607950

Kunii M (2014a) Mesoscale data assimilation for a local severe rainfall event with the NHM-LETKF system. Weather Forecast 29:1093–1105

Kunii M (2014b) The 1000-member ensemble Kalman filtering with the JMA nonhydrostatic mesoscale model on the K computer. J Meteorol Soc Jpn 91:623–633. doi:10.2151/jmsj.2014-607

Kusunoki S, Arakawa O (2015) Are CMIP5 models better than CMIP3 models in simulating precipitation over East Asia? J Clim 28:5601–5621

Lee JY, Wang B (2014) Future change of global monsoon in the CMIP5. Climate Dynam 42:101–119

Liebmann B, Smith CA (1996) Description of a complete (interpolated) outgoing longwave radiation dataset. Bull Am Meteorol Soc 77:1275–1277

Lloyd ID, Vecchi GA (2011) Observational evidence for oceanic controls on hurricane intensity. J Clim 24:1138–1153. doi:10.1175/2010JCLI3763.1

Madden R, Julian P (1972) Description of global-scale circulation cells in the tropics with a 40-50 day period. J Atmos Sci 29:1109–1123

Manganello JV, Hodges KI, Kinter JL III, Cash BA, Marx L, Jung T, Achuthavarier D, Adams JM, Altshuler EL, Huang B, Jin EK, Stan C, Towers P, Wedi N (2012) Tropical cyclone climatology in a 10-km global atmospheric GCM: toward weather-resolving climate modeling. J Clim 25:3867–3893. doi:10.1175/JCLI-D-11-00346.1

Manganello JV, Hodges KI, Dirmeyer B, Kinter JL III, Cash BA, Marx L, Jung T, Achuthavarier D, Adams JM, Altshuler EL, Huang B, Jin EK, Towers P, Wedi N (2014) Future changes in the western North Pacific tropical cyclone activity projected by a multidecadal simulation with a 16-km global atmospheric GCM. J Clim 27:7622–7646. doi: 10.1175/JCLI-D-13-00678.1

Miura H, Satoh M, Nasuno T, Noda AT, Oouchi K (2007a) A Madden-Julian oscillation event realistically simulated by a global cloud-resolving model. Science 318:1763–1765. doi:10.1126/science.1148443

Miura H, Satoh M, Tomita H, Nasuno T, Iga S, Noda AT (2007b) A short-duration global cloud-resolving simulation with a realistic land and sea distribution. Geophys Res Lett 34:L02804. doi:10.1029/2006GL027448

Miyakawa T, Satoh M, Miura H, Tomita H, Yashiro H, Noda AT, Yamada Y, Kodama C, Kimoto M, Yoneyama K (2014) Madden-Julian oscillation prediction skill of a new-generation global model demonstrated using a supercomputer. Nat Commun 5:3769. doi:10.1038/Ncomms4769

Miyamoto Y, Kajikawa Y, Yoshida R, Yamaura T, Yashiro H, Tomita H (2013) Deep moist atmospheric convection in a subkilometer global simulation. Geophys Res Lett 40(18):4922–4926. doi:10.1002/grl.50944

Miyamoto Y, Yoshida R, Yamaura T, Yashiro H, Tomita H, Kajikawa Y (2015) Does convection vary in different cloud disturbances? Atmos Sci Lett 16(3):305–309. doi:10.1002/asl2.558

Molinari J, Vollaro D (2012) A subtropical cyclonic gyre associated with interactions of the MJO and the midlatitude jet. Mon Weather Rev 140:343–357. doi:10.1175/MWR-D-11-00049.1

3

Atmospheric radioactivity over Tsukuba, Japan: a summary of three years of observations after the FDNPP accident

Yasuhito Igarashi[1*], Mizuo Kajino[1], Yuji Zaizen[1], Kouji Adachi[1] and Masao Mikami[1,2]

Abstract

A severe accident occurred in March 2011 at the Fukushima Dai-ichi nuclear power plant (FDNPP) operated by the Tokyo Electric Power Company (TEPCO), causing serious environmental pollution over a wide range covering eastern Japan and the northwestern Pacific. This accident created a large mark in the atmospheric radionuclide chronological record at the Meteorological Research Institute (MRI). This paper reports the impacts from the FDNPP accident over approximately 3 years in Tsukuba, Ibaraki (approximately 170 km southwest from the accident site), as a typical example of the atmospheric pollution from the accident. The monthly atmospheric ^{90}Sr and ^{137}Cs depositional fluxes in March 2011 reached approximately 5 Bq/m^2/month and 23 kBq/m^2/month, respectively. They are 3–4 and 6–7 orders of magnitude higher, respectively, than before the accident. Sr-90 pollution was relatively insignificant compared to that of ^{137}Cs. The ^{137}Cs atmospheric concentration reached a maximum of 38 Bq/m^3 during March 20–21, 2011. After that, the concentrations quickly decreased until fall 2011 when the decrease slowed. The pre-FDNPP accident ^{137}Cs concentration levels were, at most, approximately 1 µBq/m^3. The average level 3 years after the accident was approximately 12 µBq/m^3 during 2014. The atmospheric data for the 3 years since the accident form a basis for considering temporal changes in the decreasing trends and re-suspension (secondary emission), supporting our understanding of radioCs' atmospheric concentration and deposition. Information regarding our immediate monitoring, modeling, and data analysis approaches for pollution from the FDNPP accident is provided in the Appendices.

Keywords: Temporal change, ^{90}Sr, ^{137}Cs, Atmospheric deposition, Atmospheric concentration, FDNPP accident

Background

We have conducted observational research on radionuclides in the environment for almost 60 years at the Meteorological Research Institute (MRI) in Japan, ever since the 1950s when the USA, Soviet Union, and others performed vigorous nuclear tests in the atmosphere. The atmosphere is the major medium into which radioactive materials were directly injected by the nuclear tests and accidents, and within it, transport, diffusion, and wet and dry removal of these materials occur. During the nuclear testing era, the major purpose of our research was to clarify the radioactive pollution situation and its major controlling

factors in the atmosphere (Hirose et al. 1986; Katsuragi 1983; Miyake 1954; Miyake et al. 1963, 1975) and hydrosphere (Miyake et al. 1955, 1962, 1988). After the Chernobyl accident, the purpose of the research gradually shifted to obtaining more data about various processes in the atmosphere (Aoyama 1988; Aoyama et al. 1986, 1987, 1991, 2006; Hirose et al. 1993, 2001; Igarashi et al. 1996, 2003, 2009) and hydrosphere (Aoyama 1995, Aoyama and Hirose 2004; Hirose et al. 1999, Hirose and Aoyama 2003; Miyao et al. 2000). Of particular interest in this study, observation of monthly radionuclide deposition (atmospheric total deposition/radioactive fallout) for ^{90}Sr (half-life, 28.8 years) and ^{137}Cs (half-life, 30.2 years) had continued for 57 years as of April 2014, although the location of the observations moved from Koenji, Tokyo, to Tsukuba in 1980 when the science city was built (Katsuragi 1983). Both radionuclides are

* Correspondence: yigarash@mri-jma.go.jp
[1]Meteorological Research Institute, 1-1 Nagamine, Tsukuba, Ibaraki 305-0052, Japan
Full list of author information is available at the end of the article

scientifically important because of their health and environmental impacts (e.g., see U.S. Department of Health and Human Services, Public Health Service, Agency for Toxic Substances and Disease Registry ATSDR2004Cs 2004; U.S. Department of Health and Human Services, Public Health Service, Agency for Toxic Substances and Disease Registry ATSDR2004Sr 2004). We continued collecting and analyzing atmospheric samples after the accident at Tokyo Electric Power Company's (TEPCO) Fukushima Dai-ichi Nuclear Power Plant (FDNPP) in Ohkuma-machi and Futaba-machi, Fukushima prefecture (37.42 °N, 140.97 °E) in March 2011.

Many authors have attempted to determine the environmental impacts of the FDNPP accident, which have gradually come to light (e.g., Aoyama et al. 2012, 2013; Hirose 2012; Kusakabe et al. 2013; Masson et al. 2011; Masumoto et al. 2012; MEXT 2011a ; MEXT and USDOE 2011; Povinec et al. 2013a, b; Tsumune et al. 2013; Yamamoto et al. 2012; Yoshida and Kanda 2012; Yoshida and Takahashi 2012). We still need to study the following issues from an atmospheric science point of view (Igarashi 2009): (1) primary source terms including emissions inventory and temporal changes (e.g., Chino et al. 2011; Katata et al. 2012, b, 2014; Maki et al. 2013; Stohl et al. 2012; Terada et al. 2012; Winiarek et al. 2012), (2) transport and diffusion (e.g., Masson et al. 2011; Morino et al. 2011; Sekiyama et al. 2015; Stohl et al. 2012; Takemura et al. 2011; Tanaka 2013; Terada et al. 2012), and (3) dry and wet removal (e.g., Adachi et al. 2013; Hirose et al. 1993; Kristiansen et al. 2012), which governed radioactive surface contamination during the early phase of the accident. In addition, the physical and chemical properties of the radioactive materials (e.g., Adachi et al. 2013; Kaneyasu et al. 2012) are important factors that influence the second and third subjects to be investigated. Here, we summarize the observations, present a time series of the atmospheric impacts of the TEPCO FDNPP accident over approximately 3 years in Tsukuba, Ibaraki, Japan, and compare the levels to the situation before the accident as very basic scientific information (Igarashi, 2009). In addition, secondary emissions from contaminated surfaces to the atmosphere (re-suspension; Igarashi 2009) have become important during the later phases. Resuspension comes from contaminated surfaces, terrestrial ecosystems, and open-field burning. These sources have undoubtedly supported atmospheric radionuclides but are not yet well understood and are thus considered briefly. Other information about the accident, related to our immediate monitoring and modeling endeavors and data analysis approaches to short-lived γ-emitters and ^{89}Sr, is summarized in the Appendices.

Methods

Atmospheric deposition samples

The monthly atmospheric total deposition/atmospheric fallout has been sampled using a weathering-resistant plastic tray (area = 4 m^2) installed on a cottage roof in an open field of the MRI in Tsukuba, Ibaraki (36.1 °N, 140.1 °E; approximately 170 km southwest of the FDNPP) since the 1980s. After April 2011, the sample size was reduced to two trays, each 1 m^2, which we considered sufficient for the levels present after the FDNPP accident. The collected samples were evaporated and concentrated into a gross quantity with a rotary evaporator (Eyela NE-12) or an evaporating dish, and the samples were saved in a polyethylene safekeeping container. Each evaporated sample, packed in a cylindrical plastic container, was measured for γ-ray emitting radionuclides ([134]Cs and ^{137}Cs) using a Ge semiconductor detector (coaxial-type from ORTEC EG&G or Eurisys) coupled with a computed spectrometric analyzer (Oxford-Tennelec Multiport or Seiko EG&G 92x). The precision, accuracy, and quality control of the measurements are described elsewhere (Otsuji-Hatori et al. 1996).

Part of the sample was then stored for future reanalysis. The remaining sample was added to concentrated nitric acid along with H_2O_2 and digested in a heating operation. Sr-90 was radiochemically recovered from the obtained sample solution, purified and finally fixed as Sr carbonate precipitate, an activity measurement source. After the source was left for several weeks to achieve ^{90}Sr and ^{90}Y radioequilibrium, its β-activity was measured using a low-background 2π gas-flow detector (Tennelec LB5100) with P10 gas (Otsuji-Hatori et al. 1996). Within several months after the FDNPP accident, [89]Sr (half-life, 50.5 days) from the accident coexisted with ^{90}Sr and affected the β-activity measurement. To remove the ^{89}Sr influence, we occasionally repeated the Sr source measurement and evaluated the radioequilibrium between ^{90}Sr and ^{90}Y, as well as the decrease in ^{89}Sr activity (see Appendix 2). When required, the influence of the ^{89}Sr activity was subtracted from the β-activity counts to obtain the ^{90}Sr activity. The activity was always decay-corrected mid-sampling. The detection limit for ^{90}Sr was approximately 7.0 mBq/sample, approximately 3.5 mBq/m^2 using a total of 30,000 s of measurement. For ^{137}Cs, the limit was approximately 16.0 mBq/sample, approximately 8.0 mBq/m^2 for an average of 120,000 s of measurement.

Atmospheric radioactive aerosols

Aerosol samples were collected weekly using a high-volume air sampler (HV; Sibata Scientific Technology Ltd., HV-1000 F) on a quartz fiber filter (Advantech QR100; 203 mm × 254 mm) (Igarashi et al. 1999a). During March 2011, the sampling frequency was intensified.

The flow rate was set at 0.7 m³/min, and the daily sucked air volume was approximately 1000 m³. After collection, the filters were compressed into pellets using a hydraulic press device. They then underwent conventional γ-ray spectrometry with Ge detectors as described above. Current detection limits for ^{134}Cs and ^{137}Cs are approximately 9.0 mBq/sample (1.3 µBq/m³) and 10 mBq/sample (1.5 µBq/m³) for approximately 1,000,000 s measurements, respectively.

The filter samples collected before the radioactive plume arrived at Tsukuba were measured at the Kyoto University Research Reactor Institute to achieve lower detection limits and avoid contamination from the FDNPP accident. This was necessary because the Ge detectors, measurement environment, and experimental materials at the MRI were somehow contaminated by the radioactive plume's passage on March 14–15 and 20–23, 2011 (see Appendix 1). To date, radioSr analysis has been performed on only a limited number of aerosol samples collected during March 2011. The results are presented in Appendix 2.

Results and discussion

Figures 1 and 2 depict the results of the atmospheric ^{90}Sr and ^{137}Cs deposition observations at the MRI for different durations. The temporal changes in monthly radionuclide depositions shown in Fig. 1 include those from the late 1950s to more recently available data, i.e., after the FDNPP accident. Figure 2 compares the amounts of atmospheric deposition after the FDNPP accident and from the late 2000s. Analyses of ^{90}Sr and ^{137}Cs deposition samples taken 6 and 8 months before the accident are ongoing to control for possible sample contamination at the MRI caused by the accident. Thus, these data are missing in Figs. 1 and 2.

Figure 3 depicts the temporal change in atmospheric activity concentrations of radioCs since March 2011. Before the FDNPP accident, it was difficult to detect ^{137}Cs below about 1 µBq/m³ in the air (the global fallout background level).

Although there were small-scale Japanese nuclear accidents in the 1990s (Igarashi et al. 1999a, 2000; Komura et al. 2000), they did not cause significant marks in the present time series of monthly ^{90}Sr and ^{137}Cs depositions. The effects of the Chernobyl accident that occurred in 1986 were more evident for ^{137}Cs than ^{90}Sr (e.g., Aoyama et al. 1991) as illustrated in Fig. 1. However, the previous maximum ^{137}Cs deposition was two orders of magnitude lower than those caused by the FDNPP accident. Thus, the impact of the FDNPP

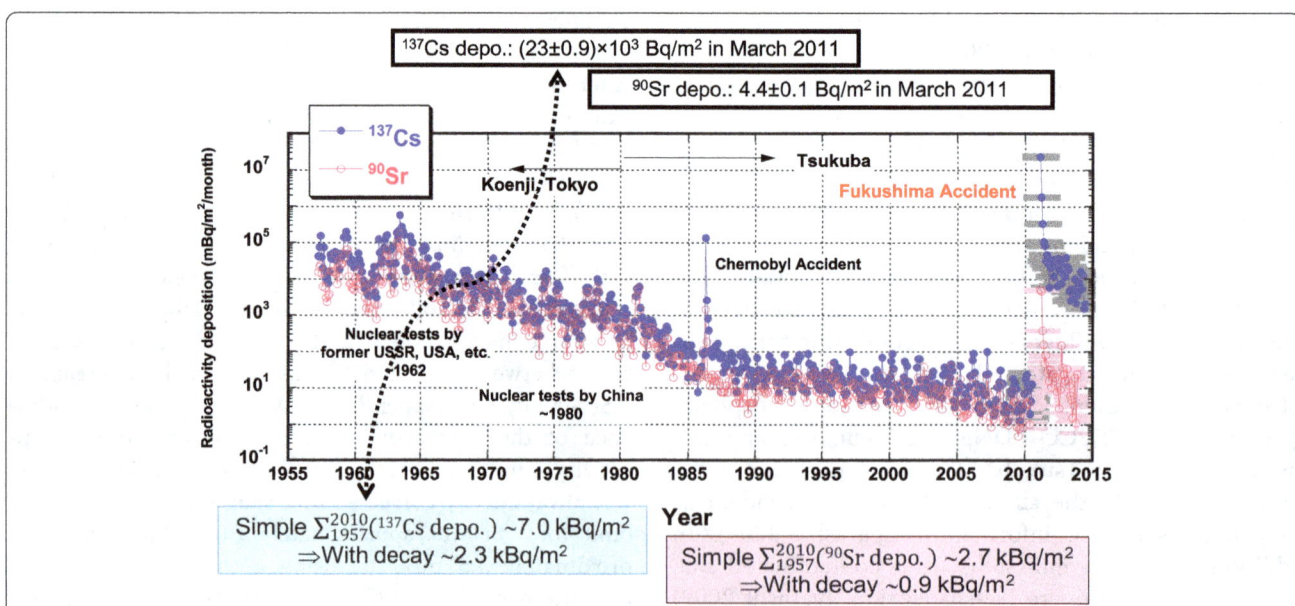

Fig. 1 Sr-90 and ^{137}Cs monthly deposition observed at the Meteorological Research Institute (MRI) from 1957 to 2014. Monthly deposition is expressed in millibecquerel per square meter on a logarithmic scale. Sr-90 and ^{137}Cs analyses from deposition samples taken 6 and 8 months before the accident, respectively, are ongoing to avoid possible sample contamination at the MRI because of the accident. Thus, these data are missing not only in Fig. 1 but also in Fig. 2. The measurement uncertainty (1σ) is shown only for the data obtained after the FDNPP accident and is reasonably small compared to the analytical data. For comparison, uncertainty for the monthly data in 2010 is also given. The effects of atmospheric nuclear bomb tests have been recorded since 1957. Until the Partial Test Ban Treaty (PTBT) became effective in 1963, the USA, Soviet Union, and UK conducted atmospheric tests. France and China continued atmospheric testing until 1974 and 1980, respectively. Since 1981, all the nuclear bomb tests have shifted underground, so additional radioSr and Cs contamination should be negligible. However, the Chernobyl accident in 1986 also affected the time series. The simple summation of the deposition from 1957 to the time before the FDNPP accident (mid-2010) and decay-corrected summations for ^{90}Sr and ^{137}Cs can be compared to the FDNPP-derived deposition

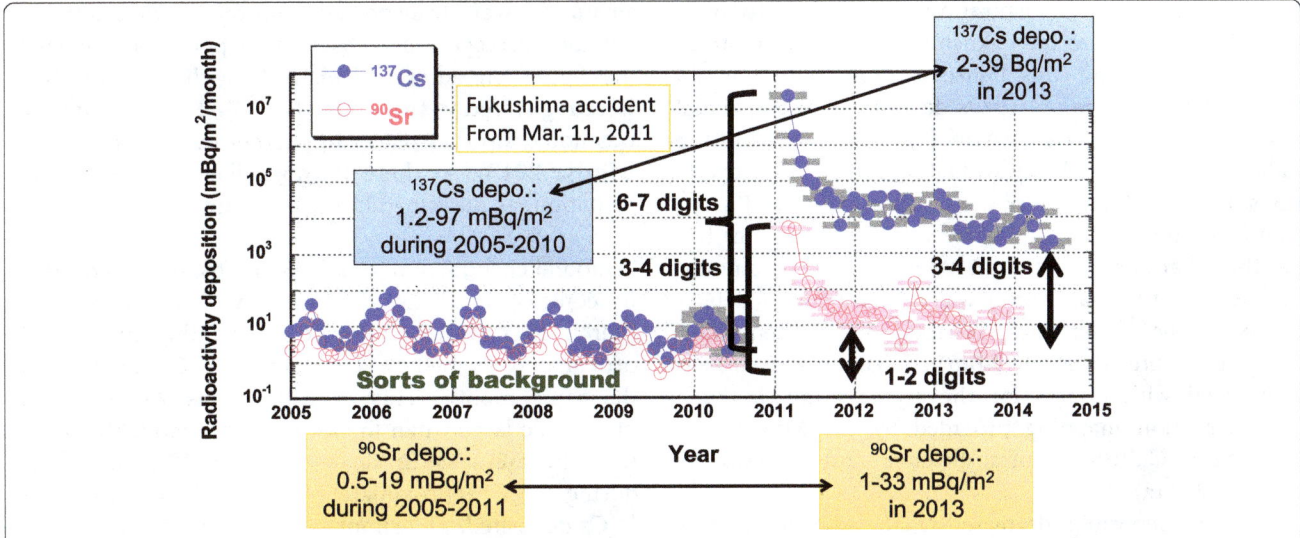

Fig. 2 Monthly ^{90}Sr and ^{137}Cs deposition levels in pre- and post-accident periods. Partial enlargement of Fig. 1. The monthly deposition is expressed in millibecquerel per square meter on a logarithmic scale. The atmospheric depositions of ^{90}Sr and ^{137}Cs in 2013 observed at the MRI were a few orders of magnitude higher than those from 2005 to 2011 before the FDNPP accident. For ^{90}Sr and ^{137}Cs, monthly depositions during 2005 to 2010 were 0.5–19 mBq/m^2/month and 1.2–97 mBq/m^2/month, whereas they were 1–33 mBq/m^2/month and 2–39 Bq/m^2/month in 2013, respectively

accident was more remarkable than any previous incident in our time series.

Temporal changes in monthly ^{137}Cs atmospheric deposition

The monthly ^{137}Cs deposition in March 2011, when the FDNPP accident occurred, was 23 ± 0.9 kBq/m^2/month, which is six to seven orders of magnitude higher than

the level before the Fukushima disaster (Figs. 1 and 2). Because the pollution source of the FDNPP accident is closer to the observation site (170 km) than it is to the weapons testing sites and Chernobyl (several thousand kilometers), the spatial representativeness of the MRI data (as an absolute value) is lower.

The cumulative ^{137}Cs deposition at the MRI was 25.5 kBq/m^2/year for the year 2011. The sum of the

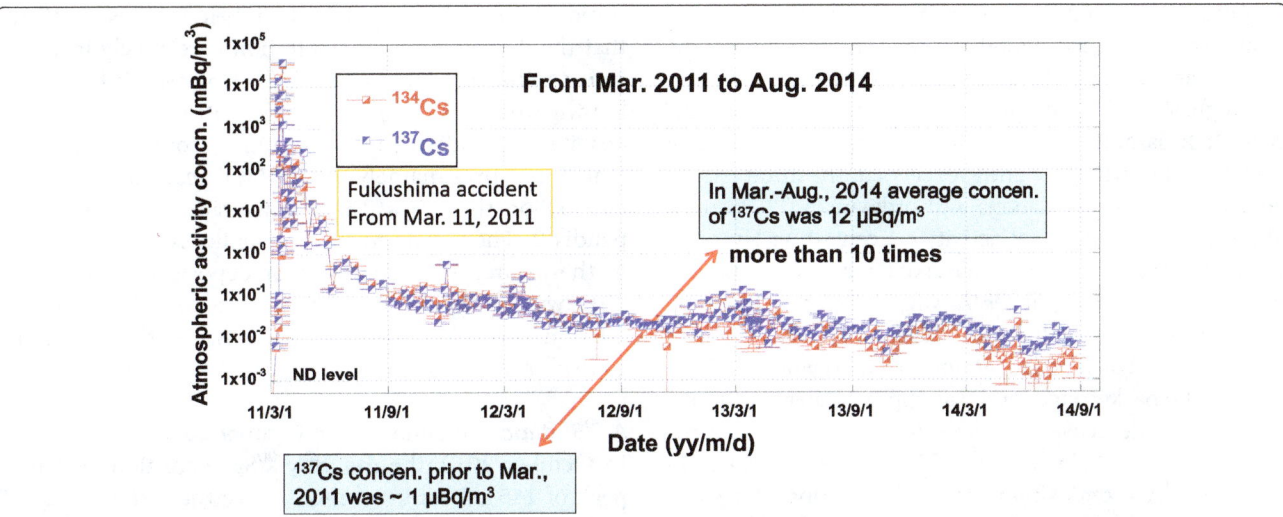

Fig. 3 Temporal change in atmospheric radioCs concentrations at the MRI before and after the FDNPP accident ("Mar.-Aug. 2014"). Activity concentration is expressed in milli becquerel per cubic meter on a logarithmic scale. The measurement uncertainty (1σ) is shown. The maximum concentration of 38 Bq/m^3 of ^{137}Cs was observed during March 20–21, 2011. After that, the radioCs concentrations rapidly decreased until fall 2011 when the decrease slowed. The levels before (approximately 1 μBq/m^3) and 3 years after the FDNPP accident (12 μBq/m^3 from March to August 2014) are also compared. A difference of at least one order of magnitude is observed between the concentration level from March to August 2014 and the level before the FDNPP accident

simple monthly ^{137}Cs depositions from 1957 to mid-2010, the time before the Fukushima disaster, is approximately 7.0 kBq/m^2 (this figure is thought to contain some error since the pre-1970s data did show individual undefined errors), as shown in Fig. 1. Considering the radioactive decay of the individual monthly ^{137}Cs depositions, this past total contribution represents 2.3 kBq/m^2. The FDNPP accident's influence was over ten times larger than that of any past event. Almost the same amount of ^{134}Cs (half-life, 2.1 years) was simultaneously deposited with the ^{137}Cs; thus, the total cesium deposition came to more than 50 kBq/m^2. This value agrees quite well with figures for the area around Tsukuba in observation mapping provided by the Ministry of Education, Culture, Sports, Science and Technology (MEXT 2011a).

Later, the deposition decreased rapidly, but the monthly ^{137}Cs deposition in 2012 and 2013 ranged from 8–36 and 2–39 Bq/m^2/month, respectively, where deposition during 2005–2010 had been in the range of 1.2–97 mBq/m^2/month, i.e., three to four orders of magnitude higher. The deposition level at the end of 2013 was still as high as values registered when atmospheric nuclear tests were conducted by China in the 1970s to the early 1980s. The deposition rate slowly decreased in the following years.

Atmospheric concentrations of radioCs

Figure 3 displays the temporal change in the atmospheric radioCs activity concentrations at the MRI in Tsukuba since the FDNPP accident. The temporal trend shows an abrupt increase (peak) of several orders of magnitude, followed by a rather rapid concentration decrease over a short period (3 to 4 months after the FDNPP accident), with a smaller decreasing rate after. The highest ^{137}Cs atmospheric concentrations (38 Bq/m^3 in a 12 h sampling period) were registered on March 20–21, 2011, which slightly exceeded the limit stipulated by Japanese regulations and ordinances (30 Bq/m^3). Although the pre-accident activity concentration level was not measured, it had been observed for a short period, from February to April 1997, which includes the time when the Power Reactor and Nuclear Fuel Development Corporation Tokai accident occurred (Igarashi et al. 1999a). The background level was approximately 1 µBq/m^3 and did not decrease far below half that value (approximately 0.5 µBq/m^3) until 2011. The decrease in monthly ^{137}Cs deposition was small during the same period (Igarashi et al. 2003, 2009). Thus, the ^{137}Cs activity concentration level registered during summer 2014 appears at least 10 times higher than that before the accident. During 2011 and 2012, small spikes were recorded from time to time (Fig. 4). In these cases, daily forward trajectory analysis suggested that the polluted air masses were transported from the accident site during the corresponding observation period as shown in the figure. In addition, relatively high concentrations were registered in the winter (Fig. 3). This phenomenon was noted at other places in northern and eastern Japan (Hirose 2013), so there is most likely a common explanation, as described in the literature.

Temporal change in monthly ^{90}Sr atmospheric deposition

In contrast to ^{137}Cs, the monthly ^{90}Sr deposition in March 2011 was 5.2 ± 0.1 Bq/m^2/month. This was approximately 1/5000 the amount of ^{137}Cs deposited in the same month. This deposition was 2–3 orders of magnitude larger than the level before the FDNPP disaster. The annual ^{90}Sr deposition was 10.6 Bq/m^2/year during 2011, approximately 1/2500 of the quantity of ^{137}Cs deposited. The simple sum of the monthly ^{90}Sr depositions from 1957 to mid-2010, before the Fukushima disaster, was approximately 2.7 kBq/m^2, as shown in Fig. 1. Taking the radioactive decay of the individual monthly ^{90}Sr depositions into account, the sum represents approximately 0.9 kBq/m^2. The FDNPP accident's impact on ^{90}Sr was very small. The most extreme monthly ^{90}Sr deposition, recorded during the global fallout era of May 1963 in Tokyo, was 170 Bq/m^2/month. The FDNPP accident's impact on the monthly ^{90}Sr deposition was less than one-thirtieth of this maximum. Therefore, it is probable that ^{90}Sr pollution over the Kanto Plain from the accident was relatively insignificant; the environmental and health impacts of ^{90}Sr are relatively minor.

In addition, the ^{137}Cs/^{90}Sr activity ratio fluctuated between approximately 400 and 5000 (Fig. 5), except for some abnormal cases described below. This confirms that the degree of radioSr pollution is relatively insignificant compared to that of radioCs. However, it is still unknown why the ^{137}Cs/^{90}Sr activity ratio varied so widely despite the radionuclides having a common accident emission source, namely, the FDNPP accident. More discussion on the ^{137}Cs/^{90}Sr activity ratio is given in Appendix 2. The reason for the variability is worth studying in the future. The monthly ^{90}Sr deposition recorded in 2012 was 10–31 mBq/m^2/month, whereas during 2005–2010, it was 0.5–19 mBq/m^2/month, a difference of up to two orders of magnitude.

A ^{90}Sr deposition anomaly in October 2012

In October 2012, the monthly ^{90}Sr deposition showed a peak of 145 ± 2 mBq/m^2/month (see the arrow in Figs. 5 and 6), which is 1–2 orders of magnitude higher than any monthly ^{90}Sr deposition registered that year, and its influence lasted a few months (Fig. 6). This small ^{90}Sr event remains puzzling. By applying forward trajectory analysis and closely examining the precipitation over Tsukuba, we believe that the ^{90}Sr may have come from

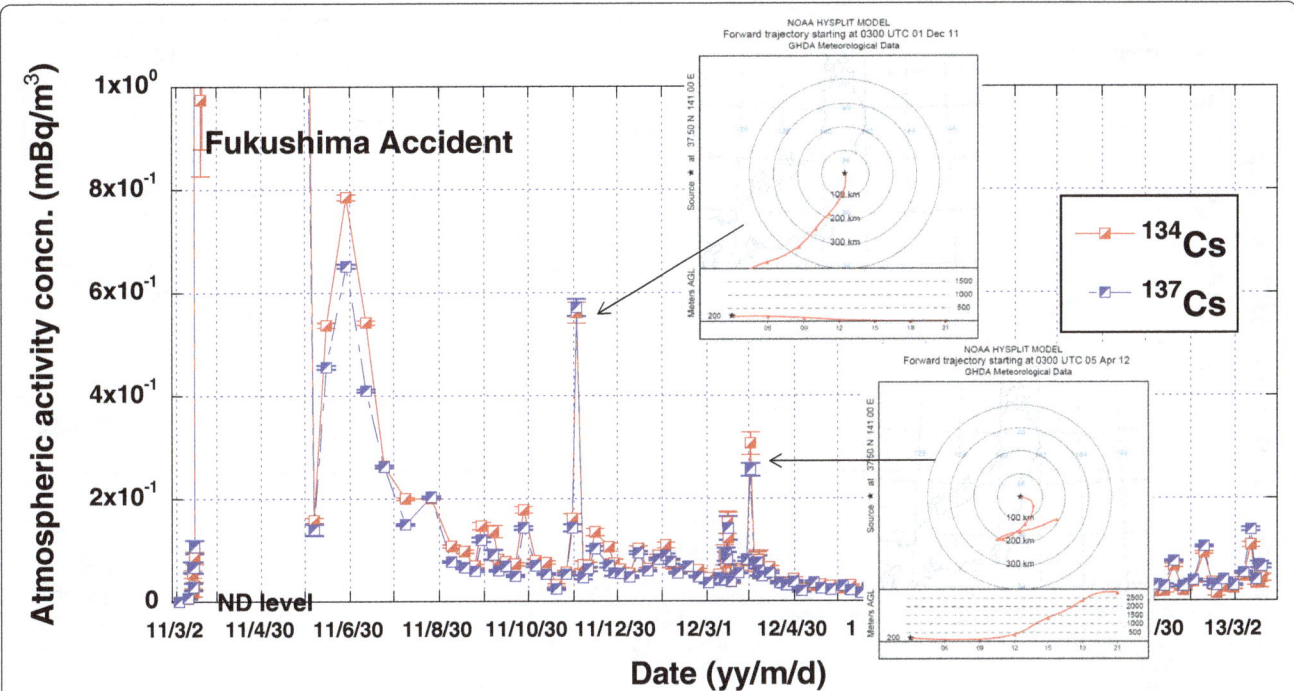

Fig. 4 Atmospheric concentration increases observed during 2011 and 2012 and their air mass trajectories. Note that the activity concentration scale is linear. The forward air mass trajectory calculated by the NOAA's HYSPLIT model is depicted for the radioCs activity concentration peaks, suggesting that the plume from the FDNPP site passed over the Tsukuba region. The shown trajectory cases are December 1, 2011 and April 5, 2012. The increases seem to be attributable to the transport of primary radioCs from the accident site

the FDNPP and encountered precipitation on October 7 and 18–19, 2012. However, this increase was not accompanied by a radioCs deposition peak, and the major radionuclide emitted by the FDNPP accident is radioCs, which is inconsistent with FDNPP accident being the source of the October anomaly.

The Japanese Radioactivity Survey data on the Internet were checked, but no consistent data were evident for the corresponding period. In addition, no such anomaly

was reported in Europe (Masson 2014, personal communication). Based on the timescale of this contamination, however, the source should be neither very local nor very small. This episode shows some similarities to the case in fall 1995 in Tsukuba (Igarashi et al. 1999b). We also assume unidentified, unreported incidents of burning and/or melting of industrial ^{90}Sr sources in the Far East region as a possible explanation, such as the Algeciras (Spain) incident in 1998 with its ^{137}Cs source of

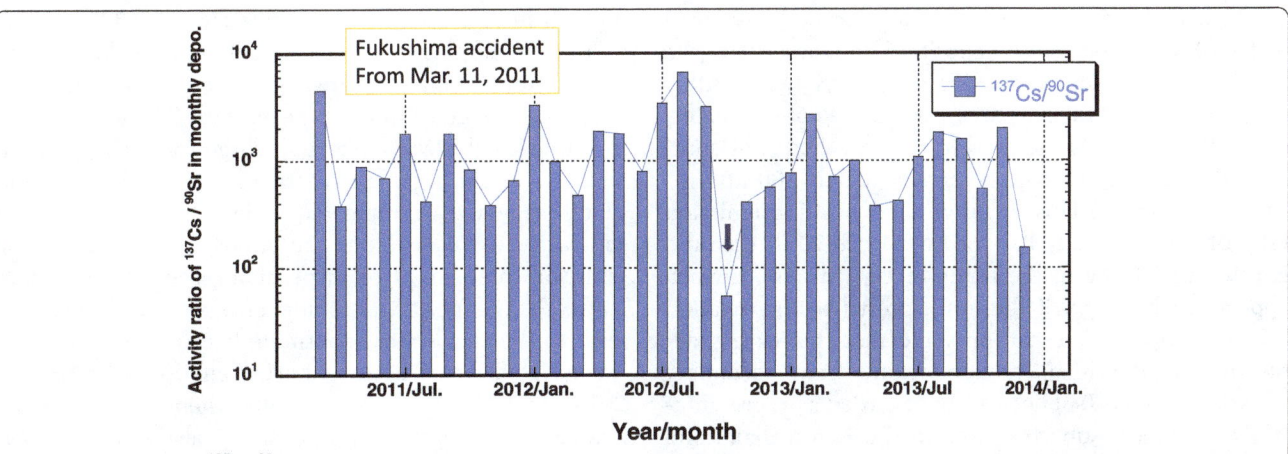

Fig. 5 Activity ratio of ^{137}Cs/^{90}Sr in monthly depositions since March 2011 at the MRI. The temporal changes do not show a clear decreasing or increasing trend. The *arrow* shows the month during which an anomalous deposition of ^{90}Sr was observed. Except for the anomaly, the ^{137}Cs/^{90}Sr activity ratio fluctuated from approximately 150 to 6700

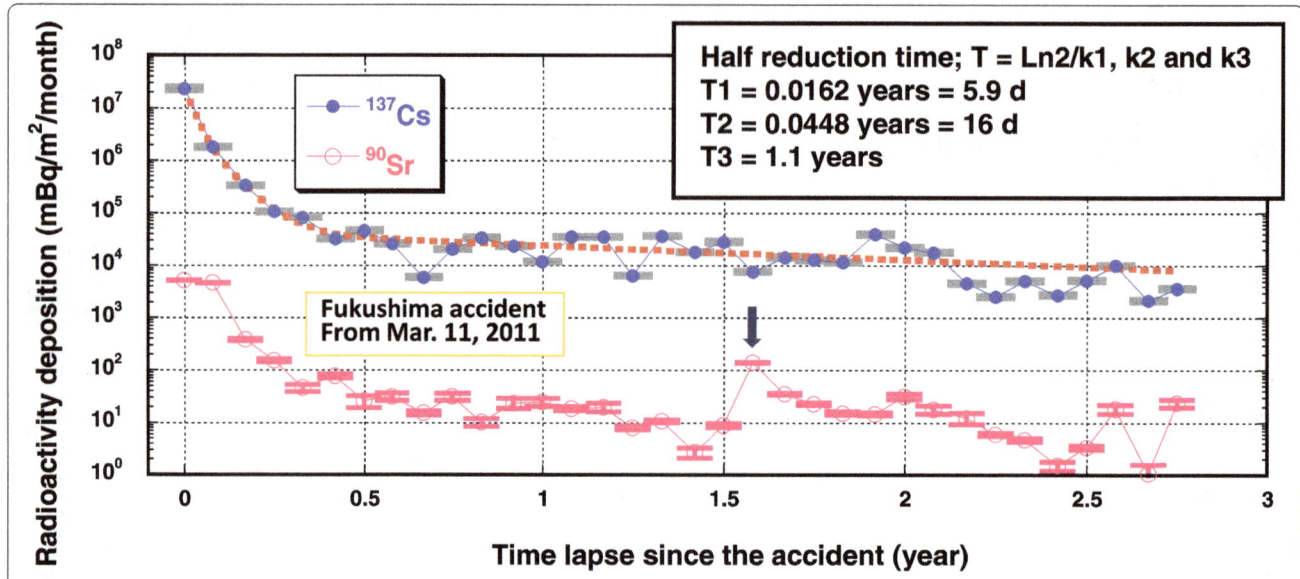

Fig. 6 Exponential fitting of the decreasing monthly ^{137}Cs deposition trend since March 2011 at the MRI. The curve is composed of three exponential functions. These are attributable to the decreasing intensity of primary emission, tropospheric aerosol residence and re-suspension. The *arrow* shows the month during which an anomalous ^{90}Sr deposition was observed. Possible causes are mentioned in the text

3.7 TBq (Estevan 2003). Sr-90 is widely used in industrial applications, such as in thickness gauges, and its activity size ranges from 740 MBq to 3.7 GBq in Japan. Because ^{90}Sr is a pure β-emitter, it is more difficult to determine the sources of its environmental pollution than it is for [137]Cs.

Decrease in monthly ^{137}Cs deposition after the FDNPP accident

Although researchers do not agree precisely on the FDNPP radioactivity emission inventory (Chino et al. 2011; Katata et al. 2012, 2012b, 2014; Maki et al. 2013; Stohl et al. 2012; Terada et al. 2012; Winiarek et al. 2012), if the ^{137}Cs emission in March 2011 is assumed to be 10 PBq/month, the deposition/emission ratio (the monthly deposition at the MRI divided by the monthly emissions from TEPCO (2012)) would be approximately 10^{-12}. If the MRI is included in the so-called "hot spot" area, the deposition could be approximately 100 kBq/m^2 (five times larger). This would give a deposition/emission ratio of approximately 10^{-11}. After March 2011, the ratio is calculated to be in the range of 10^{-10} to 10^{-9}, which appears to be large, if the emission-deposition relation above is correct. We can presume that this excess deposition at the MRI, Tsukuba came from secondary emissions. Thus, Tsukuba can be regarded as representative of a typical suburban area in the Kanto Plain, and the relative trend of temporal changes there can be considered comparable to surface contamination levels for similar geographical domains. The temporal trends

(holding time constant) may also be spatially representative, although this potential is limited.

To study the decreasing trend in monthly ^{137}Cs atmospheric deposition caused by the FDNPP accident and to make future projections, a curve was fitted on the temporal trends using multiple components. A drawing software was employed, and the fitting operation was put through 100 iterations, each time changing the initial value so that the calculation results would converge, as shown in Fig. 6. A trinomial exponential function of the form $a \times (e^{-k \times t})$ was applied to fit the data (where a is a constant and k is an inverted time scale; $\text{Ln}2/T_{1/2}$), and the individual half-times (T1, T2, and T3 in Fig. 6) were approximately 5.9 (± 11 %) days, 16 (± 18 %) days, and 1.1 (± 32 %) years, respectively. The relative uncertainty is shown in parentheses. These appear to correspond to the time scale of (1) the reduction in the original FDNPP accident surge (primary emission source), (2) the tropospheric transportation and diffusion of the radioactive plume (equivalent to the removal of radioactive aerosols from the atmosphere), and (3) the emission intensity of re-suspension (secondary emission sources). We posit that some primary radiological release to the atmosphere continues because the FDNPP is not isolated from the neighboring environment (Hirose 2013; TEPCO 2012). The results, then, cannot be assumed to be completely free of primary release. However, the first and second terms can be reasonable estimates corresponding to the primary emission and tropospheric aerosol residence, respectively.

The second term is almost identical to figures obtained by other recent studies (e.g., Hirose 2012, 2013; Kristiansen et al. 2012). Hirose (2013) analyzed radioCs deposition data obtained during 2011–2012 from several places over the Kanto Plain and Fukushima prefecture, Japan. According to his report, "The apparent half-lives at Ichihara, Tokyo, Utsunomiya, Hitachinaka and Maebashi were 11.9, 10.6, 13.5, 11.5 and 12 d, respectively." Hirose (2012) states that "the residence times of aerosols in the troposphere, which are in the range of 5–30 d, have been determined by natural and anthropogenic radionuclides, which depend on particle size and altitude (Ehhalt, 1973)." Hirose (2012) also argues "the temporal change of the Fukushima-derived ^{137}Cs revealed that the apparent atmospheric residence time of the Fukushima-derived ^{137}Cs in sites within 300 km from the Fukushima Dai-ichi NPP is about 10 d." This long residence time might reflect the Fukushima radioactive plume's circulation over the Northern Hemisphere, which takes about 20 days (Hernández-Ceballos et al. 2012). As shown in Fig. 8a in the Appendix 1, the third Fukushima plume's arrival over the Kanto Plain was observed from March 28–31, 2011. It was well reconstructed by the aerosol transport model. Other observations over the Kanto Plain also revealed this transport event (e.g., Amano et al. 2012; Haba et al. 2012). However, we cannot clearly determine whether this concentration peak is due to delayed primary emission (e.g., Terada et al. 2012), hemispheric circulation, or a combination of both. This is because the current model simulation uses the emission inventory, which is also based on atmospheric monitoring results (e.g., Terada et al. 2012). Regarding this connection, Kristiansen et al. (2012) investigated the ^{131}I and ^{137}Cs removal times from the atmosphere using global-scale monitoring data. Their estimated ^{137}Cs removal times were in the range of 10.0–13.9 days, which is closer to our present result. They also noted the difference from the typical values of 3–7 days obtained by aerosol model simulations, suggesting that the aerosol transportation models need improvement. We would like to add that the deposition results should be interpreted to reflect not only the surface air but also the air column up to at least the mixed layer. Therefore, the deposition may be affected by large-scale transportation, in contrast to indications obtained from the surface concentration only. For further reference, based on the monthly emission of radioCs until the end of 2011 estimated by TEPCO 2012, the primary emission decrease can be fitted using two exponential laws with half-time constants of 2.3 days (±2 %) and 48 days (±23 %).

The third term's half-time of 1.1 years for the MRI data, despite its relatively large associated uncertainty, appears to reveal the total re-suspension of radioCs from contaminated surfaces. This value is too large to correspond to any primary releases from the FDNPP in the early phases. In addition, it agrees with the value for the re-suspension "descending trend" due to the Chernobyl accident reported by Garger et al. (2012), which was 300 days. It was possible to fit a two-term exponential curve to the present ^{137}Cs data by fixing the 1.1-year half-time, obtaining a value of 7.8 days for the first term. When compared with the triple exponential (three-term) model, the fitting distance (defined by the ratio of the calculation to the observation) for the double exponential (two-term) model was larger for elapsed times of 2–12 months, although there were exceptions. The mean and standard deviation for the two- and three-term fit distances are 2.50 ± 2.02 and 1.54 ± 1.14, respectively. The medians are 1.82 and 1.09, respectively, suggesting that the three-term model fits better. Although we do not provide an illustration here, we found that fitting with three-term functions for the decrease in monthly ^{90}Sr deposition after the disaster was also possible. Therefore, we preferred fitting with a trinomial exponential function to reproduce the deposition flux of radionuclides from the FDNPP accident. Again, the primary emissions of radioCs to the atmosphere are anticipated to continue at a non-negligible level (less than 7.2 GBq/ month is assumed in TEPCO's latest press release (in Japanese) at http://www.tepco.co.jp/life/custom/faq/images/d150129-j.pdf) because the FDNPP is not isolated from the surrounding environment (Hirose 2013). These delayed primary emissions of approximately 7 GBq are 6–7 orders of magnitude lower than the emissions in March 2011 (e.g., 15 PBq for ^{137}Cs; NISA 2011). If the primary emission deposits were delayed in a fashion similar to those from March 2011, recent MRI records after the FDNPP accident would correspondingly be 6–7 orders of magnitude lower than the peak value caused by the accident (see Fig. 2). Therefore, we consider that the present decrease in the third term reflects secondary emission (re-suspension) trends over the Kanto Plain moderately well. In future, we plan to confirm this by applying different evaluation methods such as transport simulations or others.

Consideration of re-suspension and its persistence

Currently, there may be interest and concern about how long it will take for the atmospheric radionuclide deposition fluxes to return to pre-FDNPP accident levels (cf. Garger et al. 2012; Hatano and Hatano 2003). Although it seems slightly arbitrary, the monthly ^{137}Cs depositions can be estimated if the fitted curve described above is extrapolated. The result of this extrapolation is illustrated in Fig. 7. This simple estimation shows that more than a decade will likely be required for the activity levels to return to pre-accident levels. Thus, re-suspension (secondary emission to the atmosphere;

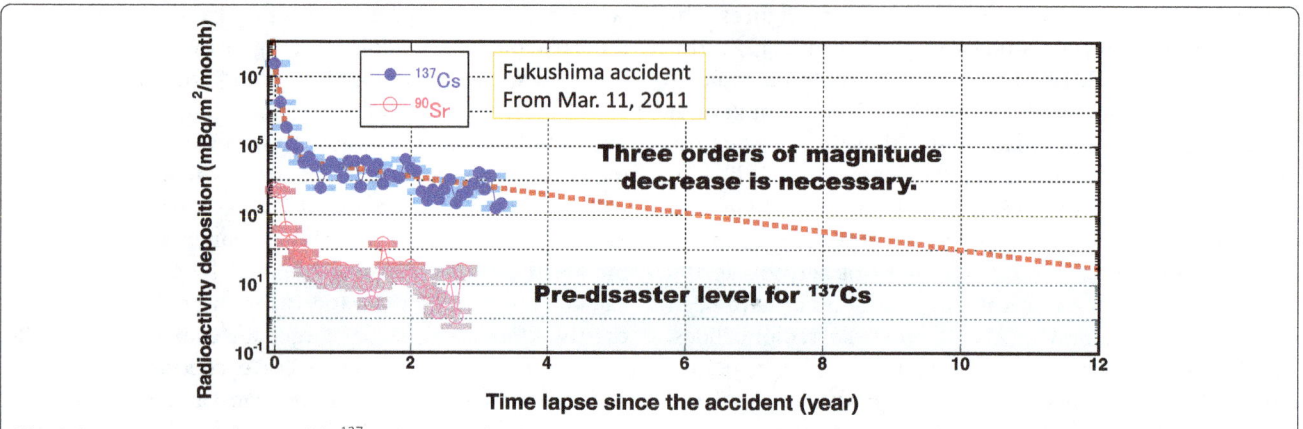

Fig. 7 Future projection for monthly ^{137}Cs deposition level using a trinomial exponential function. The present simple estimation shows that more than a decade would be necessary for the ^{137}Cs atmospheric deposition level to return to pre-accident levels

e.g., Igarashi 2009) must be scrutinized with long-term monitoring. Because it seems natural that radionuclide emission flux would be proportional to surface pollution density, there could be radioCs fluxes several orders of magnitude higher than those measured in Tsukuba in areas nearer the FDNPP site whose Cs surface pollution is several orders of magnitude higher than in Tsukuba. Therefore, elucidating the secondary emission processes of the FDNPP radionuclides remains an imminent scientific challenge, especially for heavily polluted areas. Secondary sources can include soil dust suspension from polluted earth surfaces, emissions from polluted vegetation and forests, and volatilization and release from combustion of polluted garbage and open field burning (e.g., Igarashi 2009). Although the main emission sources are not yet well understood, this elucidation must be performed as soon as possible.

Conclusions

The authors conducted atmospheric monitoring of airborne radioSr and Cs and their deposition at the MRI in Tsukuba, Japan. The monitoring period encompasses the FDNPP accident and the subsequent few years. The monthly ^{137}Cs deposition at the MRI was $(23 \pm 0.9) \times 10^3$ Bq/m^2/month in March 2011, which is 6–7 orders of magnitude higher than pre-accident levels. Almost equal amounts of [134]Cs and ^{137}Cs were deposited, causing surface pollution of more than 50 kBq/m^2 in Tsukuba in 2011, in close agreement with the Ministry of Education, Culture, Sports, Science and Technology, Japan (MEXT)'s airborne mapping. Deposition of ^{90}Sr was 5.2 ± 0.1 Bq/m^2/month in March 2011, which is less than 0.02 % of the total ^{137}Cs deposition in that month. The level of ^{90}Sr deposition was 3–4 orders of magnitude higher than pre-accident levels and did not reach the level registered during the 1960s after nuclear tests; the effects from ^{90}Sr will not be as large as from

radioCs. During 2013, the Fukushima fallout decreased by 3–4 orders from its magnitude at the time of the accident, yet some becquerel per square meter of monthly deposition continues. This corresponds to the level in the 1970s and early 1980s when China performed atmospheric nuclear tests. During 2013, the ^{137}Cs concentration remained at a level of tens of micro becquerel per cubic meter. Because re-suspension (secondary emission) will continue over a long time, it is necessary to monitor its future trends and variability. An apparent decrease in atmospheric radioCs deposition was fitted by trinomial exponentials, giving information regarding the reducing trend of airborne radionuclide persistence through re-suspension into the atmosphere. Extrapolation of the decreasing rate suggests that it would take at least a decade for the activity to return to pre-disaster period levels. Further monitoring efforts are essential.

Appendix

Appendix 1 Temporal changes in radioactive aerosol concentrations and plume transport from the FDNPP accident over Tsukuba in March 2011
Introduction
The heat and blast at the FDNPP accident resulted in the leakage of a huge amount of anthropogenic radionuclides, near the levels of the Chernobyl accident in 1986, into the environment (IAEA 2006; Janžekovič and Križman 2011; NISA 2011), as seen on both the domestic and Northern Hemispheric scale (Hernández-Ceballos et al. 2012; Masson et al. 2011; Takemura et al. 2011; Tanaka 2013). The transport of the radioactive plume and its deposition over the Pacific Ocean (Aoyama et al. 2013; Honda et al. 2012), North America (e.g., Schwantes et al. 2012; Zhang et al. 2011), and Europe (e.g., Masson et al. 2011) as well as within the Japanese

territories (Hirose 2012; Kinoshita et al. 2011; Morino et al. 2011; Terada et al. 2012; Tsuruta et al. 2014) has been well depicted by many researchers. The pattern of domestic pollution of the land by local fallout was made fairly clear by the creation of a contamination map based on many university investigations (Kinoshita et al. 2011; Tanihata 2013) and airborne surveys by Japan's MEXT and the USA's NASA/DOE (MEXT and USDOE 2011; Sanada et al. 2014; Torii et al. 2013; USDOE 2013). The transport of the radioactive plume and its subsequent deposition over the capital area (the Kanto Plain; Amano et al. 2012; Haba et al. 2012; Tsuruta et al. 2014) has been reported and monitored in Tsukuba (Doi et al. 2013; Kanai 2012). The MRI in Tsukuba suffered almost no electricity outage soon after the earthquake. Thus, aerosol sampling at the observation field continued from before the FDNPP accident through its aftermath. Here, we add our independent observations of the temporal changes in atmospheric radionuclide concentrations over Tsukuba covering all of March 2011, with our specific transport model simulation for reference.

Experiment

Intensified aerosol sampling

Aerosol samples were collected onto quartz fiber filters using a high-volume sampler, as described in the body of the paper; the only change was the duration of sampling, from 1 day to 6 h—which was altered as soon as the accident was made public. The total sucked air volume was thus between 250 and 1000 m^3.

Activity measurement

After collection, the filters were treated in the same manner as usual and measured with Ge detectors, as described previously. The filter samples collected before the radioactive plume's arrival at Tsukuba were measured at the Kyoto University Research Reactor Institute (KURRI) to lower the detection limits. This was necessary because the Ge detector and the laboratory environment at the MRI building were contaminated by the radioactive plume on March 14–15 and 20–22, increasing the background levels. Before the compression procedure, portions of the filter were punched out (33 mmϕ × 4 pieces), of which one piece was selected for radioSr analysis, as noted in Appendix 2.

Transport modeling

The Eulerian chemical transport model RAQM2 (Kajino et al. 2012; Adachi et al. 2013; Sekiyama et al. 2015) was used to simulate radioactive plume transport from the FDNPP accident over the Kanto Plain. The JMA/MRI non-hydrostatic meteorological model (NHM; Saito et al. 2007) was used to simulate the meteorological

field to calculate the transport and deposition processes of radionuclides using RAQM2. The horizontal domain and its grid resolution (3 km) were common to both NHM and RAQM2, with 50 vertical layers from the surface up to 22 km for NHM and 20 layers to 10 km for RAQM2. The JMA's Meso-Regional Objective Analysis (MANAL), which has a horizontal resolution of 5 km, was used to define the boundary conditions for NHM. The calculated domains cover southern Tohoku and the central part of Honshu. Details of the transport (advection, diffusion, and convective transport) and deposition schemes (dry and wet (in cloud and below cloud, grid-scale and subgrid-scale)) are described in Kajino et al. (2012) and Sekiyama et al. (2015).

We simulated five species of particulate radionuclides (volatile and reactive ^{131}I (I_2), volatile and non-reactive ^{131}I (CH_3I), non-volatile ^{131}I, ^{134}Cs, and ^{137}Cs). We conducted dispersion and deposition simulation of radioCs in two very different forms—hygroscopic submicrons vs. hydrophobic supermicrons—in a previous study (Adachi et al. 2013) and showed that the deposition regions were significantly different. However, because the proportions of hygroscopic and hydrophobic radioCs in emissions have never been estimated, we assumed the hygroscopic submicron aerosols to be the carriers of radionuclides and used dimensions equivalent to the geometric mean of the dry diameter $D_{g,n,\mathrm{dry}} = 102$ nm, geometric standard deviation $\sigma_g = 1.6$, particle density $\rho_p = 1.83$ g/cm^3, and hygroscopicity $\kappa = 0.4$ (Petters and Kreidenweis 2007; Adachi et al. 2013). The emission inventories of ^{131}I and ^{137}Cs were taken from Katata et al. (2014). RAQM2 incorporates aerosol dynamic processes, such as nucleation, condensation/volatilization, and coagulation, within and among different aerosol categories, but the size distribution of the aerosols was assumed to remain unchanged in this simulation.

Results and discussion

Particulate fission products and radioCs

The detected γ-emitting radionuclides were ^{99}Mo-^{99m}Tc (half-life, 65.9–6 hours), ^{129m}Te (33.6 days), ^{131}I (8.02 days), ^{132}Te-^{132}I (3.20 days–2.3 hours), ^{133}I (20.8 hours), ^{134}Cs (2.07 years), ^{136}Cs (13.2 days), and ^{137}Cs (30.0 years) as shown in Fig. 8a in the Appendix 1. Note that gaseous iodine was not captured by the present sampling. The ^{90}Sr results are also plotted in the figure (for analytical details, please refer to Appendix 2). There were two significant transport events that brought the radioactive plume toward the Kanto Plain in March 2011. One was during March 14–15 and the other occurred during March 20–22. Plume transport is determined by temporal changes in emission intensity and the wind field near the ground surface, which have been addressed by many authors (e.g., Katata et al. 2012, 2014; Morino et al. 2011; Terada et al. 2012). The releasing sources are attributed to a venting operation at an individual reactor vessel, reactor core damage, buildings damaged by a hydrogen explosion, and continuous release through a reactor building (see, e.g., TEPCO 2012; Katata et al. 2014). The activity concentrations of these radionuclides were consistent with those described in previous reports regarding Tsukuba (e.g., Doi et al. 2013; Kanai 2012). The March 7–12, 12–13, 13, and 13–14 samples exhibited detectable levels of radioCs and ^{131}I, for which we cannot totally rule out the possibility of sample contamination despite their measurement at KURRI. The two events exhibited different radionuclide compositions, reflecting different source at the accident site. Although the ^{134}Cs/^{137}Cs ratio was unity for both transport events, the activity ratios were ^{131}I/^{137}Cs ≈ 5 and ^{132}Te/^{137}Cs ≈ 8 during the first event and ^{131}I/^{137}Cs ≈ 2.5 and ^{132}Te/^{137}Cs ≈ 1 during the second event. Te-132 was significant during the first transport event. Because the melting point of metallic Te is 450 °C, whereas that of Cs is only 28 °C, the finding may suggest a higher temperature for the source in the earlier phase. For comparison, ^{90}Sr data are included in Fig. 8a in the Appendix 1; the details of the measurements are given in Appendix 2.

After the FDNPP accident, unlike in Chernobyl, no radioRu was found (Aoyama et al. 1986, 1987). This may be because of the different accident scenarios; the melting temperature of metallic Ru is very high (approximately 2500 °C).

Another notable point is the magnitude of the concentration drop between the first and second plume events. RadioCs and ^{132}Te concentrations were 4–5 orders of magnitude lower for the second plume than the concentration peaks, and those for ^{131}I were 2–3 orders of magnitude lower. This difference appeared to be caused by either the re-suspension of radioI or the contamination of our materials and instruments. The latter seems unlikely, however, because the filter samples were treated identically and the maximum contamination levels would be those found for the March 7–14 samples (measured at the KURRI). We gave sufficient attention to reducing contamination during sampling and sample handling. Nevertheless, the entire environment was contaminated, and therefore, it was difficult to avoid entirely. In any case, the volatile nature of iodine (the boiling point of CH$_3$I is 42 °C, while the melting point of I$_2$ is 113 °C) is likely part of the cause. Therefore, immediate re-suspension of radioI should be given more attention. This is briefly addressed below.

Transport model simulation

The aerosol simulation model captures the events that transported the radioactive plume to the Kanto Plain very well (see Fig. 8b and 9 in the Appendix 1). The transport of the plume from the southern Tohoku district is not considered very exceptional (the MRI is approximately 170 km southwest from the accident site). Aoyama et al. (1999) and Igarashi et al. (1999a) analyzed the radioactive plume over the Kanto Plain from the earlier PNC accident in Tokai, Ibaraki, in 1997. Igarashi et al. (2000a,b) conducted continuous observations at the MRI of ^{85}Kr, of which the local source was the Tokai nuclear fuel reprocessing plant approximately 60 km northeast of Tsukuba. They noted the incidence of plume transport from a point source in northern Ibaraki over the Kanto Plain with a northeasterly wind, a prevalent weekly wind pattern occurring during the spring in Japan. Similar meteorological situations appeared to occur on March 14–15 and March 20–22, 2011 over the Kanto Plain. Notably, the drop in activity concentration between the plume advections is evident in the simulation results (Fig. 8b and c in the Appendix 1) despite only primary emissions coming from the FDNPP accident. The reality of the observations differed from the simulations (Fig. 8a in the Appendix 1). As described above, contamination in the observation procedures cannot be totally ruled out, but by coupling the model and observations, it is possible to evaluate the immediate resuspension of the atmospheric Fukushima radionuclides (see section below).

Finally, we argue that aerosol transport modeling is an indispensable tool for the assessment of accident effects. However, many uncertainties remain, especially concerning the emission inventory, wet and dry deposition, and cloud processes. Data and information are collected to improve the transport model schemes, and comparison of different models has been performed to contribute to an accurate evaluation of the source term and transport and deposition processes (SCJ 2014).

Estimation of immediate re-suspension factor

The quantity of the deposited radionuclides that could return again to the air (re-suspension) is notable. Maximum re-suspension is known to occur just after radioactive plume passage (hereafter, we call this immediate re-suspension). Thus, as a primary approach, immediate re-suspension factors were roughly estimated with modeled amounts deposited in the Kanto Plain by the first plume and the observed minimum activity concentration between the two plume events, i.e., March 17 09JST to March 20 09JST. We assumed mass closure between resuspension from the contaminated surface and outflow by horizontal advection and turbulence vertical mixing as below.

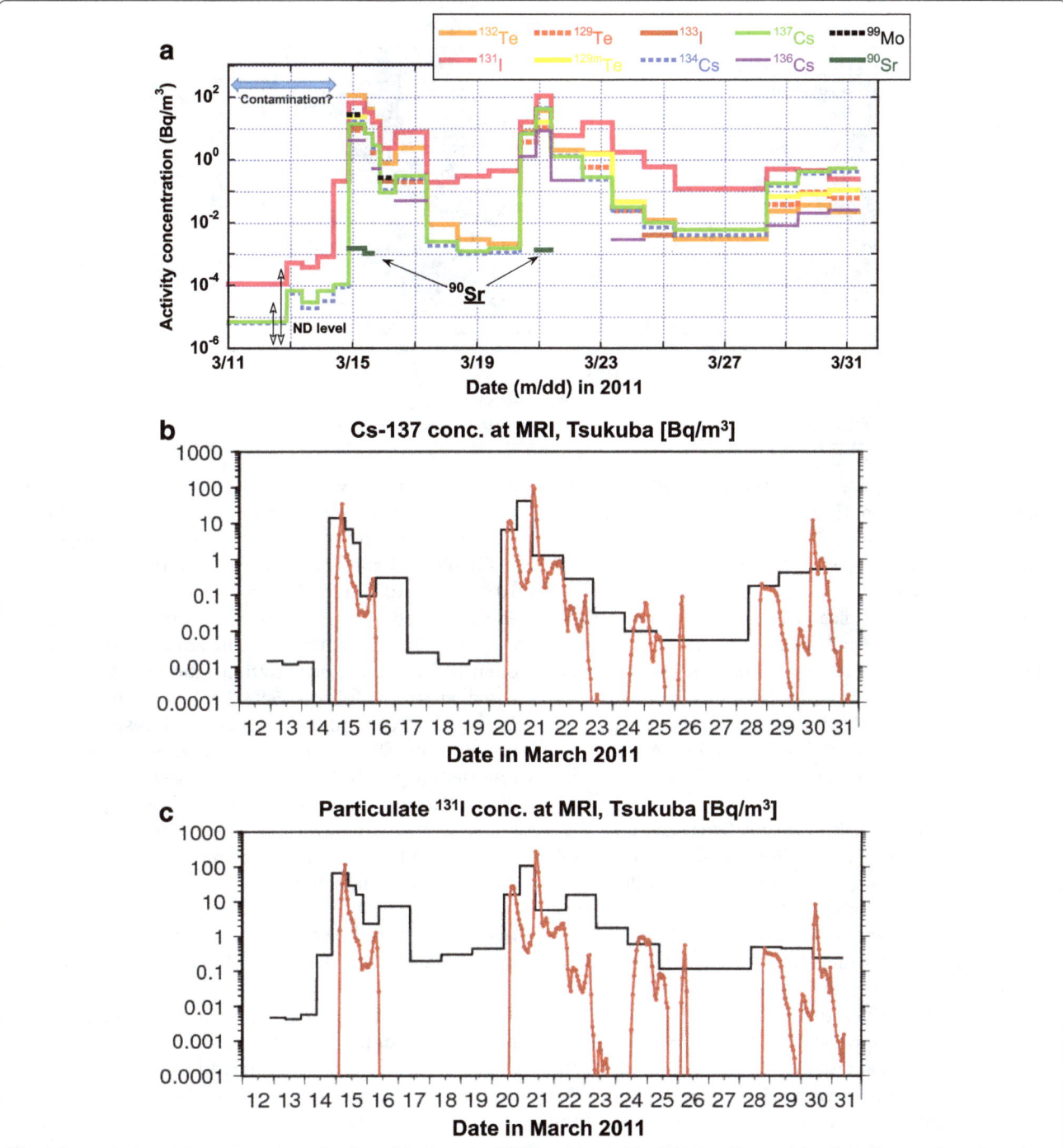

Fig. 8 Atmospheric activity concentrations of radionuclides from the FDNPP accident in March 2011. **a** Observed data from filter samples collected at the MRI, Tsukuba, Japan; **b** comparison of observed (*black*) and simulated results (*red*) for ^{137}Cs; and **c** similar to **b** but for particulate ^{131}I. The abscissa is expressed in dates in March 2011 and is labeled at the start of the day in **a** and the middle of the day in **b** and **c**. Contamination of the filter samples cannot be totally ruled out for the period before March 14 in **a**, which is depicted by the *left-right pointing double arrow*

The continuity equation is expressed as

$$\partial C/\partial t = \nabla(K_{\mathrm{dif}}\nabla C) - \nabla(UC) - \lambda C + \Phi,$$

in which C is concentration, K_{dif} indicates three-dimensional diffusion terms, U denotes the wind field, λ is

the decay constant, and Φ is a re-suspension term for individual radionuclides. On the other hand, the concentration increase in one unit of time from re-suspension is expressed as

$$\Delta C/\Delta t = \Phi = k_i \times D_i \times (\Delta x \Delta y / \Delta x \Delta y \Delta z),$$

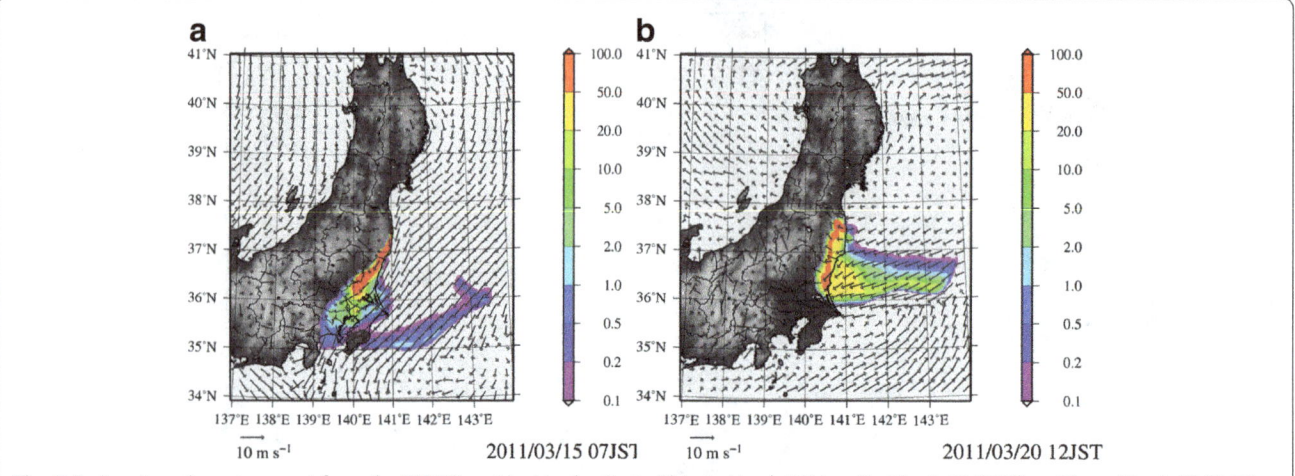

Fig. 9 Radioactive plume transport from the FDNPP accident in the Kanto Plain in March 2011. **a** On March 15 07JST and **b** on March 20 12JST. The figures show the simulated surface ^{137}Cs concentration in *shaded colors* with the model topography in *grayscale*

in which k_i and D_i are a re-suspension factor (/s) and surface contamination (Bq/m^2) for individual radionuclides, respectively. Also, Δx, Δy, and Δz are the horizontal and vertical lengths of the space where the mass closure is obtained.

We can disregard radioactive decay, horizontal diffusion, and convective wind. Balancing the mass between inflow and outflow, we finally obtain the following relationship:

$$(k_i \cdot D_i)/(\Delta z) = (\Delta K_z/\Delta z) \times (\Delta C/\Delta z) + (\Delta u/\Delta x + \Delta v/\Delta y) \times C_i,$$

in which i indicates the radionuclides, namely, ^{137}Cs and ^{131}I; D_i indicates the modeled total (gas + aerosol) cumulative deposition (Bq/m^2) by March 17 09JST; k_i is the re-suspension factor (s^{-1}); U and K_z are the modeled space- and time-averaged horizontal wind speed (m/s) and vertical turbulent diffusivity (m^2/s), respectively; C_i indicates the time-averaged observed concentrations of the radionuclides (9.75×10^{-4} and 3.14×10^{-1} Bq/m^3 for ^{137}Cs and ^{131}I, respectively); and Δx, Δy, and Δz are the horizontal and vertical distances in space over which the above mass closure is obtained. To obtain the horizontal and vertical gradient terms on the right-hand side of the equation, the concentrations outside the space are assumed to be zero (no inflow into the space).

The re-suspension factors for ^{137}Cs and ^{131}I are 7.0×10^{-6} /s and 5.3×10^{-4} /s, respectively, for the smallest volume of the RAQM2 model grid ($\Delta x = 3$ km, $\Delta y = 3$ km, and $\Delta z = 100$ m). Those for ^{137}Cs and ^{131}I varied from 1.6×10^{-6} /s to 1.5×10^{-5} /s (6.1×10^{-6} /s on average) and from 5.3×10^{-4} /s to 1.3×10^{-3} /s (4.6×10^{-4} /s on average), respectively, for the various horizontal spaces plus neighboring zero, one, or two RAQM2 grids from the grid where the MRI is located (i.e., Δx, $\Delta y = 3$,

9, or 15 km) and vertical spaces plus zero, one, or two RAQM2 grids from the bottom ($\Delta z = 100$, 200, or 400 m).

In summary, the immediate re-suspension factors k_i of ^{137}Cs and ^{131}I are estimated to be on the order of 10^{-6}–10^{-5} /s and 10^{-4}–10^{-3} /s, respectively, and that of ^{131}I is approximately two orders of magnitude larger than that of ^{137}Cs. These values are converted correspondingly, often quoting the concentration ratio over the contaminated surface as follows: 5.8×10^{-6} – 1.7×10^{-5} and 4.4×10^{-4} – 1.3×10^{-3} /m) for ^{137}Cs and ^{131}I, respectively. The present data do not display the large deviation hitherto reported (e.g., 10^{-6}–10^{-4} /m; Maxwell and Anspaugh 2011). Because those values are based on rough assumptions, further studies based on surface flux measurements need to be conducted to more accurately estimate the re-suspension factors.

Appendix 2 RadioSr in the aerosol samples collected during March 2011

Introduction

There are several reports containing estimates of the radioactive contamination from the FDNPP accident, presented in the form of mapped images produced from the results of investigations of radionuclides in the soil (e.g., MEXT 2011a ; Sanada et al. 2014; Torii et al. 2013) and in the form of air dose rate figures produced from aircraft observations. Among the radionuclides, radioSr is an important indicator of contamination. The former Nuclear and Industrial Safety Agency (NISA) in Japan reported the following emission estimates within the atmosphere: ^{89}Sr (half-life, 50.5 days) as 2.0×10^{15} Bq and ^{90}Sr (half-life, 28.8 years) as 1.4×10^{14} Bq (NISA 2011). Nevertheless, there have been no reports on ^{89}Sr and ^{90}Sr in air samples because of analytical difficulty. The detection of nine different γ-emitting radionuclides, including ^{99}Mo, is described in Appendix 1. However, ^{89}Sr and ^{90}Sr emit no γ-rays with their radioactive decay, making it impossible to determine their presence by γ-spectrometry. To evaluate their radioactive pollution levels, the aerosol components were radiochemically extracted from the HV filter sample to analyze the radioSr and assess the emission ratios of ^{137}Cs, ^{89}Sr, and ^{90}Sr.

Experiment
Sub-HV filter sample for Sr analysis
HV filter samples from the γ-spectrometry measurements noted earlier were used for the radioSr analysis. Approximately 2 % of the filter area was punched out (as circles) and provided for this analysis, which was performed on sub-filter samples collected during March 2011 (Table 1 in the Appendix 2).

Analysis of radioSr
To dissolve the aerosols on the filter, 100–200 ml of concentrated nitric acid was added and heated on a 200 °C hotplate, then 1–5 ml of hydrogen peroxide solution was added to accelerate the decomposition of any organic matter. This was followed by further thermolysis for more than an hour. The obtained solution was subjected to separation, which was conducted through radiochemical analysis comprising several precipitation separations, such as oxalate, fuming nitric acid, hydroxide, carbonate, and barium chromate precipitations. The last separation was repeated twice, which allowed the Sr fraction to be freed from radioBa and Ra isotopes. The final strontium carbonate deposit was β-counted with the low-background 2π gas-flow counter described earlier (Tennelec LB5100).

Estimating the activity ratio of ^{89}Sr and ^{90}Sr
The atmospheric aerosol sample contained ^{89}Sr and ^{90}Sr, indicating that the total β-activity must be deconvoluted. The measurement sensitivity of the gas-flow counter was confirmed for possible energy independence; therefore, the temporal change in the β-counting rate of a purified ^{90}Sr (maximum β-ray energy 0.546 MeV) source and ^{90}Y (maximum β-ray energy 2.24 MeV) growth from the parent nuclide was observed in five specimens of the MRI reference fallout samples (Otsuji-Hatori et al. 1996) that contained no ^{89}Sr. The following equation was then applied to find the counting efficiency of ^{90}Sr and ^{90}Y:

$$N_{\text{total}} = A_{\text{Sr-90}} \times m_1 + A_{Y-90} \times (1-e^{-\lambda t}) \times m_2.$$

N_{total} is the total counting rate (cpm); A stands for each nuclide's β-activity (dpm); λ is the decay constant of ^{90}Y; t is the elapsed time; and m_1 and m_2 are the counting efficiencies of ^{90}Sr and ^{90}Y, respectively. The β-ray energy emitted by ^{90}Y is approximately 4 times that of ^{90}Sr, and the average values of m_1 and m_2 from the five specimens were 27.3 ± 1.8 % and 24.8 ± 3.7 %, respectively.

Table 1 Temporal variation of ^{90}Sr activity concentration in the air over Tsukuba

Sampling start date and time (JST)	End date and time (JST)	^{90}Sr activity concentration (mBq/m^3)
March 12 21 pm	March 13 9 am	nd
March 13 21 pm	March 14 9 am	nd
March 14 9 am	March 14 21 pm	nd
March 14 21 pm	March 15 9 am	1.50 ± 0.13
March 15 9 am	March 15 15 pm	1.04 ± 0.095
March 15 15 pm	March 15 21 pm	nd
March 15 21 pm	March 16 9 am	nd
March 16 9 am	March 17 8 am	nd
March 17 9 am	March 18 8 am	nd
March 18 8 am	March 19 9 am	nd
March 19 9 am	March 20 8 am	nd
March 20 9 am	March 20 21 pm	nd
March 20 21 pm	March 21 9 am	1.32 ± 0.13
March 21 9 am	March 22 9 am	nd
March 22 9 am	March 23 8 am	nd
March 23 9 am	March 24 9 am	nd
March 24 9 am	March 25 9 am	nd
March 25 9 am	March 28 9 am	nd
March 28 9 am	March 29 9 am	nd
March 29 9 am	March 30 9 am	nd

Although the "nd" measurements change, depending mainly on the sample volume, the average level was approximately 0.2 mBq/m^3
nd not detected

There were no statistically significant differences. Thus, the β-activities of radioSr were interpreted to have the same counting efficiency regardless of the β-energy. The activity ratio of ^{89}Sr and ^{90}Sr was elucidated from the value traced back to the date of sample collection as well as the fixed date when the strontium carbonate precipitated. The activity was always decay corrected in the middle of the sampling time. The current detection limit for radioSr in air at that time was approximately 230 μBq/m^3.

Results and discussion
Estimation of ^{90}Sr in the aerosol sample
We will now quantify and describe the radioSr found in the air over Tsukuba. The radioactivity in Tsukuba indicated a two-fold concentration increase in March 2011, as shown in Fig. 8 in the Appendix 1. The amount of radioSr in the sample was smaller than what was anticipated based on past experience (e.g., Aoyama et al. 1991). ^{90}Sr was unable to be detected except when plume transport occurred. From March 14 9 pm (JST) to March 15 9 am, from March 15 9 am to 3 pm, and March 20 9 pm to March 21 9 am, the results were 1.5 ± 0.13, 1.0 ± 0.10, and 1.3 ± 0.13 mBq/m^3, respectively. For the other samples, the radioSr was

lower than the detection limits (Table 1 in the Appendix 2). The ^{90}Sr activity results shown here were calculated based on β-counts made long enough after the events that the contribution of ^{89}Sr could be negligible (less than 5 % of ^{90}Sr activity). For example, we waited at least 200 days after chemical separation (separation was performed after December 2011). The accompanying uncertainty was estimated from the average of the relative β-count uncertainties in the five latest individual measurements.

The activity ratio of ^{137}Cs/^{90}Sr in the aerosol samples, which was in the range of 4700–23,000, is very large compared with the activity ratio of radioactive fallout, which was 1.63 during the 1960–1970s; this indicates a clear difference in the data before and after the FDNPP accident. Furthermore, the MRI's estimated ^{137}Cs/^{90}Sr ratio for the Chernobyl radionuclides in May 1986 in Japan was 96 (Aoyama et al. 1991), which indicates that the Fukushima radionuclide composition was dominated by radioCs. In the activity peak on March 14–15, the ratio was 4700–6000, and the peak on March 20–21 was 23,000 times higher with ^{137}Cs, which also shows that the composition of the radioactive plume differed between the earlier and later dates during the course of the FDNPP accident.

The measured ^{137}Cs/^{90}Sr activity ratio in Tsukuba was more than 40 times higher than the emission assessment by NISA 2011 for the FDNPP accident (^{137}Cs: ^{90}Sr = 15: 0.14). The IAEA (2006) had estimated that the amount of ^{90}Sr emitted (approximately 10 PBq) for the Chernobyl accident was only 12 % that of ^{137}Cs (approximately 85 PBq), yet in reality, the atmosphere/precipitation observations in Japan showed approximately the amount of ^{90}Sr to be only 1/100 that of ^{137}Cs (Aoyama et al. 1991), indicating that less than 1/10 of the emitted ^{90}Sr was transported. Thus, the 8000 km long-range transportation from Chernobyl produced the radionuclide separation. With that in mind, it could be possible that fractionation

Table 2 Curve fitting results with assumed ^{89}Sr over ^{90}Sr activity ratio

March 14–15, 2011		March 20–21, 2011	
^{89}Sr:^{90}Sr	σ	^{89}Sr:^{90}Sr	σ
9:0.14	0.3356	8:0.14	0.4550
10:0.14	0.3340	9:0.14	0.4165
11:0.14	0.3493	10:0.14	0.5432
12:0.14.	0.3795	11:0.14	0.5869

The results for two air filter samples collected in March 2011. The minimum standard deviation σ suggests the best estimate

caused by particle size deviation (Hirose et al. 1993) occurred in the FDNPP plume. The plume was transported less than a few hundred kilometers in the present case, but fractionation could be very effective.

^{89}Sr/^{90}Sr activity ratio

The emissions estimated by NISA 2011 showed that the ^{89}Sr proportion was 14 times higher than that of ^{90}Sr after the nuclear accident, which indicated that the radioactivity estimate would be 1/3 that of ^{90}Sr after a year. The results from the aerosol sample observations suggest the presence of ^{89}Sr; therefore, the temporal change in the β-counts was fitted based on emission estimates by the former NISA (^{89}Sr:^{90}Sr = 2:0.14). Figure 10 in the Appendix 2 shows the fitted results of the aerosol sample measurements for March 14–15. As shown in the figure, the sample counting values exhibited a large decay after 40 days of fixation as strontium carbonate, which indicates that the amount of coexisting ^{89}Sr was relatively large. Therefore, appropriately different ratios were examined instead of the 2:0.14 ratio, which could not be fitted. Therefore, the emitted ratio for the sample collected on March 14–15 was 10:0.14 for ^{89}Sr:^{90}Sr. The peak data for March 20–21 indicated that a ratio of 9:0.14 fit perfectly. Table 2 in

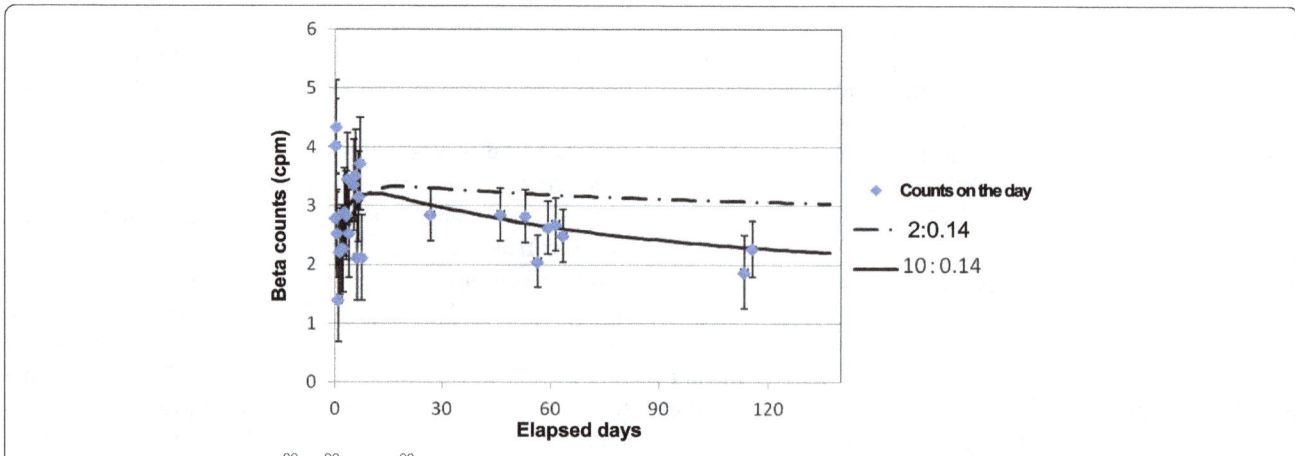

Fig. 10 Deconvolution of the ^{89}Sr, ^{90}Sr and ^{90}Y activities. It is possible to deconvolute radionuclides by measuring the temporal change in the total β-activity (cpm) of the purified radioSr source (March 14–15, 2011 sample). Elapsed days means the time after the radiochemical separation. An initial activity ratio of ^{89}Sr/^{90}Sr was assumed and applied to the curve fit as 2/0.14 and 10/0.14.

Table 3 Efficiency of ^{137}Cs extracted from air filter samples by heated concentrated nitric acid

Sampling date and time (JST)	Before ext. ^{137}Cs (Bq)	After ext. ^{137}Cs (Bq)	Extraction efficiency (%)
March 14 21 pm–March15 9 am	243 ± 0.7	72 ± 0.4	70.4
March 15 9 am–March 15 15 pm	41 ± 0.3	7.4 ± 0.12	82.0
March 15 15 pm–March 15 21 pm	20 ± 0.2	6.1 ± 0.11	69.5
March 15 21 pm–March 16 9 am	19 ± 0.2	0.51 ± 0.04	97.3
March 16 9 am–March 17 9 am	9.2 ± 0.19	0.52 ± 0.04	94.3
March 19 9 am–March 20 21 pm	94.2 ± 0.6	0.20 ± 0.03	99.8
March 20 21 pm–March 21 9 am	423 ± 0.9	1.2 ± 0.05	99.7
March 21 9 am–March 22 9 pm	30.8 ± 0.24	0.15 ± 0.03	99.5

Air filter samples were subjected to heated conc. nitric acid extraction for Sr analysis. Cs-137 was measured to confirm the extraction efficiency. Samples shown as "nd" before extraction were excluded from the table. Some samples exhibited significantly lower extraction efficiencies of 70–80 %. Insoluble and refractory radioactive particles must have been incorporated in these samples

the Appendix 2 shows these fitting results. Therefore, the emission ratio of ^{89}Sr/^{90}Sr for both March 14–15 and 20–21 was approximately 70 (10:0.14), which was five times bigger than what NISA 2011 had estimated.

The MEXT has reported ^{89,90}Sr in approximately 50 soil samples within 80 km of the FDNPP (MEXT 2011b). The decay data are corrected as of June 2011, and the activity ratio was reported to be in the range of 1.9–6.5 (average: 4). Another decay correction as of March 11, 2011 gives ^{89}Sr/^{90}Sr ratios of 7–24 with an average of 15. The ratio is not consistent with our results, and the fluctuation was large. The cause of the discrepancy and fluctuation is still unknown. The most likely explanation is that stable Sr, already present in reactor materials or seawater components, absorbed neutrons and formed ^{89}Sr. The extent of and fluctuation in mixing (inhomogeneity) might produce the discrepancy.

Efficiency of acid extraction of ^{137}Cs from filter specimens

The rates at which ^{137}Cs could be extracted from the filter and aerosol samples using acid are shown in Table 3 in the Appendix 2. The samples collected on March 14–15 and 20–21 have different extraction rates, indicating that the ^{137}Cs in the sample from the March 14–15 was refractory to some extent (20–30 %), even in a heated solution of nitric acid. This is possibly because of the difference in the physical and chemical nature of the radioactive aerosol. Thus, it is possible that the current radioSr concentration has been slightly underestimated (20–30 %) because of the low water dissolution rate of the radioactive material, especially for the March 14–15 sample.

As shown here, observations of the radioactive plume over Tsukuba at different times demonstrated that the ^{89}Sr/^{90}Sr ratio was almost constant, but the ^{137}Cs/^{90}Sr ratio and the extraction efficiency of ^{137}Cs with nitric acid differed. Moreover, it was shown earlier that the activity ratios among other γ-emitters differed (see Appendix 1).

These findings confirm that the characteristics of the aerosol particles that carried major radionuclides from the first plume differed from later advected radioactive plumes. Adachi et al. (2013) addressed this sort of contrast in the characteristics of the two plumes' radioactive aerosols in detail, and Abe et al. (2014) added more information. They documented the discovery of insoluble, glassy spherules containing radioCs and assumed that the major fraction came from the first event. Indeed, no such particles were detected in the later event. This should also affect the ratio of ^{137}Cs/^{90}Sr in the air, and evidence regarding this will be obtained in future work. In conclusion, the present results support the previous findings of less ^{90}Sr contamination than radioCs contamination from the FDNPP accident and indicate the necessity of further investigations of radioSr in the atmospheric environment.

Abbreviations

ATSDR: US Department of Health and Human Services, Public Health Service, Agency for Toxic Substances and Disease Registry; DOE: US Department of Energy; FDNPP: Fukushima Dai-ichi nuclear power plant; HYSPLIT: Hybrid Single Particle Lagrangian Integrated Trajectory Model; IAEA: International Atomic Energy Agency; JMA: Japan Meteorological Agency; KURRI: Kyoto University Research Reactor Institute; MANAL: Meso-regional objective analysis; MEXT: Ministry of Education, Culture, Sports, Science and Technology, Japan; MRI: Meteorological Research Institute, Japan; NASA: National Aeronautics and Space Administration, USA; NHM: The JMA/MRI non-hydrostatic meteorological model; NISA: The former Nuclear and Industrial Safety Agency, Japan Carriage Return; RAQM2: Regional Air Quality Model 2; SCJ: Science Council of Japan Carriage Return; TEPCO: Tokyo Electric Power Company.

Competing interests

The authors declare that they have no competing interests.

Authors' contributions

YI designed and supervised the study and summarized the manuscript. MK conducted the transport simulation and wrote that part of the manuscript. Both YI and MK analyzed the data and helped in their interpretation. YZ helped conduct the sampling. KA and MM provided important suggestions for summarizing the work. They collaborated with the corresponding author in the preparation of the manuscript. All the authors read and approved the final manuscript.

Authors' information

YI received his PhD degree in chemistry from the University of Tsukuba in 1987. From 1987 to 1991, he was at the National Institute of Radiological Sciences and studied radiochemical analysis and radioecology. He moved to the MRI in 1991 because of his scientific ambition to be involved in more global issues. His current interests are atmospheric aerosols and their precursors, including Asian dust and $PM_{2.5}$, and their possible influences on climate, general environmental change, and other phenomena. He is working enthusiastically and is a member of several organizations, including the Japan Association of Aerosol Science and Technology, the Geochemical Society of Japan, the Japan Society of Nuclear and Radiochemical Sciences, the Meteorological Society of Japan, the Japan Radioisotope Association, the Japan Society of Analytical Chemistry, and the Japan Geoscience Union. He considers collaboration between observational researchers and modelers as a basic requisite in pursuing the geo- and environmental sciences.

MK received his PhD from the Graduate School of Science at Kyoto University in 2005. Since then, he has been engaged in the development of chemical transport models coupled with regional-scale meteorological models at the Disaster Prevention Research Institute, Kyoto University, Research Center for Advanced Science and Technology, University of Tokyo, and currently, the MRI. His main research interest is elemental processes of chemistry and microphysics of airborne particles and its impacts on air quality, ecosystem, and climate.

YZ graduated Meteorological College in 1984 and worked at the Nagasaki Marine Observatory, Japan Meteorological Agency from 1984 to 1989. He has been studying atmospheric aerosols at the MRI since 1989 and received his PhD from Nagoya University in 2005.

KA received his PhD from Kobe University in 2005, worked at Arizona State University between 2005 and 2011 as a postdoctoral/faculty research associate, and is currently studying atmospheric aerosols at the MRI.

MM received his PhD from Tohoku University in 1996, has worked at the MRI from 1985 to 2015, and finally holds the position of Senior Researcher for Research Affairs at the institute. Currently, he is working at Japan Meteorological Business Support Center.

Acknowledgements

The authors are deeply indebted to the following part-time and temporary staff members at the MRI: Chitsuko Takeda, Tokyo Nuclear Service Co. Ltd., and Kazue Inukai and Keiko Kamioka, currently at ATOX Co. Ltd. for the difficult work of analyzing samples and preparing samples measurements under an unusually severe accident situation; Hiroshi Sakou, Toru Kimura, and Sakae Mayama, ATOX Co. Ltd. for sampling, sample preparation, and general analysis; Wakari Iwai and Kazuma Nabeshima for radiochemical analysis of Sr isotopes and sample preparation, respectively; and Yuriko Kamiya, Kayo Yanagida, and Rina Mori for sampling, logistical support, figure preparation, and manuscript editing. Takeshi Ito helped with the trajectory analysis in the revised manuscript. The authors are also grateful to the following MRI academic colleagues: Hiroaki Naoe, Michio Aoyama (currently at Fukushima University), and Hiroshi Takahashi (currently at the Japan Meteorological Agency) for their help with sampling. Additionally, they acknowledge the assistance of Naoyuki Osada (currently at Okayama University) and Yuichi Oki (Kyoto University Research Reactor Institute) with low-background γ-ray measurements. The authors benefitted from discussions with Kazuyuki Kita (Ibaraki University) and Yuko Hatano (University of Tsukuba) regarding the re-suspension issue. Olivier Masson, IRSN, France, kindly gave critical comments on an early version of the manuscript, for which the authors are grateful. This study was financially supported by the former MEXT and current Nuclear Regulation Authority "Japanese Radioactivity Survey" fund and partially by the MEXT "Kakenhi" (a Grant-in-Aid for Scientific Research on Innovative Areas under the A01-1 and A01-2 research teams in the "Interdisciplinary Study on Environmental Transfer of Radionuclides from the Fukushima Daiichi NPP Accident (ISET-R; leader: Professor Yuichi Onda, University of Tsukuba)"; grant nos. 24110002 and 24110003) and the JSPS "Kakenhi" (leader Dr. Tsuyoshi T Sekiyama, MRI; grant no. 24340115). We gratefully acknowledge NOAA's ARL for providing the HYSPLIT transport and dispersion model and the READY website used in this study. The present paper was written and organized based on previous proceedings (Iwai et al. 2012; Igarashi et al. 2013a) and presentations at the ICAS, AMS, EGU, domestic meetings, and other settings (Igarashi et al. 2011, 2013b, 2013c) and the MRI home page (MRI 2011).

Author details

[1]Meteorological Research Institute, 1-1 Nagamine, Tsukuba, Ibaraki 305-0052, Japan. [2]Present address: Japan Meteorological Business Support Center, Chiyoda-ku, Tokyo 101-0054, Japan.

References

Abe Y, Iizawa Y, Terada Y, Adachi K, Igarashi Y, Nakai I. Detection of uranium and chemical state analysis of individual radioactive microparticles emitted from the Fukushima nuclear accident using multiple synchrotron radiation X-ray analyses. Anal Chem. 2014;86:8521–5. doi:10.1021/ac501998d.

Adachi K, Kajino M, Zaizen Y, Igarashi Y. Emission of spherical cesium-bearing particles from an early stage of the Fukushima nuclear accident. Sci Rep. 2013;3:2554. doi:10.1038/srep02554.

Amano H, Akiyama M, Chunlei B, Kawamura T, Kishimoto T, Kuroda T, et al. Radiation measurements in the Chiba Metropolitan Area and radiological aspects of fallout from the Fukushima Dai-ichi Nuclear Power Plants accident. J Environ Radioact. 2012;111:42–52. doi:10.1016/j.jenvrad.2011.10.019.

Aoyama M. Evidence of stratospheric fallout of caesium isotopes from Chernobyl accident. Geophys Res Lett. 1988;15:327–30. doi:10.1029/GL015i004p00327.

Aoyama M, Hirose K. The temporal and spatial variation of ^{137}Cs concentration in the Western North Pacific and its marginal seas during the period from 1979 to 1988. J Environ Radioact. 1995;29:57–74. doi:10.1016/0265-931X(94)00050-7.

Aoyama M, Hirose K. Artificial radionuclides database in the Pacific Ocean: HAM database. Sci World J. 2004;4:200–15. doi:10.1100/tsw.2004.15.

Aoyama M, Hirose K, Suzuki Y, Inoue H, Sugimura Y. High level radioactive nuclides in Japan in May. Nature. 1986;321:819_20. doi:10.1038/321819a0.

Aoyama M, Hirose K, Sugimura Y. Deposition of gamma-emitting nuclides in Japan after the reactor-IV accident at Chernobyl. J Radioanal Nucl Chem. 1987;116:291–306. doi:10.1007/BF02035773.

Aoyama M, Hirose K, Sugimura Y. The temporal variation of stratospheric fallout derived from the Chernobyl accident. J Environ Radioact. 1991;13:103–15. doi:10.1016/0265-931X(91)90053-I.

Aoyama M, Ohara T, Komura K. Donen Tokai jiko niyoru houshasei sesiumu no Kanto heiya heno hirogari (Transport and diffusion of radioactive caesium to the Kanto plain by the PNC Tokai accident). Kagaku. 1999;69(1):16–21 (in Japanese).

Aoyama M, Hirose K, Igarashi Y. Re-construction and updating our understanding on the global weapons tests ^{137}Cs fallout. J Environ Monitor. 2006;8:431–8. doi:10.1039/B512601K.

Aoyama M, Tsumune D, Hamajima Y. Distribution of ^{137}Cs and ^{134}Cs in the North Pacific Ocean: impacts of the TEPCO Fukushima-Daiichi NPP accident. J Radioanal Nucl Chem. 2012;296(1):535–9. doi:10.1007/s10967-012-2033-2.

Aoyama M, Uematsu M, Tsumune D, Hamajima Y. Surface pathway of radioactive plume of TEPCO Fukushima NPP1 released ^{134}Cs and ^{137}Cs. Biogeosciences. 2013;10:3067–78. doi:10.5194/bg-10-3067-2013.

Chino M, Nakayama H, Nagai H, Terada H, Katata G, Yamazawa H. Preliminary estimation of release amounts of ^{131}I and ^{137}Cs accidentally discharged from the Fukushima Daiichi Nuclear Power Plant into atmosphere. J Nucl Sci Technol. 2011;48:1129–34. doi:10.1080/18811248.2011.9711799.

Doi T, Masumoto K, Toyoda A, Tanaka A, Shibata Y, Hirose K. Anthropogenic radionuclides in the atmosphere observed at Tsukuba: characteristics of the radionuclides derived from Fukushima. J Environ Radioact. 2013;122:55–62. doi:10.1016/j.jenvrad.2013.02.001.

Ehhalt DH. Turnover times of ^{137}Cs and HTO in the troposphere and removal rates of natural particles and vapor. J Geophys Res. 1973;78:7076–86.

Estevan MT. Consequences of the Algeciras accident, and the Spanish system for the radiological surveillance and control of scrap and the products of its processing. In: Security of radioactive sources, proceedings of an international conference. Vienna Austria: IAEA; 2003. p. 357–62. 10–13 March 2003.

Garger EK, Kuzmenko YI, Sickinger S, Tschiersch J. Prediction of the ^{137}Cs activity concentration in the atmospheric surface layer of the Chernobyl exclusion zone. J Environ Radioact. 2012;110:53–8. doi:10.1016/j.jenvrad.2012.01.017.

Haba H, Kanaya J, Mukai H, Kambara T, Kase M. One-year monitoring of air-borne radionuclides in Wako, Japan, after the Fukushima Dai-ichi nuclear power plant accident in 2011. Geochem J. 2012;46(4):271–8. Special Issue: Fukushima Review.

Hatano Y, Hatano N. Formula for the resuspension factor and estimation of the date of surface contamination. Atmos Environ. 2003;37:3475–80. doi:10.1016/S1352-2310(03)00410-2.

Hernández-Ceballos MA, Hong GH, Lozano RL, Kim YI, Lee HM, Kim SH, et al. Tracking the complete revolution of surface westerlies over Northern Hemisphere using radionuclides emitted from Fukushima. Sci Total Environ. 2012;438:80–5.

Hirose K. Fukushima Dai-ichi nuclear power plant accident: summary of regional radioactive deposition monitoring results. J Environ Radioact. 2012;111:13–7. doi:10.1016/j.jenvrad.2011.09.003.

Hirose K. Temporal variation of monthly ^{137}Cs deposition observed in Japan: effects of the Fukushima Daiichi nuclear power plant accident. Appl Radiat Isot. 2013;81:325–9. doi:10.1016/j.apradiso.2013.03.076.

Hirose K, Aoyama M. Present background levels of surface ^{137}Cs and ^{239,240}Pu concentrations in the Pacific. J Environ Radioact. 2003;69(1–2):53–60. doi:10.1016/S0265-931X(03)00086-9.

Hirose K, Sugimura Y, Katsuragi Y. ^{90}Sr and $^{239+240}$Pu in the surface air in Japan: their concentrations and size distributions. Pap Met Geophys. 1986;37(4):255–69.

Hirose K, Takatani S, Aoyama M. Wet deposition of radionuclides derived from the Chernobyl accident. J Atmos Chem. 1993;17:61–71. doi:10.1007/BF00699114.

Hirose K, Amano H, Baxter MS, Chaykovskaya E, Chumichev VB, Hong GH, et al. Anthropogenic radionuclides in seawater in the East Sea/Japan Sea: results of the first-stage Japanese-Korean-Russian expedition. J Environ Radioact. 1999;43:1–13.

Hirose K, Igarashi Y, Aoyama M, Miyao T. Long-term trends of plutonium fallout observed in Japan. In: Kudo A (ed) Radioactivity in the environment Vol 1. pp 251–266. Plutonium in the environment proceedings of the second international symposium, Osaka, Japan, November 9–12, 1999. Amsterdam: Elsevier; 2001. doi:10.1016/S1569-4860(01)80018-8.

Honda MC, Aono T, Aoyama M, Hamajima Y, Kawakami H, Kitamura M, et al. Dispersion of artificial caesium-134 and −137 in the western North Pacific one month after the Fukushima accident. Geochem J. 2012;46:e1–9.

Igarashi Y. Anthropogenic radioactivity in aerosol: a review focusing on studies during the 2000s. Jpn J Health Phys. 2009;44(3):313–23.

Igarashi Y, Otsuji–Hatori M, Hirose K. Recent deposition of ^{90}Sr and ^{137}Cs observed in Tsukuba. J Environ Radioact. 1996;31:157–69. doi:10.1016/0265-931X(96)88491-8.

Igarashi Y, Aoyama M, Miyao T, Hirose K, Komura K, Yamamoto M. Air concentration of radiocaesium in Tsukuba, Japan following the release from the Tokai waste treatment plant: comparisons of observations with predictions. Appl Radiat Isotopes. 1999a;50:1063–73. doi:10.1016/S0969-8043(98)00129-8.

Igarashi Y, Aoyama M, Miyao T, Hirose K, Tomita M. Anomalous ^{90}Sr deposition during the fall, 1995 at MRI, Tsukuba, Japan. J Radioanal Nucl Chem. 1999b;239:539–42. doi:10.1007/BF02349065.

Igarashi Y, Miyao T, Aoyama M, Hirose K, Sartorius H, Weiss W. Radioactive noble gases in the surface air monitored at MRI, Tsukuba, before and after the JCO accident. J Environ Radioact. 2000a;50:107–18.

Igarashi Y, Sartorius H, Miyao T, Weiss W, Fushimi K, Aoyama M, et al. ^{85}Kr and ^{133}Xe monitoring at MRI, Tsukuba and its importance. J Envrion Radioactiv. 2000b;48:191–202. doi:10.1016/S0265-931X(99)00076-4.

Igarashi Y, Aoyama M, Hirose K, Miyao T, Nemoto K, Tomita M, et al. Resuspension: decadal monitoring time series of the anthropogenic radioactivity deposition in Japan. J Radiat Res. 2003;44:319–28.

Igarashi Y, Inomata Y, Aoyama M, Hirose K, Takahashi H, Shinoda Y, et al. Possible change in Asian dust source suggested by atmospheric anthropogenic radionuclides during the 2000s. Atmos Environ. 2009;43:2971–80.

Igarashi Y, Kajino M, Osada N, Oki Y, Takeda C. Aerosol radioactivity observed in Tsukuba during March 2011. Kyoto: ICAS2011 (IUPAC International Congress on Analytical Sciences); 2011.

Igarashi Y, Zaizen Y, Adachi K, Kajino M. Mikami M. Atmospheric pollution by the Fukushima accident: two years observations in Tsukuba. In: Bessho K, Tagami K, Takamiya K, Miura T, editors. Proceedings the 14th workshop on environmental radioactivity, pp 35–39 (in Japanese with English abstract). Tsukuba: High Energy Accelerator Research Organization; 2013a.

Igarashi Y, Kajino M, Zaizen Y, Mikami M. Observations of atmospheric radionuclides from the Fukushima nuclear accident in Tsukuba, Japan, American Meteorological Society 93rd annual meeting, Austin, Texas, Jan 2013. 2013b.

Igarashi Y, Kajino M, Zaizen Y, Adachi K, Mikami M, Kita K, et al. Atmospheric radionuclides from the Fukushima nuclear accident: two years observations in Tsukuba, Japan, the EGU general assembly 2013, Vienna, Austria, Apr 2013. 2013c.

International Atomic Energy Agency. Environmental consequences of the Chernobyl accident and their remediation: twenty years of experience report of the Chernobyl Forum Expert Group "Environment". Vienna, Austria: 2006. ISBN 92-0-114705-8, ISSN 1020-6566

Iwai W, Igarashi Y, Nabeshima K. Observation of radioactivity of the atmospheric aerosol (radioactive Sr). In: Bessho K, Tagami K, Takamiya K, Miura T, editors. Proceedings the 13th workshop on environmental radioactivity. Tsukuba: High Energy Accelerator Research Organization; 2012. p. 102–7 (in Japanese with English abstract).

Janžekovič H, Križman MJ. Comparison of discharges of the nuclear accidents in Japan 2011 and Chernobyl 1986. In: Proceedings of the international conference nuclear energy for New Europe, Bovec, Slovenia, Sept 12–15, 2011. 2011.

Kajino M, Inomata Y, Sato K, Ueda H, Han Z, An J, et al. Development of the RAQM2 aerosol chemical transport model and predictions of the Northeast Asian aerosol mass, size, chemistry, and mixing type. Atmos Chem Phys. 2012;12:11833–56. doi:10.5194/acp-12-11833-2012.

Kanai Y. Monitoring of aerosols in Tsukuba after Fukushima Nuclear Power Plant incident in 2011. J Environ Radioact. 2012;111:33–7. doi:10.1016/j.jenvrad.2011.10.011.

Kaneyasu N, Ohashi H, Suzuki F, Okuda T, Ikemori F. Sulfate aerosol as a potential transport medium of radiocesium from the Fukushima nuclear accident. Environ Sci Technol. 2012;46:5720–6.

Katata G, Ota M, Terada H, Chino M, Nagai H. Atmospheric discharge and dispersion of radionuclides during the Fukushima Dai-ichi Nuclear Power Plant accident. Part I: source term estimation and local-scale atmospheric dispersion in early phase of the accident. J Environ Radioact. 2012a;109:103–13.

Katata G, Terada H, Nagai H, Chino M. Numerical reconstruction of high dose rate zones due to the Fukushima Dai-ichi Nuclear Power Plant accident. J Environ Radioact. 2012b;111:2–12.

Katata G, Chino M, Kobayashi T, Terada H, Ota M, Nagai H, et al. Detailed source term estimation of the atmospheric release for the Fukushima Daiichi Nuclear Power Station accident by coupling simulations of atmospheric dispersion model with improved deposition scheme and oceanic dispersion model. Atmos Chem Phys Discuss. 2014;14:14725–832. doi:10.5194/acpd-14-14725-2014.

Katsuragi Y. A study of ^{90}Sr fallout in Japan. Pap Met Geophys. 1983;33(4):277–91.

Kinoshita N, Sueki K, Sasa K, Kitagawa J, Ikarashi S, Nishimura T, et al. Assessment of individual radionuclide distributions from the Fukushima nuclear accident covering central-east Japan. Proc Natl Acad Sci U S A. 2011;108(49):19526–9. doi:10.1073/pnas.1111724108.

Komura K, Yamamoto M, Muroyama T, Murata Y, Nakanishi T, Hoshi M, et al. The JCO criticality accident at Tokai-mura, Japan: an overview of the sampling campaign and preliminary results. J Environ Radioact. 2000;50:3–14.

Kristiansen NI, Stohl A, Wotawa G. Atmospheric removal times of the aerosol-bound radionuclides ^{137}Cs and ^{131}I measured after the Fukushima Dai-ichi nuclear accident: a constraint for air quality and climate models. Atmos Chem Phys. 2012;12:10759–69.

Kusakabe M, Oikawa S, Takata H, Misonoo J. Spatiotemporal distributions of Fukushima-derived radionuclides in nearby marine surface sediments. Biogeosciences. 2013;10:5019–30. doi:10.5194/bg-10-5019-2013.

Maki T, Tanaka TY, Kajino M, Sekiyama TT, Igaraashi Y, Mikami M. Radioactive nuclei emission analysis from Fukushima Daiichi Nuclear Power Plant by inverse model, Manuscript for 93rd American Meteorological Society annual meeting, Austin Texas U.S. 2013. (https://ams.confex.com/ams/93Annual/webprogram/Paper216873.html Accessed date: October 14, 2015)

Masson O. Private communication, Radioprotection Division, Environmental Radioactivity Study and Monitoring Department, Institute for Radiological Protection and Nuclear Safety, France. 2014.

Masson O, Baeza A, Bieringer J, Brudecki K, Bucci S, Cappai M, et al. Tracking of air-borne radionuclides from the damaged Fukushima Dai-ichi nuclear reactors by European networks. Environ Sci Technol. 2011;45:7670–7.

Masumoto Y, Miyazawa Y, Tsumune D, Kobayashi T, Estournel C, Marsaleix P, et al. Oceanic dispersion simulation of Cesium-137 from Fukushima Dai-ichi Nuclear Power Plant. Elements. 2012;8:207–12.

Maxwell RM, Anspaugh LR. An improved model for prediction of resuspension. Health Phys. 2011;101:722–30. doi:10.1097/HP.0b013e31821ddb07.

Meteorological Research Institute (MRI). Tokyo Denryoku Fukushima Dai-ichi Genshiryoku Hatsudensho Jiko nitomonau houshaseibusshitsu no iryuukakusan nitsuite (On the transport and diffusion of radioactive materials by the Fukushima Dai-ichi Nuclear Power Station accident)" (in Japanese). 2011. http://www.mri-jma.go.jp/Topics/H23/H23_tohoku-taiheiyo-oki-eq/1107fukushima.html.

Ministry of Education, Culture, Sports, Science and Technology (MEXT). Monbu Kagaku-sho oyobi Ibaraki-ken niyoru kokuki monitaringu no sokutei kekka no shuusei nitsuite (On the correction of the aerial monitoring results conducted by MEXT and Ibaraki prefecture) dated August 31, 2011 (in Japanese). 2011a. http://radioactivity.nsr.go.jp/ja/contents/5000/4933/24/1940_0831.pdf.

Ministry of Education, Culture, Sports, Science and Technology (MEXT). Monbu Kagaku-sho ni yoru, purutoniumu, sutoronchiumu no kakushu bunseki no kekka nitsuite (On the analytical results for Pu and Sr by MEXT) dated September 30, 2011 (in Japanese). 2011b. http://radioactivity.nsr.go.jp/ja/contents/6000/5048/24/5600_110930_rev130701.pdf.

Ministry of Education, Culture, Sports, Science and Technology, Results of Air-borne Monitoring by the Ministry of Education, Culture, Sports, Science and Technology and the U.S. Department of Energy (MEXT & USDOE). 2011. dated May 6, 2011.

Miyake Y. The artificial radioactivity in rain water observed in Japan, from autumn 1954 to spring 1955. Pap Met Geophys. 1954;6(1):26–31.

Miyake Y, Sugiura Y, Kameda K. On the distribution of radioactivity in the sea: around Bikini Atoll in June, 1954. Pap Met Geophys. 1955;5:253–62.

Miyake Y, Saruhashi K, Katsuragi Y, Kanazawa T. Penetration of ^{90}Sr and ^{137}Cs in deep layers of the Pacific and vertical diffusion rate of deep water. J Radiat Res. 1962;3(3):141–7.

Miyake Y, Samhashi K, Katsuragi Y, Kanazawa T, Tsunogai S. Deposition of Sr-90 and Cs-137 in Tokyo through the end of July 1963. Pap Met Geophys. 1963;14:58–65.

Miyake Y, Katsuragi Y, Sugimura Y. Plutonium fallout in Tokyo. Pap Met Geophys. 1975;26(1):1–8.

Miyake Y, Saruhashi K, Sugimura Y, Kanazawa T, Hirose K. Contents of ^{137}Cs, plutonium and americium isotopes in the Southern Ocean waters. Pap Met Geophys. 1988;39:95–113.

Miyao T, Hirose K, Aoyama M, Igarashi Y. Trace of the recent deep water formation in the Japan Sea deduced from historical ^{137}Cs data. Geophys Res Lett. 2000;27(22):3731–4.

Morino Y, Ohara T, Nishizawa M. Atmospheric behavior, deposition, and budget of radioactive materials from the Fukushima Daiichi nuclear power plant in March 2011. Geophys Res Lett. 2011;38:L00G11. doi:10.1029/2011GL048689.

Nuclear and Industrial Safety Agency (NISA). Houshasei bussitsu houshutsuryo deta no ichibu ayamari nitsuite (On a partial mistake in emission inventory estimate of radioactive materials) dated October 20, 2011 (in Japanese). 2011. http://warp.ndl.go.jp/info:ndljp/pid/6086248/www.meti.go.jp/press/2011/10/20111020001/20111020001.pdf.

Otsuji-Hatori M, Igarashi Y, Hirose K. Preparation of a reference fallout material for activity measurements. J Environ Radioact. 1996;31:143–55.

Petters MD, Kreidenweis SM. A single parameter representation of hygroscopic growth and cloud condensation nucleus activity. Atmos Chem Phys. 2007;7:1961–71. doi:10.5194/acp-7-1961-2007.

Povinec PP, Gera M, Holý K, Hirose K, Lujaniené G, Nakano M, et al. Dispersion of Fukushima radionuclides in the global atmosphere and the ocean. Appl Radiat Isot. 2013a;81:383–92.

Povinec PP, Aoyama M, Biddulph D, Breier R, Buesseler K, Chang CC, et al. Cesium, iodine and tritium in NW Pacific waters: a comparison of the Fukushima impact with global fallout. Biogeosciences. 2013b;10:5481–96. doi:10.5194/bg-10-5481-2013.

Saito K, Ishida J, Aranami K, Hara T, Segawa T, Narita M, et al. Nonhydrostatic atmospheric models and operational development at JMA. J Meteorol Soc Jpn. 2007;85B:271–304. doi:10.2151/jmsj.85B.271.

Sanada Y, Sugita T, Nishizawa Y, Kondo A, Torii T. The aerial radiation monitoring in Japan after the Fukushima Daiichi nuclear power plant accident. Progress in Nuclear Science and Technology. 2014;4:76–80.

Schwantes JM, Orton CR, Clark RA. Analysis of a nuclear accident: fission and activation product releases from the Fukushima Daiichi Nuclear Facility as remote indicators of source identification, extent of release, and state of damaged spent nuclear fuel. Environ Sci Technol. 2012;46(16):8621–7. doi:10.1021/es300556m.

Sectional Committee on Nuclear Accident Committee on Comprehensive Synthetic Engineering, Science Council of Japan (SCJ). Report "A review of the model comparison of transportation and deposition of radioactive materials released to the environment as a result of the Tokyo Electric Power Company's Fukushima Daiichi Nuclear Power Plant Accident" dated September 2, 2014. 2014. http://www.scj.go.jp/en/report/index.html.

Sekiyama TT, Kunii M, Kajino M, Shimbori T. Horizontal resolution dependence of atmospheric simulations of the Fukushima nuclear accident using 15-km, 3-km, and 500 m grid models. J Meteor Soc Japan. 2015;93(1):49–64. doi:10.2151/jmsj.2015-002.

Stohl A, Seibert P, Wotawa G, Arnold D, Burkhart JF, Eckhardt S, et al. Xenon-133 and caesium-137 releases into the atmosphere from the Fukushima Dai-ichi Nuclear Power Plant: determination of the source term, atmospheric dispersion, and deposition. Atmos Chem Phys. 2012;12:2313–43. doi:10.5194/acp-12-2313-2012.

Takemura T, Nakamura H, Takigawa M, Kondo H, Satonuma T, Miyasaka T, et al. A numerical simulation of global transport of atmospheric particles emitted from the Fukushima Daiichi Nuclear Power Plant. Scientific Online Letters on the Atmosphere (SOLA). 2011;7:101–4.

Tanaka TY. Numerical simulation of global dispersion of radionuclides. Wind Engineers JAWE. 2013;38(4):388–95 (in Japanese).

Tanihata I. Sampling and mapping of soil contamination and what we have learn from it. Radioisotopes. 2013;62:724–40 (in Japanese).

Terada H, Katata G, Chino M, Nagai H. Atmospheric discharge and dispersion of radionuclides during the Fukushima Dai-ichi Nuclear Power Plant accident. Part II: verification of the source term and analysis of regional-scale atmospheric dispersion. J Environ Radioact. 2012;112:141–54. doi:10.1016/j.jenvrad.2012.05.023.

Tokyo Electric Power Company (TEPCO). Estimation of radioactive material released to the atmosphere during the Fukushima Daiichi NPS Accident dated May 2012. 2012. http://www.tepco.co.jp/en/press/corp-com/release/betu12_e/images/120524e0205.pdf.

Torii T, Sugita T, Okada CE, Reed MS, Blumenthal DJ. Enhanced analysis methods to derive the spatial distribution of ^{131}I deposition on the ground by air-borne surveys at an early stage after the Fukushima Daiichi Nuclear Power Plant accident. Health Phys. 2013;105(2):192–200.

Tsumune D, Tsubono T, Aoyama M, Uematsu M, Misumi K, Maeda Y, et al. One-year, regional-scale simulation of ^{137}Cs radioactivity in the ocean following the Fukushima Dai-ichi Nuclear Power Plant accident. Biogeosciences. 2013;10:5601–17. doi:10.5194/bg-10-5601-2013.

Tsuruta H, Oura Y, Ebihara M, Ohara T, Nakajima T. First retrieval of hourly atmospheric radionuclides just after the Fukushima accident by analyzing filter-tapes of operational air pollution monitoring stations. Sci Rep. 2014;4:6717. doi:10.1038/srep06717.

U.S. Department of Energy (USDOE). 2013. http://energy.gov/situation-japan-updated-12513

U.S. Department of Health and Human Services, Public Health Service, Agency for Toxic Substances and Disease Registry (ATSDR2004Cs). Toxicological Profile for Cesium. 2004. http://www.atsdr.cdc.gov/ToxProfiles/tp157.pdf.

U.S. Department of Health and Human Services, Public Health Service, Agency for Toxic Substances and Disease Registry (ATSDR2004Sr). Toxicological Profile for Strontium. 2004. http://www.atsdr.cdc.gov/ToxProfiles/tp159.pdf.

Winiarek V, Bocquet M, Saunier O, Mathieu A. Estimation of errors in the inverse modeling of accidental release of atmospheric pollutant: Application to the reconstruction of the cesium-137 and iodine-131 source terms from the Fukushima Daiichi Power Plant. J Geophys Res. 2012;117, D05122. doi:10.1029/2011JD016932.

Yamamoto M, Takada T, Nagao S, Koike T, Shimada K, Hoshi M, et al. An early survey of the radioactive contamination of soil due to the Fukushima Dai-ichi Nuclear Power Plant accident, with emphasis on plutonium analysis. Geochem J. 2012;46:341–53.

Yoshida N, Kanda J. Tracking the Fukushima radionuclides. Science. 2012;336:1115–6. doi:10.1126/science.1219493.

Yoshida N, Takahashi Y. Land-surface contamination by radionuclides from the Fukushima Daiichi Nuclear Power Plant accident. Elements. 2012;8:201–6. doi:10.2113/gselements.8.3.201.

Zhang W, Bean M, Benotto M, Cheung J, Ungar K, BAhier B. Development of a new aerosol monitoring system and its application in Fukushima nuclear accident related aerosol radioactivity measurement at the CTBT radionuclide station in Sidney of Canada. J Environ Radioactiv. 2011;102:1065–9. doi:10.1016/j.jenvrad.2011.08.007.

Moon JY, Wang B, Ha KJ, Lee JY (2013) Teleconnections associated with Northern Hemisphere summer monsoon intraseasonal oscillation. Climate Dynam 40: 2761–2774. doi:10.1007/s00382-012-1394-0

Murakami T, Matsumoto J (1994) Summer monsoon over the Asian continent and western North Pacific. J Meteorol Soc Jpn 72:719–745

Murakami H, Wang Y, Yoshimura H, Mizuta R, Sugi M, Shindo E, Adachi T, Yukimoto S, Hosaka M, Kusunoki S, Ose T, Kitoh A (2012) Future changes in tropical cyclone activity projected by the new high-resolution MRI-AGCM. J Clim 25:3237–3260. doi:10.1175/JCLI-D-11-00415.1

Murakami H, Vecchi GA, Underwood S, Delworth TL, Wittenberg AT, Anderson WG, Chen JH, Gudgel RG, Harris LM, Lin SJ, Zeng F (2015) Simulation and prediction of category 4 and 5 hurricanes in the high-resolution GFDL HiFLOR coupled climate model. J Clim 28:9058–9079. doi:10.1175/JCLI-D-15-0216.1

Nakano M, Sawada M, Nasuno T, Satoh M (2015) Intraseasonal variability and tropical cyclogenesis in the western North Pacific simulated by a global nonhydrostatic atmospheric model. Geophys Res Lett 42:565–571. doi:10.1002/2014gl062479

Nakano M, Wada A, Sawada M, Yoshimura H, Onishi R, Kawahara S, Sasaki W, Nasuno T, Yamaguchi M, Iriguchi T, Sugi M, Takeuchi Y (2017) Global 7-km mesh nonhydrostatic Model Intercomparison Project for improving TYphoon forecast (TYMIP-G7): Experimental design and preliminary results. Geosci Model Dev 10:1363-1381. doi:10.5194/gmd-2016-184

Nakazawa T (2006) Madden-Julian oscillation activity and typhoon landfall on Japan 2004. SOLA 2:136–139. doi:10.2151/sola.2006-035

Nasuno T, Satoh M (2011) Properties of precipitation and in-cloud vertical motion in a global nonhydrostatic aquaplanet experiment. J Meteorol Soc Jpn 89: 413–439. doi:10.2151/jmsj.2011-502

Noda AT, Satoh M, Yamada Y, Kodama C, Seiki T (2014) Responses of tropical and subtropical high-cloud statistics to global warming. J Clim 27:7753–7768. doi: 10.1175/JCLI-D-14-00179.1

Noda AT, Yamada Y, Kodama C, Miyakawa T, Seiki T, Satoh M (2015) Cold and warm rain simulated using a global nonhydrostatic model without cumulus parameterization, and their responses to a warmer atmospheric condition. J Meteorol Soc Jpn 93:181–197. doi:10.2151/jmsj.2015-010

Ogata T, Ueda H, Inoue T, Hayasaki M, Yoshida A, Watanabe S, Kira M, Ooshiro M, Kumai A (2014) Projected future changes in the Asian monsoon: a comparison of CMIP3 and CMIP5 model results. J Meteorol Soc Jpn 92:207–225. doi:10.2151/jmsj.2014-302

Ohno T, Satoh M (2015) On the warm core of a tropical cyclone formed near the tropopause. J Atmos Sci 72:551–571. doi:10.1175/JAS-D-14-0078.1

Oishi Y, Imamura F, Sugawara D (2015) Near-field tsunami inundation forecast using the parallel TUNAMI-N2 model: application to the 2011 Tohoku-Oki earthquake combined with source inversions. Geophys Res Lett 42:1083–1091. doi:10.1002/2014GL062577

Oouchi K, Yoshimura J, Yoshimura H, Mizuta R, Kusunoki S, Noda A (2006) Tropical cyclone climatology in a global-warming climate as simulated in a 20 km-mesh global atmospheric model: frequency and wind intensity analyses. J Meteorol Soc Jpn 84:259–276

Oouchi K, Noda AT, Satoh M, Miura H, Tomita H, Nasuno T, Iga S (2009a) A simulated preconditioning of typhoon genesis controlled by a boreal summer Madden-Julian oscillation event in a global cloud-system-resolving model. SOLA 5:65–68. doi:10.2151/sola.2009-017

Oouchi K, Noda AT, Satoh M, Wang B, Xie SP, Takahashi HG, Yasunari T (2009b) Asian summer monsoon simulated by a global cloud-system-resolving model: Diurnal to intra-seasonal variability. Geophys Res Lett 36:L11815. doi:10.1029/2009gl038271

Palmer T (2014) Build high-resolution global climate models. Nature 515:338–339

Petch JC, Brown AR, Gray MEB (2002) The impact of horizontal resolution on the simulations of convective development over land. Q J R Meteorol Soc 128:2031–2044. doi:10.1256/003590002320603511

Prein AF, Langhans W, Fosser G, Ferrone A, Ban N, Goergen K, Keller M, Tölle M, Gutjahr O, Feser F, Brisson E, Kollet S, Schmidli J, van Lipzig NPM, Leung R (2015) A review on regional convection-permitting climate modeling: Demonstrations, prospects, and challenges. Rev Geophys 53:323–361

Roberts MJ, Vidale PL, Mizielinski MS, Strachan J, Hodges K, Bell R, Camp J (2015) Tropical cyclone in the UPSCALE ensemble of high resolution global climate models. J Clim 28:574–596. doi:10.1175/JCLI-D-14-00131.1

Rogers RF, Reasor P, Lorsolo S (2013) Airborne Doppler observations of the inner-core structural differences between intensifying and steady-state tropical cyclones. Mon Weather Rev 141:2970–2991

Saito K, Tsuyuki T, Seko H, Kimura F, Tokioka T, Kuroda T, Duc L, Ito K, Oizumi T, Chen G, Ito J, SPIRE Field 3 Mesoscale NWP group (2013) Super high-resolution mesoscale weather prediction. J Phys Conf Ser 454:012073. doi:10.1088/1742-6596/454/1/012073

Sato T, Miura H, Satoh M, Takayabu YN, Wang Y (2009) Diurnal cycle of precipitation in the tropics simulated in a global cloud-resolving model. J Clim 22:4809–4826. doi: 10.1175/2009jcli2890.1

Satoh M, Matsuno T, Tomita H, Miura H, Nasuno T, Iga S (2008) Nonhydrostatic Icosahedral Atmospheric Model (NICAM) for global cloud resolving simulations. J Comp Phys 227:3486–3514. doi:10.1016/j.jcp.2007.02.006

Satoh M, Iga S, Tomita H, Tsushima Y, Noda AT (2012a) Response of upper clouds due to global warming tested by a global atmospheric model with explicit cloud processes. J Clim 25:2178–2191. doi:10.1175/JCLI-D-11-00152.1

Satoh M, Oouchi K, Nasuno T, Taniguchi H, Yamada Y, Tomita H, Kodama C, Kinter J, Achuthavarier D, Manganello J, Cash B, Jung T, Palmer T, Wedi N (2012b) The intra-seasonal oscillation and its control of tropical cyclones simulated by high-resolution global atmospheric models. Clim Dyn 39:2185–2206. doi:10.1007/s00382-011-1235-6

Satoh M, Tomita H, Yashiro H, Miura H, Kodama C, Seiki T, Noda AT, Yamada Y, Goto D, Sawada M, Miyoshi T, Niwa Y, Hara M, Ohno Y, Iga S, Arakawa T, Inoue T, Kubokawa H (2014) The Non-hydrostatic Icosahedral Atmospheric Model: Description and development. Prog Earth Planet Sci 1:18. doi: 10.1186/s40645-014-0018-1

Satoh M, Yamada Y, Sugi M, Kodama C, Noda AT (2015) Constraint on future change in global frequency of tropical cyclones due to global warming. J Meteorol Soc Jpn 93:489–500. doi:10.2151/jmsj.2015-025

Seko H, Kunii M, Yokota S, Tsuyuki T, Miyoshi T (2015) Ensemble experiments using a nested LETKF system to reproduce intense vortices associated with tornadoes of 6 May 2012 in Japan. Prog Earth Planet Sci 2:42. doi:10.1186/s40645-015-0072-3

Sperber K, Annamalai H, Kang IS, Kitoh A, Moise A, Turner A, Wang B, Zhou T (2013) The Asian summer monsoon: an intercomparison of CMIP5 vs. CMIP3 simulations of the late 20th century. Clim Dyn 41:2711–2744. doi:10.1007/s00382-012-1607-6

Stevens B, Bony S (2013) What are climate models missing? Science 340:1053–1054. doi:10.1126/science.1237554

Sugi M, Noda A, Sato N (2002) Influence of the global warming on tropical cyclone climatology: An experiment with the JMA global model. J Meteorol Soc Jpn 80:249–272. doi: 10.2151/jmsj.80.249

Sugi M, Murakami H, Yoshimura J (2012) On the mechanism of tropical cyclone frequency changes due to global warming. J Meteorol Soc Jpn 90A:397–408. doi:10.2151/jmsj.2012-A24

Taniguchi H, Yanase W, Satoh M (2010) Ensemble simulation of cyclone Nargis by a global cloud-system-resolving model - Modulation of cyclogenesis by the Madden-Julian oscillation. J Meteorol Soc Jpn 88:571–591

Terai M, Yashiro H, Sakamoto K, Iga S, Tomita H, Satoh M, Minami K (2014) Performance optimization and evaluation of a global climate application using a 440 m horizontal mesh on the K computer, Abstract presented at the 2014 International Conference for High Performance Computing. Networking, Storage and Analysis, Ernest N. Morial Convention Center, New Orleans, 16-21 November 2014

Terasaki K, Sawada M, Miyoshi T (2015) Local ensemble transform Kalman Filter experiments with the Nonhydrostatic Icosahedral Atmospheric Model NICAM. SOLA 11:23–26. doi:10.2151/sola.2015-006

Tomita H (2008) New microphysical schemes with five and six categories by diagnostic generation of cloud ice. J Meteorol Soc Jpn 86:121–142

Tomita H, Satoh M (2004) A new dynamical framework of nonhydrostatic global model using the icosahedral grid. Fluid Dyn Res 34:357–400

Tomita H, Miura H, Iga S, Nasuno T, Satoh M (2005) A global cloud-resolving simulation: Preliminary results from an aqua planet experiment. Geophys Res Lett 32:L08805. doi: 10.1029/2005GL022459

Top 500 (2011a) Top 500 List June 2011., https://www.top500.org/lists/2011/06/. Accessed 10 Mar 2017

Top 500 (2011b) Top 500 List November 2011., https://www.top500.org/lists/2011/11/. Accessed 10 Mar 2017

Tsuboi S, Ando K, Miyoshi T, Peter D, Komatitsch D, Tromp J (2016) A 1.8 trillion degrees-of-freedom, 1.24 petaflops global seismic wave simulation on the K computer. Int J High Perform Comput Appl 30(4):411–422. doi:10.1177/1094342016632596

Tsushima Y, Iga S, Tomita H, Satoh M, Noda AT, Webb M (2014) High cloud increase in a perturbed SST experiment with a global nonhydrostatic model including explicit convective processes. J Adv Model Earth Syst 6:571–585. doi:10.1002/2013MS000301

Vecchi GA, Soden BJ (2007) Global warming and the weakening of the tropical circulation. J Clim 20:4316–4340. doi:10.1175/JCLI4258.1

Vitart F, Robertson AW, Anderson DLT (2012) Subseasonal to Seasonal Prediction Project: Bridging the gap between weather and climate. WMO Bull 61:23–28

Vitart F, Ardilouze C, Bonet A, Brookshaw A, Chen M, Codorean C, Déqué M, Ferranti L, Fucile E, Fuentes M, Hendon H, Hodgson J, Kang H, Kumar A, Lin H, Liu G, Liu X, Malguzzi P, Mallas I, Manoussakis M, Mastrangelo D, MacLachlan C, McLean P, Minami A, Mladek R, Nakazawa T, Najm S, Nie Y, Rixen M, Robertson A et al (2017) The Subseasonal to Seasonal (S2S) Prediction Project Database. Bull Am Meteorol Soc 98:163–173. doi:10.1175/BAMS-D-16-0017.1

Wada A, Uehara T, Ishizaki S (2014) Typhoon-induced sea surface cooling during the 2011 and 2012 typhoon seasons: observational evidence and numerical investigations of the sea surface cooling effect using typhoon simulations. Prog Earth Planet Sci 1:11. doi:10.1186/2197-4284-1-11

Walsh KJE, Camargo SJ, Vecchi GA, Daloz AS, Elsner J, Emanuel K, Horn M, Lim Y-K, Roberts M, Patricola C, Scoccimarro E, Sobel AH, Strazzo S, Villarini G, Wehner M, Zhao M, Kossin JP, LaRow T, Oouchi K, Schubert S, Wang H, Bacmeister J, Chang P, Chauvin F, Jablonowski C, Kumar A, Murakami H, Ose T, Reed KA, Saravanan R et al (2015) Hurricanes and climate: the U.S. CLIVAR working group on hurricanes. Bull Am Meteorol Soc 96:997–1017. doi:10.1175/BAMS-D-13-00242.1

Walsh KJE, Mcbride JL, Klotzbach PJ, Balachandran S, Camargo SJ, Holland G, Knutson TR, Kossin JP, Lee T, Sobel A, Sugi M (2016) Tropical cyclone and climate change. WIREs Clim Change 7:65–89. doi:10.1002/wcc.371

Wang B, Rui H (1990) Synoptic climatology of transient tropical intraseasonal convection anomalies: 1975-1985. Meteorol Atmos Phys 44:43–61

Wang B, Xie X (1997) A model for the boreal summer intraseasonal oscillation. J Atmos Sci 54:72–86

Wang H, Wang Y (2014) A numerical study of typhoon Megi (2010). Part I: rapid intensification. Mon Weather Rev 142:29–48. doi:10.1175/MWR-D-13-00070.1

Wang B, Ding QH, Fu XH, Kang IS, Jin K, Shukla J, Doblas-Reyes F (2005) Fundamental challenge in simulation and prediction of summer monsoon rainfall. Geophys Res Lett 32:L15711. doi:10.1029/2005gl022734

Wehner M, Reed KA, Stone D, Collins WD, Bacmeister J (2015) Resolution dependence of future tropical cyclone projections of CAM5.1 in the US CLIVAR Hurricane Working Group idealized configurations. J Clim 28:3905–3925. doi:10.1175/JCLI-D-14-00311.1

Wheeler MC, Hendon HH (2004) An all-season real-time multivariate MJO index: development of an index for monitoring and prediction. Mon Weather Rev 132:1917–1932

Yamada Y, Oouchi K, Satoh M, Tomita H, Yanase W (2010) Projection of changes in tropical cyclone activity and cloud height due to greenhouse warming: global cloud-system-resolving approach. Geophys Res Lett 37:L07709. doi:10.1029/2010GL042518

Yamamoto K, Uno A, Murai H, Tsukamoto T, Shoji F, Matsui S, Sekizawa R, Sueyasu F, Uchiyama H, Okamoto M, Ohgushi N, Takashina K, Wakabayashi D, Taguchi Y, Yokokawa M (2014) The K computer operations: experiences and statistics. Procedia Comput Sci 29:576–585. doi:10.1016/j.procs.2014.05.052

Yamaura T, Kajikawa Y, Tomita H, Satoh M (2013) Possible impact of a tropical cyclone on the northward migration of the Baiu frontal zone. SOLA 9:89–93. doi:10.2151/sola.2013-020

Yanase W, Taniguchi H, Satoh M (2010) The genesis of tropical cyclone Nargis (2008): environmental modulation and numerical predictability. J Meteorol Soc Jpn 88:497–519. doi:10.2151/jmsj.2010-314

Yashiro H, Kajikawa Y, Miyamoto Y, Yoshida R, Yamaura R, Tomita H (2016a) Resolution dependency of diurnal precipitation cycle simulated by global cloud resolving model. SOLA 12:272–276

Yashiro H, Terai M, Yoshida R, Iga S, Minami K, Tomita H (2016b) Performance analysis and optimization of Nonhydrostatic ICosahedral Atmospheric Model (NICAM) on the K Computer and TSUBAME2.5. In: Proceedings of the Platform for Advanced Scientific Computing Conference (PASC'16). ACM, New York, p Article 3. doi:10.1145/2929908.2929911

Yashiro H, Terasaki K, Miyoshi T, Tomita H (2016c) Performance evaluation of a throughput-aware framework for ensemble data assimilation: the case of NICAM-LETKF. Geosci Model Dev 9:2293–2300. doi:10.5194/gmd-9-2293-2016

Yokokawa M, Shoji F, Uno A, Kurokawa M, Watanabe T (2011) The K computer: Japanese next-generation supercomputer development project. In: IEEE/ACM International Symposium on Low Power Electronics and Design (ISLPED) 2011. IEEE, New York, pp 371–372. doi:10.1109/ISLPED.2011.5993668

Yokota S, Seko H, Kunii M, Yamauchi H, Niino H (2016) The tornadic supercell on the Kanto Plain on 6 May 2012: polarimetric radar and surface data assimilation with EnKF and ensemble-based sensitivity analysis. Mon Weather Rev 144:3133–3157. doi:10.1175/MWR-D-15-0365.1

Yoneyama K, Zhang C, Long CN (2013) Tracking pulses of the Madden-Julian oscillation. Bull Am Meteorol Soc 94:1871–1891

Zarzycki CM (2016) Tropical cyclone intensity errors associated with lack of two-way ocean coupling in high-resolution global simulations. J Clim 29:8589–8610

Zou LW, Zhou TJ (2015) Asian summer monsoon onset in simulations and CMIP5 projections using four Chinese climate models. Adv Atmos Sci 32:794–806

4

Resolution dependence of deep convections in a global simulation from over 10-kilometer to sub-kilometer grid spacing

Yoshiyuki Kajikawa[1*], Yoshiaki Miyamoto[1], Ryuji Yoshida[1], Tsuyoshi Yamaura[1], Hisashi Yashiro[1] and Hirofumi Tomita[1,2]

Abstract

The success of sub-kilometer global atmospheric simulation opens the door for resolving deep convections, which are fundamental elements of cloudy disturbances that drive global circulation. A previous study found that the essential change in the simulated convection properties occurred at a grid spacing of about 2 km as a global mean. In grid-refinement experiments, we conducted further comprehensive analysis of the global-mean state and the characteristics of deep convection, to clarify the difference of the essential change by location and environment. We found that the essential change in convection properties was different in the location and environment for each cloudy disturbance. The convections over the tropics show larger resolution dependence than convections over mid-latitudes, whereas no significant difference was found in convections over land or ocean. Furthermore, convections over cloudy disturbances [(i.e., Madden-Julian oscillation (MJO), tropical cyclones (TCs)] show essential change of convection properties at about 1 km grid spacing, suggesting resolution dependence. As a result, convections not categorized as cloudy disturbances make a large contribution to the global-mean convection properties. This implies that convections in disturbances are largely affected organization processes and hence have more horizontal resolution dependence. In contrast, other categorized convections that are not involved in major cloudy disturbances show the essential change at about 2 km grid spacing. This affects the latitude difference of the resolution dependence of convection properties and hence the zonal-mean outgoing longwave radiation (OLR). Despite the diversity of convection properties, most convections are resolved at less than 1 km grid spacing. In the future, longer integration of global atmosphere, to 0.87 km grid spacing, will stimulate significant discussion about the interaction between the convections and cloudy disturbances.

Keywords: Deep convection, High resolution, Global simulation, NICAM, Cloudy disturbance, Resolution dependence

Background

The rapid increase of computer capabilities has enabled meteorological and climatological researchers to increase horizontal and vertical resolutions in the numerical model (Simmons et al. 1989; Mizuta et al. 2006; Saito et al. 2006; Kodama et al. 2015). The demand for high-resolution simulation not only from regional-model researchers but also from global-model researchers has become intense. In the atmospheric general circulation model (AGCM), one of the key issues is to explicitly resolve deep convection for the following reasons.

Deep convection is a minimum element for the organization of cloud systems, including cloudy disturbances, and plays an essential role in driving the atmosphere, through the transportation of energy in the troposphere from the tropics to the polar region (Webster 1972; Gill 1982; Emanuel and Raymond 1993). Cloudy disturbances are sometimes associated with natural disasters

* Correspondence: ykaji@riken.jp
[1]RIKEN Advanced Institute for Computational Science, 7-1-26 Minatojima-minami-machi, Chuo-ku, Kobe, Hyogo 650-0047, Japan
Full list of author information is available at the end of the article

because of related heavy rainfall; hence, it is important to enrich our understanding of the processes and mechanisms of cloudy disturbances. One of the important processes that characterize cloudy disturbances is the interaction between deep convection and cloudy disturbances through the hierarchical structure of cloud cluster and super cloud cluster (SCC). SCC is the eastward-moving convection area near the equator with a horizontal scale of 2000–4000 km (Nakazawa 1988; Mapes and Houze 1993). However, it has been difficult to obtain observational data that is spatiotemporal enough to examine that process (Stephens et al. 2010; Mrowiec et al. 2012). Numerical simulations have been used to compensate the deficiency and gaps in observational data. The ability to express deep convection, with its effect on larger-scale phenomenon, is crucial for better simulating global circulation and associated organizations of cloud systems and disturbances.

However, because there is a large gap in spatiotemporal scale between convection on the order of 10^0–10^1 km and cloudy disturbances on the order of 10^3 km, it has been challenging to globally simulate both phenomena and comprehensively discuss their interaction. In previous decades, due to the low horizontal resolution of the model, clouds have been expressed as parameterization. Various types of cumulus parameterization were established and have been used in the AGCM [e.g., (Arakawa and Schubert 1974; Kain and Fritsch 1990; Yoshimura et al. 2015)]. The variety of these parameterizations has been a source of uncertainties in model results.

It is impossible for a conventional GCM to represent the hierarchy of cloud organization from elemental convective clouds. Thus, comprehensive understanding, including the impact of organized clouds on general circulation, was not realistic. The new Nonhydrostatic Icosahedral Atmospheric Model (NICAM) (Tomita and Satoh 2004; Satoh et al. 2008; Satoh et al. 2014) is designed to conduct global simulation without cumulus parameterization. Previous studies have shown the usefulness of the global nonhydrostatic model without cumulus parameterization for large-scale organized convective systems and disturbances, such as tropical cyclones (Fudeyasu et al. 2008; Yamaura et al. 2013; Miyamoto et al. 2014; Nakano et al. 2015), the Madden-Julian oscillation (MJO) (Miura et al. 2007; Miyakawa et al. 2014), and monsoon onset (Kajikawa et al. 2015). However, in the early 2010s, the horizontal grid spacing with 3.5 km was limited. Although this resolution was the best possible in those days for qualitatively acceptable results for the cloud disturbance expression, the impact of higher resolution on model results was expected for the interaction between different spatiotemporal scale phenomena.

Because of these considerations, Miyamoto et al. (2013) (hereafter MY13) successfully conducted the first-ever global atmosphere simulations with sub-kilometer grid spacing. They further stepped forward to resolve the convection for the entire Earth. Through a set of grid-refinement experiments from over 10 km to sub-kilometer, they found that the simulated convection core averaging over the globe is expressed not by a single grid point but by multiple grid points in the sub-kilometer grid-spacing simulation. They also showed that the expression of convection core was drastically changed between 3.5 and 1.7 km grid spacing. Furthermore, Miyamoto et al. (2015) (hereafter MY15) conducted detailed analysis for convections simulated in the finest grid spacing. They explained that the statistical properties of deep convection are significantly different in various cloudy disturbances, such as MJOs or tropical cyclones (TCs).

In short, we have been in a transition stage of improving the representation of cloud processes, including the feedbacks, with computer resources (Randall et al. 2003), and deep convection is one of the important components of climate modeling (Stevens and Bony 2013). The success of sub-kilometer global atmospheric simulation in MY13 and MY15 opens the door to the next stage of global research for deep convection and the cloudy disturbances that arise from its organization, by truly resolving the deep convection.

However, several issues remain. The primary issue is the dependence of the essential change of convection properties on the location and environment of the grid spacing under consideration. MY13 showed the change in the sense of global mean and the subsequent paper, and MY15 clarified the different convection aspects between the main cloud disturbances under the highest resolution simulation only. MY13 did not discuss the dependence on the cloud disturbance, while MY15 did not discuss the dependence on the resolution. In this study, we investigate the primary issue more comprehensively, considering both MY13 and MY15. The specific question is as follows: In what area does convection make the larger contribution to the resolution dependence of detected convection properties averaging over the globe as shown in MY13? Another important question is as follows: What environmental condition is effective in producing the diversity of convection properties, i.e., what is the resolution relationship between the number of deep convections, areas of deep convection, vertical mass flux outgoing longwave radiation (OLR), and precipitation? In this paper, we perform comprehensive analysis to address these questions and resolve the issues by describing an overview of sub-kilometer global simulation with a set of grid-refinement experiments.

Methods

We used a set of global atmospheric simulation results of a grid-refinement experiment. The simulation was conducted using NICAM. The number of vertical layers was 94, and the grid interval gradually expanded with height. The height of the lowest level was 36 m, and the average resolutions in the boundary layer and in the troposphere were about 80 and 250 m, respectively. The height of the top of the atmosphere was 39,291 m. The detailed description of the dynamical core is summarized in Tomita and Satoh (2004). Physical processes such as radiation process, microphysical process, boundary-layer turbulence, and surface flux were solved using the parameterizations of Sekiguchi and Nakajima (2008), Tomita (2008), Noda et al. (2010), and Louis (1979). Cumulus parameterization was not used in any experiment. The horizontal grid spacing in the series of experiments was set at 0.87, 1.7, 3.5, 7.0, and 14 km. Hereafter, the simulations are referred to (as in MY13) as Δx, where x is one of the horizontal grid-spacing values. For example, grid spacing for 0.87 km is $\Delta 0.87$. In addition, we used 20,480 nodes as the maximum for conducting $\Delta 0.87$. The detailed computational performance is described in the Appendix.

The initial conditions and integration of each experiment are illustrated in Fig. 1. The initial conditions in each experiment were constructed using the results of a one-step coarser resolution after a 3 day integration, beginning 2012082200UTC, except for $\Delta 0.87$, which was initialized with the two-step coarser resolution of $\Delta 3.5$, as shown. This integration was initialized by linearly

interpolated data from the final analysis of the National Center for Atmospheric Research (Kalnay et al. 1996). For example, the initial condition of $\Delta 14$ was obtained from the 3 day integration of $\Delta 28$. The simulation period was 12 h, from 2012082500UTC to 2012082512UTC. We used the data at 2012082512UTC in each experiment for the following analysis.

To define the convection in the simulation and validate such data for our analysis, we applied the following method to extract the convection in the simulated results, as in MY13. First, we defined the convective grids by optical thickness (>35) and cloud top pressure (<400 hPa) based on the cloud categorization scheme of the International Satellite Cloud Climatology Project (ISCCP) (Rossow and Schiffer 1999). Then, we could detect the convection core in the convective grids as the grid at which the vertical velocity averaged in the troposphere is greater than that in all neighboring grids. This diagnosis, using the local peak of vertical velocity, enables us to avoid the controversial discussion on the threshold dependences of detected convections. Figure 2 shows the time series of the number of convective grid points in each simulation, which was not shown in MY13. The number of convective grid points is sensitive to the initial adjustment, abruptly increasing at the beginning of the integration. A 3 day integration of the observational data seems to be long enough for the number of deep convections to become almost constant. Twelve-hour simulations initialized by the simulation results of a one-step coarser resolution are also long enough to obtain a constant number of deep

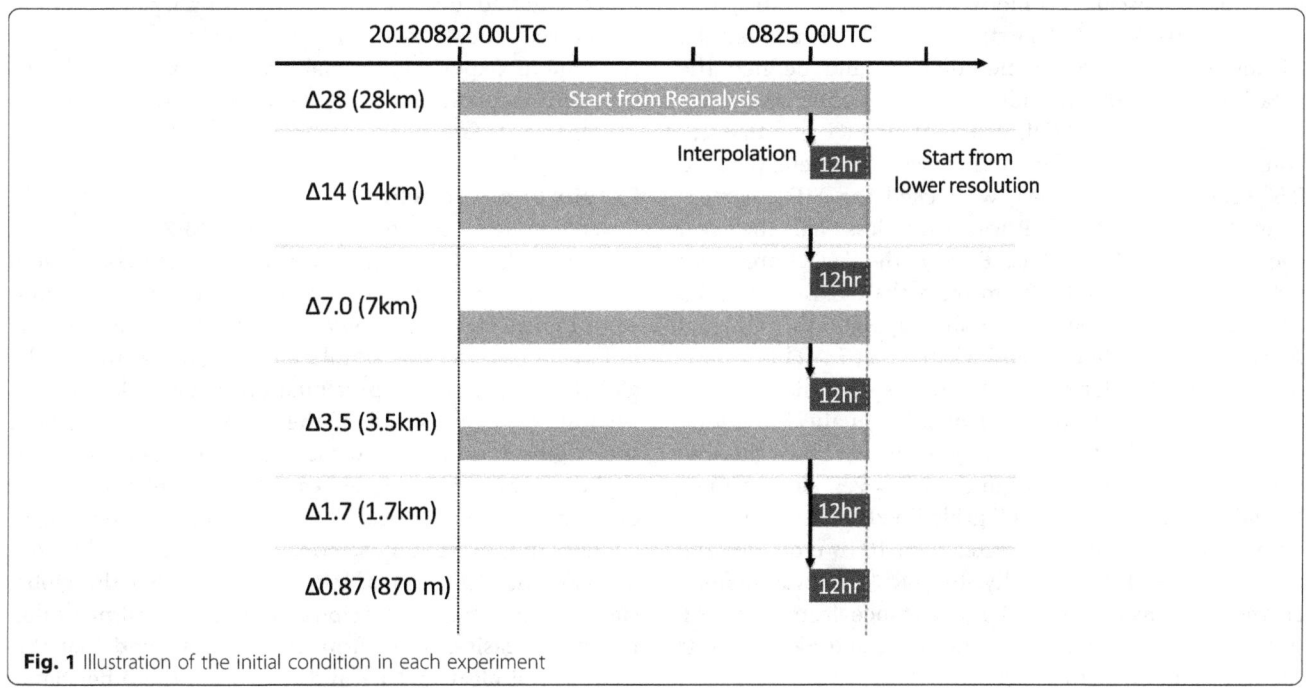

Fig. 1 Illustration of the initial condition in each experiment

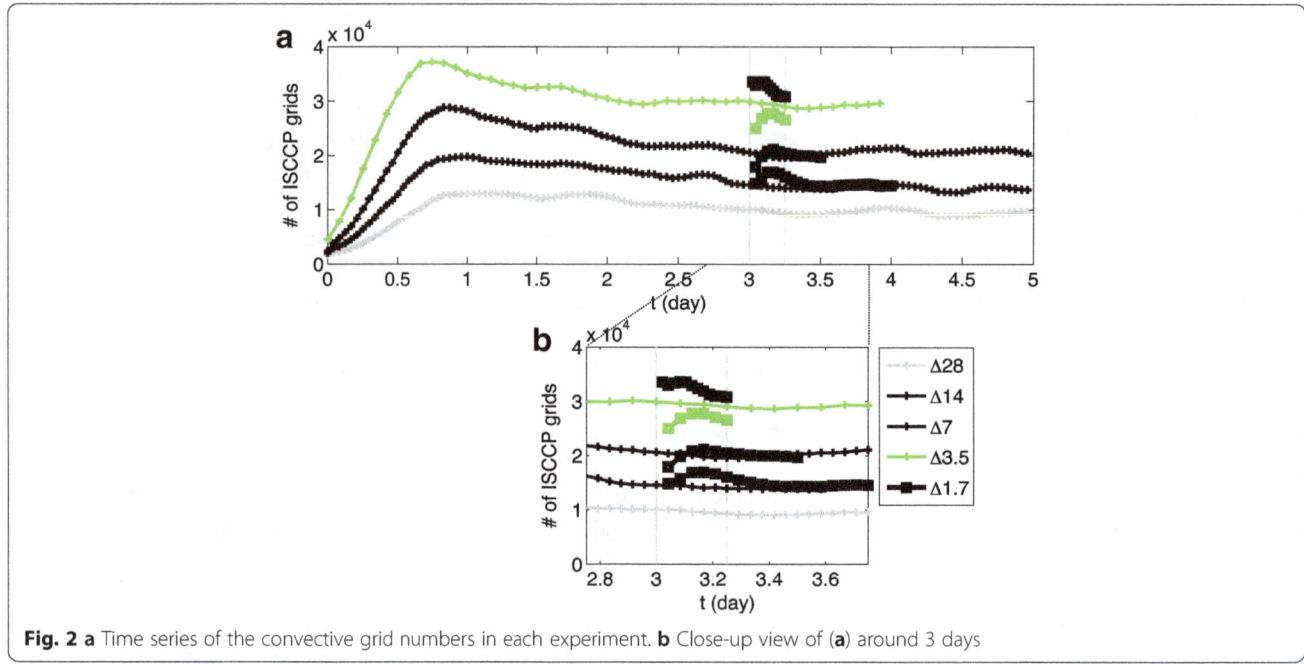

Fig. 2 a Time series of the convective grid numbers in each experiment. **b** Close-up view of (**a**) around 3 days

convections. Hence, the spin-up can be ignored in the data snapshot that was used.

Since we focus on the resolution dependence of the convection properties in each cloudy disturbance, we extracted the area of MJOs, TCs, mid-latitudinal low disturbances (MDL), and fronts (FRT), as in MY15. Although the method is the same as that used in MY15, we review it here to aid the following discussion. The MJO area is defined as the grid where the reconstructed OLR with first and second modes of boreal summer intraseasonal oscillation index (Kikuchi et al. 2012) was less than –10 W m^{-2}. To extract the TCs, we applied the Miyamoto et al. (2014) methodology and defined the area inside the 600 km radius from the center as the TC area. To detect the MDL centers and FRT, we first reconstructed the simulation data with coarsened 2.5° × 2.5° grid resolution. Then, we picked the MDL centers as grids at which the SLP was 5 hPa less than the areal average in the 10° radius. Finally, the MDL area was defined as inside of 1000 km from the MDL center. To extract the FRT region, we first applied the thermal frontal locator (Renard and Clarke 1965). This represented a third-order differential of the equivalent potential temperature at the 1500 m level in the horizontal direction. We detected the grid where the thermal frontal locator was >10–13 in the 2.5° × 2.5° data as a potential FRT grid. If the FRT grids lined up continuously with maximum distance greater than 10°, we defined the area as the FRT line. Finally, the FRT area was defined as the area inside the 200 km distance from the FRT line. In this study, we merge the MDL and FRT areas as the mid-latitude disturbance (MLD) area.

Figure 3 shows the horizontal distribution of the clouds in Δ14, Δ3.5, and Δ0.87, with the close-up region over the western North Pacific in the same experiment. The large-scale structure of the clouds, including cloud clusters, tropical cyclones, and mid-latitude disturbances, is almost unchanged among the grid-refinement experiments. The convection reasonably represents the global aspect in higher resolution. By comparison, among three simulations focusing on the specific area (Fig. 3d–f), it is remarkable that the Δ0.87 provides a more detailed description of the smaller-scale cloud structure. These available datasets from the grid-refinement experiments enabled us to statistically investigate the resolution dependence of the convection properties.

Results and discussion

Resolution dependence in the global field

Before we discuss the characteristics of deep convection, we need to confirm the consistency of the background environment for the simulated convections among the different grid-spacing simulations. Figure 4 shows the global mean of OLR, precipitation, zonal velocity, and vertical mass flux in all of the simulation results, with their global standard deviations. Global-mean OLR slightly increases on finer-resolution simulation. However, no meteorological variables showed significant change; they were almost constant among the different grid-spacing simulations. It is of interest that the global standard deviation of precipitation and vertical mass flux had a decreasing trend from Δ14 to Δ3.5 and that this trend is not clear in Δ1.7 and Δ0.87. On the other hand,

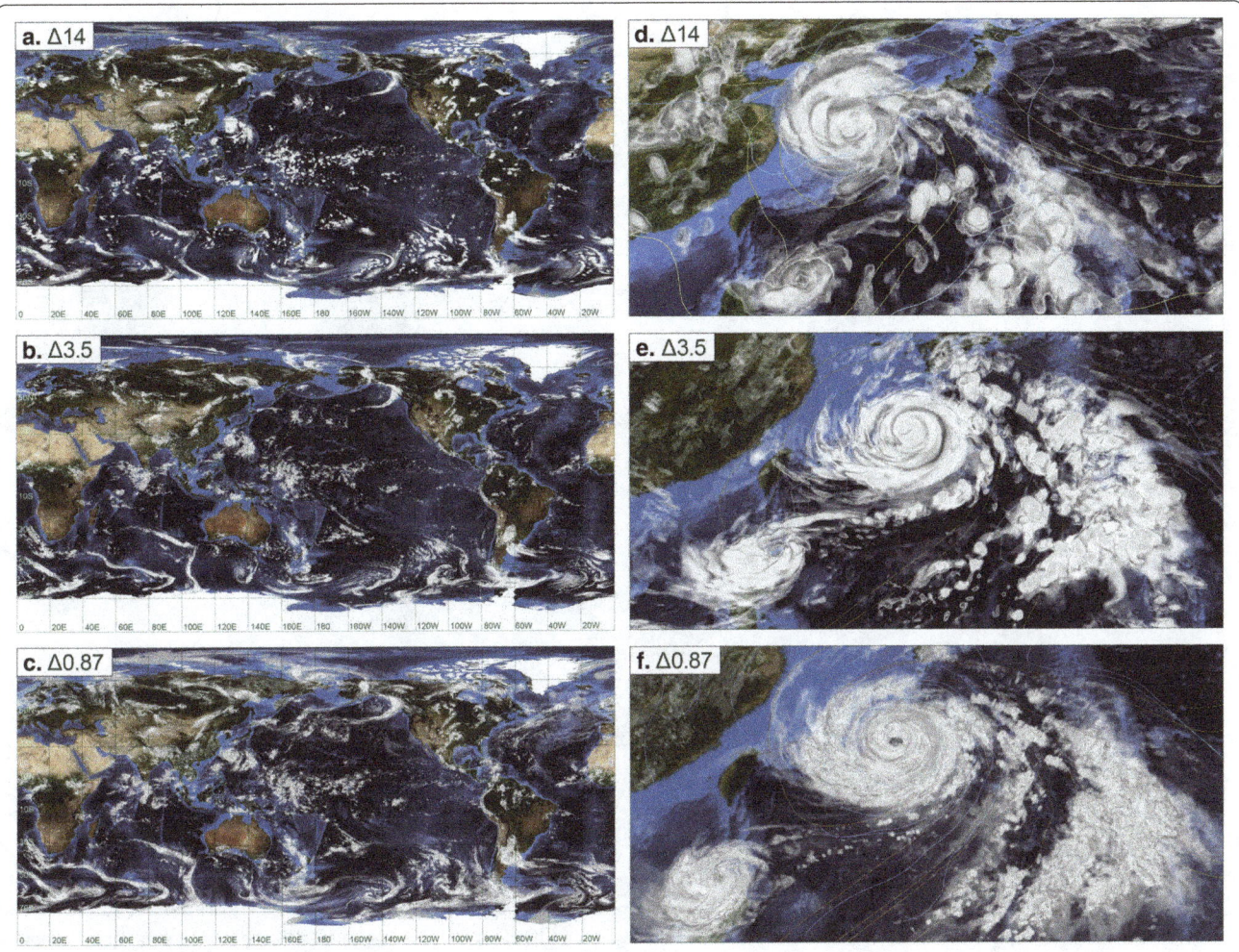

Fig. 3 Global view of cloud distribution simulated by (**a**) Δ14 km, (**b**) Δ3.5 km, and (**c**) Δ0.87 km (**d–f**) focusing over the western North Pacific region of (**a–c**)

the standard deviation of zonal wind does not show such a trend. When compared to existing observation data, the global-mean OLR in the simulations is around 240–250 W m^{-2}, which is relatively larger than the observed OLR at around 235 W m^{-2}, [obtained from daily-mean-interpolated OLR data from the National Oceanic and Atmospheric Association satellite (Liebmann and Smith 1996)]. The global-mean precipitation of around 0.15 mm h^{-1} was relatively larger than the daily-mean precipitation on August 25, 2012 [provided by the Global Precipitation Climatology Project (Adler et al. 2003) (around 0.109)]. Since these variables in Fig. 4 are based not on a specific time average but on a snapshot, slight differences are conceivable. The potential reason for no significant change is that the integration time of each simulation was 3.5 days in total, which is not enough time to respond to the interactions among different-scale atmospheric phenomena.

Figure 5 shows the zonal-mean OLR and the number of detected convections in each simulation. Zonal-mean

OLR has variability among the horizontal grid spacing. This variability is large over the mid-high latitudes (especially 30 N–90 N) and small over the tropics and arid regions (30 S–10 S) in the southern hemisphere (Fig. 5a). Since we used snapshot data for this analysis, the location of the simulated convection and cloudy disturbances is different in each simulation, especially the synoptic disturbances over the mid-high latitudes, which causes the relatively large variability. Overall, zonal-mean OLR in Δ0.87 is higher than in other resolution experiments, especially over the mid-high latitudes. Zonal-mean OLR in each simulation is consistent with observation, except for the area between 30 S and 10 S. The OLR in the simulation is about 30 W m^{-2} larger than observations in the peak. This strongly affects the positive bias of the global-mean OLR in Fig. 4a.

How is the relatively larger OLR in Δ0.87 and Δ1.7 produced? MY13 pointed out that convection is resolved in simulations with finer than 2 km grid spacing. This is linked with the OLR difference. However, we also need

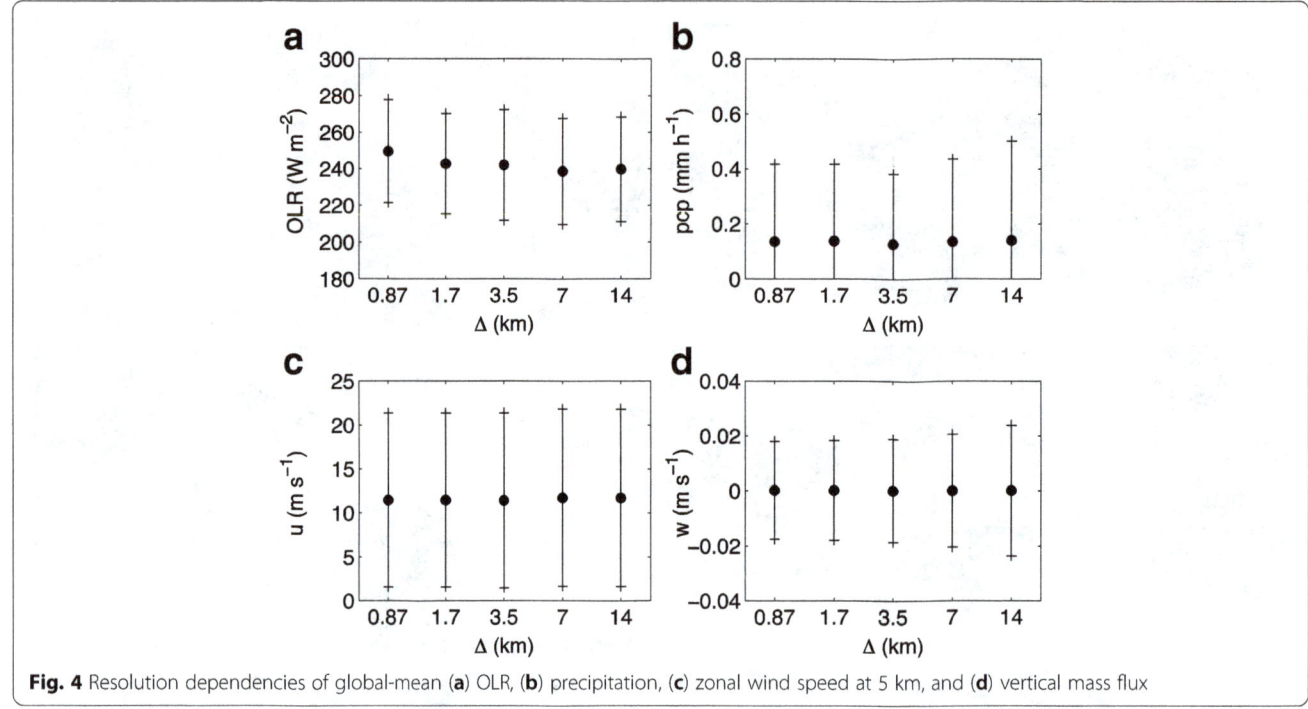

Fig. 4 Resolution dependencies of global-mean (**a**) OLR, (**b**) precipitation, (**c**) zonal wind speed at 5 km, and (**d**) vertical mass flux

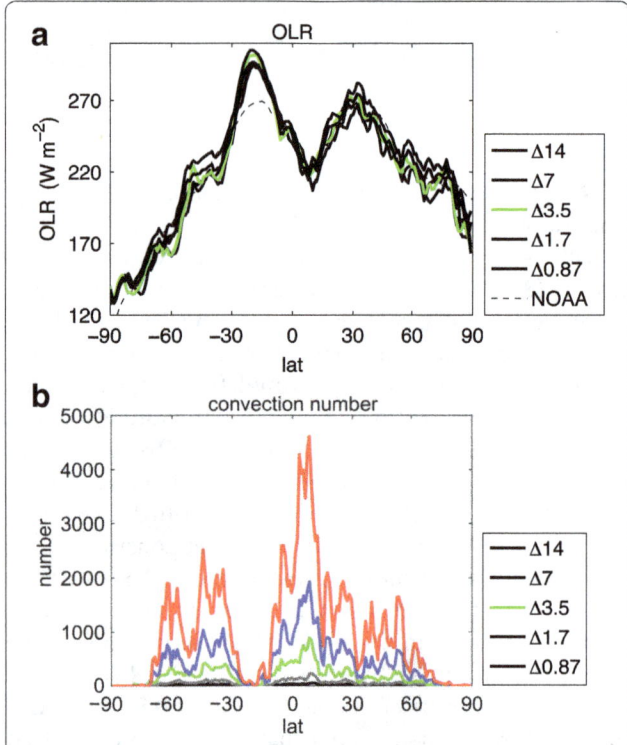

Fig. 5 Resolution dependencies of zonal-mean (**a**) OLR and (**b**) the number of detected convection cores

to investigate the resolution dependence of the other clouds in the simulation. Figure 6 shows the area of each ISCCP cloud type over the globe, defined by cloud top pressure and cloud optical thickness (Rossow and Schiffer 1999), in the grid-refinement experiments. The area of deep convection that we focused on in this study is abruptly decreased between $\Delta 3.5$ and $\Delta 1.7$, whereas the area of clear sky is increased with the same grid-spacing change. This suggests that the simulations in $\Delta 1.7$ and $\Delta 0.87$ express the proper size of the convection core with the multiple grid points and distinguish the non-cloudy sky around the convection as well. It is of interest that the areas of cirrus also increased from $\Delta 14$ and $\Delta 0.87$, implying that the simulations with higher spatial resolution are reasonably able to express the anvil associated with deep convections. According to the resolution dependence of the cloud area, increasing the zonal-mean OLR over the mid-high latitudes in the finer-resolution simulation can be explained by the increase of the area of clear sky. On the other hand, in the tropics, the zonal-mean OLR may be canceled out between the increased area of clear sky and that of cirrus with deeper convection. Further analysis is needed to quantitatively discuss the cause of resolution dependence in the other types of clouds.

The area of low cloud (e.g., stratocumulus) gradually decreases with higher resolution, especially from $\Delta 14$ and $\Delta 3.5$. Two possible reasons of this low-cloud trend are suggested. The low-cloud cells are likely to be resolved with finer resolution as well as deep convection, decreasing the low-cloud area and increasing the clear-sky area. The upward motion in the low-cloud core may

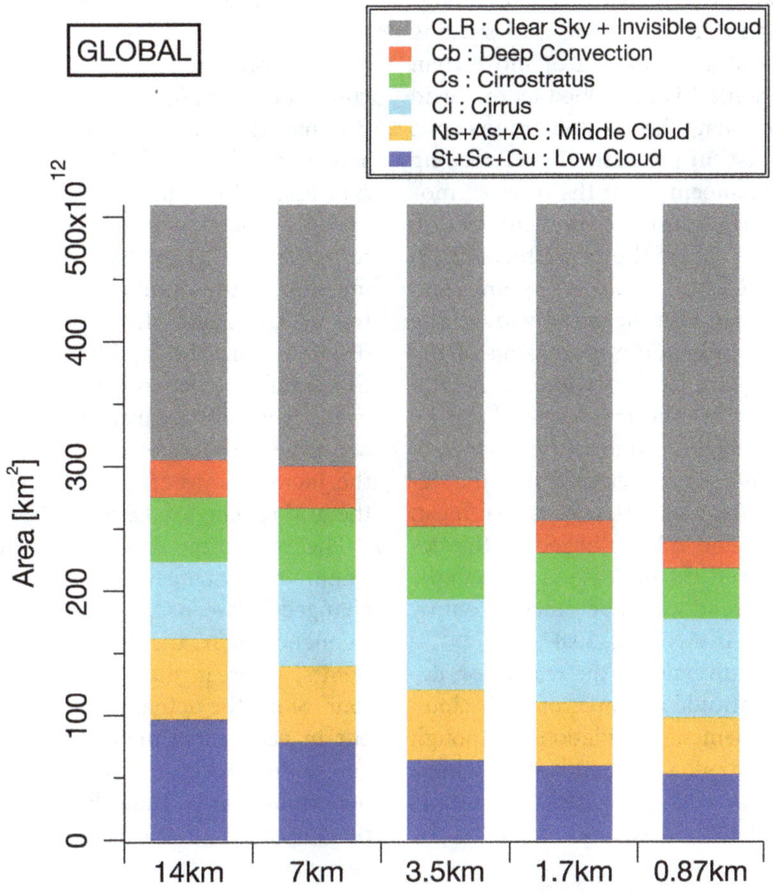

GLOBAL

Legend:
- CLR : Clear Sky + Invisible Cloud
- Cb : Deep Convection
- Cs : Cirrostratus
- Ci : Cirrus
- Ns+As+Ac : Middle Cloud
- St+Sc+Cu : Low Cloud

Area [km²] vs resolution: 14km, 7km, 3.5km, 1.7km, 0.87km

Fig. 6 Resolution dependence of total area of clear sky, deep convection, cirrostratus, cirrus, middle cloud, and low cloud, averaged over the globe

be accelerated with resolving the convection core. The faster upward motion in this model provides faster conversion from cloud to rain (Tomita 2008; Sato et al. 2015). This would reflect the decrease of the low-cloud area. On the other hand, drastically increasing the horizontal and vertical resolutions with the large-eddy simulation (LES) technique gives better representation of stratocumulus (e.g., Sato et al. 2015). Their results and our results suggest that some parameterization or tuning of microphysics is still needed in the current global modeling stage.

The area of deep convection over the globe has significantly decreased, between $\Delta 3.5$ and $\Delta 1.7$, which is consistent with the change of the convection number and the distance to the nearest convection core, as pointed out in MY13. The active convection is often accompanied by rainfall. In Fig. 7, we examine the resolution dependence of the precipitation rate at the grid of the convection core and show the frequency of the precipitation. The vertical axis denotes the logarithmic axis of the frequency. The precipitation over the convection core in $\Delta 14$, $\Delta 7.0$, and $\Delta 3.5$ is confined to less than 20 mm h⁻¹. Strong precipitation is shown in $\Delta 1.7$

(around 50 mm h⁻¹) and $\Delta 0.87$ (more than 100 mm h⁻¹). The ratio of strong precipitation, which is in excess of 20 mm h⁻¹, is drastically increased from $\Delta 3.5$ to $\Delta 1.7$. This is consistent with the idea that convection is resolved using multiple grid points in simulations and enhancement of the upward motion with the grid

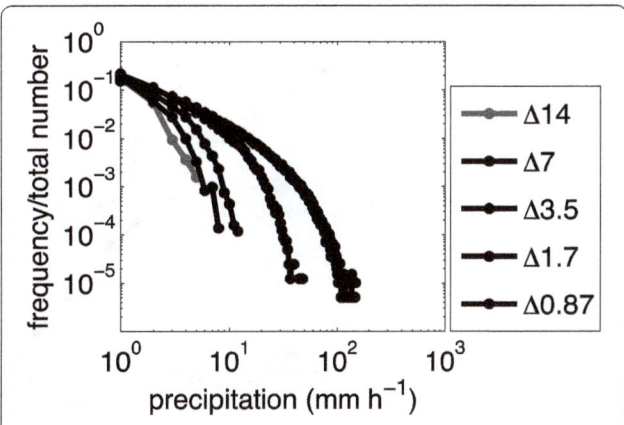

Fig. 7 Probability function of precipitation at the grid of the simulated convection core at each horizontal resolution

spacing of less than 2 km. The global-mean precipitation is almost constant in the different grid-spacing simulations. In the case of grid spacing of more than 2 km, convection with rainfall is described as one grid for the larger area, and then, the rainfall intensity is relatively weak. In contrast, in the case of grid spacing of less than 2 km, the enhancement of the upward motion provides strong precipitation at the convection core grid. In addition, we suggest the two effects of (1) resolving the convection with multiple grid points and (2) enhancement of the convection activity over the tropics. Since the former induce the increasing of the area of clear sky and OLR and the latter induce the decrease of OLR, the OLR over the tropics tends to be less variable among the simulations due to the balance of the two effects (Fig. 5a). We also speculated that the decreasing trend in the standard deviation of global-mean precipitation between Δ14 and Δ3.5 is linked to the decrease of total cloud area and the increase of the sample grid number, whereas this trend would be canceled out by the increasing heavy rainfall in Δ1.7 and Δ0.87.

In short, we confirm the diversity of the resolution dependence of OLR in the latitude and area of each cloud type among the grid-refinement simulations, although the global mean is consistent among the simulations. We also found that the convection resolved by multiple grid points in Δ1.7 and Δ0.87 was accompanied by stronger rainfall.

Resolution dependence on convection properties

First, we briefly introduce the results in MY13 and describe the resolution dependence of the simulated convection properties averaging over the globe. MY13 found that the convection core is resolved with multiple grid points when the horizontal grid space is less than 2 km. Figure 8 shows the resolution dependence of (a) the number of convections and (b) grid distance to the nearest convection core over the globe. The number of convection core in Δ3.5 and Δ7.0 is about four times larger than the number in Δ7.0 and Δ14; hence, those are shown on the $\log\Delta^4$ line starting from the number in Δ14 (dashed line in Fig. 8a). This indicates that the simulated convection in double resolution is simply an interpolated result of the original. However, the rate of increase in the number of convection core is decreased from Δ3.5 to Δ1.7. In other words, the number tends to distance from the $\log\Delta^4$ line in Δ1.7. If the number of the simulated convection core reaches the appropriate value, it is constant regardless of the grid spacing. Clear convergence was not confirmed even in Δ0.87; however, the increase rate of the convection is changed between the grid spacing of 3.5 and 1.7 km.

The resolution dependence in the histogram of the minimum distance between convection cores is also changed between Δ3.5 and Δ1.7 (Fig. 8b). The peak of frequency in Δ14, Δ7.0, and Δ3.5 appears in four grind points, whereas that in Δ1.7 and Δ0.87 is larger than four. Since the actual length of four grid points is different in each experiment, we suggest that the simulated convection distance in coarser resolution experiments is determined not physically but numerically; it may depend on the numerical discretization method. Hence, we speculated that the realistic distance between convection cores appears to be larger than the effective resolution in Δ1.7 and Δ0.87.

To investigate the horizontal diversity of the above features as a global mean, we have further analyzed the resolution dependence of the simulated convection properties, considering the perspective of the land-ocean difference (4.1), the latitude difference (4.2), and the

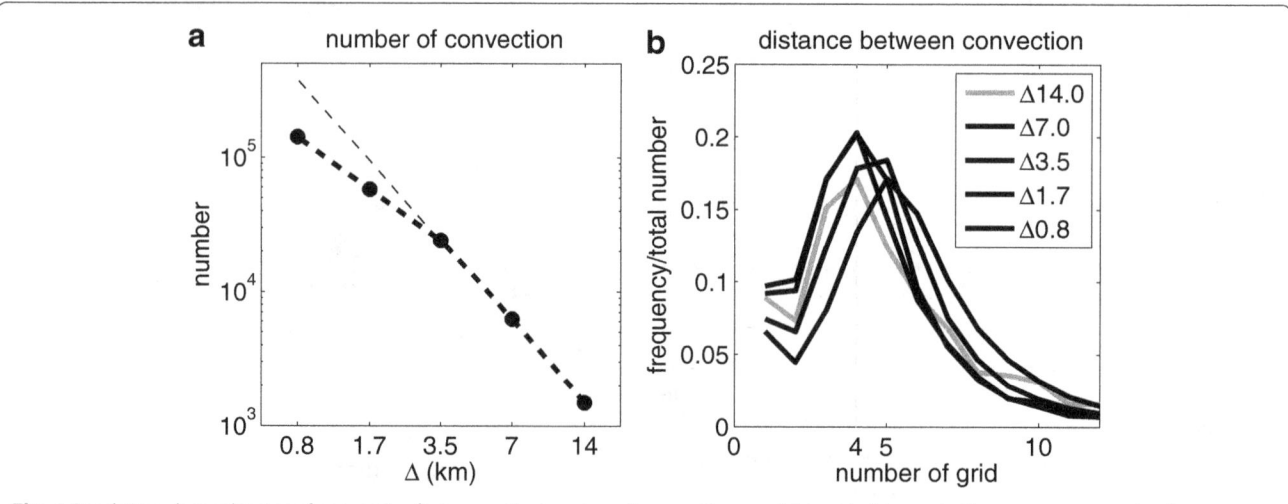

Fig. 8 Resolution dependencies of convective features: **a** the number of convections and (**b**) grid distance to the nearest convective feature. The *thin dashed line* in (**a**) indicates a log Δ^4 crossing at the point of Δ14 as a reference (Miyamoto et al 2013)

difference in cloudy disturbances (4.3), and we picked up the convections over the other specific region (4.4) in the following sections.

Land and ocean difference

Different surface conditions are one of the potential factors for characterizing convection. We investigated convection properties over land and ocean separately. Figure 9 shows the number of convections and grid distance to the nearest convection core in each experiment as similar as Fig. 8 with vertical velocity of composited convections based on the detected convection cores in Δ0.87 over land and ocean, respectively. The number of convections over ocean is larger than those over land because many deep convection cores are found over the tropical ocean. The number of convections drastically changes between Δ3.5 and Δ1.7, over both land and ocean. The grid distance between convections also changes at around the 2 km grid spacing; the peak of frequency tends to be larger from four grids. This is consistent with the global mean (Fig. 8). The vertical velocity of the convection core is slightly larger in the convection over land than that over ocean. In addition, a decrease of the increasing rate of the convection number occurred in the simulation of Δ3.5 over land. However, both convection properties over land and ocean are very consistent with the global-mean feature (Fig. 8), and in a qualitative sense, little significant difference in the convection feature over land and over ocean is found in this study.

Latitudinal difference

The location of the convection, especially in latitude, should be connected to the thermal condition affecting the property of convection. To examine the latitude difference of convection properties and its resolution dependence, we divided the convection over the globe into four groups based on the gaps in its zonal-mean distribution (Fig. 5): (1) 70 S–30 S, (2) 15 S–15 N, (3) 15 N–30 N, and (4) 30 N–70 N. Figure 10 shows the change in the number of detected convections and grid distance to the nearest convection core with the horizontal grid space in each region. Significant difference between the tropics and the mid-latitude region is found. The increasing rate of the convection number in the mid-latitude begins to decrease between Δ3.5 and Δ1.7.

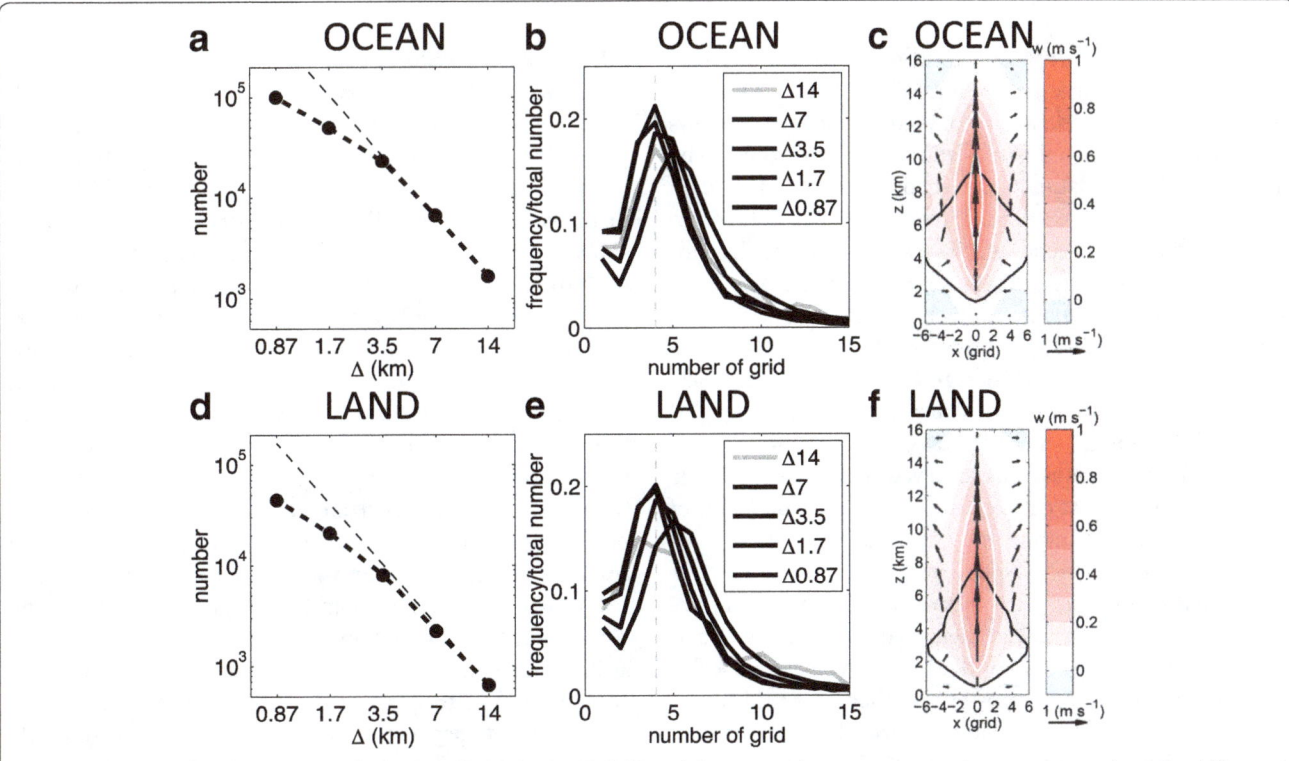

Fig. 9 Resolution dependence of (**a**) the number of convections and (**b**) grid distance to the nearest convection core detected over land. The *thin dashed line* in (**a**) denotes a log Δ⁴ based on the value of Δ14 as a reference. **c** Radius-height cross section for composites of vertical velocity (w: *shaded*) and velocity vector of radial and vertical velocity (vector) for the simulated convection core detected over ocean in Δ0.87. The *white contour* denotes 0.3 and 0.6 ms⁻¹, and the *black contour* indicates the mixing ratio of hydrometeors weighted by air density with an interval of 0.5 g m⁻³. **d–f** The same as (**a–c**) but for the convection detected over land

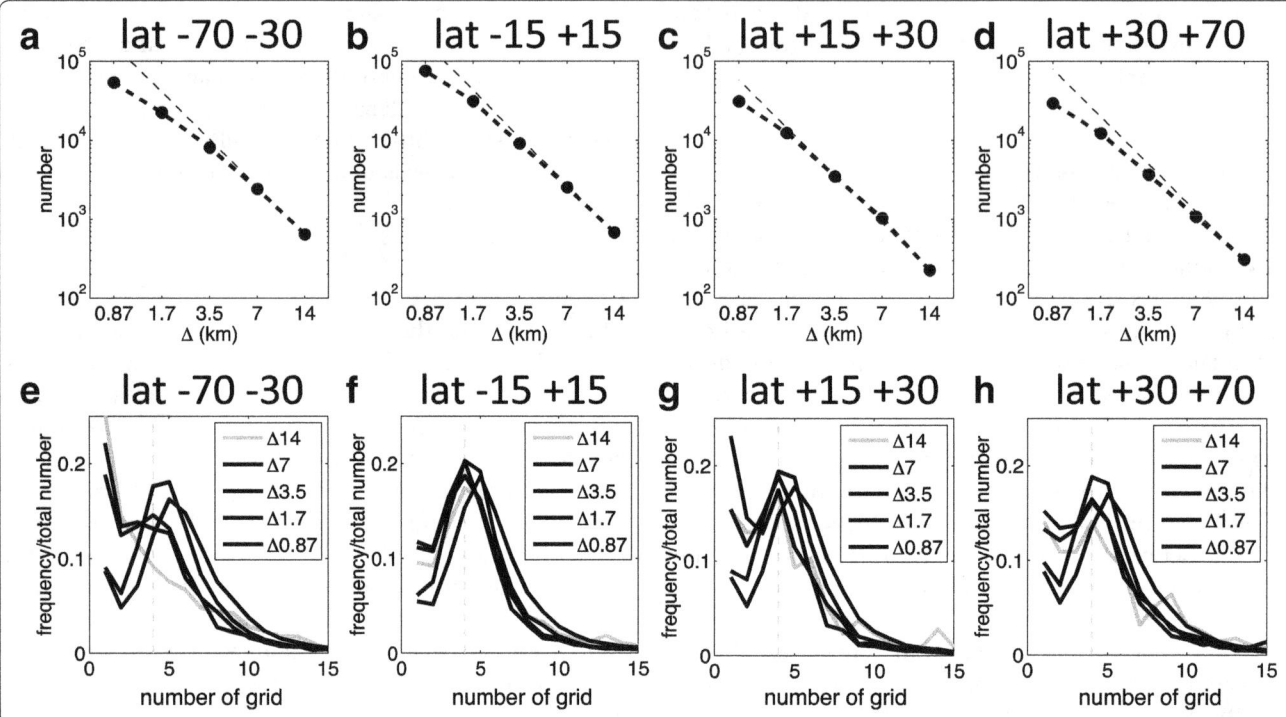

Fig. 10 Resolution dependence of (**a**) the number of convections and (**e**) grid distance to the nearest convection core detected over the latitudinal area between 70 S and 30 S. Figure structure is the same as Fig. 8. The same as (**a**) and (**e**) but for the area between 15 S and 15 N (**b**, **f**); 15 N and 30 N (**c**, **g**); and 30 N and 70 N (**d**, **h**)

On the other hand, the convection number over the tropics tends to converge between Δ0.87 and Δ1.7.

The grid distance between convection cores also shows the difference between the mid-latitude and the tropics. The distance of the mid-latitude convection becomes larger than four grids in Δ1.7 and Δ0.87. Interestingly, the distance to the mid-latitude convection in the coarse resolution has another peak around one grid. This indicates the convections occur close to one another. When the convection core is resolved with multiple grid points in Δ1.7 and Δ0.87, the number of convections with one grid distance is dramatically decreased and the number of convections with larger grid distance is increased as well. This also supports that convections in the mid-latitudes are resolved and have a realistic distance to the nearest convection cores in the simulation of less than 2 km. In contrast, the grid distance between convections over the tropics is distributed widely, with the peak at four to five grids. Since the increasing ratio of the convection number does not change significantly around Δ1.7 and Δ3.5, the grid distance also does not change drastically. However, the distance becomes larger than four grid points in Δ0.87. Meanwhile, the change of grid distance between convection cores can be seen from Δ1.7 to Δ0.87 in the whole area. It also supports that most of the convections would be resolved with multiple grid points in Δ0.87. In short, the resolution dependence of the convection shows different trends in the tropics and the mid-latitude area.

Different cloudy disturbances

MY15 showed the diversity of convection properties in various disturbances in Δ0.87. In this study, we investigated the resolution dependence of the convection number in cloudy disturbances: MJO, TC, and mid-latitude disturbances, as in MY15. Each disturbance region is shown in Fig. 11. In this study, we also add the category of "other," which is defined as all convections that are not located in any of the above cloudy disturbances. Figure 12 shows the resolution dependence of the number of detected convections in each cloudy disturbance, "other," and global accumulation. The global accumulation is the same as in Fig. 8. The trend of the number of convections does not converge at about 2 km except for the convections categorized as "other" in this study. The number of convections in each cloudy disturbance in Δ0.87 is about 10^4, while those categorized as "other" is about 10^5, which is similar to the global accumulation. Therefore, the convection categorized as "other" makes a large contribution to the global aspects, while the contribution of the convection in the cloudy disturbance to the global aspect is limited. The difference in the convection numbers between global accumulation and each cloudy disturbance clearly shows that the convergent trend is weaker in MJO and TC. This would affect the latitudinal difference in the "Latitudinal difference" section. It is important that the trend of the convection

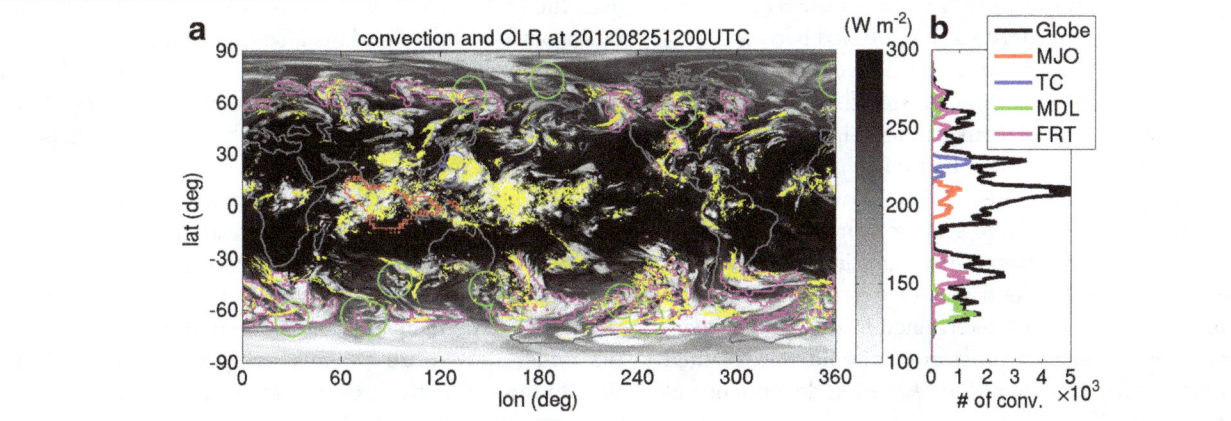

Fig. 11 a Horizontal distribution of the extracted convection core in Δ0.87 (*yellow*) and the defined area of MJO (*red*), TC (*blue*), MDL (*green*), and FRT (*magenta*) disturbances, respectively. **b** Latitudinal profile of the number of the extracted convective grid points in each disturbance in Δ0.87 (Miyamoto et al., 2015)

number in the organized cloudy disturbance has less convergence around 2 km but tends to converge from Δ1.7 to Δ0.87 at least. This supports that most of the organized convection in the tropics would be resolved in Δ0.87.

Specific area

We also checked the convection properties and resolution dependence in the specific area where the convection is active. We selected the three regions: (1) mid-Pacific (150 E–150 W, 15 S–15 N), (2) Maritime Continent (90 E–150 E, 15 S–10 N), and (3) Tibetan Plateau (70 E–100 E, 20 N–35 N), based on the lower OLR in the surrounding areas and the large number of detected convections (about 10^4) (see Fig. 11). These three areas can be categorized as the tropics or Asian monsoon region. The essential change of the convection properties starts between Δ1.7 and Δ0.87 over the mid-Pacific and Maritime Continent

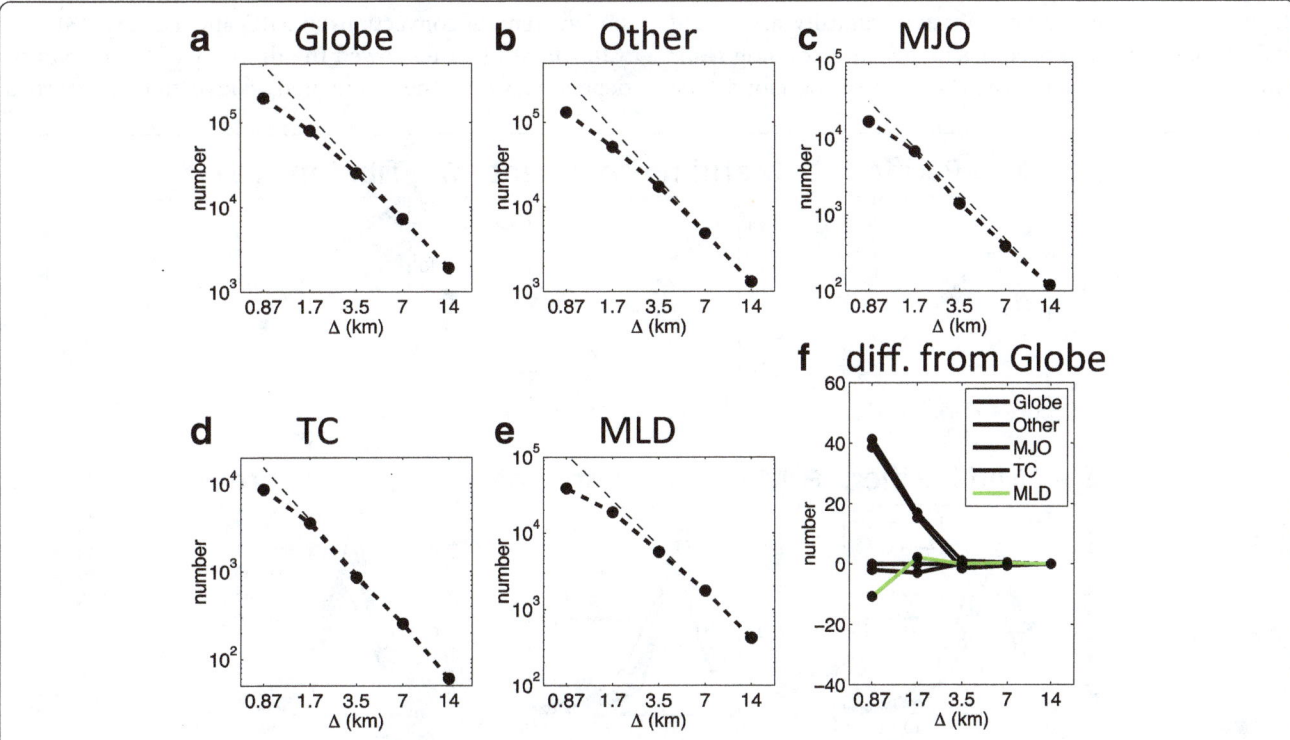

Fig. 12 Resolution dependence of the number of detected convections in (**a**) the globe, (**b**) other, (**c**) MJO, (**d**) TC, (**e**) MLD (MDL and FRT), and (**f**) difference from globe after notarizing based on the convection number of Δ14 in each disturbance. The *thin dashed line* in (**a–e**) denotes a log Δ^4 based on the value of each Δ14 as a reference

(Fig. 13). The number of the convection distances around one to two grids is also drastically decreased between Δ3.5 and Δ1.7. On the other hand, the convections over the Tibetan Plateau do not show a significant convergent trend for the number of convections. Since the convections over the Tibetan Plateau and mid-Pacific seem to be in the organized cloud system (Fig. 11), it is implied that convections forced by large-scale disturbances tend to have a less significant change of properties, even in Δ1.7. Although the detailed cloud system over the mid-Pacific and Tibetan Plateau is not determined in this study, environmental conditions over those areas should be explained for a better understanding of the resolution dependence of the convection.

Conclusions

We comprehensively investigated the simulated convection properties and global-mean field by focusing on their resolution dependence, based on grid-refinement experiments from 14 to 0.87 km grid spacing by using NICAM. The convergence trend for the number of convections is confirmed to occur between Δ3.5 and Δ1.7 (Miyamoto et al. 2013). The global mean of vertical mass flux, precipitation, and zonal wind at 5 km are conserved in different resolution simulations (Fig. 4). Global-mean OLR is slightly increased in Δ1.7 and Δ0.87, and this trend is more remarkable over the mid-latitude area (Fig. 5). Global-mean precipitation has no resolution dependence, while the precipitation intensity associated with deep convection becomes higher in increasing resolutions (Fig. 7). Interestingly, the ratio of the cloud type

over the globe is different between simulations. The area of low and middle clouds is also decreased with increasing the resolution as well as deep convection, and those of clear sky are increased particularly in Δ1.7 and Δ0.87 (Fig. 6). These differences reflect the resolution dependence of zonal-mean OLR (Fig. 5).

MY13 pointed out that the essential change of the number of convection cores and distance to the nearest convective core occurred around the 2 km grid spacing as a global mean (Fig. 8). We further investigated the resolution dependence of the simulated convection from the wider and more various perspectives than MY13. We found that the results in MY13 were different between latitudinal regions and cloudy disturbances (Figs. 10 and 12), although the trend of the convection number and grid distance between convection cores is not significantly different between over land and ocean (Fig. 9). The convergence trend for the number of convections in the mid-latitude area is more predominant than that in the tropics (Fig. 10). The essential change of the convection properties around the 2 km grid spacing is not clear, even in Δ1.7, for the convections in the cloudy disturbances categorized by MY15. In contrast, the convection properties that are not detected in the categorized cloudy disturbance change drastically between Δ3.5 and Δ1.7, which is similar to global accumulations (Fig. 12). Moreover, the convections in the cloud cluster over the mid-Pacific and Maritime Continent show a similar trend of convections in MJO and TC (Fig. 13).

In this study, we showed the diversity of the resolution dependency of the simulated convection properties.

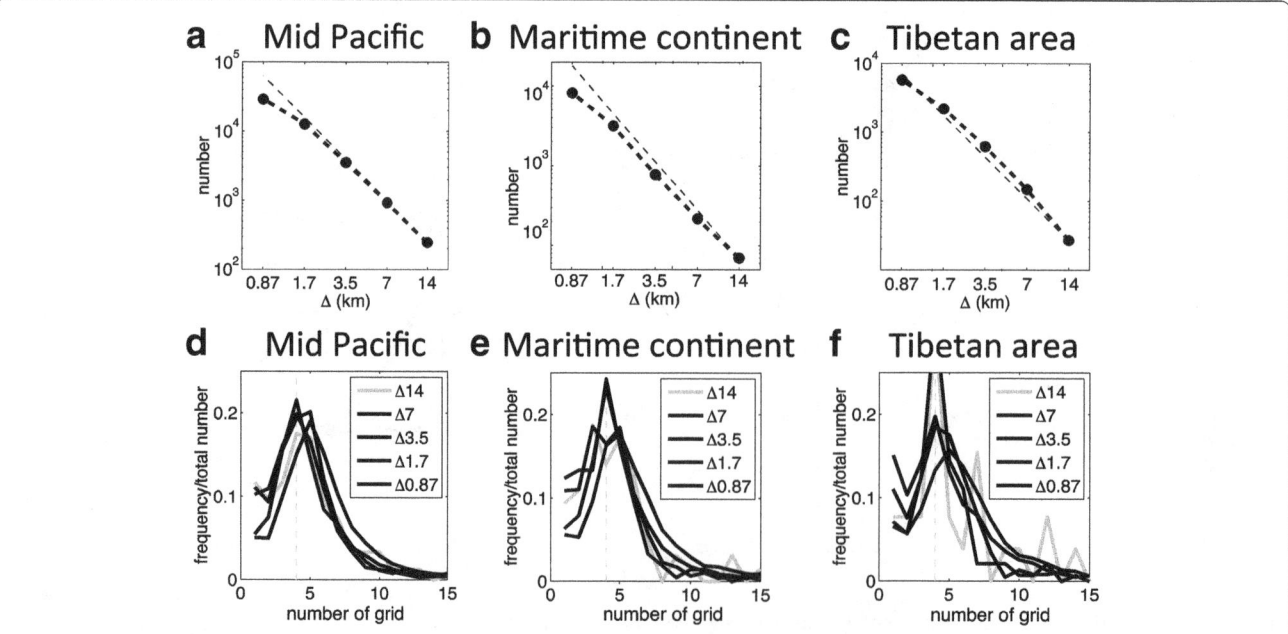

Fig. 13 Resolution dependence of (**a**) the number of convections over the mid-Pacific. **b** The same as (**a**) but for over the Maritime Continent. **c** The same as (**a**) but for over the Tibetan area. **d–f** Grid distance to the nearest convection core detected. The same area as (**a–c**)

Convections over the mid-latitudes and convections detected other than in the cloudy disturbances show the essential change of properties more clearly around the 2 km grid spacing, largely contributing the global mean. We speculate that this difference is related to the strength of forcing under the tropical cloudy disturbance. In fact, larger CAPE in the MJO area and larger low-level convergence in the TC area are confirmed in MY15. It is also speculated that the convection naturally arises as many as possible in such area under the strong forcing. Therefore, the convections are tightly packed in the cloudy disturbance, the size of convection core is relatively smaller than other categorized convections, and the convection core is not resolved by multiple grid points, even in $\Delta 1.7$. In contrast, where the environmental atmosphere allows convection to occur freely, the simulated convection, likely the isolated convective cloud, is relatively larger than that in tropical cloudy disturbances and it can be resolved by multiple grid points in $\Delta 1.7$ and $\Delta 0.87$. Since this difference in the relationship between cloudy disturbances appears to not link with the surface condition, it does not affect the convection property difference between convection over land and ocean.

We found a difference in resolution dependency in the simulated convection property. It is important that the convections, even in cloudy disturbances, show a convergent trend for the number and are resolved not by a single grid but by multiple grid points between $\Delta 1.7$ and $\Delta 0.87$, at least, despite the existence of the above difference. This is a noteworthy aspect for a

series of grid-refinement experiments. It recalls the further high spatial resolution for better simulations of tropical convections. This would bring a better understanding of tropical cloudy disturbances, based on the hierarchical structure of convections. Hence, longer time integration of the global atmosphere in the 0.87 km grid spacing in the future will provide significant discussion about the interaction between convections and cloudy disturbances.

Appendix

In the highest resolution run of $\Delta 0.87$, we achieved the greatest computational performance for weak scaling on the K computer. Figure 14 from Terai et al. (2014) shows a computational performance and scalability of NICAM on the K computer. At this scaling test, we increased the number of grids while increasing the number of nodes. We achieved 10 % of the peak performance using five nodes (40 cores) with 56 km of horizontal grid spacing. The performance was maintained to 8 % with 81,920 nodes (655,360 cores) and 0.44 km of horizontal grid spacing. Elapsed time per step was increased ~15 % from 80 nodes ($\Delta 14$ run) to 20,480 nodes ($\Delta 0.87$ run) due to the imbalance of the number of computations. This load imbalance is mainly related to computation in cloud microphysics and the spatial inhomogeneity of cloud distribution. In this study, the run of $\Delta 0.87$ with 20,480 nodes was conducted as the highest resolution of the production run, by taking the total computational time and resources into account.

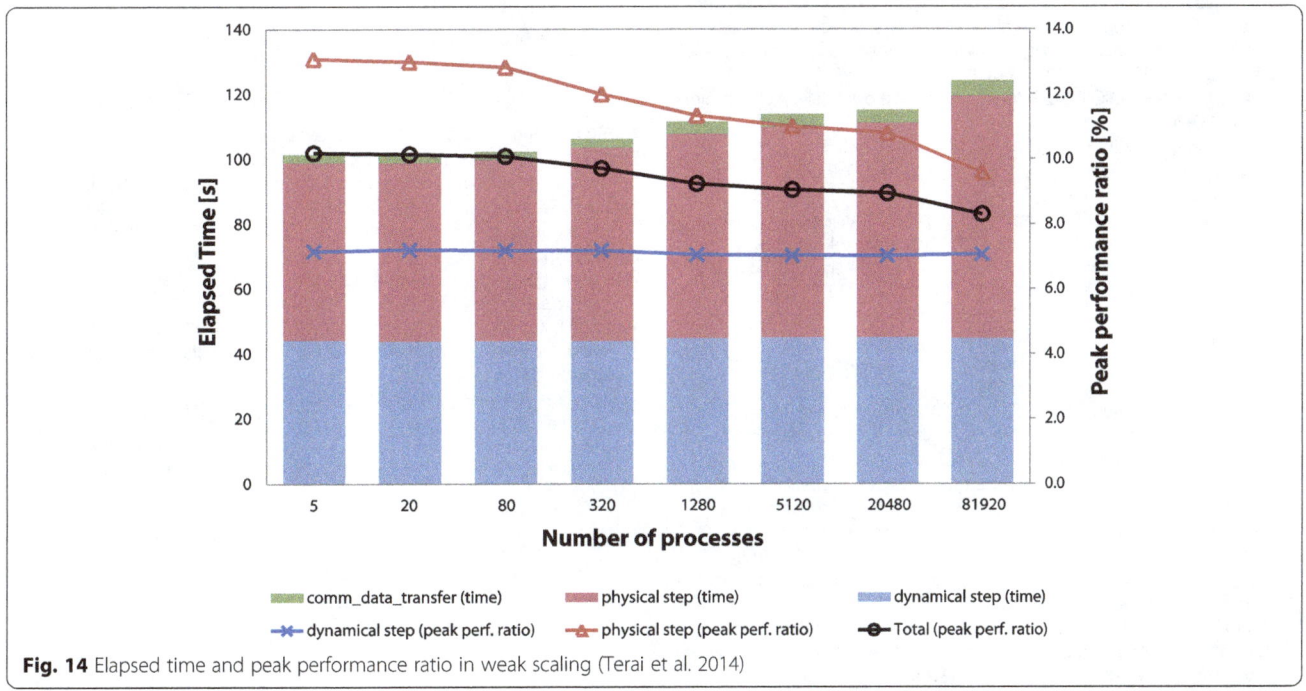

Fig. 14 Elapsed time and peak performance ratio in weak scaling (Terai et al. 2014)

Acknowledgements

The authors are grateful to the editor of the Progress in Earth and Planetary Science and anonymous reviewers. Special thanks are due to Drs. S. Nishizawa Y. Sato, and S. Iga in the RIKEN Advanced Institute for Computer Science (AICS) for the valuable comments. The simulations were performed using the K computer at the RIKEN AICS under the supported by Strategic Programs for Innovative Research (SPIRE) Field 3 (Projection of Planet Earth Variations for Mitigating Natural Disasters).

Authors' contributions

YK conceived the idea, organized this study, and wrote the manuscript. YM carried out the simulations, analyzed the simulated results, and collaborated with YK to draft the manuscript. RY and TY supported to conduct the simulation and analysis with arranging the necessary environment on the K computer. HY led the implementation of NICAM on the K computer. HT coordinated the project and collaborated with YK in the construction of the manuscript. All co-authors provided guidance for the analysis and commented on the manuscript. All authors read and approved the final manuscript.

Competing interests

The authors declare that they have no competing interests.

Author details

[1]RIKEN Advanced Institute for Computational Science, 7-1-26 Minatojima-minami-machi, Chuo-ku, Kobe, Hyogo 650-0047, Japan. [2]Japan Agency for Marine-Earth Science and Technology, 2-15, Natsushima-cho, Yokosuka, Kanagawa 237-0061, Japan.

References

Adler RF et al (2003) The version-2 global precipitation climatology project (GPCP) monthly precipitation analysis (1979-present). J Hydrometeorol 4:1147–1167

Arakawa A, Schubert WH (1974) Interaction of a cumulus cloud ensemble with large-scale environment. Part 1. J Atmos Sci 31:674–701

Emanuel KA, Raymond DJ (1993) The Representation of cumulus convection in numerical modelsAmerican Meteorological Society, Boston, Mass, p 246

Fudeyasu H, Wang YQ, Satoh M, Nasuno T, Miura H, Yanase W (2008) Global cloud-system-resolving model NICAM successfully simulated the lifecycles of two real tropical cyclones. Geophys Res Lett 35: doi: 10.1029/2008gl036003.

Gill AE, (1982) Atmosphere-ocean dynamics. Academic Press, New York, p 662

Kain JS, Fritsch JM (1990) A one-dimensional entraining/detraining plume model and its application in convective parameterization. J Atmos Sci 47:2784–2802

Kajikawa Y, Yamaura T, Tomita H, Satoh M (2015) Impact of tropical disturbance on the Indian summer monsoon onset simulated by a global cloud-system-resolving model. SOLA 11:80–84. doi:10.2151/sola.2015-020

Kalnay E et al (1996) The NCEP/NCAR 40-year reanalysis project. Bull Amer Meteorol Soc 77:437–471

Kikuchi K, Wang B, Kajikawa Y (2012) Bimodal representation of the tropical intraseasonal oscillation. Clim Dynam 38:1989–2000. doi:10.1007/s00382-011-1159-1

Kodama C et al (2015) A 20-year climatology of a NICAM AMIP-type simulation. J Meteorol Soc Jpn 93:393–424. doi:10.2151/jmsj.2015-024

Liebmann B, Smith CA (1996) Description of a complete (interpolated) outgoing longwave radiation dataset. Bull Amer Meteorol Soc 77:1275–1277

Louis JF (1979) A parametric model of vertical eddy fluxes in the atmosphere. Bound-Lay Meteorol 17:187–202

Mapes BE, Houze RA (1993) Cloud clusters and superclusters over the oceanic warm pool. Mon Weather Rev 121:1398–1415

Miura H, Satoh M, Nasuno T, Noda AT, Oouchi K (2007) A Madden-Julian oscillation event realistically simulated by a global cloud-resolving model. Science 318:1763–1765. doi:10.1126/science.1148443

Miyakawa, T., and Coauthors (2014) Madden-Julian oscillation prediction skill of a new-generation global model demonstrated using a supercomputer. Nat Commun 5, doi: 10.1038/Ncomms4769

Miyamoto Y, Kajikawa Y, Yoshida R, Yamaura T, Yashiro H, Tomita H (2013) Deep moist atmospheric convection in a subkilometer global simulation. Geophys Res Lett 40:4922–4926. doi:10.1002/Grl.50944

Miyamoto Y, Satoh M, Tomita H, Oouchi K, Yamada Y, Kodama C, Kinter J (2014) Gradient wind balance in tropical cyclones in high-resolution global experiments. Mon Weather Rev 142:1908–1926. doi:10.1175/Mwr-D-13-00115.1

Miyamoto Y, Yoshida R, Yamaura T, Yashiro H, Tomita H, Kajikawa Y (2015) Does convection vary in different cloud disturbances? Atmos Sci Lett 16:305–309

Mizuta R et al (2006) 20-km-mesh global climate simulations using JMA-GSM model—mean climate states. J Meteorol Soc Jpn 84:165–185

Mrowiec AA, and Coauthors (2012) Analysis of cloud-resolving simulations of a tropical mesoscale convective system observed during TWP-ICE: vertical fluxes and draft properties in convective and stratiform regions. J Geophys Res-Atmos 117 doi: 10.1029/2012jd017759

Nakano M, M Sawada, T Nasuno, M Satoh (2015) Intraseasonal variability and tropical cyclogenesis in the western North Pacific simulated by a global nonhydrostatic atmospheric model, 2014GL062479, 10.1002/2014gl062479

Nakazawa T (1988) Tropical super clusters within intraseasonal variations over the western Pacific. J Meteorol Soc Jpn 66:823–839

Noda AT, Oouchi K, Satoh M, Tomita H, Iga S, Tsushima Y (2010) Importance of the subgrid-scale turbulent moist process: cloud distribution in global cloud-resolving simulations. Atmos Res 96:208–217

Randall D, Khairoutdinov M, Arakawa A, Grabowski W (2003) Breaking the cloud parameterization deadlock. Bull Amer Meteorol Soc 84:1547–1564. doi:10.1175/Bams-84-11-1547

Renard RJ, Clarke LC (1965) Experiments in numerical objective frontal analysis. Mon Weather Rev 93:547–556

Rossow WB, Schiffer RA (1999) Advances in understanding clouds from ISCCP. Bull Amer Meteorol Soc 80:2261–2287

Saito K et al (2006) The operational JMA nonhydrostatic mesoscale model. Mon Weather Rev 134:1266–1298

Sato Y, S Nishizawa, H Yashiro, Y Miyamoto, Y Kajikawa, H Tomita (2015) Component-level intercomparison: proof of concept using the case of shallow cumulus simulation, 2, 10.1186/s40645-015-0053-6.

Satoh, M., and Coauthors (2014) The non-hydrostatic icosahedral atmospheric model: description and development., 1, doi:10.1186/s40645-014-0018-1

Satoh M, Matsuno T, Tomita H, Miura H, Nasuno T, Iga S (2008) Nonhydrostatic icosahedral atmospheric model (NICAM) for global cloud resolving simulations. J Comput Phys 227:3486–3514. doi:10.1016/j.jcp.2007.02.006

Sekiguchi M, Nakajima T (2008) A k-distribution-based radiation code and its computational optimization for an atmospheric general circulation model. J Quant Spectrosc Radiat Trans 109:2779–2793

Simmons AJ, Burridge DM, Jarraud M, Girard C, Wergen W (1989) The ECMWF medium-range prediction models development of the numerical formulations and the impact of increased resolution. Meteorol Atmos Phys 40:28–60

Stephens GL et al (2010) Dreary state of precipitation in global models. J J Geophys Res-Atmos 115:D24211. doi:10.1029/2010JD014532

Stevens B, Bony S (2013) What are climate models missing? Science 340:1053–1054. doi:10.1126/science.1237554

Terai M, H Yashiro, K Sakamoto, S Iga, H Tomita, M Satoh, K Minami (2014) Performance optimization and evaluation of a global climate application using a 440m horizontal mesh on the K computer., Abstract from International Conference for High Performance Computing, Networking, Storage and Analysis, SC14, New Orleans, United States

Tomita H (2008) New microphysical schemes with five and six categories by diagnostic generation of cloud ice. J Meteorol Soc Jpn 86a: 121-142

Tomita H, Satoh M (2004) A new dynamical framework of nonhydrostatic global model using the icosahedral grid. Fluid Dyn Res 34:357–400. doi:10.1016/j.fluiddyn.2004.03.003

Webster PJ (1972) Response of tropical atmosphere to local, steady forcing. Mon Weather Re 100:518

Yamaura T, Kajikawa Y, Tomita H, Satoh M (2013) Possible impact of a tropical cyclone on the northward migration of the Baiu frontal zone. SOLA 9:89–93. doi:10.2151/sola.2013-020

Yoshimura H, Mizuta R, Murakami H (2015) A spectral cumulus parameterization scheme interpolating between two convective updrafts with semi-lagrangian calculation of transport by compensatory subsidence. Mon Weather Rev 143:597–621. doi:10.1175/Mwr-D-14-00068.1

5

Impacts of cloud microphysics on trade wind cumulus: which cloud microphysics processes contribute to the diversity in a large eddy simulation?

Yousuke Sato[*], Seiya Nishizawa, Hisashi Yashiro, Yoshiaki Miyamoto, Yoshiyuki Kajikawa and Hirofumi Tomita

Abstract

This study investigated the impact of several cloud microphysical schemes on the trade wind cumulus in the large eddy simulation model. To highlight the differences due to the cloud microphysical component, we developed a fully compressible large eddy simulation model, which excluded the implicit scheme and approximations as much as possible. The three microphysical schemes, the one-moment bulk, two-moment bulk, and spectral bin schemes were used for sensitivity experiments in which the other components were fixed. Our new large eddy simulation model using a spectral bin scheme successfully reproduced trade wind cumuli, and reliable model performance was confirmed. Results of the sensitivity experiments indicated that precipitation simulated by the one-moment bulk scheme started earlier, and its total amount was larger than that of the other models. By contrast, precipitation simulated by the two-moment scheme started late, and its total amount was small. These results support those of a previous study. The analyses revealed that the expression of two processes, (1) the generation of cloud particles and (2) the conversion from small droplets to raindrops, were crucial to the results. The fast conversion from cloud to rain and the large amount of newly generated cloud particles at the cloud base led to evaporative cooling and subsequent stabilization in the sub-cloud layer. The latent heat released at higher layers by the condensation of cloud particles resulted in the development of the boundary layer top height.

Keywords: Large eddy simulation; Shallow clouds

Background

The effect of clouds is one of the most uncertain factors in climate projection and numerical weather prediction. Shallow clouds (e.g., stratus, stratocumulus, shallow cumulus) play particularly important roles in the energy budget of the earth through radiation process because they cover a broad area of the earth (e.g., Randall et al. 1984). The 5th Intergovernmental Panel on Climate Change (IPCC) report suggested that the uncertainties with regards to shallow clouds should be reduced for reliable assessments (IPCC AR5, Stocker et al. 2013).

In global scale models (e.g., general circulation model (GCM)) and regional models with coarse grid spacing, shallow clouds are usually expressed by parameterizations

(e.g., Tiedtke 1993; Considine et al. 1997; Kain 2004), but these parameterizations have not been able to effectively simulate the shallow cloud cover observed from satellites (e.g., Chepfer et al. 2008; Naud et al. 2010).

To improve the expression of shallow cloud, the results of large eddy simulation (LES) models have been utilized. For example, Bretherton and Park (2009) used the results of an LES model to develop a new moist turbulent parameterization. Suzuki et al. (2004) and Posselt and Lohmann (2008) introduced an autoconversion parameterization that was typically used in an LES model (Khairoutdinov and Kogan 2000) into a global scale model. Many studies using LES models have been conducted to determine the characteristics of shallow cloud and improve large-scale modeling (e.g., Wang and Feingold 2009; Xue et al. 2008, Savic-Jovcic and Stevens 2008, Yamaguchi and Randall 2012). However, the

* Correspondence: yousuke.sato@riken.jp
RIKEN Advanced Institute for Computational Science, 7-1-26
Minatojima-Minami-machi, Chuo-ku, Kobe, Hyogo 650-0047, Japan

results of LES models are diverse, as indicated by several LES model intercomparison studies targeting shallow cloud (e.g., Stevens et al. 2005; Ackerman et al. 2009; van Zanten et al. 2011; Siebesma et al. 2003). It has been suggested that the difference in the microphysical schemes used in LES models is one of the reasons for the diversity in the results (Ackerman et al. 2009; van Zanten et al. 2011). However, it is difficult to investigate the exact effect of the different cloud microphysical schemes (i.e., the effect on the results that comes from changing the cloud microphysical scheme while keeping all other components fixed), because each LES model uses a different scheme, not only in cloud microphysics but also many other components (e.g., governing equation, turbulent scheme, advection scheme, and others). Thus, it is also difficult to determine which microphysical processes contribute most to the diversity of the LES results. Sensitivity experiments, which change only the cloud microphysical scheme, are required to better understand the exact effects of the cloud microphysical scheme.

The kinematic driver (KiD) model developed by Shipway and Hill (2012) enables us to conduct sensitivity simulations by changing only the cloud microphysics. Using KiD, we can consider the exact effects of the differences in cloud microphysical schemes. However, KiD ignores feedbacks to the atmosphere, even though the feedbacks of microphysics can affect the microphysical properties of shallow clouds and the turbulent structure of the boundary layer (e.g., Stevens et al. 1998; Wang et al. 2010). It is necessary to consider the feedback of microphysical processes on the dynamics when determining which process causes the diversity in the results of LES models. We should use the model that excludes approximation and implicit schemes as much as possible, because these features also affect the cloud microphysical properties simulated by the model.

This study developed a model that satisfies these requirements. Using the model, we attempted to reproduce the diversity in the results of the LES model used in van Zanten et al. (2011) and to determine impact of the various cloud microphysical schemes. Three types of cloud microphysical schemes (one-moment bulk, two-moment bulk, and spectral bin schemes) were considered in this study, because these three schemes have been used in previous intercomparison studies.

Of the several types of shallow cloud, we focused on trade wind cumulus because their variability in the results of LES models was larger than that of other types of shallow clouds. For example, the variability of surface precipitation in an intercomparison of trade wind cumulus (van Zanten et al. 2011) was larger than that in an intercomparison of stratocumulus (Ackerman et al. 2009). van Zanten et al. (2011) proposed that one of the reasons for the diversity in the microphysical properties

of trade wind cumuli was the different cloud microphysical models. They interpreted that a simple (one-moment bulk) microphysical scheme produced large amounts of precipitation (i.e., Table 3 of van Zanten et al. (2011)) and liquid water simulated by one-moment bulk schemes tended to be distributed in the lower layer. By contrast, the liquid water was located in the higher layer, and the precipitation flux was small in most of the two-moment schemes (i.e., Figure 6a of van Zanten et al. (2011)). In this study, we confirmed the validity of their interpretation through a simulation using our new fully compressible LES model and determined the main processes contributing to the diversity in the results of LES model intercomparison studies.

Methods

Experimental setup, dynamic framework, turbulence model, and external forcing

This section describes the common parts of the model with its experimental setup (i.e., the dynamic framework, turbulence model, and external forcing). The different parts of the microphysical schemes are highlighted in the subsequent section.

The dynamic model used in this study is an LES model that is included in the Scalable Computing for Advanced Library and Environment library (SCALE). Henceforth, we call this LES model SCALE-LES. The details of SCALE-LES are found at http://scale.aics.riken.jp/.

A fully compressible system is adopted for the governing equation of SCALE-LES. The prognostic variables are the three-dimensional momentum (ρu, ρv, ρw), total density (ρ), mass-weighted potential temperature ($\rho \theta$), and mass concentration of tracers (ρq_s), where q_s includes specific humidity, ratio of mass, and number concentration ratio of hydrometeors to total mass. Explicit time integration is used in all directions. Furthermore, the fourth-order central difference scheme is adopted for spatial discretization to avoid the numerically implicit diffusion that would be induced by odd-ordered difference schemes. The three-step Runge–Kutta scheme is adopted. To retain stability of the model, fourth-order superviscosity/diffusion is applied for all prognostic variables. Using SCALE-LES, developed as described above, we exclude the effects of approximation in the governing equation system and implicit diffusion as far as possible, enabling a consideration of the effects of the target component (i.e., cloud microphysics in this study).

To guarantee monotonicity, the flux-corrected transport (FCT) scheme (Zalesak 1979) is applied for all prognostic variables except for density. The effects of sub-grid scale turbulence are calculated using a Smagorinsky-type scheme based on Brown et al. (1994) and Scotti et al. (1993).

The experimental setup is almost the same as that of the previous Global Energy and Water Exchange project (GEWEX) Cloud System Study (GCSS) intercomparison of Rain in Cumulus over the Ocean (RICO) (van Zanten et al. 2011). The calculation domain covers 12.8×12.8 km^2 horizontally with a double periodic boundary condition and 4.0 km vertically. The horizontal and vertical grid intervals are 100 and 40 m, respectively. Rayleigh damping is applied for three-dimensional momentum over the upper atmosphere of $z > 3.5$ km. The strength of the numerical diffusion is set as 1.25×10^5 m^4 s^{-1}, which is determined by the sensitivity experiment (the results of the sensitivity experiment are shown in Appendix 1 of this paper). The simulations are conducted for 24 h with time steps (Δt) of 0.05, 0.15, and 0.5 s for dynamics, cloud microphysics, and other physics. The Δt for cloud microphysics is determined as the largest time step to avoid artificial noise, and the ratio of Δt for dynamics to Δt for other physics is set to 10 based on the sensitivity experiments (see Appendix 2 of this paper for details of the sensitivity experiment).

The external forcing of the radiation, the surface flux, and the large-scale horizontal advection are applied in the same way as in van Zanten et al. (2011). The forcing of the large-scale subsidence is applied for all prognostic variables including density, except for u and v. In nature, large-scale subsidence generates a divergence of total density, which makes the air mass flow out from the domain. However, it cannot flow out in the compressible model using the configuration of van Zanten et al. (2011) due to the periodicity in the lateral boundary condition. Although this problem can be ignored in the anelastic and Boussinesq systems due to the fixed density, it is necessary to consider this problem in the compressible system. Although no description is available for this problem in van Zanten et al. (2011), the density of each layer should be reduced according to the divergence. The density reduction rate and the equation system with large-scale forcing are given below.

Large-scale subsidence (w_{LS}) was given by van Zanten et al. (2011) as:

$$
w_{\mathrm{LS}} = \begin{cases} -\dfrac{0.005}{2260} z & (z < 2260\,\mathrm{m}) \\ -0.005 & (z \ge 2260\,\mathrm{m}) \end{cases}
$$

where z is the height. Instead of this formulation, we give the subsidence formulated directly to vertical momentum as:

$$
\rho w_{\mathrm{LS}} = \begin{cases} -\dfrac{0.005}{2260} z & (z < 2260\,\mathrm{m}) \\ -0.005 & (z \ge 2260\,\mathrm{m}) \end{cases}
$$

Consequently, the density reduction rate (D) is given as:

$$
D \equiv \frac{\partial(\rho w_{\mathrm{LS}})}{\partial z} = \begin{cases} -\dfrac{0.005}{2260} & (z < 2260\,\mathrm{m}) \\ 0 & (z \ge 2260\,\mathrm{m}) \end{cases}
$$

The continuous equation modified with the subsidence term is given by

$$
\frac{\partial \rho}{\partial t} + \frac{\partial(\rho u)}{\partial x} + \frac{\partial(\rho v)}{\partial y} + \frac{\partial(\rho(w + w_{\mathrm{LS}}))}{\partial z} = D \tag{1}
$$

The density reduction derived from the divergence by the large-scale subsidence is added in the right-hand side (rhs) of Eq. (1). Note that this equation is identical to the equation without large-scale subsidence. The Lagrangian conservation equation of the scalar quantities (ϕ) is given as:

$$
\rho \frac{\partial \phi}{\partial t} + \rho u \frac{\partial \phi}{\partial x} + \rho v \frac{\partial \phi}{\partial y} + \rho(w + w_{\mathrm{LS}}) \frac{\partial \phi}{\partial z} = 0 \tag{2}
$$

The prognostic variable of SCALE-LES is a mass-weighted value, and the equation of mass-weighted values is derived from Eqs. (1) and (2) as:

$$
\frac{\partial(\rho \phi)}{\partial t} + \frac{\partial(\rho u \phi)}{\partial x} + \frac{\partial(\rho v \phi)}{\partial y} + \frac{\partial(\rho(w + w_{\mathrm{LS}})\phi)}{\partial z} = D\phi \tag{3}
$$

As shown in the rhs of Eq. (3), the scalar quantities (ϕ) flow out from each layer of the system by subsidence. The equation for the vertical momentum is modified, as is that of the scalar quantities. The vertical flux $\rho w_{\mathrm{LS}}\phi$ at the top boundary can be determined so that such additional convergence of the flux is canceled with $D\phi$ at the top layer.

Three microphysical schemes

To reproduce the diversity in the results of the RICO study, three types of cloud microphysical scheme are used for this study: the one-moment bulk microphysical scheme (Tomita 2008), the two-moment bulk scheme (Seiki and Nakajima 2014), and the spectral bin microphysical scheme (Suzuki et al. 2010). The one-moment bulk scheme and the two-moment bulk scheme are based on Berry (1968) and Seifert and Beheng (2001), respectively. Both of the original bulk schemes were used in the RICO study.

The essential difference among the three schemes is their treatment of the size distribution of the number of hydrometeor particles. The one-moment bulk scheme expresses this value by the Marshall–Palmer distribution, with the assumption of a constant total number of particles. By this assumption, only the mass concentration is needed to determine the size distribution. Although the two-moment bulk scheme conceptually treats the size distribution almost the same way, it differs from the one-moment bulk scheme in the assumption about the type of size distribution function and in the process of its determination. The two-moment bulk scheme assumes the generalized gamma distribution as the size distribution function, and the size distribution itself is determined not only by the mass concentration but also by the number concentration. In this sense, the two-moment bulk scheme is more sophisticated than the one-moment bulk scheme. The spectral bin scheme is an intrinsically sophisticated method compared with the others. It explicitly predicts the size distribution. To compensate for the detailed expression of size distribution, the spectral bin scheme requires about four- to fivefold larger number of prognostic variables compared with the other two schemes.

Although the three microphysical schemes treat both warm and ice phase cloud, the ice phase was not calculated because the temperature at the model top is greater than 273.15 K. In this case, the categorization of hydrometeors for each scheme is as follows. The one-moment and two-moment bulk schemes address two types of hydrometeors: cloud droplet and raindrop for warm rain process. The spectral bin scheme addresses only a type of water drop, which covers the particle size of both cloud droplets and raindrops. For the spectral bin scheme, the radius of newly generated cloud particles by nucleation is set to 3 μm (Suzuki et al. 2006). The size distribution of hydrometeors is discretized to 33 bins; its configuration has been established by several previous studies (e.g., Khain and Sednev 1996; Iguchi et al. 2008). The center of mass of ith bin (m_i) is set by using the i-1-th bin (m_{i-1}) as $m_i = 1.874$ m_{i-1}. m_1 is the mass of cloud particles whose radius is 3 μm.

All three schemes consider the generation of cloud droplets, condensation, evaporation, and sedimentation of cloud hydrometeors. Although the two-moment bulk scheme also considers the breakup of cloud droplets, this difference is minor based on sensitivity experiments examining the breakup process (figure not shown). Sedimentation is calculated by the first-order upwind scheme for all three schemes.

Since the generation of new cloud droplets is one of the critical processes in this experiment, the difference in this process among the schemes should be noted. The mass of newly generated cloud droplets is calculated by saturation adjustment in the one-moment bulk scheme, which was also used in some one-moment bulk schemes in the RICO study. By contrast, it is calculated by the nucleation schemes in the two-moment bulk and spectral bin schemes. The number concentration of the cloud droplets $(N_{c,nucl})$ generated by the nucleation process is calculated as follows (e.g., Pruppacher and Klett 1997):

$$N_{c,nucl} = N_0 S_w{}^k, \tag{4}$$

where S_w is supersaturation over water. The constants N_0 and k are set as $N_0 = 100 \times 10^6$ m^{-3} and $k = 0.462$. This scheme was also used in several models used in the RICO study. The growth of cloud droplets into raindrops is another key issue for this experiment, as well as the underlying creation process. In the bulk scheme, this is expressed as autoconversion and accretion. To investigate the strength of the impact of the autoconversion and accretion processes, we conducted the same simulation with autoconversion and accretion ratios that were twice as large (0.067-fold smaller) as the two-moment (one-moment) scheme. The autoconversion rate (P_{auto}) in the one-moment bulk scheme is calculated as in Tomita (2008), which was based on Berry (1968):

$$P_{auto} = \frac{1}{\rho} \left[16.7 \times (\rho q_c)^2 \left(5 + \frac{3.6 \times 10^{-5} N_{c,T08}}{D_d \rho q_c} \right) \right]$$
$$[\text{kg kg}^{-1}\text{s}^{-1}], \tag{5}$$

where $D_d = 0.1456 - 5.964 \times 10^{-2} \log (N_{c,T08}/2000)$, q_c is the cloud water mixing ratio, and $N_{c,T08}$ is the number concentration of cloud droplets. In Tomita (2008), $N_{c,T08}$ was set as 50 cm^{-3}, but in this study, $N_{c,T08}$ was set as 70 cm^{-3} based on the experimental setup in the RICO study (van Zanten et al. 2011). Another autoconversion scheme from Khairoutdinov and Kogan (2000) was implemented into the one-moment bulk scheme because it had performed better than the scheme of Berry (1968) in a GCM (Suzuki et al. 2004). The scheme was also adopted in some models used in the RICO study. P_{auto} in the Khairoutdinov and Kogan (2000) scheme is calculated as:

$$P_{auto} = 1350 \times q_c^{2.47} \times N_{c,T08}^{-1.79} \quad [\text{kgkg}^{-1}\text{s}^{-1}] \tag{6}$$

Using the schemes shown above, we attempted to reproduce the diversity of the LES results in the RICO study.

Results and discussion
Basic performance of SCALE-LES
The validity of SCALE-LES was confirmed through comparison with a previous study (van Zanten et al. 2011). The results of the spectral bin scheme were regarded as a reference solution of SCALE-LES, because it is the most sophisticated scheme of the three. First, the results of the reference solution were compared with the previous study. As shown in Figs. 1 and 2, the results of our model (SCALE-LES) are within the range between the maximum and minimum of the intercomparison study in terms of temporal evolution and vertical profile

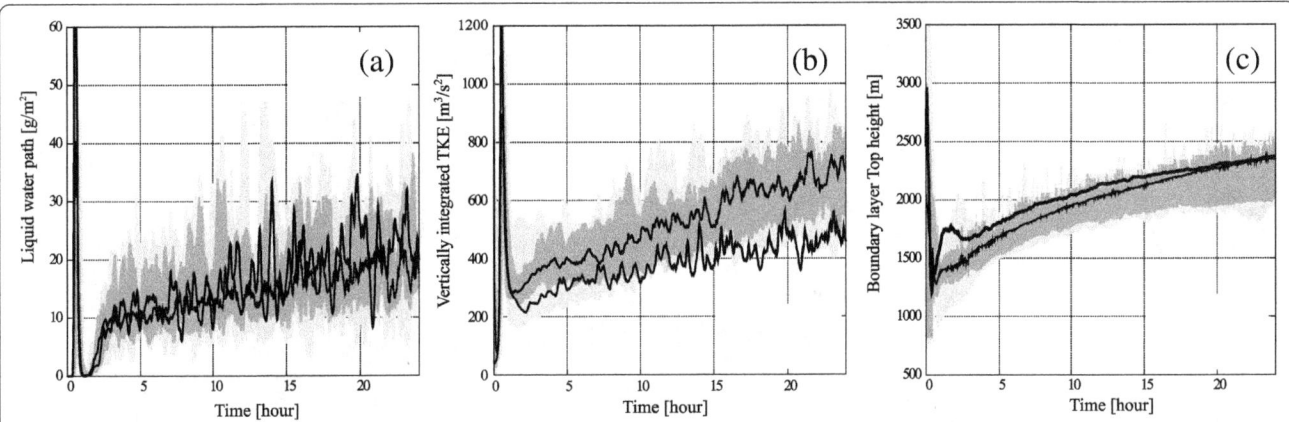

Fig. 1 Comparison of the time evolution between SCALE-LES and a previous intercomparison study. Time evolution of the **a** liquid water path, **b** vertically integrated turbulence kinetic energy, and **c** boundary layer top height averaged over the entire calculation domain simulated by (*black line*) SCALE-LES, with the spectral bin scheme. The *blue line*, *dark gray shading*, and *light gray shading* indicate the median, range between the first and third quartiles, and range between the maximum and minimum values, respectively, of the previous intercomparison study (van Zanten et al. 2011)

for several quantities. This indicates that our model could reproduce the shallow cumulus simulated by the LES models used in the previous study.

As well as the physical performance, the computational performance and the scalability of SCALE were investigated. The elapsed time for a time step (Δt) and the performance efficiency are shown in Table 1. The elapsed time and performance of SCALE-LES do not change when it is used with a large number of Message Passing Interface (MPI) processes (e.g., over 10,000 MPI processes). This indicates that SCALE-LES has excellent scalability and a reasonable performance in massive parallel computing.

Impacts of cloud microphysical scheme on simulation results

Second, we show the results of the same simulation (RICO) by using three microphysical schemes for investigating the impacts of the cloud microphysical scheme. Figure 3 indicates the differences among the three schemes. The precipitation flux simulated by the two-moment bulk scheme is small, and its peak value is distributed in the upper layer. By contrast, the precipitation flux by the one-moment bulk scheme is large and the peak value locates in the lower layer. The precipitation by the spectral bin scheme is between the other two (Fig. 3a). The trend is consistent with the previous study (Figure 6a of van Zanten et al. 2011).

The impacts of the different cloud microphysical schemes on the vertical distribution of the precipitation flux, the liquid water mixing ratio (q_l), and cloud fraction are clearly shown in Fig. 3a–c. The q_l in the one-moment bulk scheme is distributed in the lower layer (the peak value is represented at $z \sim 900$ m). Table 2 shows the surface precipitation averaged during the last

4 h for the three schemes. The precipitation amount in the one-moment bulk scheme is the largest, and the precipitation begins earliest among the three. The surface precipitation over 0.1 W m^{-2} starts at 1.8 h in the one-moment bulk scheme, at 10.05 h in the two-moment bulk scheme, and at 2.63 h in the spectral bin scheme. The liquid water simulated in the two-moment bulk scheme is located in the upper layer (the peak value is located at $z \sim 2400$ m), and only trace amounts of precipitation reach the surface (Table 2). The cloud fraction of the one-moment scheme is located in the lower layer and is small in the upper layer. On the other hand, the positive cloud fraction in the two-moment bulk scheme reaches a higher layer. The spectral bin scheme simulates an intermediate value between the other two, with the same trend as the RICO study. These results are consistent with the result of the previous study (van Zanten et al. 2011).

The large amount of precipitation in the one-moment bulk scheme carries a large amount of liquid water to the lower layer. As shown in Fig. 3d, the total water mixing ratio (q_t) of the lower layer (i.e., $z < 1500$ m) in the one-moment bulk scheme is larger than that in the others. Despite the difference in the vertical distribution of liquid water, the liquid water path (LWP) shows the same value (Fig. 3e). By contrast, the cloud coverage of the one-moment scheme is smaller than in the other scheme (Fig. 3f). This implies that the amount of cloud that extends horizontally at the top of boundary layer is small in the one-moment scheme, which is because of the liquid water removed from the cloud layer earlier by precipitation. This is also found in Fig. 3c, where the peak of the cloud fraction does not appear around the top of the boundary layer (i.e., $z \sim 2000$ m).

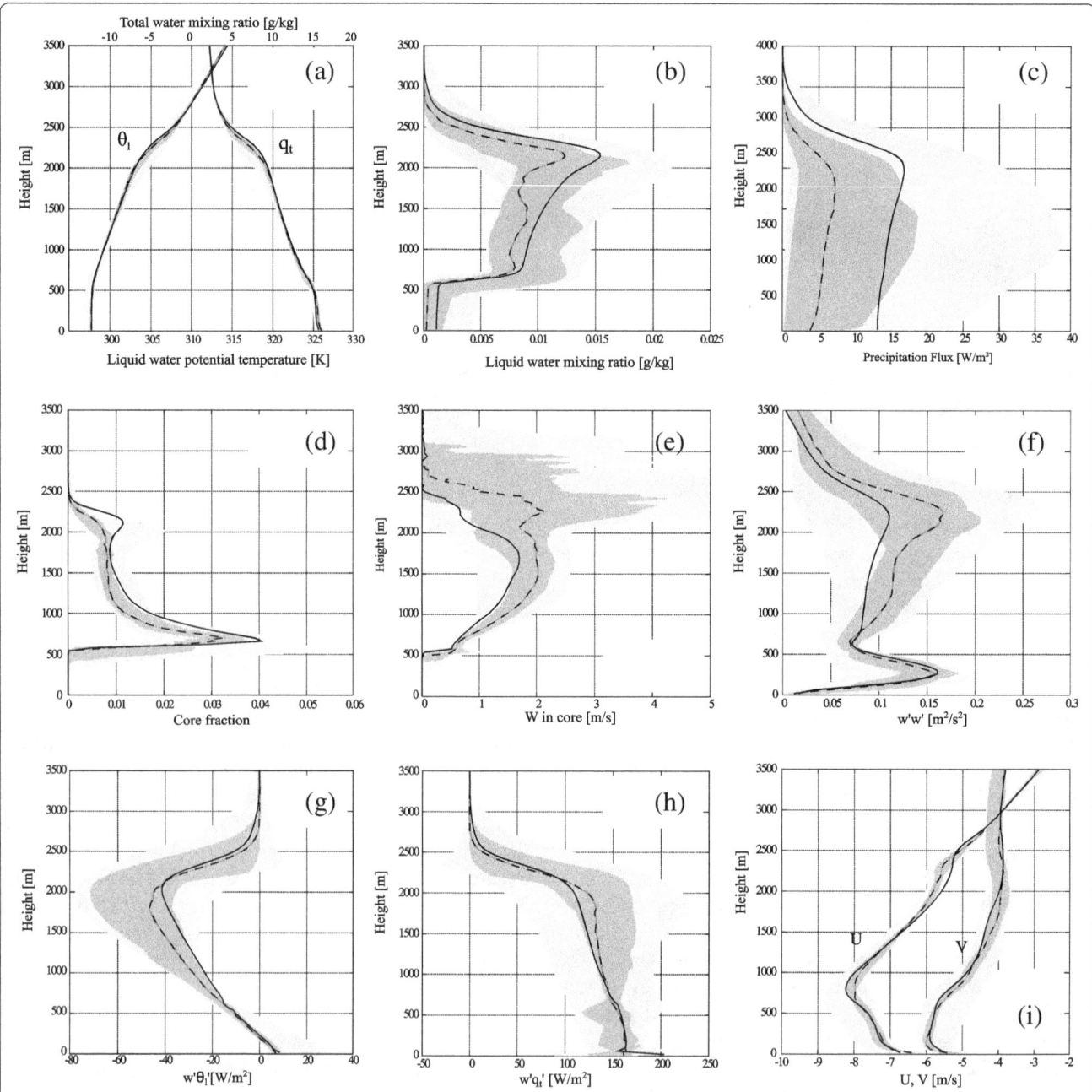

Fig. 2 Comparison of vertical profiles between SCALE-LES and a previous intercomparison study. Horizontally averaged profile of the **a** liquid water potential temperature (θ_l) and total water mixing ratio (q_t), **b** liquid water mixing ratio (q_l), **c** precipitation flux, **d** cloud fraction, **e** vertical velocity in cloud core, **f** variance of resolved w', **g** $w'\theta_l'$, **h** $w'q_t'$, and **i** horizontal wind velocity, averaged during the last 4 h. The *solid line* indicates the results of SCALE-LES. The *dashed line*, *heavy gray shading*, and *light gray shading* indicate the median, range between the first and third quartiles, and range between the maximum and minimum values, respectively, of the previous intercomparison study (van Zanten et al. 2011)

The impacts of cloud microphysics appear not only on the precipitation and vertical distribution of liquid water, but also the turbulent properties such as the turbulence kinetic energy (TKE), the variance in resolved vertical velocity (w'), and boundary layer top height. The boundary layer top height in the two-moment bulk scheme is the highest among the three schemes, whereas the one-moment bulk scheme simulates the lowest boundary layer height. This trend in boundary layer height continues to the end of the simulation time, as shown in Fig. 4a, and the difference among the schemes gradually becomes large. The variance in w' (Fig. 4c) and TKE (Fig. 4d) in the two-moment bulk scheme attributes a larger value to the upper layer. By contrast, those in the one-moment scheme are smaller in the upper layer. Vertically integrated TKE tends to be large and small in the

Table 1 Computational performance of SCALE-LES

Number of core	16	2048	8192	32,768	131,072	663,552[a]
(Number of MPI process)	(2)	(256)	(1024)	(4096)	(16,384)	(82,944)
Elapsed time (s step^{-1})	2.528	2.172	2.443	2.017	1.995	2.113
Performance efficiency (%)	5.5	6.3	5.6	6.8	6.9	6.5

Elapsed time (s step^{-1}) and performance efficiency (%) of SCALE-LES measured using the two-moment bulk microphysical scheme. The performance was measured through an experiment in which the number of grids for each MPI process was the same (weak scaling test) on the K computer
[a]This test used all nodes of the K computer

two-moment and the one-moment bulk schemes, respectively (Fig. 4b).

From the results shown above, we concluded that the turbulent properties of the boundary layer as well as the cloud microphysical properties of cumulus are significantly affected by the different microphysical schemes when other components are unchanged. These results

support the proposal of van Zanten et al. (2011) that the use of different cloud microphysical schemes is one of the main reasons for the diversity among LES models. We have confidence in this conclusion because direct effects other than the cloud microphysical schemes were excluded in our experiments.

Reasons for the differences among the results of the three schemes

The impacts of the cloud microphysical schemes are clearly indicated by the differences in the boundary layer top height, vertical distribution of liquid water, and the precipitation flux, as shown in the previous section. The reasons for these differences will be discussed in this section.

Since the impacts of each cloud microphysical scheme originate from the expression of the liquid water in each scheme, an examination of the tendency of q_l and potential temperature (θ) in the cloud microphysical process

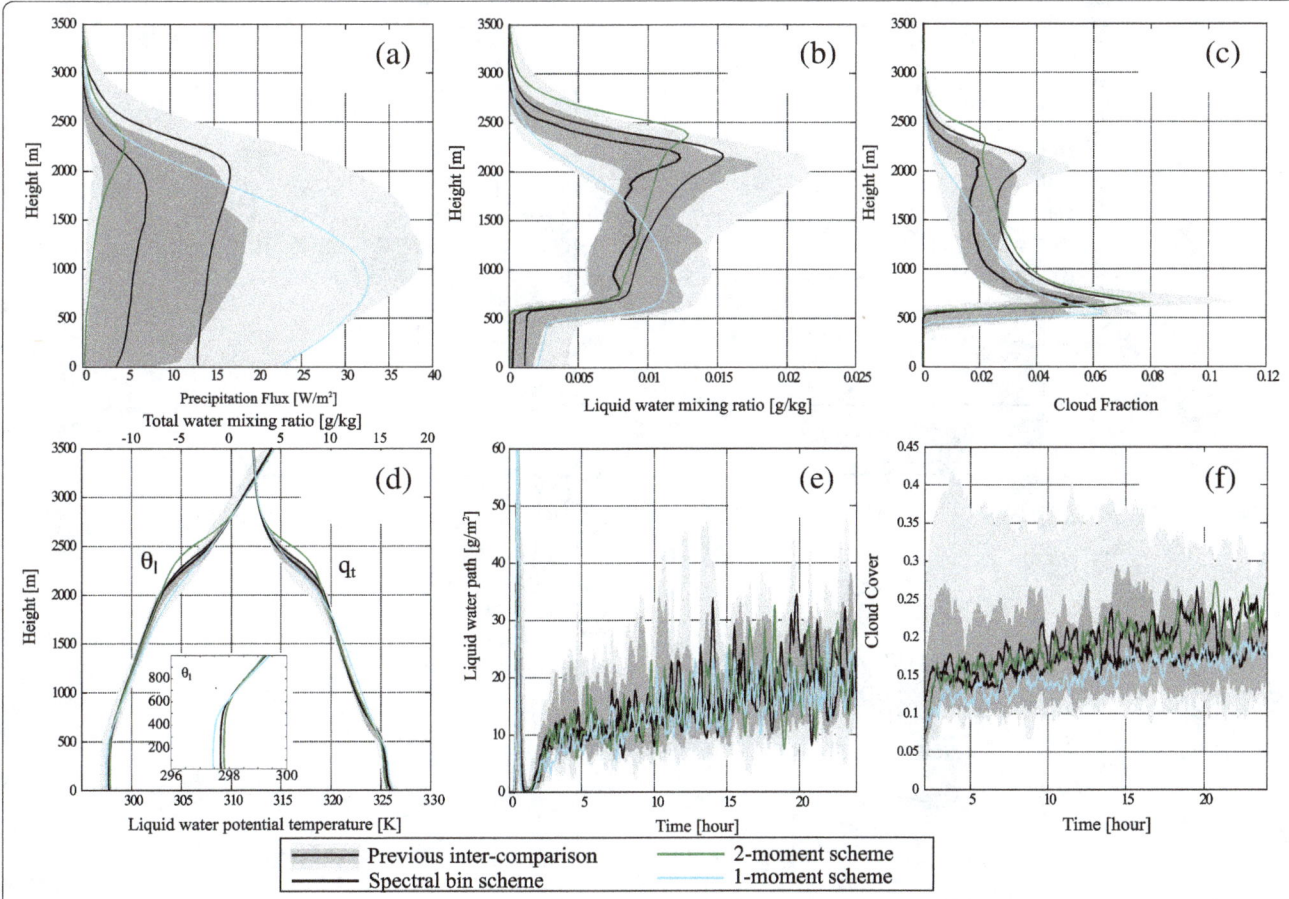

Fig. 3 Comparison of the three cloud microphysical schemes. Horizontally averaged profile of the **a** precipitation flux, **b** liquid water mixing ratio, **c** cloud fraction, **d** total water mixing ratio and liquid water potential temperature, and time evolution of **e** the liquid water path and **f** cloud cover. The *red*, *green*, and *sky-blue lines* show results of the spectral bin scheme, the two-moment bulk scheme, and the one-moment bulk scheme, respectively, and the *black line*, *dark gray shading*, and *light gray shading* indicate the median, range between the first and third quartiles, and range between the maximum and minimum values, respectively, of the previous intercomparison study (van Zanten et al. 2011). The *small figure* in **d** indicates the extension of the profile of liquid water potential temperature below 1000 m

Table 2 Comparison of surface precipitation flux. The surface precipitation flux averaged over the whole calculation domain during the last 4 h of calculation

Scheme	One-moment bulk	Two-moment bulk	Spectral bin
Precipitation flux (W m^{-2})	22.89	0.174	13.17

is helpful for understanding the differences among the results. We first investigate the difference during $t = 3$–4 h of the calculation, because the effects of feedbacks of cloud physics to dynamics is not large and it is easy to interpret the difference. The tendencies at each height averaged during $t = 3$ h to $t = 4$ h are shown in Fig. 5a, b. The results from $t = 0$ h to $t = 3$ h were removed to ignore the effects of spin-up. The effect of sedimentation was omitted from the tendencies. The generation of liquid water at the cloud base in the one-moment bulk scheme is more active than that in the others (Fig. 5a). This is attributed to the difference in the mechanism for generating

cloud particles. In the one-moment bulk scheme, newly generated cloud particles are calculated by saturation adjustment, whereas in the other schemes, they are calculated based on Eq. (4). Because the saturation adjustment does not permit supersaturation, it can generate larger amounts of liquid water at the cloud base than that can be generated by the scheme based on Eq. (4), which allows for supersaturation. This large amount of liquid water generation results in a large heat release at the cloud base (Fig. 5b) and subsequently a strong vertical velocity (Fig. 5c).

In addition to the large amount of liquid water generated in the one-moment bulk scheme, it is clear from the particle size distribution that the conversion from cloud to rain is fast. Size distributions at the lower part of cloud (i.e., $z = 1000$ m) are shown in Fig. 5d. The generation of drizzle and raindrops (i.e., particles over 40 µm in radius) in the lower part of the cloud is active in the one-moment bulk scheme, whereas small cloud

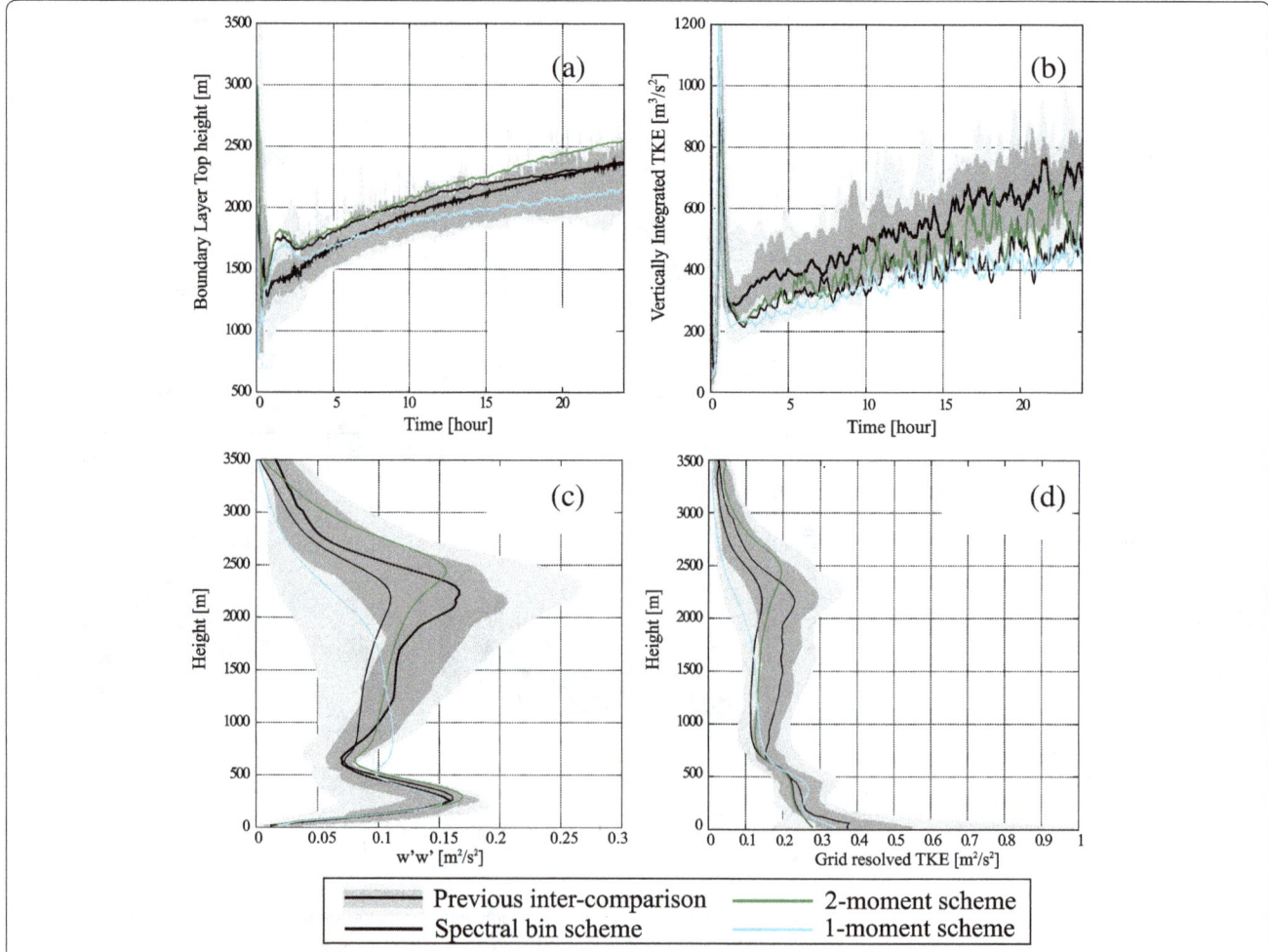

Fig. 4 Comparison of turbulence properties between the three schemes. Time evolution of the **a** boundary layer top height and **b** vertically integrated turbulence kinetic energy (grid resolved + sub-grid scale) and horizontally averaged profile of **c** the variance of resolved w and **d** grid-resolved turbulence kinetic energy averaged during the last 4 h of calculation

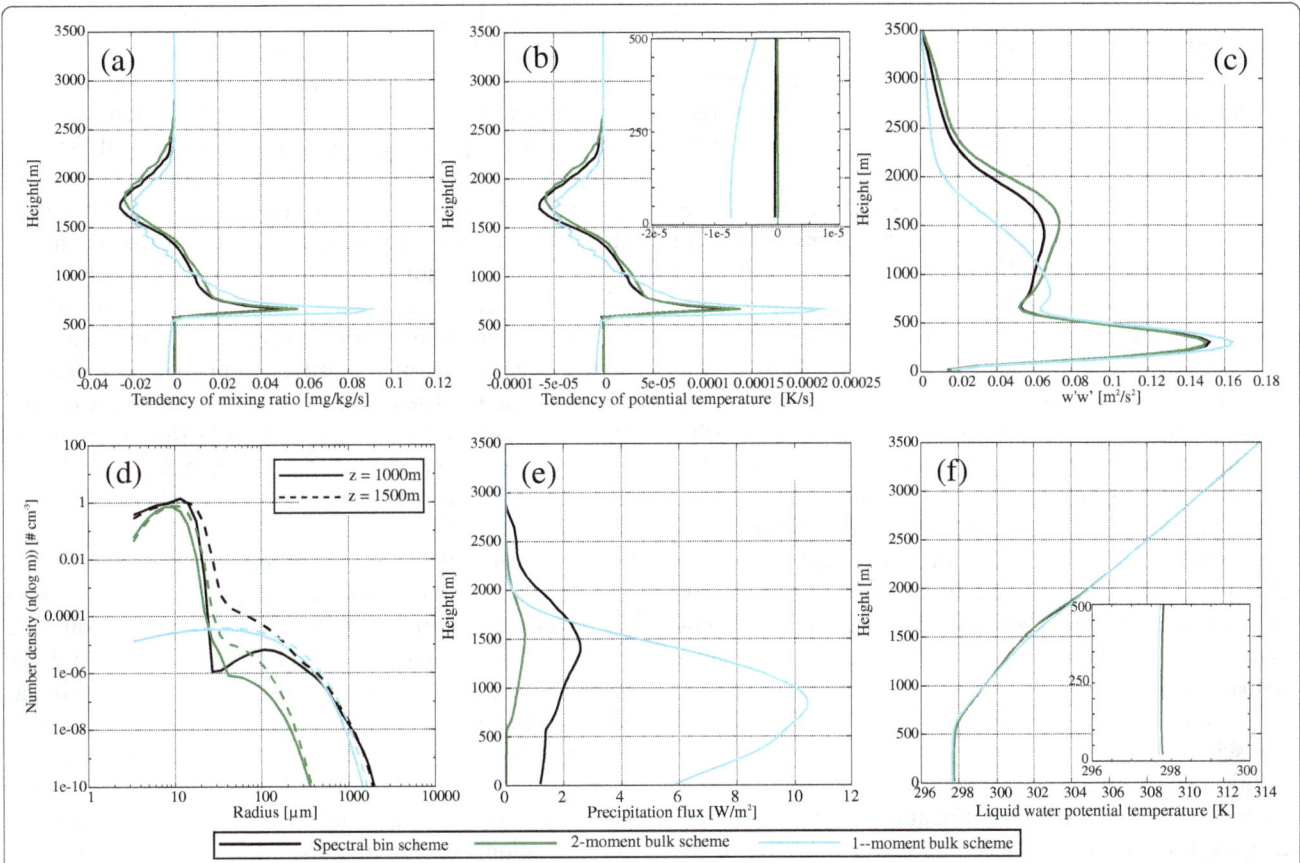

Fig. 5 Profiles and size distribution function averaged during $t = 3 \sim 4$ h. Horizontally averaged profile of the **a** tendency of the liquid water, **b** tendency of potential temperature, **c** variance of w', **e** precipitation flux, and **f** liquid water potential temperature averaged during $t = 3$ to 4 h. **d** Number density distribution (n (log m)) averaged over the whole cloudy grid at (*solid*) $z = 1000$ m and (*dashed*) $z = 1500$ m, where n and m are the number concentration and mass of liquid water, respectively. The *red*, *green*, and *sky-blue lines* show results of the spectral bin, two-moment bulk, and one-moment bulk schemes, respectively. The extended figures in **b** and **f** show the tendencies of potential temperature and liquid water potential temperature below 500 m, respectively

particles are dominant in the two-moment bulk schemes. This indicates that the conversion from cloud to rain in the one-moment bulk scheme is faster than that in the others, and larger numbers of raindrops are generated in the lower part of the cloud by the one-moment bulk scheme.

The generation of large raindrops leads to the fast terminal velocity of the hydrometeors and a large precipitation flux in the one-moment bulk scheme (Fig. 5e), which, in turn, leads to the large precipitation flux at the surface and the fast onset of surface precipitation. Figure 5a, b also show that the peak negative tendency of q_l and θ near the cloud top ($z \sim 1700$ m), which corresponds to the evaporation of cloud droplets at the top of the boundary layer, is located in a lower layer in the one-moment scheme than in the others. This indicates that the large precipitation volume and large cloud size in the one-moment bulk scheme restrain the cloud particles from reaching the upper layer. Therefore, the boundary

layer top height of the one-moment bulk scheme becomes lower (Fig. 4a).

The large amount of raindrops in the one-moment scheme creates feedback for the thermodynamic structure below the cloud. The water loading due to raindrops leads to active evaporative cooling below the cloud (shown in the negative tendency of q_l and θ below the cloud as shown in Fig. 5a, b and the lower potential temperature below the cloud shown in Fig. 5f). This evaporative cooling stabilizes the boundary layer and suppresses the heat transfer from the ground to the upper part of the boundary layer, which is shown in the fact that the positive tendency of θ in the one-moment bulk scheme does not reach $z > 1200$ m but reaches $z \sim 1400$ m in the other scheme. This supports Stevens et al. (1998), who indicated that precipitation suppresses cloud growth and entrainment. This suppression can limit the development of the boundary layer and results in a more stable boundary layer.

The difference between the two-moment bulk and the spectral bin schemes in the tendency of q_l and θ are relatively small. However, the difference in size distribution function between the two schemes is clearly apparent (Fig. 5d). A larger amount of raindrops ($r > 100$ μm) were present in the spectral bin scheme than in the case in the two-moment bulk scheme. It is indicated that the conversion from cloud droplets to raindrops is slow in the two-moment bulk scheme. Because the amount of large raindrops is small in the two-moment bulk scheme (shown as green lines in Fig. 5d), liquid water is carried to the upper layer more easily than in the other schemes. The liquid water evaporates at the top of the boundary layer. This is shown by the negative tendency of q_l and by θ locating in a higher layer in the two-moment bulk scheme than in the others (Fig. 5a, b).

The presence of larger particles in the spectral bin scheme subsequently increases the rate of collisions with other particles, which leads to more rapid growth of particles and earlier precipitation in the spectral bin scheme than in the two-moment bulk scheme. This provides feedback due to the large amount of precipitation, which is the same feedback that occurs in the one-moment bulk scheme.

With the feedback, the differences among the three schemes increase with the integration time. The difference in the boundary layer top height in the three schemes becomes large as the integration time increases (Fig. 4a), and the difference in the liquid water potential temperature below clouds during the last 4 h (Fig. 3d) is larger than at $t = 3–4$ h (Fig. 5f). The differences in the tendencies of θ and q_l shown above also become clear (figure not shown).

In summary, the one-moment bulk scheme creates a larger amount of precipitation, because the saturation adjustment was adopted in the one-moment bulk scheme, and raindrops are subsequently produced by the fast conversion from cloud to raindrops. This results in earlier onset and larger amounts of surface precipitation. The evaporative cooling by raindrops, which occurs actively below the cloud, stabilized the boundary layer in the one-moment bulk scheme.

The two-moment scheme creates raindrops more slowly, resulting in a smaller amount of precipitation compared with the other schemes. The smaller amount of precipitation results in an active latent heat release in the higher layer (shown in Fig. 5b) and a high boundary layer. The spectral bin scheme shows an intermediate rate of conversion and creates an intermediate amount of precipitation, with values between those of the two-moment and one-moment bulk schemes.

The large number of cloud particles newly generated at the cloud bottom by saturation adjustment, the difference in the speed of conversion from cloud to rain, and the

difference in the timing of the surface precipitation all originate from the variety of microphysical schemes used in this study. The differences in each scheme would not always appear when the results of other one-moment, two-moment, and spectral bin schemes are compared. By contrast, the evaporative cooling and subsequent stabilization of the sub-cloud layer and the suppression of the development of boundary layer height, which appeared in the results of the one-moment scheme used in this study, can be expected if the fast conversion from cloud to rain or the active generation of cloud particles at the cloud base occurs as a result of natural phenomena, regardless of which scheme is used. The active latent heat release and high boundary layer, which appeared in the two-moment scheme used in this study, can also be expected. The results are commonly expected for trade wind cumulus.

Speed of conversion from cloud to rain

The analyses in the previous sections hint that the performance of the bulk microphysical schemes could be modified by changing the nucleation schemes and the conversion speed from cloud to rain. In the bulk scheme, autoconversion and accretion are the main processes involved in the conversion from cloud to rain. To obtain information for the modification of the parameterization of these two processes, a comparison of the autoconversion and accretion rates among the three schemes is helpful. Figure 6 shows the autoconversion and accretion rates

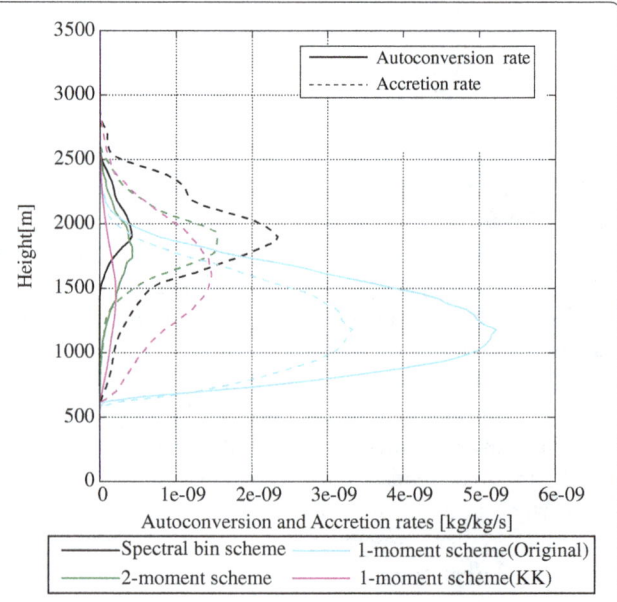

Fig. 6 Autoconversion and accretion rate averaged during $t = 3~4$ h. The horizontally averaged (*solid line*) autoconversion rate and (*dashed line*) accretion rate simulated by the spectral bin scheme (*red*), two-moment scheme (*green*), one-moment scheme (*sky blue*), and one-moment scheme with the Khairoutdinov and Kogan (2000) scheme (*pink*), averaged during $t = 3~4$ h

averaged during $t = 3 \sim 4$ h. The autoconversion rate of the spectral bin scheme is regarded as the rate of increasing mass of raindrops (defined as liquid particles whose radius is larger than 40 μm) generated by the coagulation between cloud particles (defined as liquid particles whose radius is smaller than 40 μm). The accretion rate of the spectral bin scheme is determined as the increasing rate of mass due to the coagulation between cloud particles and raindrops. Figure 6 shows that the autoconversion rate of the one-moment scheme is about 15 times larger than that of the spectral bin scheme. By contrast, the accretion rate of the two-moment scheme is 1.5~2 times smaller than that of the spectral bin scheme. Based on these results, the sensitivity of these two processes in each scheme was investigated and is discussed in the next section.

Sensitivity of the autoconversion to the one-moment bulk scheme

As shown above, the one-moment bulk scheme overestimates the conversion speed from cloud droplet to raindrop. We first investigated the difference between the original autoconversion scheme of Tomita (2008) and that of Khairoutdinov and Kogan (2000) shown in Eq. (6). The results of the one-moment bulk scheme with the Khairoutdinov and Kogan (2000) scheme are more similar to the results of the spectral bin scheme and the previous study than those calculated by Eq. (5) (Fig. 7). This indicates that the conversion speed calculated by Eq. (5) is too fast for shallow clouds because the validity of the one-moment bulk scheme with Eq. (5) was confirmed through experiments with deep convective clouds (Tomita 2008).

In addition to the experiment with the Khairoutdinov and Kogan (2000) scheme, other sensitivity experiments

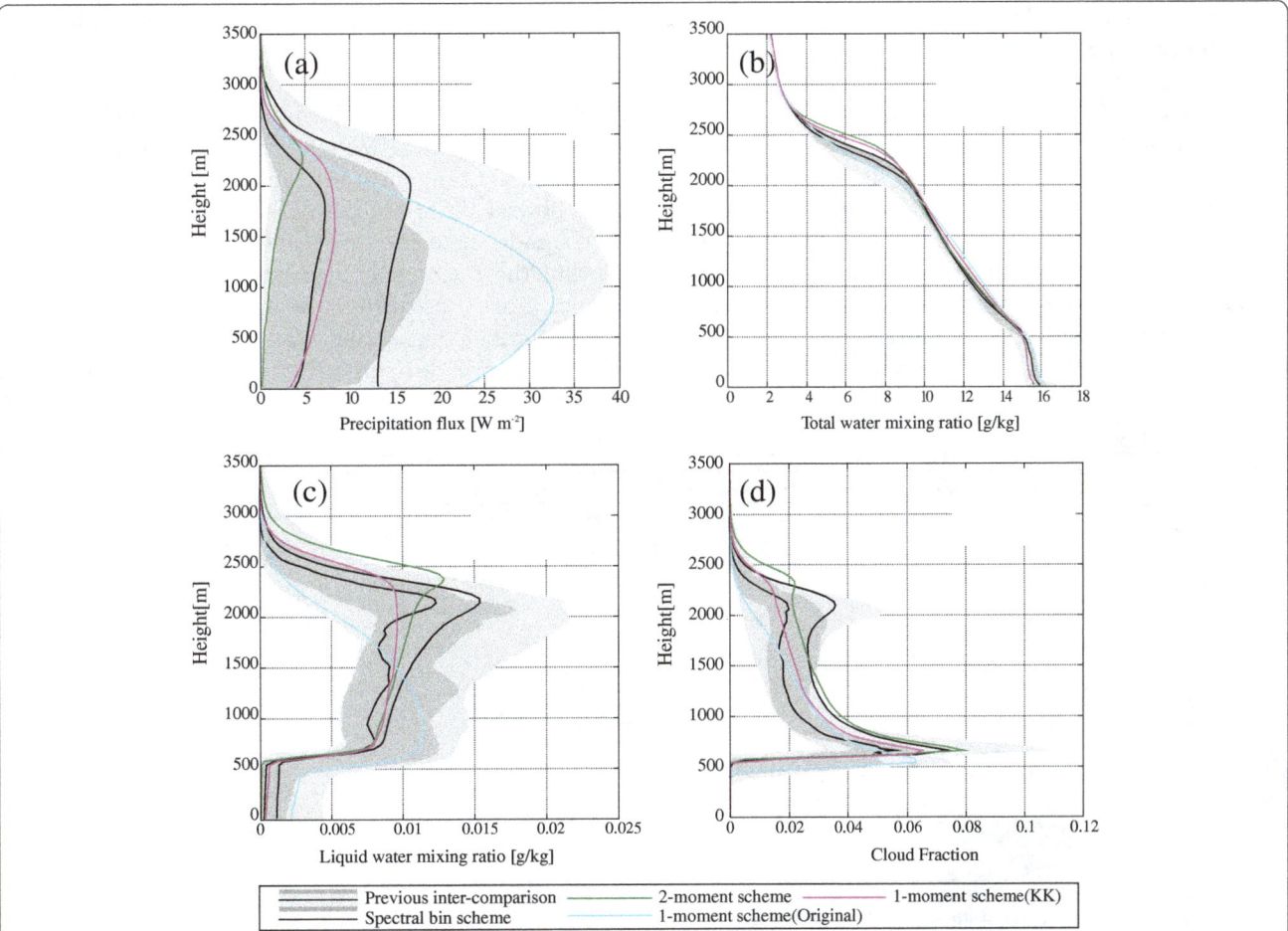

Fig. 7 Cloud microphysical properties simulated by Khairoutdinov and Kogan's auto conversion scheme. Horizontally averaged profile of the **a** precipitation flux, **b** total water mixing ratio (q_t), **c** liquid water mixing ratio, and **d** cloud fraction averaged during the last 4 h. The *red, green, sky-blue,* and *pink lines* show results by the spectral bin scheme, the two-moment bulk scheme, the one-moment bulk scheme, and one-moment bulk scheme using the Khairoutdinov and Kogan (2000) autoconversion rate, respectively. The *black line, dark gray shading,* and *light gray shading* indicate the median, range between the first and third quartiles, and range between the maximum and minimum values, respectively, of the previous intercomparison study (van Zanten et al. 2011)

were conducted by reducing the autoconversion rate. Based on the analyses in Fig. 6, a 0.067-fold (i.e., 1/15) smaller autoconversion rate than the default value was used for the sensitivity experiment. Another sensitivity experiment with an accretion rate that was 0.067-fold smaller than the default value was also conducted. Figure 8 shows the profile of q_l and the precipitation flux, which were calculated using the smaller autoconversion and accretion rate. It can be seen that a small autoconversion rate results in liquid water locating in the upper layer, whereas it does not locate in the upper layer when the accretion rate was reduced to the same extent as the autoconversion rate. Hence, the autoconversion process is more sensitive to the production of raindrops.

Sensitivity of the accretion to the two-moment bulk scheme

Contrasting the one-moment bulk scheme, the two-moment bulk scheme underestimates the conversion speed from cloud to raindrop. We speculate that the faster conversion from cloud droplet to raindrop in the two-moment scheme produced similar results using the spectral bin scheme. Sensitivity experiments were conducted by changing both the autoconversion and accretion rates of the two-moment scheme. Based on the analyses in Fig. 6, a twofold increase in the accretion rate was used, and a twofold increase of the autoconversion rate was also used for the sensitivity experiment. Figure 9 shows the results of these experiments. In the two-moment bulk scheme, both the precipitation rate and q_l increase when the autoconversion rate is doubled. However, doubling the autoconversion rate does not result in

a peak value of q_l and a precipitation flux in the lower layer as simulated by the one-moment bulk scheme. By contrast, the precipitation flux simulated when the accretion rate is doubled is considerably larger than both the default and when the autoconversion rate is doubled. This indicates that the accretion process was the major contributor to the creation of liquid water in the lower layer in the two-moment bulk scheme.

In short, the sensitivity of the accretion and the autoconversion processes to the cloud microphysical properties differ between the schemes.

Component level intercomparison

In this study, we investigated the exact effects of the various cloud microphysical schemes on model simulations of trade wind cumulus. If we change components other than the cloud microphysical scheme (e.g., dynamical core, turbulence scheme, advection scheme), the response of the cloud microphysical properties would change as suggested by van Zanten et al. (2011). It is necessary to conduct sensitivity experiments by changing each component while keeping all other components fixed, as we did when targeting the cloud microphysical scheme in this study. We refer to these sensitivity experiments as a "Component level intercomparison".

Grabowski (2014) suggested a piggyback approach to better understand the exact effects of cloud microphysics and the interaction between microphysics and dynamics. This approach can be also applied to the other components. Using a component level intercomparison and the piggyback approach, we can discuss the effects

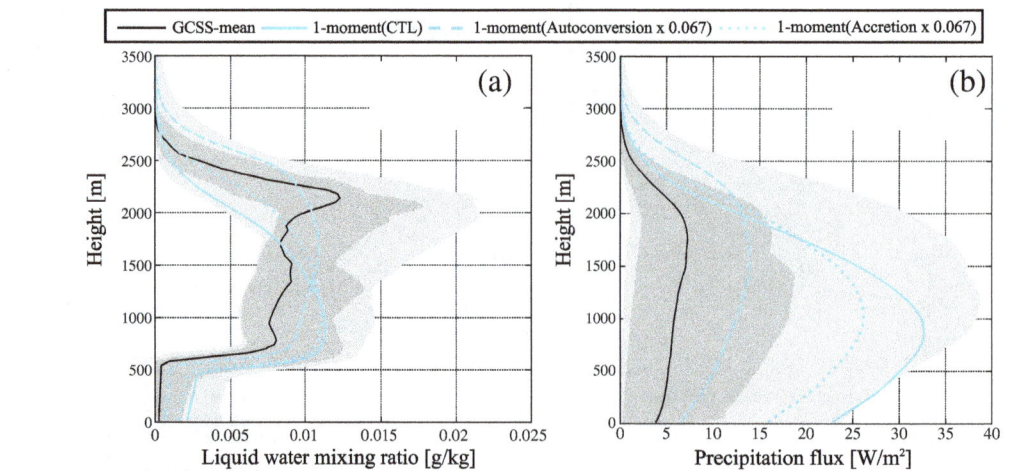

Fig. 8 Cloud microphysical properties simulated in the sensitivity experiment with a varying conversion ratio in the one-moment scheme. Horizontally averaged profile of the **a** liquid water mixing ratio (q_l) and **b** precipitation flux averaged during the last 4 h. The *solid sky-blue*, *dashed sky-blue*, and *dotted sky-blue lines* show the results using the one-moment bulk scheme with the default autoconversion rate, one-moment bulk scheme with an autoconversion rate 0.067-fold (i.e., 1/15) smaller, and one-moment bulk scheme with accretion ratio 0.067-fold smaller, respectively. The *black line*, *dark gray shading*, and *light gray shading* indicate the median, range between the first and third quartiles, and range between the maximum and minimum values, respectively, for a previous intercomparison study (van Zanten et al. 2011)

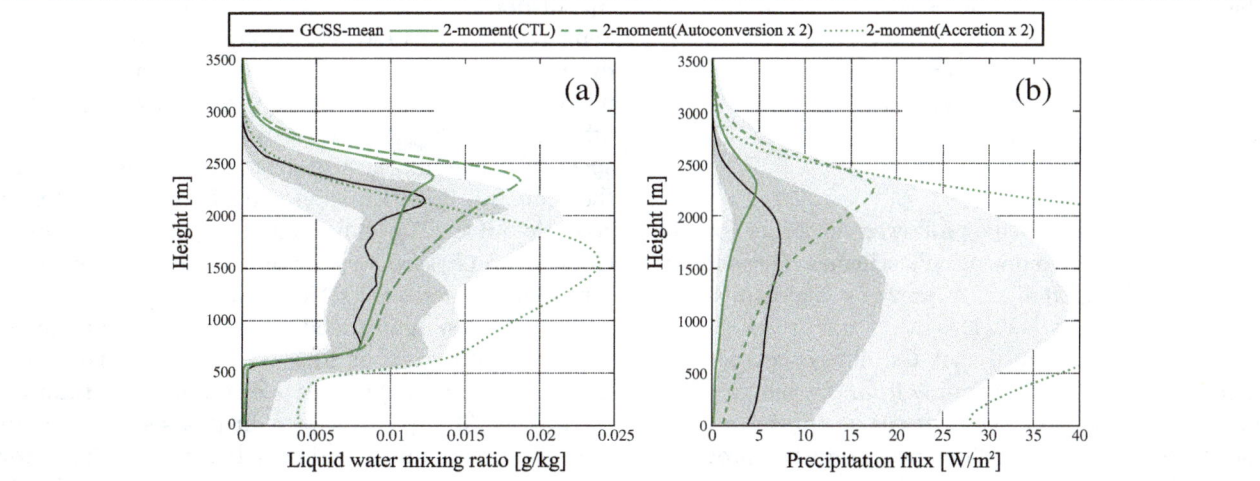

Fig. 9 Cloud microphysical properties simulated in the sensitivity experiment with a changing conversion ratio in the two-moment scheme. The horizontally averaged profile of the **a** liquid water mixing ratio (q_l) and **b** precipitation flux averaged during the last 4 h. The *solid green, dashed green,* and *dotted green lines* show the results using the two-moment bulk scheme with the default autoconversion rate, two-moment bulk scheme with an autoconversion rate twice as large as the default, and two-moment bulk scheme with an accretion ratio twice as large as the default. The *black line, dark gray shading,* and *light gray shading* indicate the median, range between the first and third quartiles, and the range between the maximum and minimum values, respectively, for a previous intercomparison study (van Zanten et al. 2011)

of each component separately and obtain knowledge that would improve the parameterization of global scale models or regional models with coarse resolution.

Conclusions

In this study, we developed a large eddy simulation model (SCALE-LES), which excludes approximations and implicit schemes as much as possible. The results of a benchmark test indicated that SCALE-LES effectively reproduces the trade wind cumuli simulated in a previous LES intercomparison study (van Zanten et al. 2011). Using SCALE-LES, we investigated the impacts of cloud

microphysical schemes on shallow cumulus and investigated which processes were critical for the diversity observed in previous LES intercomparison studies. Three types of cloud microphysical scheme, the one-moment bulk scheme of Tomita (2008), the two-moment bulk scheme of Seiki and Nakajima (2014), and the one-moment spectral bin scheme of Suzuki et al. (2010), were implemented with SCALE-LES for the sensitivity experiments.

The results indicated that the precipitation at the surface increases, in order, from the two-moment bulk scheme to the spectral bin scheme and the one-moment bulk scheme. Surface precipitation begins first in the

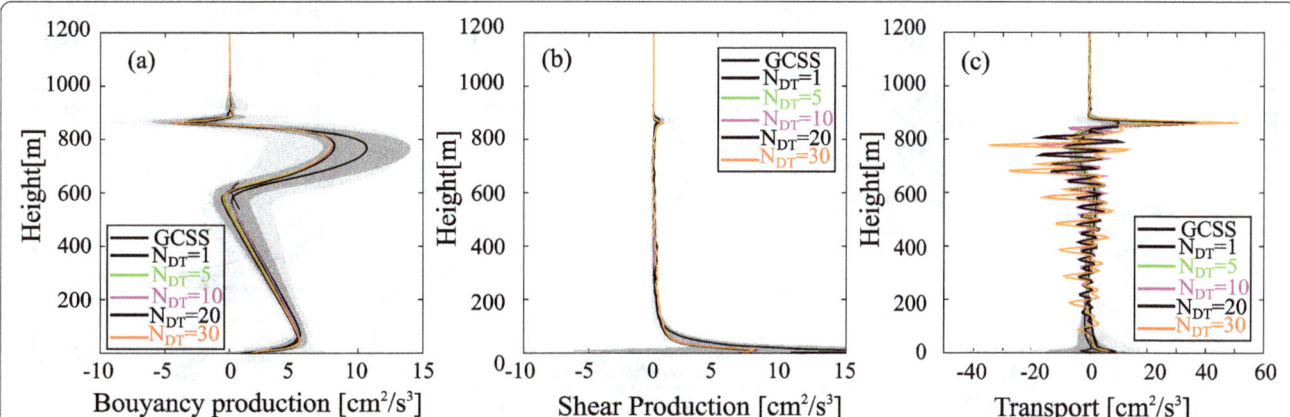

Fig. 10 Cloud microphysical properties in the sensitivity experiment of N_{DT} ($=\Delta t_{microphy}/\Delta t_{dyn}$). Hourly averaged profile of the **a** buoyancy production, **b** shear production, and **c** transport terms averaged over the entire calculation domain during the last hour of each simulation. The *red, green, pink, black,* and *orange lines* represent the results of $N_{DT} = 1, 5, 10, 20,$ and 30, respectively. The *blue line, dark gray shading,* and *light gray shading* indicate the mean, standard deviation, and the range between the maximum and minimum values, respectively, from a previous intercomparison study (Stevens et al. 2005)

Table 3 List of non-dimensional coefficient for sensitivity experiment

Non-dimensional coefficient (γ)	$\gamma = 10^{-3}$	$\gamma = 10^{-5}$	$\gamma = 10^{-7}$
The strength of the numerical filter	1.25×10^{5}	1.25×10^{3}	1.25×10^{1}

The value of the numerical filter (m^4 s^{-1}) for each experiment

one-moment bulk scheme, followed in order by the spectral bin and two-moment bulk schemes. These results support the suggestion of a previous intercomparison study (van Zanten et al. 2011)

Our analyses confirmed that the differences between the schemes were derived mainly from the generation of cloud particles and the speed of conversion from cloud droplets to raindrops. The differences in the two processes originated from the differences in the microphysical schemes used. By contrast, the phenomena generated by this variety, i.e., evaporative cooling and stabilization below the cloud and a low boundary layer, active latent heat release, and a high boundary layer can be expected regardless of the scheme used, if the active conversion from cloud to rain and the active generation of new cloud particles occur in nature.

The sensitivity of the autoconversion and accretion processes to the cloud microphysical properties simulated by the bulk microphysical schemes was also investigated. In the two-moment bulk scheme the accretion process was more sensitive to the cloud microphysical properties, whereas the autoconversion was more sensitive in the one-moment bulk scheme. These results indicate that the tuning method of the microphysical scheme differs from scheme to scheme, and a component level intercomparison is useful to obtain the exact method for each scheme.

Appendix 1

Sensitivity of the ratio of the physical time step to the dynamical time step. In this study, the time step of dynamics (Δt_{dyn}), cloud microphysics ($\Delta t_{\mathrm{microphy}}$), and other physics ($\Delta t_{\mathrm{phy}}$) were set as 0.05, 0.15, and 0.5 s, respectively. The time step of dynamics was determined by the Courant–Friedrichs–Lewy (CFL) condition for the acoustic wave. The time step for physics (except for cloud microphysics) was set as $10 \times \Delta t_{\mathrm{dyn}}$ based on the sensitivity experiment (Nishizawa et al., 2015). The $\Delta t_{\mathrm{microphy}}$ was determined in sensitivity experiments examining the ratio of $\Delta t_{\mathrm{microphy}}$ to Δt_{dyn}. The results of the sensitivity experiments examining the ratio N_{DT} ($= \Delta t_{\mathrm{microphy}} / \Delta t_{\mathrm{dyn}}$) are shown in this section. For this sensitivity test, the experimental setup of the second Dynamics and Chemistry of Marine Stratocumulus Research Flight 1 (DYCOMS-II RF01) (Stevens et al. 2005) was used. In this study, the experimental setup of the RICO study was used in most cases, but the effects of the acoustic wave appeared more clearly in the DYCOMS-II case (figure not shown). Consequently, the ratio (N_{DT}) was determined using the DYCOMS-II RF01 experimental setup. For this sensitivity experiment, Δt_{dyn} was set as 0.01 s and N_{DT} was swept from 1 to 30 (i.e., $\Delta t_{\mathrm{microphy}}$ was set from 0.01 to 0.3 s).

The results of the sensitivity experiment indicated that the effects of N_{DT} were mostly small (figures not shown), except for the TKE budget. Figure 10 shows the profile of the buoyancy production term, shear production term, and transport term of TKE production. The transport term was quite noisy when N_{DT} was large ($N_{\mathrm{DT}} > 10$). Noise is also present in the shear production term. This noise was derived from the acoustic wave that was generated at every time step of the microphysical processes. When N_{DT} was

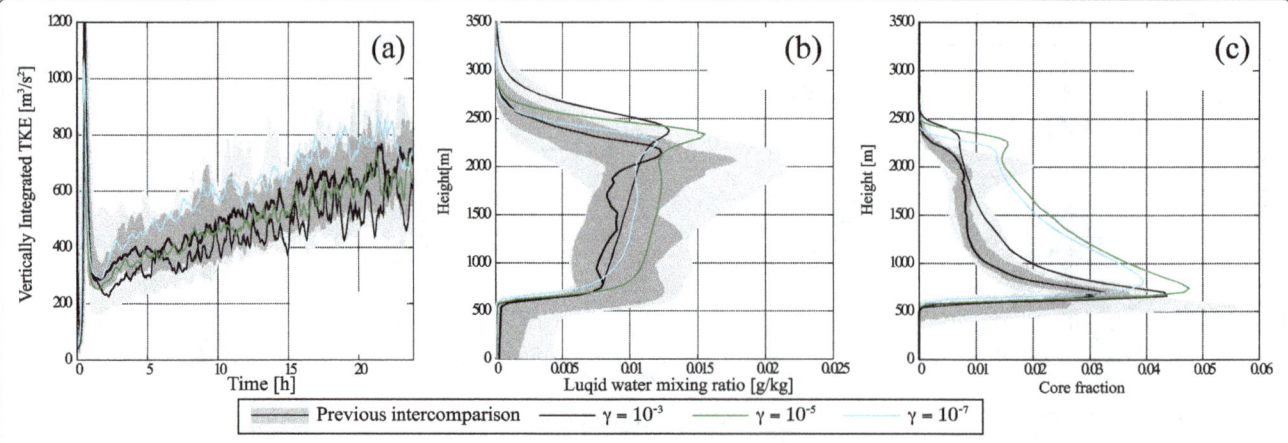

Fig. 11 Results of a sensitivity experiment of the strength of numerical diffusion. Time evolution of the **a** vertically integrated turbulence kinetic energy averaged over the whole calculation domain and the profile of the **b** liquid water mixing ratio and **c** cloud fraction averaged over the whole calculation domain during the last 4 h. The *red*, *green*, and *sky-blue line* shows the results with the coefficient (γ) as 10^{-3}, 10^{-5}, and 10^{-7}, respectively. The *black line*, *thick gray shade*, and *thin gray shade* indicate the median, range between the first and third quartiles, and range between the maximum and minimum values of a previous intercomparison study, respectively (van Zanten et al. 2011)

large, the variation of θ at each step of the microphysical processes was also large. The large variation in θ generates an acoustic wave, which produced the noise. This indicates that N_{DT} should be smaller than 5 to render the effects of the acoustic wave negligibly. From these results and for safety, N_{DT} was set as 3 for the RICO experiment (i.e., $\Delta t_{\mathrm{microphy}} = 3 \times \Delta t_{\mathrm{dyn}}$).

Appendix 2

Numerical filter. This section describes the sensitivity experiments of the strength of the numerical diffusion. The nth ordered superviscosity/diffusion is defined as:

$$\frac{\partial}{\partial x}\left[\nu\rho\frac{\partial^{n-1}f}{\partial x^{n-1}}\right] \quad f\in\{u,v,w,\theta,q_s\}.$$
$$\frac{\partial}{\partial x}\left[\nu\frac{\partial^{n-1}\rho}{\partial x^{n-1}}\right]$$

The ν is written as:

$$\nu = (-1)^{\frac{n}{2}+1}\gamma\frac{\Delta x^n}{2^n\Delta t}$$

where γ is a non-dimensional coefficient, Δx is the grid spacing, and n is the order of the numerical filter. This study adopted a fourth-order numerical diffusion (i.e., $n = 4$) to dampen the artificial noise. n was set as 4 because of the computational efficiency.

To investigate the sensitivity of the strength of the numerical diffusion, simulations of the RICO study were conducted while changing the strength of the numerical diffusion. Based on the rough estimation of Nishizawa et al. (2015), the non-dimensional coefficient (γ) should be smaller than $O(10^{-3})$–$O(10^{0})$ for $n = 4$. Thus, γ was changed from 10^{-3} to 10^{-7}. The strength of the numerical diffusion for each γ is shown in Table 3. The two-moment bulk scheme was used for the sensitivity experiment. As a result of the numerical diffusion, one-dimensional sinusoidal two-grid noise will decay to $1/e$ with a $1/\gamma$ time step.

Figure 11 shows the results of the sensitivity experiment. The vertically integrated TKE increased with a weakening of the numerical filter, even though the TKE of all experiments was located in the range between the maximum and minimum of the previous studies. This is because a strong numerical diffusion can efficiently dampen small-scale turbulence. As well as the TKE, the liquid water mixing ratio of all the sensitivity experiments was also located in the range of previous intercomparison studies regardless of the strength of the numerical diffusion. However, the cloud fraction simulated with a small numerical diffusion (i.e., $\gamma \leq 10^{-5}$) was much larger than that of the intercomparison study, and the cloud fraction was completely outside the range of the inter comparison studies. The cloud fraction, core fraction, and variance of w' were also outside the range of the inter comparison studies when γ was smaller than 10^{-5} (figure not shown). The temporal evolution of the cloud fraction indicates that the large cloud fraction, with small numerical diffusion, seems to originate from artificial noise. However, it is difficult to identify the reason for the large cloud fraction being artificial noise. The same experiment must be conducted with a fine grid resolution to divide all of the elements of the wave into a physically meaningful wave and artificial noise. However, computational limitations prevented us from conducting these experiments. Consequently, γ was determined using the results of the intercomparison study (van Zanten et al. 2011) as a reference solution, and γ was set as 10^{-3} (i.e., the strength of the numerical diffusion is 1.25×10^{5} m^4 s^{-1}).

Abbreviations
DYCOMS: Dynamics and Chemistry of Marine Stratocumulus; GCSS: GEWEX Cloud System Study; GEWEX: Global Energy and Water Exchange project; KiD: the kinematic driver; LES: large eddy simulation; LWP: liquid water path; MPI: Message Passing Interface; RICO: Rain in Cumulus over the Ocean; SCALE: Scalable Computing for Advanced Library and Environment; TKE: turbulence kinetic energy.

Competing interests
The authors declare that they have no competing interests.

Authors' contributions
YS implemented the cloud microphysical schemes in the SCALE library, designed this study, conducted the numerical simulations, analyzed the results of the simulations, and developed the manuscript. SN, HY, and YM developed the main frames of the SCALE library. YK collaborated with the corresponding author in the creation of the manuscript. HT proposed the development of the SCALE library. All authors read and approved the final manuscript.

Acknowledgement
Part of the results is obtained by the K computer at the RIKEN Advanced Institute for Computational Science. This work was supported by FOCUS Establishing Supercomputing Center of Excellence. SCALE-LES developed by Team-SCALE of the RIKEN Advanced Institute for Computational Sciences. The data from GCSS intercomparison studies used in several figures were downloaded from http://www.knmi.nl/samenw/rico/.

References
Ackerman AS, van Zanten MC, Stevens B, Savic-Jovcic V, Bretherton CS, Chlond A, Golaz J-C, Jiang H, Khairoutdinov M, Krueger SK, Lewellen DC, Lock A, Moeng C-H, Nakamura K, Petters MD, Snider JR, Weinbrecht S, Zulauf M (2009) Large-eddy simulations of a drizzling, stratocumulus-topped marine boundary layer. Mon Weather Rev 137:1083–1110. doi:10.1175/2008MWR2582.1
Berry EX (1968) Modification of the warm rain process. In: Proceeding of First Conference on Weather Modification., pp 81–85
Bretherton CS, Park S (2009) A new moist turbulence parameterization in the Community Atmosphere Model. J Clim 22:3422–3448. doi:10.1175/2008JCLI2556.1
Brown AR, Derbyshire SH, Mason PJ (1994) Large-eddy simulation of stable atmospheric boundary layers with a revised stochastic subgrid model. Q J R Meteorol Soc 120:1485–1512. doi:10.1002/qj.49712052004
Chepfer H, Bony S, Winker D, Chiriaco M, Dufresne JL, Sèze G (2008) Use of CALIPSO lidar observations to evaluate the cloudiness simulated by a climate model. Geophys Res Lett 35:1–6. doi:10.1029/2008GL034207
Considine G, Curry JA, Wielicki B (1997) Modeling cloud fraction and horizontal variability in marine boundary layer clouds. J Geophys Res 102:13517. doi:10.1029/97JD00261

Grabowski WW (2014) Extracting microphysical impacts in large-eddy simulations of shallow convection. J Atmos Sci 71:4493–4499. doi:10.1175/JAS-D-14-0231.1

Iguchi T, Nakajima T, Khain AP, Saito K, Takemura T, Suzuki K (2008) Modeling the influence of aerosols on cloud microphysical properties in the east Asia region using a mesoscale model coupled with a bin-based cloud microphysics scheme. J Geophys Res 113:D14215. doi:10.1029/2007JD009774

Kain, JS (2004) The Kain–Fritsch convective parameterization: An update. J Appl Meteorol. doi:10.1175/1520-0450(2004)043<0170:TKCPAU>2.0.CO;2

Khain AP, Sednev I (1996) Simulation of precipitation formation in the Eastern Mediterranean coastal zone using a spectral microphysics cloud ensemble model. Atmos Res 43:77–110. doi:10.1016/S0169-8095(96)00005-1

Khairoutdinov M, Kogan Y (2000) A new cloud physics parameterization in a large-eddy simulation model of marine stratocumulus. Mon Weather Rev 128:229–243. doi:10.1175/1520-0493(2000)128<0229:ANCPPI>2.0.CO;2

Naud CM, Del Genio AD, Bauer M, Kovari W (2010) Cloud vertical distribution across warm and cold fronts in cloudsat-CALIPSO data and a general circulation model. J Clim 23:3397–3415. doi:10.1175/2010JCLI3282.1

Nishizawa S, Yashiro H, Sato Y, Miyamoto Y, Tomita H (2015) Influence of grid aspect ratio on planetary boundary layer turbulence in large-eddy simulations. Geosci Model Dev Discuss 8:6021–6094. doi:10.5194/gmdd-8-6021-2015

Posselt R, Lohmann U (2008) Introduction of prognostic rain in ECHAM5: design and single column model simulations. Atmos Chem Phys 8:2949–2963. doi:10.5194/acpd-7-14675-2007

Pruppacher, HR, and Klett, JD, 1997, Microphysics of Clouds and Precipitation, 2nd ed., Kluwer Academic Publisher, Dordrecht, The Netherlands, 954pp.

Randall DA, Coakley JA, Lenschow DH, Fairall CW, Kropfli RA (1984) Outlook for research on subtropical marine stratification clouds. Bull Am Meteorol Soc 65:1290–1301. doi:10.1175/1520-0477(1984)065<1290:OFROSM>2.0.CO;2

Savic-Jovcic V, Stevens B (2008) The structure and mesoscale organization of precipitating stratocumulus. J Atmos Sci 65:1587–1605. doi:10.1175/2007JAS2456.1

Scotti A, Meneveau C, Lilly DK (1993) Generalized Smagorinsky model for anisotropic grids. Phys Fluids A Fluid Dyn 5:2306–2308. doi:10.1063/1.858537

Seifert A, Beheng KD (2001) A double-moment parameterization for simulating autoconversion, accretion and self collection. Atmos Res 59–60:265–281. doi:10.1016/S0169-8095(01)00126-0

Seiki T, Nakajima T (2014) Aerosol effects of the condensation process on a convective cloud simulation. J Atmos Sci 71:833–853. doi:10.1175/JAS-D-12-0195.1

Shipway BJ, Hill AA (2012) Diagnosis of systematic differences between multiple parametrizations of warm rain microphysics using a kinematic framework. Q J R Meteorol Soc 138:2196–2211. doi:10.1002/qj.1913

Siebesma A, Bretherton CS, Brown A, Chlond A, Cuxart J, Duynkerke P, Jiang H, Khairoutdinov M, Lewellen D, Moeng C-H, Sanchez E, Stevens B, Stevens DE (2003) A large-eddy simulation intercomparison study of shallow cumulus convection. J Atmos Sci 60:1201–1219. doi:10.1175/1520-0469(2003)060<1201:AALESIS>2.0.CO;B2

Stevens B, Cotton WR, Feingold G, Moeng C-H (1998) Large-eddy simulations of strongly precipitating, shallow, stratocumulus-topped boundary layers. J Atmos Sci 55:3616–3638. doi:10.1175/1520-0469(1998)055<3616:LESOSP>2.0.CO;2

Stevens B, Moeng C-H, Ackerman AS, Bretherton CS, Chlond A, de Roode S, Edwards J, Golaz J-C, Jiang H, Khairoutdinov M, Kirkpatrick MP, Lewellen DC, Lock A, Müller F, Stevens DE, Whelan E, Zhu P (2005) Evaluation of large-eddy simulations via observations of nocturnal marine stratocumulus. Mon Weather Rev 133:1443–1462. doi:10.1175/MWR2930.1

Stocker, TF, Qin D, Plattner G-K, Tignor M, Allen SK, Boschung J, Nauels A, Xia Y, Bex V, and Midgley PM (2013), IPCC, 2013: Climate Change 2013: The Physical Science Basis, Cambridge University Press, Cambridge, United Kingdom and New York, NY, USA.

Suzuki K, Nakajima T, Nakajima TY, Khain A (2006) Correlation pattern between effective radius and optical thickness of water clouds simulated by a spectral bin microphysics cloud model. SOLA 2:116–119. doi:10.2151/sola.2006-030

Suzuki K, Nakajima T, Nakajima TY, Khain AP (2010) A study of microphysical mechanisms for correlation patterns between droplet radius and optical thickness of warm clouds with a spectral bin microphysics cloud model. J Atmos Sci 67:1126–1141. doi:10.1175/2009JAS3283.1

Suzuki K, Nakajima T, Numaguti A, Takemura T, Kawamoto K, Higurashi A (2004) A study of the aerosol effect on a cloud field with simultaneous use of GCM modeling and satellite observation. J Atmos Sci 61:179–194. doi:10.1175/1520-0469(2004)061<0179:ASOTAE>2.0.CO;2

Tiedtke M (1993) Representation of clouds in large-scale models. Mon Weather Rev 121:3040–3061. doi:10.1175/1520-0493(1993)121<3040:ROCILS>2.0.CO;2

Tomita H (2008) New microphysical schemes with five and six categories by diagnostic generation of cloud ice. J Meteorol Soc Japan 86A:121–142. doi:10.2151/jmsj.86A.121

van Zanten MC, Stevens B, Nuijens L, Siebesma AP, Ackerman AS, Burnet F, Cheng A, Couvreux F, Jiang H, Khairoutdinov M, Kogan Y, Lewellen DC, Mechem D, Nakamura K, Noda A, Shipway BJ, Slawinska J, Wang S, Wyszogrodzki A (2011) Controls on precipitation and cloudiness in simulations of trade-wind cumulus as observed during RICO. J Adv Model Earth Syst 3:M06001. doi:10.1029/2011MS000056

Wang H, Feingold G (2009) Modeling mesoscale cellular structures and drizzle in marine stratocumulus. Part II: the microphysics and dynamics of the boundary region between open and closed cells. J Atmos Sci 66:3257–3275. doi:10.1175/2009JAS3120.1

Wang H, Feingold G, Wood R, Kazil J (2010) Modelling microphysical and meteorological controls on precipitation and cloud cellular structures in Southeast Pacific stratocumulus. Atmos Chem Phys 10:6347–6362. doi:10.5194/acp-10-6347-2010

Xue H, Feingold G, Stevens B (2008) Aerosol effects on clouds, precipitation, and the organization of shallow cumulus convection. J Atmos Sci 65:392–406. doi:10.1175/2007JAS2428.1

Yamaguchi T, Randall DA (2012) Cooling of entrained parcels in a large-eddy simulation. J Atmos Sci 69:1118–1136. doi:10.1175/JAS-D-11-080.1

Zalesak ST (1979) Fully multidimensional flux-corrected transport algorithms for fluids. J Comput Phys 31:335–362. doi:10.1016/0021-9991(79)90051-2

6

A numerical experiment on the formation of the tropopause inversion layer associated with an explosive cyclogenesis: possible role of gravity waves

Shigenori Otsuka[1]*, Megumi Takeshita[2] and Shigeo Yoden[3]

Abstract

The tropopause inversion layer (TIL) is a persistent layer with high static stability. Although some mechanisms for the formation of the TIL have been proposed, the time evolution of the TIL under realistic conditions especially when factoring in the contribution of small-scale processes such as gravity waves is not well understood. To gain an understanding of this factor, we conducted a numerical experiment on an explosive cyclogenesis in mid-latitudes using a nonhydrostatic regional atmospheric model. Although the TIL in the model is consistent with previous observations in the sense that it is stronger in the negative vorticity areas, the relationship is clear only in the development and mature stages of a cyclone, suggesting that the evolution of the cyclone plays an important role in the formation of the TIL. To ascertain the effects of gravity waves on the TIL, vertical convergence at the tropopause is analyzed. Histograms of maximum buoyancy frequency squared $\left(N_{max}^2\right)$ within the TIL show that regions of vertical convergence have higher N_{max}^2, in addition to regions with high $\partial^2 w/\partial z^2$, implying that waves having downward phase propagation also play an important role in the dynamical formation of the TIL. This tendency is clearer in regions of negative relative vorticity at the tropopause. By taking account of the fact that the gravity wave activities associated with the cyclone and the jet streak are enhanced during the development and mature stages of the cyclone, vertical convergence due to gravity waves associated with synoptic weather systems can be seen to be a key process in the formation of the negative correlation between the strength of the TIL and the local relative vorticity at the tropopause.

Keywords: Tropopause inversion layer; Extratropical cyclone; Gravity wave

Background

The tropopause inversion layer (TIL) is a layer with enhanced static stability just above the tropopause, and it has been extensively studied over the past decade (e.g., Birner et al. 2002; 2006). The TIL is observed at all latitudes from the tropics to the polar regions (e.g., Randel et al. 2007; Tomikawa et al. 2009; Son et al. 2011). The TIL exhibits a dependence on the relative vorticity around the tropopause, in that anticyclones exhibit substantially stronger inversion than cyclones (Zängl and Wirth 2002; Randel et al. 2007.

This vorticity dependence has also been reproduced in idealized numerical simulations (e.g., Son and Polvani 2007). In general, two major mechanisms for TIL formation have been proposed: dynamical formation (e.g., Wirth 2003) and radiative formation (e.g., Randel et al. 2007; Randel and Wu 2010). Wirth (2003, 2004) applied a theory of conservative balanced dynamics to axisymmetric cyclonic and anticyclonic vortices on a mid-latitude f plane to explain the dependence of the TIL on vorticity. Idealized baroclinic life cycle experiments also support the formation of the TIL by dry dynamics (Wirth and Szabo 2007; Müller and Wirth 2009; Erler and Wirth 2011). Birner (2010) reported that a static stability forcing associated with stratospheric residual circulation represents the main cause for the zonal mean TIL in winter mid-latitudes. Erler and Wirth 2011 showed that wave

*Correspondence: shigenori.otsuka@riken.jp
[1] RIKEN Advanced Institute for Computational Science, 7-1-26 Minatojima-minami-machi, Chuo-ku, 650-0047 Kobe, Japan
Full list of author information is available at the end of the article

breaking in baroclinic life cycle experiments is important for producing asymmetry between cyclonic and anticyclonic vorticies, resulting in net production of the TIL - even in zonal mean statistics. As for radiative formation, Randel et al. 2007 performed one-dimensional radiative transfer calculations, in which strong gradients in both ozone and water vapor near the tropopause contribute to the formation of the TIL. Randel and Wu 2010 reported that the radiative influence of water vapor provides a primary mechanism for the summer inversion layer in the polar region. These formation processes depend on latitude and season; for example, dynamical formation is considered to be important in winter mid-latitudes, whereas radiative formation is dominant in polar summer (Birner 2010; Miyazaki et al. 2010b; Randel and Wu 2010).

In addition to these processes, Müller and Wirth (2009) stated that gravity waves, turbulence, and deep convection may be important for the formation of the TIL. Of these processes, however, the effect of gravity waves on the formation of the TIL is not well understood yet. Previous studies focused on idealized theories (Wirth 2003, 2004), and researchers performed numerical experiments using idealized systems (e.g., Erler and Wirth 2011; Müller and Wirth 2009; Son and Polvani 2007; Wirth and Szabo 2007), and analyzed them from a global point of view (e.g., Birner 2010; Miyazaki et al. 2010b; Randel et al. 2007). Thus, time-dependent small-scale processes around the TIL are not fully understood yet. Gravity wave activity around the TIL is considered to be important in terms of mixing. Miyazaki et al. 2010a reported that the degree of mixing around the TIL due to small-scale processes is high, and the local Richardson number tends to be less than 0.25, indicating higher possibilities of gravity wave breaking around the TIL. Mixing around the tropopause leads to material exchange between the stratosphere and the troposphere.

Motivated by these studies, the objective of this work is as follows. The first question is how the TIL evolves in a realistic life cycle of an extratropical cyclone in winter mid-latitudes, in which dynamical formation mechanisms seem to be dominant. The second question is how small-scale processes such as gravity waves contribute to the TIL. To address these questions, a case study on a realistic explosive extratropical cyclogenesis is conducted using a mesoscale regional atmospheric model. Although this case study provides only limited statistics on the small-scale processes around the TIL, it will fill in the gap in previous studies.

The rest of this paper is organized as follows. The 'Methods' section describes the numerical model. The 'Results' section gives the results of the numerical experiment and analyses for the TIL. In the 'Discussion'

section, implications of the results are discussed, and the 'Conclusions' section provides our conclusions.

Methods

The numerical model used in this study is the Japan Meteorological Agency Nonhydrostatic Model (Saito et al. 2006, 2007). We set up a single computational domain with a 20 km horizontal resolution. The domain has 207 × 200 grid points centered at 39.0°N, 139.46°E on a Lambert projection. See Figure 1 for the coverage of the domain. The model employs a terrain-following coordinate system with 210 layers from the surface to 25 km above mean sea level. The vertical resolution is about 150 m around the tropopause. Model outputs are interpolated to pressure levels that have almost the same vertical grid spacing as that of the original data. Time integration is performed over 48 h from 0000 UTC 19 February, 2009 with a time step of 10 s. During this period, an explosive cyclogenesis occurred around Japan.

The cumulus parameterization scheme is a modified Kain-Fritsch. The cloud microphysics scheme is a 6-class bulk microphysics. The GSM0412 radiation scheme (Yabu et al. 2005) and an improved Mellor-Yamada Level-3 planetary boundary layer scheme (Nakanishi and Niino 2004; 2006) are used. The time constant for the nonlinear diffusion term is 1,200 s and that for the linear diffusion term is 600 s. The National Centers for Environmental Prediction (NCEP) Global Tropospheric Analyses (final analyses, FNL) with a horizontal resolution of 1° × 1° and a time interval of 6 h are used as the initial and boundary conditions of the model.

Water vapor and ozone are important species for the radiative formation of the TIL as noted in the 'Background' section. In our experiment, water vapor is a prognostic variable in the model and contributes to the formation of the TIL. The ozone distribution is taken from three-dimensional monthly mean climatology interpolated in time and does not follow the temporal change of the local tropopause.

The two horizontal axes of the model domain are denoted by x (almost eastward) and y (almost northward). Note that the zonal wind u in this paper denotes the wind component in the x direction, and the meridional wind v denotes that in the y direction.

Results

Synoptic evolution

This subsection presents an overview of the cyclogenesis during the period of the experiment. Figure 1a,b,c,d,e shows the horizontal distributions of temperature (color) and geopotential height (contour) at 950 hPa for the entire computational domain at 12-h intervals. The explosive cyclogenesis is captured by this simulation as follows. At $t = 12$ h, a cyclone starts to develop over Japan. The

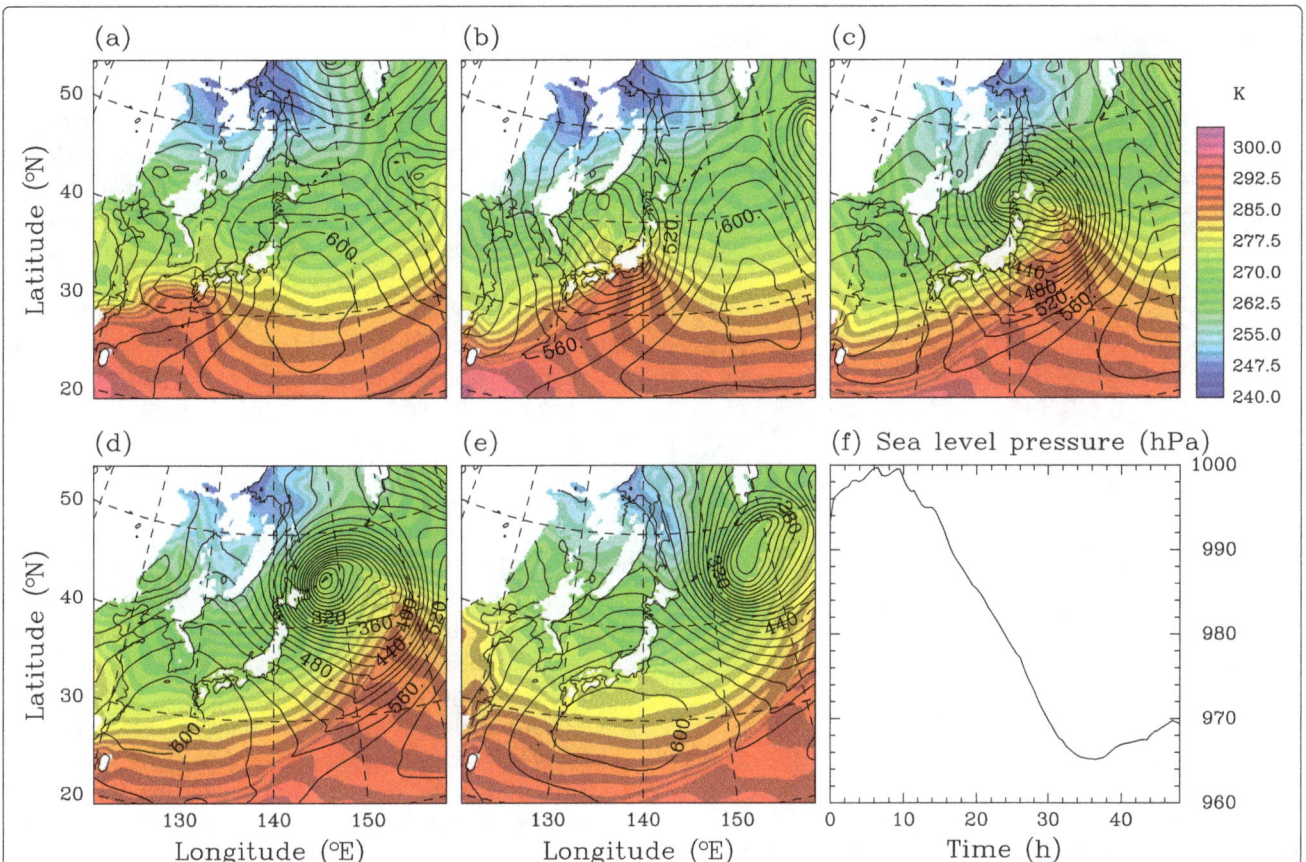

Figure 1 Temperature and geopotential height at 950 hPa. Horizontal maps of temperature (color) and geopotential height (contours) at 950 hPa for $t =$ **(a)** 0, **(b)** 12, **(c)** 24, **(d)**, 36, and **(e)** 48 h. The white color denotes regions where the surface pressure is below 950 hPa. **(f)** Time series of minimum sea level pressure in the computational domain.

horizontal gradient of temperature becomes steeper, leading to frontogenesis. At $t = 36$ h, the cyclone moves to the east of Hokkaido Island, at which time the central pressure drops to the lowest in the cyclone's life. At $t = 48$ h, the cyclone moves to the east of the Kamchatka Peninsula, and the sharp core of the cyclone starts to decay. The model domain covers the cyclone throughout the period of the experiment. The domain also covers an anticyclone to the east of the cyclone until $t = 24$ h and another anticyclone to the west from $t = 24$ h. Asymmetry between the cyclone and the anticyclones is large; the cyclone is deeper and compact, and the anticyclones are wider. Although this asymmetry may create some bias in domain-averaged statistics on the TIL, the main purpose of the paper is to describe the time evolution of the TIL with fine-scale structures, and global or zonal mean statistics are beyond the scope of this study.

Figure 1f shows a time series of minimum sea level pressure within the computational domain. As the cyclone develops, the pressure starts to decrease from $t = 10$ h. The pressure becomes the lowest, 965 hPa, at around

$t = 36$ h and then starts to increase. The pressure decreases nearly 35 hPa in 24 h between 35 and 45°N, which satisfies the criterion for a 'bomb' cyclone by put forward by Sanders and Gyakum (1980). As the position and the central pressure of the cyclone at $t = 36$ h are almost the same as those in the JMA analysis (Japan Meteorological Agency 2009), we consider that the model simulates the cyclone realistically. In this paper, we define the development stage of the cyclone as the period from 10 to 36 h; its typical snapshot is at $t = 24$ h, and the mature stage is at $t = 36$ h.

Figure 2 shows vertical cross sections along $x = 131$ in the model (almost meridional sections crossing the center of the surface cyclone) of the zonal wind (u, color) and potential temperature (θ, contour) at $t = 36$ h. The tropopause detected according to the definition by the World Meteorological Organization (WMO) is shown by the dots. The axis of the subtropical jet is located around 1,600 km from the southern boundary (145°E, 35°N), 300 hPa. The center of the surface cyclone is located at 2,700 km (147°E, 44°N), where the zonal wind reverses. The tropopause is located above 100 hPa at equatorward

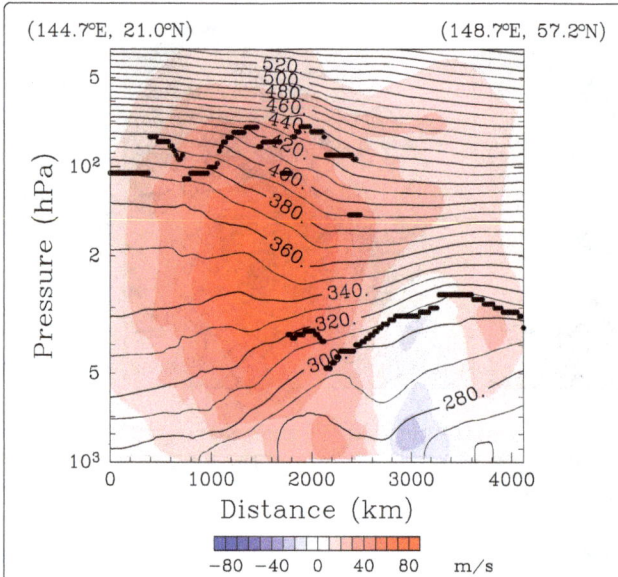

Figure 2 Zonal wind, potential temperature, and tropopause along 146°E at $t = 36$ h. Vertical cross sections of the zonal wind u (color), potential temperature (contours), and tropopause (dots) along $x = 131$ (about 146°E, crossing the center of the surface cyclone) at $t = 36$ h. Pressure levels greater than the surface pressure are shown in white.

h. When multiple tropopauses are detected at the same horizontal grid point, the lowest level is used in this plot. The contour line for 80 m s^{-1}, which shows the jet axis, lies from 132°E, 32°N to 154°E, 38°N. The gap of Z_{TP} is located around the jet axis, $Z_{TP} > 14$ km to the south and $Z_{TP} < 14$ km to the north. The lowest Z_{TP} appears around 140 to 150°E, 40 to 45°N, where the center of the surface cyclone is located. In the following analyses, only regions where $Z_{TP} < 14$ km are used.

Figure 3b shows the vertical component of relative vorticity at Z_{TP} (hereafter ζ_{TP}) at $t = 36$ h. Horizontal winds at Z_{TP} are also shown by arrows. A spiral pattern of positive and negative vorticities associated with the cyclone is discernible. The highest ζ_{TP} is obtained around 145 to 150°E, 40 to 45°N, which is located to the north of the jet exit region and to the east of the surface cyclone. The lowest ζ_{TP} is obtained around 160 to 165°E, 45°N. The region of $\zeta_{TP} > 0$ along the northern edge of the jet streak (southern edge of the analysis region) is caused by the horizontal wind shear of the jet streak. A spiral band of $\zeta_{TP} > 0$ extending to the northwestern corner of the domain coincides with the northeastern edge of strong westerlies at the tropopause level. Within the analysis region, the area of positive ζ_{TP} is greater than that of negative ζ_{TP}. Thus, data are binned by ζ_{TP} and normalized by the sample size to ensure that the conclusions are not biased by the difference of positive and negative ζ_{TP} areas.

latitudes of about 45°N, and around 300 to 500 hPa at latitudes poleward of the jet. Near the jet, a multiple tropopause appears, exhibiting tropopause folding.

Figure 3a shows the horizontal distributions of the tropopause height according to the WMO definition (Z_{TP}, color) and the wind speed at 300 hPa (contour) at $t = 36$

Relationship between TIL and relative vorticity at tropopause

As previous studies reported the existence of a negative correlation between ζ_{TP} and the strength of the TIL, we

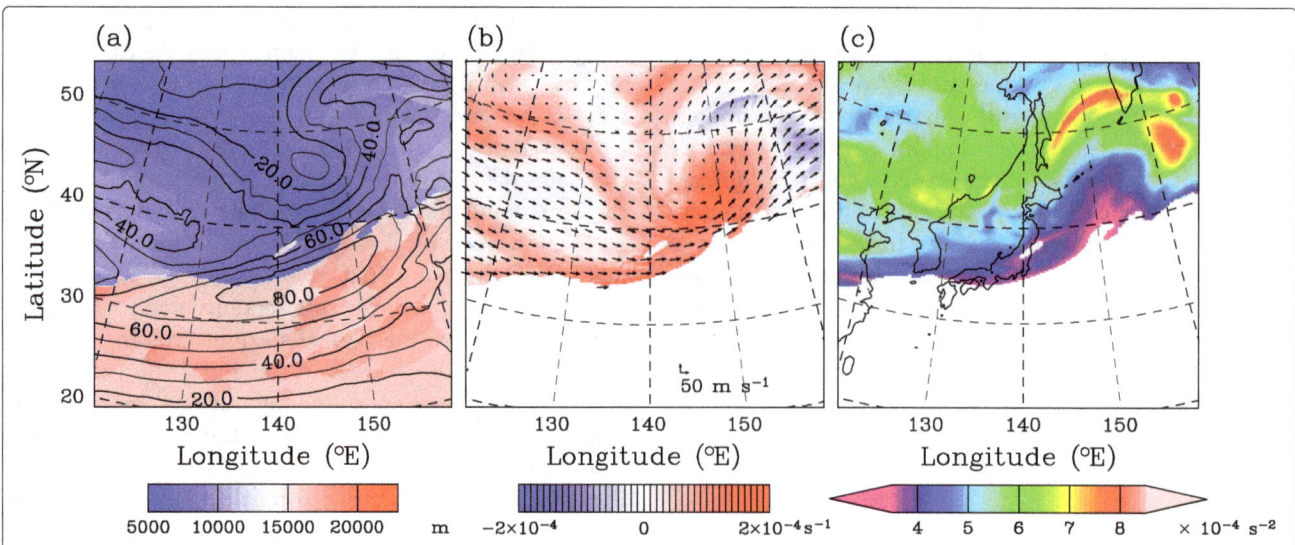

Figure 3 Tropopause height, wind speed at 300 hPa, ζ_{TP}, horizontal winds at Z_{TP}, and N_{max}^2. (a) Height of the lowermost tropopause (Z_{TP}, color) and wind speed at 300 hPa (contours, m s^{-1}), **(b)** ζ_{TP} (color) and horizontal winds at Z_{TP} (arrows), and **(c)** N_{max}^2 at $t = 36$ h. **(b)** and **(c)** are only for the tropopause below 14 km.

will examine the relationship between these two quantities. In order to examine the dependence on ζ_{TP}, 22,631 temperature profiles within regions with $Z_{TP} < 14$ km at $t = 36$ h are binned by ζ_{TP} and averaged in each bin (Figure 4a). The bin width is 5×10^{-5} s^{-1}. The vertical coordinate is the height above the tropopause. The mean profiles for negative ζ_{TP} (dark blue lines) show clear temperature inversions with amplitudes of 4 to 5 K and thicknesses of 1.5 to 2 km just above the tropopause, whereas the profiles for positive ζ_{TP} (red lines) do not show clear inversions. This result is consistent with previous observational and numerical studies (e.g., Randel et al. 2007; Son and Polvani 2007). Note that the absence of inversion layers in the mean temperature profile for $\zeta_{TP} \geq 1.5 \times 10^{-4}$ s^{-1} seems partly due to the vertical resolution of the model.

Figure 4b is the same as Figure 4a but for the buoyancy frequency squared (N^2). The tropospheric N^2 (4 km below the tropopause) is about 1.5×10^{-4} s^{-2}, whereas the stratospheric N^2 (4 km above the tropopause) is about 3 to 4×10^{-4} s^{-2}. The profiles for $\zeta_{TP} < 0$ show a sharp local maximum just above the tropopause and a sharp local minimum just below the tropopause (except for $-5 \times 10^{-5} \leq \zeta_{TP} < 0$ s^{-1}), whereas the profiles for $\zeta_{TP} > 0$ show a smooth transition from the troposphere to the stratosphere without sharp peaks. As ζ_{TP} increases, the sharpness of the N^2 profiles decreases. Following Birner et al. (2006), the maximum N^2 within the temperature inversion (denoted by N^2_{max}) is used as a measure of the strength of the TIL in the following analyses. In this study, N^2_{max} is searched for within 4 km above the local tropopause.

The horizontal map of N^2_{max} at $t = 36$ h is shown in Figure 3c. Regions with high N^2_{max} are located to the north of the surface cyclone, whereas regions with low N^2_{max} are located around the south edge of the analysis region. The distribution of N^2_{max} around the cyclone is similar to that in the idealized simulation with dry dynamics by Wirth and Szabo (2007) (their Figure three), implying common mechanisms between these two simulations. A comparison of Figure 3b,c shows that regions with high (low) N^2_{max} coincide with negative (positive) ζ_{TP}. Thus, in general, a negative correlation exists between ζ_{TP} and N^2_{max} in the horizontal distributions at $t = 36$ h. However, the regions of $\zeta_{TP} < 0$ in the eastern part of the domain coincide with N^2_{max} greater than 8×10^{-4} s^{-2} (red regions in Figure 3c), whereas those with similar negative values of ζ_{TP} in the western part coincide with N^2_{max} less than 7×10^{-4} s^{-2} (green and blue regions in Figure 3c), implying the existence of other factors that determine N^2_{max}.

The relationship between ζ_{TP} and N^2_{max} is further examined along the life cycle of the cyclone. Figure 5 shows two-dimensional histograms of ζ_{TP} (the horizontal axis) and N^2_{max} (the vertical axis) at 6-h intervals. From $t = 0$ to 18 h, the negative correlation between ζ_{TP} and N^2_{max} is not clear. The correlation coefficient between ζ_{TP} and N^2_{max} is -0.28 at $t = 0$ h and increases to -0.64 at $t = 18$ h. At the initial time ($t = 0$), the TIL might be underestimated due to the coarser resolution of the NCEP FNL, which is used as the initial condition. However, it is difficult to distinguish the spin-up of the TIL from the evolution of the TIL due to the synoptic evolution. Thus, we mainly analyze $t = 24$ to 36 h.

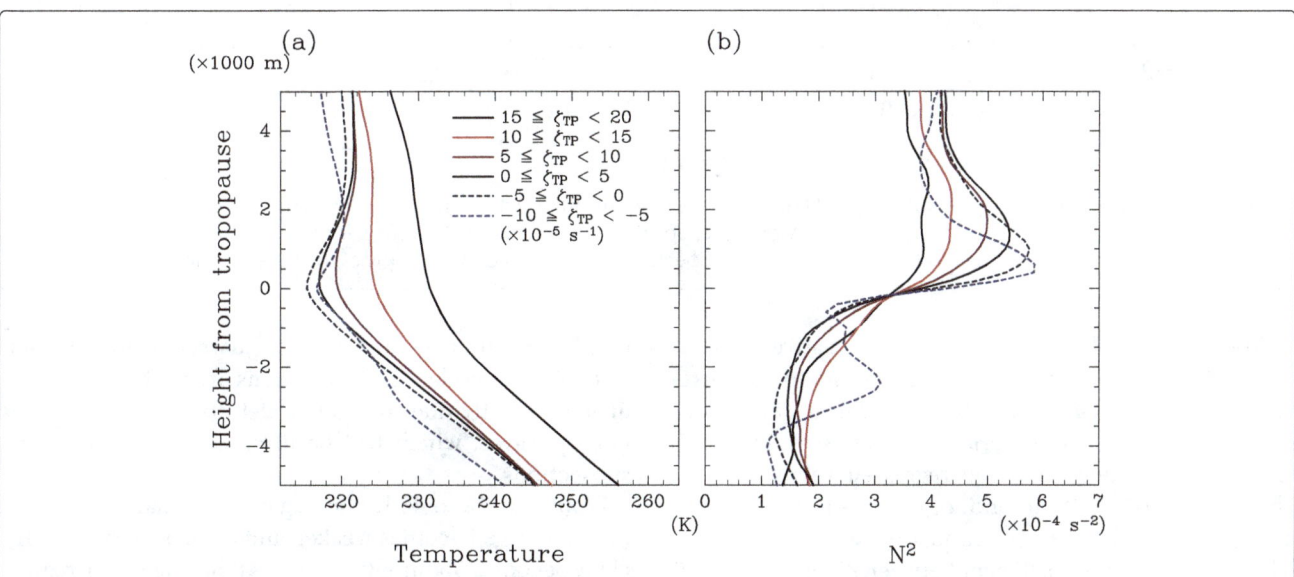

Figure 4 Mean profiles of temperature and N^2 for each category of ζ_{TP}. Mean profiles of **(a)** temperature and **(b)** N^2 for each category of ζ_{TP} at $t = 36$ h for the tropopause below 14 km. Vertical axis is the height from the tropopause.

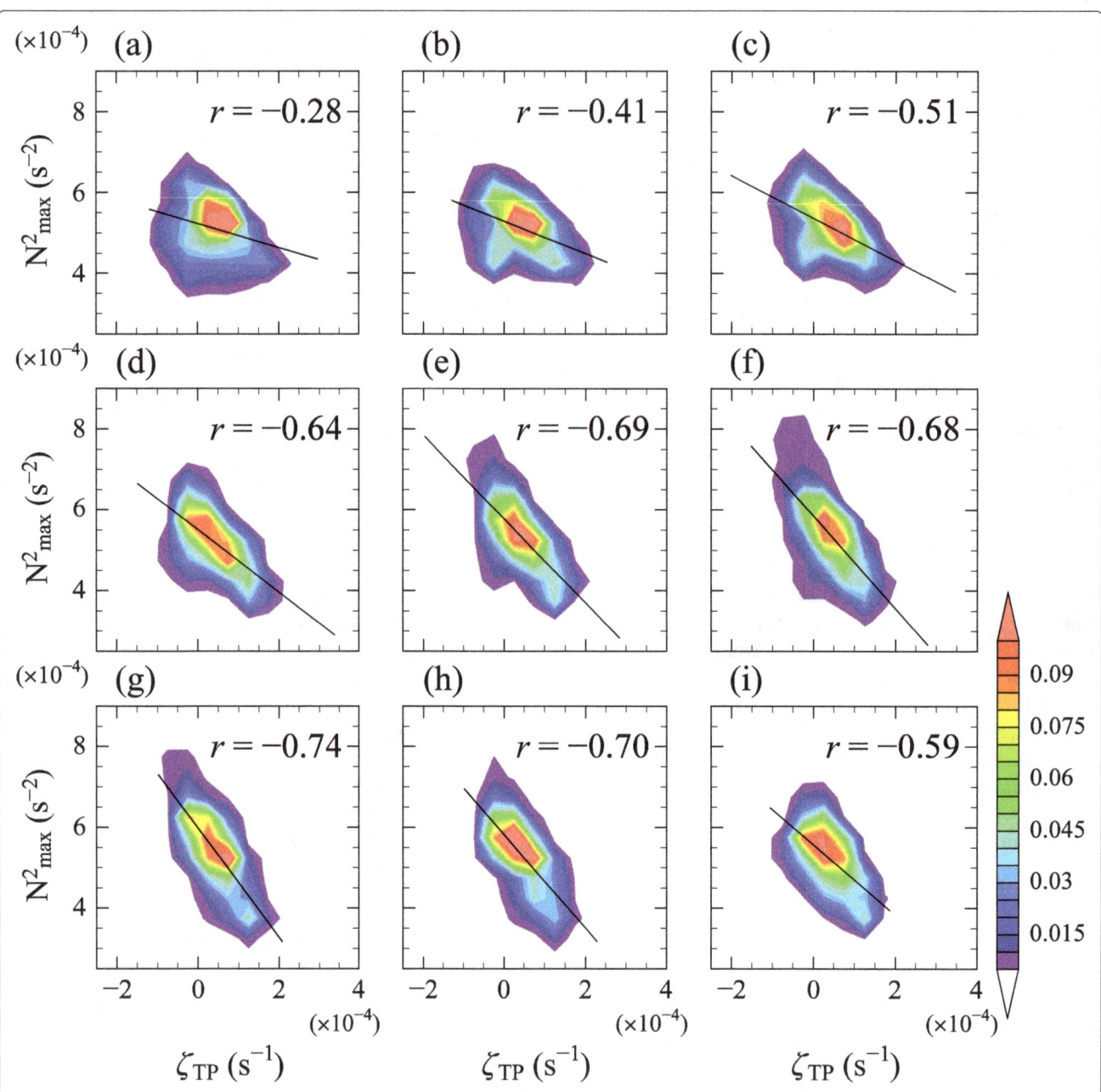

Figure 5 Two-dimensional histograms of N_{max}^2 and ζ_{TP} for every 6 h. **(a)** 0 h. **(b)** 6 h. **(c)** 12 h. **(d)** 18 h. **(e)** 24 h. **(f)** 30 h. **(g)** 36 h. **(h)** 42 h. **(i)** 48 h. Two-dimensional histograms of ζ_{TP} and N_{max}^2 for every 6 h for the tropopause below 14 km. Bin widths of ζ_{TP} and N_{max}^2 are 5×10^{-5} s^{-1} and 5×10^{-5} s^{-2}, respectively. Numbers in each panel show the correlation coefficients r between ζ_{TP} and N_{max}^2. Black lines show the linear regression lines.

From $t = 24$ to 42 h, a clear negative correlation between ζ_{TP} and N_{max}^2 is obtained with a correlation coefficient of about -0.7. The linear regressions of N_{max}^2 as functions of ζ_{TP} (black lines in Figure 5) show steeper slopes, and the regressed values become about 7×10^{-4} s^{-2} around $\zeta_{TP} = -1 \times 10^{-4}$ s^{-2}, and about 5×10^{-4} s^{-2} around $\zeta_{TP} > 2 \times 10^{-4}$ s^{-1}. The proportionality coefficient between ζ_{TP} and the mean N_{max}^2 for $\zeta_{TP} < 0$ is about -1, which is somewhat smaller than that for the reference run in Erler and Wirth

(2011), presumably due to the coarser spatial resolution of our model. The deviations from the regression lines are partly due to small-scale disturbances such as gravity waves, which will be discussed in the following subsections.

From $t = 42$ to 48 h, the negative correlation between ζ_{TP} and N_{max}^2 becomes weaker, and the correlation coefficient becomes about -0.6 at $t = 48$ h. The linear regression slope becomes gentler. In summary, the negative correlation between ζ_{TP} and N_{max}^2 is clearer during the development and mature stages of the cyclone, whereas it is unclear before and after that period.

Analysis of gravity waves around the tropopause

Figure 6 shows $\partial w/\partial z$ at 260 hPa every 3 h from $t = 24$ to 36 h. Red colors show vertical convergence, and blue colors show vertical divergence. An arc-shaped wave packet with east-west wave fronts over the Sea of Okhotsk-Northwestern Pacific propagates northward. At $t = 24$ h, the wave fronts extend eastward from the southern tip of Sakhalin. The wave fronts tilt gradually toward the southwest-northeast direction, and at $t = 36$ h, they extend almost in a south-north direction off the east coast of middle Sakhalin. In the region of the arc-shaped wave

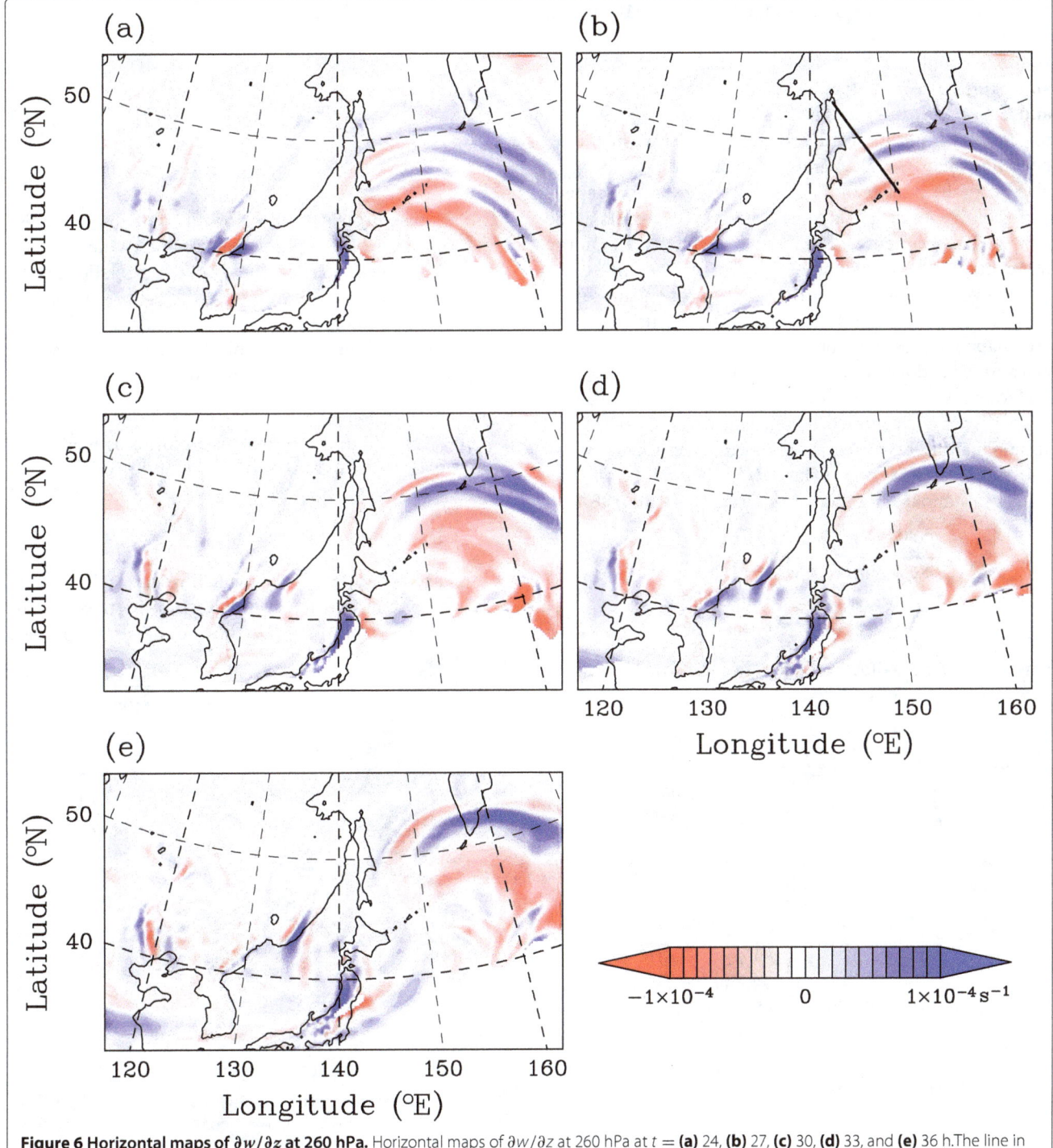

Figure 6 Horizontal maps of $\partial w/\partial z$ at 260 hPa. Horizontal maps of $\partial w/\partial z$ at 260 hPa at $t =$ **(a)** 24, **(b)** 27, **(c)** 30, **(d)** 33, and **(e)** 36 h. The line in **(b)** shows the position of the vertical section in Figure 7.

packet, $N_{\max}^2$ is the highest, where ζ_{TP} is negative as shown in Figures 3b,c at the tropopause. The wave packets over the Asian continent around (120°E, 40°N), (130°E, 40°N), and (135°E, 45°N) are likely mountain waves. As the wind field changes with time, the mountain waves also change their structure.

The arc-shaped waves are robustly simulated in sensitivity experiments with different horizontal resolutions (15 and 20 km), vertical resolutions (50 layers and 210 layers), numerical integration time steps (10 and 20 s), time constants for nonlinear numerical diffusion (1,200 and 2,400 s), or time constants for linear numerical diffusion (600 and 2,400 s). If the number of vertical layers is reduced to 50, the amplitude of the arc-shaped waves in $\partial w/\partial z$ becomes slightly weaker, and the associated peak in $N_{\max}^2$ also becomes smaller (not shown). The resolution dependency of the peak $N_{\max}^2$ is consistent with the results of the idealized experiment by Müller and Wirth (2009). On the other hand, the weakening of the TIL around the center of the cyclone also becomes gentle. Although the small-scale structures are different, the time evolution of the arc-shaped waves and the negative correlation between ζ_{TP} and $N_{\max}^2$ does not change much.

Here, temporal modulation of local N^2 by propagating gravity waves or stationary disturbances is discussed in a simple manner because the regions of clear gravity waves coincide with the regions of large $N_{\max}^2$ (Figures 3c and 6e). If we use standard notations (e.g., Andrews et al. 1987),

$$\frac{\partial N^2}{\partial t} \equiv \frac{\partial}{\partial t}\left(\frac{g}{\theta}\frac{\partial \theta}{\partial z}\right)$$
$$= \frac{g}{\theta}\left(\frac{\partial}{\partial z} - \frac{1}{\theta}\frac{\partial \theta}{\partial z}\right)\left(\frac{\partial \theta}{\partial t}\right), \quad (1)$$

where g is the gravitational acceleration. If the process is adiabatic,

$$\frac{\partial N^2}{\partial t} = \frac{g}{\theta}\left(\frac{\partial}{\partial z} - \frac{1}{\theta}\frac{\partial \theta}{\partial z}\right)(-\boldsymbol{u}\cdot\nabla\theta)$$
$$= \frac{g}{\theta}\left(-\frac{\partial \boldsymbol{u}}{\partial z}\cdot\nabla\theta - \boldsymbol{u}\cdot\nabla\frac{\partial \theta}{\partial z} + \frac{1}{\theta}\frac{\partial \theta}{\partial z}\boldsymbol{u}\cdot\nabla\theta\right), \quad (2)$$

where $\boldsymbol{u} \equiv (u, v, w)$ denotes three-dimensional winds. If θ is assumed to be uniform in the horizontal direction,

$$\frac{\partial N^2}{\partial t} \approx \frac{g}{\theta}\left[-\frac{\partial w}{\partial z}\frac{\partial \theta}{\partial z} + w\left\{\frac{1}{\theta}\left(\frac{\partial \theta}{\partial z}\right)^2 - \frac{\partial^2 \theta}{\partial z^2}\right\}\right]. \quad (3)$$

Vertically propagating disturbances
If $w = A\sin(mz - \omega t)$,

$$\frac{\partial N^2}{\partial t} = A\frac{g}{\theta}\left[-m\cos(mz - \omega t)\frac{\partial \theta}{\partial z}\right.$$
$$\left. + \sin(mz - \omega t)\left\{\frac{1}{\theta}\left(\frac{\partial \theta}{\partial z}\right)^2 - \frac{\partial^2 \theta}{\partial z^2}\right\}\right], \quad (4)$$

where A is a constant, m is the vertical wave number, and ω is the frequency. Assuming that $\partial \theta/\partial z$ is a positive constant and $\partial \theta/\partial z \ll m\theta$,

$$\frac{\partial N^2}{\partial t} \approx -AmN^2\cos(mz - \omega t). \quad (5)$$

This gives

$$N^2 = \int \frac{\partial N^2}{\partial t}dt$$
$$\approx \frac{AmN^2}{\omega}\sin(mz - \omega t) + \text{const}, \quad (6)$$
$$\frac{\partial w}{\partial z} = Am\cos(mz - \omega t), \quad (7)$$
$$\frac{\partial^2 w}{\partial z^2} = -Am^2\sin(mz - \omega t). \quad (8)$$

If the phase propagation is downward ($m/\omega < 0$), N^2 and $\partial^2 w/\partial z^2$ become in phase. For simplicity, the feedback from the changes in N^2 to the dispersion relation of the wave is considered to be free from the variation in N^2 in this paper. Nonlinearity in the TIL (e.g., Miyazaki et al. 2010a) is beyond the scope of this paper, which focuses on a diagnosis of linear wave propagation.

Stationary disturbances
If $w = A\sin(mz)$ and the same assumption is made with regard to θ and N^2 as above, from Equation 3,

$$N^2 \approx -AmN^2\cos(mz)\cdot t + \text{const}, \quad (9)$$
$$\frac{\partial w}{\partial z} = Am\cos(mz). \quad (10)$$

In this case, $-\partial w/\partial z$ and N^2 are in phase.

Figure 7 shows the vertical cross section of $\partial w/\partial z$ (color) and N^2 (contour) along the line shown in Figure 6b

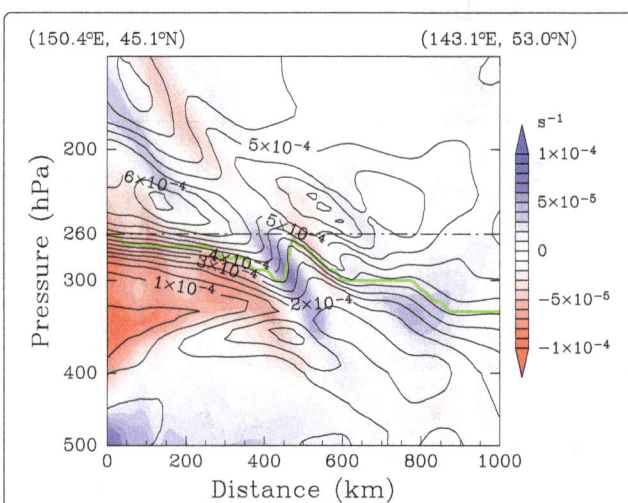

Figure 7 Vertical cross sections of $\partial w/\partial z$ and N^2. Vertical cross sections of $\partial w/\partial z$ (color, s^{-1}) and N^2 (contour, s^{-2}) at $t = 27$ h along the line shown in Figure 6b. Tropopause is denoted by the green line. The dashed-dotted line shows the 260 hPa level.

at $t = 27$ h. The color coding is the same as that in Figure 6. The green line shows the tropopause. A wave pattern is located at around 0 to 800 km in the horizontal direction and 300 to 200 hPa in the vertical direction, which corresponds to the arc-shaped wave packet in Figure 6.

The dispersion relation of the wave packet around the tropopause at 147°E, 49°N, $t = 27$ h is examined because the wave packet is clearer at this point and time. The region has a strong wind shear and a gap in N^2, showing a nonlinear nature at a later time. From Figures 6b and 7, the wavelengths in the x, y, and z directions are 5.0×10^2 km, 1.8×10^2 km, and 2.0–3.0 km, respectively. From Figure 7, N^2 is about 5×10^{-4} s^{-2}. Using the Coriolis parameter at 50°N, $f = 1.1 \times 10^{-4}$ s^{-1}, the dispersion relation for an inertia-gravity wave (e.g., Gill 1982),

$$\hat{\omega} = \omega - \mathbf{k} \cdot \mathbf{u} = \sqrt{\frac{N^2(k^2 + l^2) + f^2 m^2}{k^2 + l^2 + m^2}}, \quad (11)$$

gives an intrinsic frequency $\hat{\omega} = 2.9$–4.1×10^{-4} s^{-1}, or a period of about 2.5–3.7×10^2 minutes. Here, ω denotes the frequency in a fixed frame of reference, k and l are the wavenumbers along the x and y directions, respectively, and $\mathbf{k} \equiv (k, l, m)$. We assume $k > 0$, $l < 0$, and $m < 0$ from the phase lines in Figures 6 and 7. Substituting environmental wind speeds of of about $(2.8 \times 10, 3.2 \times 10, 1.8 \times 10^{-2})$ m s^{-1}, ω becomes -5.4 to -4.0×10^{-4} s^{-1} or a period of about 1.9 to 2.6×10^2 min. In the model, the period at a fixed point is about 2.0×10^2 min with northward phase propagation, or $\omega = -5.2 \times 10^{-3}$ s^{-1}, which is close to the value estimated from Equation 11. Thus, this wave satisfies the dispersion relation for an inertia-gravity wave.

From a hodograph analysis of high-pass filtered horizontal winds, it is also confirmed that the wave is an inertia-gravity wave that has a downward phase speed (not shown). Although detailed analysis of gravity waves is beyond the scope of this paper, several discussions will be presented later.

Relationship between TIL and gravity waves

In this subsection, the relationship between the TIL and the gravity waves is further examined to ascertain if the vertical convergence shown in Figure 7 is the dominant factor in the strengthening of the TIL during the development and mature stages of the cyclone.

In Figure 7, $\partial w/\partial z$ is in quadrature with N^2 in the stratosphere; the local maxima of N^2 are located between local maxima of $\partial w/\partial z$ above and the local minima below. From Equations 6 to 7, this implies that the wave pattern for N^2 is produced by the downward propagating wave disturbance of w, which is consistent with the analysis in the previous subsection that indicated that the wave with

a downward phase speed is an inertia-gravity wave. These high-N^2 regions are detected as TILs.

It is important to confirm that the amplitudes of the wave patterns in N^2 and w are consistent with each other. From Equation 6, the amplitude of N^2 becomes AmN^2/ω. In Figure 7, the amplitude of $\partial w/\partial z$, which becomes Am, is about 5×10^{-5} s^{-1}. Substituting $Am = 5 \times 10^{-5}$ s^{-1}, $N^2 = 5 \times 10^{-4}$ s^{-2}, and $\omega = 3.1 \times 10^{-4}$ s^{-1} into Equation 6, the amplitude of the wave pattern of N^2 is estimated to be 8.3×10^{-5} s^{-2}, whereas the actual amplitude in Figure 7 is about 10^{-4} s^{-2}. These are the same order of magnitude as each other. As the amplitude of the wave pattern for N^2 is proportional to N^2, the wave pattern is clearly seen only within the TIL (Figure 7).

Figure 8 shows histograms of N^2_{max} in the TIL, using the hourly outputs from $t = 24$ to 36 h. Horizontal grid points within the analysis regions of 13 snapshots (in total 286,836 grid points) are categorized by the sign of either (a) ζ_{TP}, (b) $\partial w/\partial z$, or (c) $\partial^2 w/\partial z^2$. The histograms of N^2_{max} are created for each group. The bin width of the histograms is 5×10^{-5} s^{-1}. Each histogram is normalized by its sample size.

First, differences due to the sign of ζ_{TP} are examined (Figure 8a). The peak for $\zeta_{\text{TP}} < 0$ appears at $N^2_{\text{max}} = 5.25 \times 10^{-4}$ s^{-2}, whereas that for $\zeta_{\text{TP}} \geq 0$ appears at $N^2_{\text{max}} = 6.25 \times 10^{-4}$ s^{-2}. This result is consistent with the negative correlation between ζ_{TP} and N^2_{max} in Figures 4 to 5.

Next, differences due to the sign of $\partial w/\partial z$ are examined in Figure 8b. Although the two histograms both peak at $N^2_{\text{max}} = 5.25 \times 10^{-4}$ s^{-2}, the distribution for $\partial w/\partial z < 0$ s^{-1} (short-dashed) appears slightly to the right of that for $\partial w/\partial z \geq 0$ s^{-1} (solid), indicating that N^2_{max} increases as $\partial w/\partial z$ decreases. This indicates that stationary disturbances enhance the TIL through vertical convergence. The difference between positive and negative $\partial w/\partial z$ is slightly clearer when the histograms are computed only for the regions of $\zeta_{\text{TP}} < 0$ (Figure 9). For $\zeta_{\text{TP}} < 0$, the peak for $\partial w/\partial z \geq 0$ is at $N^2_{\text{max}} = 5.75 \times 10^{-4}$ s^{-2}, and that for $\partial w/\partial z < 0$ is at $N^2_{\text{max}} = 6.25 \times 10^{-4}$ s^{-2} (Figure 9a). For $\zeta_{\text{TP}} \geq 0$, the two histograms peak at the same N^2_{max} (Figure 9b). Note that the time integration of $\partial w/\partial z$ strengthens N^2 when the stationary disturbances are dominant. Thus, local $\partial w/\partial z$ at a specific location and time may have little correlation with N^2_{max}.

Finally, differences due to the sign of $\partial^2 w/\partial z^2$ are examined in Figure 8c. The two histograms both peak at $N^2_{\text{max}} = 5.25 \times 10^{-4}$ s^{-2}. However, the distribution for $\partial^2 w/\partial z^2 \geq 0$ is located to the right of that for $\partial^2 w/\partial z^2 < 0$, indicating that N^2_{max} increases as $\partial^2 w/\partial z^2$ increases. This implies that waves with a downward phase propagation are important for the strengthening of the TIL. The difference between positive and negative $\partial^2 w/\partial z^2$ is clearer when the histograms are computed only for the

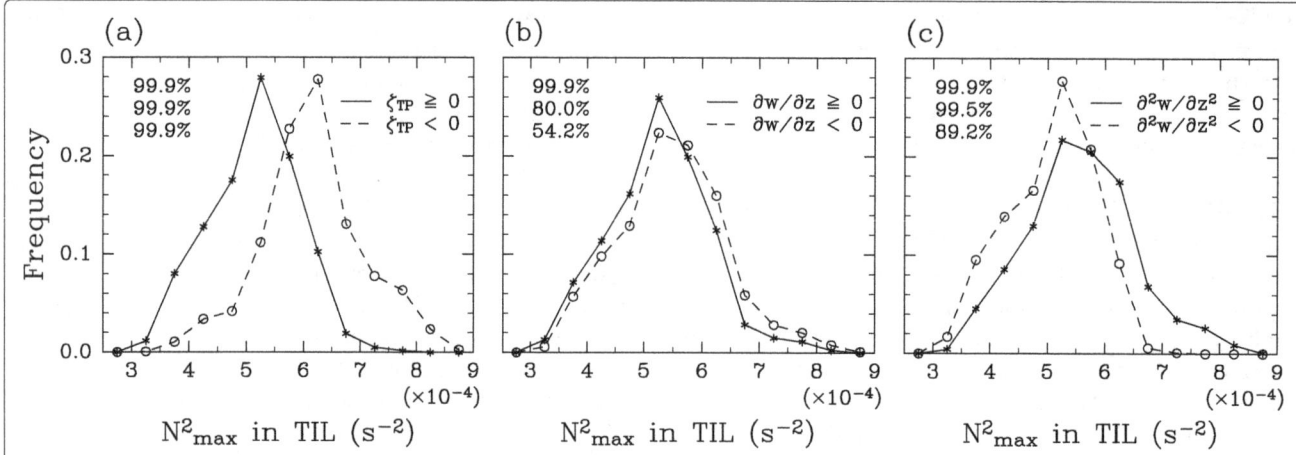

Figure 8 Histograms of N^2_{max} for different ζ_{TP}, $\partial w/\partial z$, or $\partial^2 w/\partial z^2$. Histograms of N^2_{max} in the TIL in the regions of either **(a)** $\zeta_{TP} < 0$ and $\zeta_{TP} \geq 0$, **(b)** $\partial w/\partial z < 0$ and $\partial w/\partial z \geq 0$, or **(c)** $\partial^2 w/\partial z^2 < 0$ and $\partial^2 w/\partial z^2 \geq 0$. Computed from hourly outputs from $t = 24$ to 36 h. The $\partial w/\partial z$ and $\partial^2 w/\partial z^2$ are computed at the same level as N^2_{max}. Each histogram is normalized by its sample size. Values on the top left are significance levels for the differences in mean N^2_{max} by the moving block bootstrap method with block sizes of (from top to bottom) $10 \times 10 \times 5$, $20 \times 20 \times 10$, and $40 \times 40 \times 10$ grid points (or, equivalently 200 km $\times$ 200 km $\times$ 5 h, 400 km $\times$ 400 km $\times$ 10 h, and 800 km $\times$ 800 km $\times$ 10 h) in the x, y, and t directions.

regions of $\zeta_{TP} < 0$ (Figure 10). For $\zeta_{TP} < 0$, the peak for $\partial^2 w/\partial z^2 < 0$ is at $N^2_{max} = 5.75 \times 10^{-4}$ s^{-2} and that for $\partial^2 w/\partial z^2 \geq 0$ is at $N^2_{max} = 6.25 \times 10^{-4}$ s^{-2} (Figure 10a). For $\zeta_{TP} \geq 0$, the two histograms peak at the same N^2_{max} (Figure 10b). The difference by $\partial^2 w/\partial z^2$ in Figure 10a is clearer than that by $\partial w/\partial z$ in Figure 9a. Although waves with upward phase propagation may also exist, their contribution seems to be smaller.

It is not a straightforward task to test the statistical significance of spatially correlated samples with appropriate estimates of correlation scales. Thus, we have adopted the moving block bootstrap method (Künsch 1989) to test the

statistical significance of the differences in mean N^2_{max}. Because the optimum block size is unknown, we choose block sizes of $10 \times 10 \times 5$, $20 \times 20 \times 10$, and $40 \times 40 \times 10$ grid points (or, equivalently 200 km $\times$ 200 km $\times$ 5 h, 400 km $\times$ 400 km $\times$ 10 h, and 800 km $\times$ 800 km $\times$ 10 h) in the x, y, and t directions. We obtained 10,000 bootstrap observations each. As a result, the differences in the means in Figure 8a,c are significant to levels higher than 99% except for Figure 8c with the largest block size (see the values at the top left of each figure). The difference in the means in Figure 8b is not as significant as that in other pairs of data except the smallest block size. Statistical tests for Figures 9

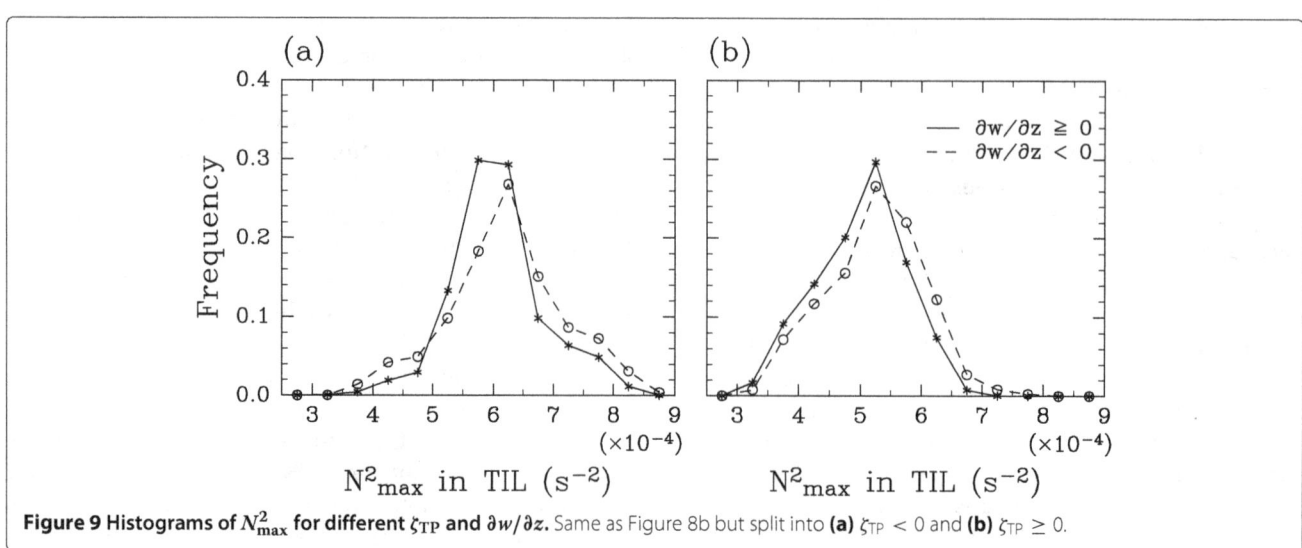

Figure 9 Histograms of N^2_{max} for different ζ_{TP} and $\partial w/\partial z$. Same as Figure 8b but split into **(a)** $\zeta_{TP} < 0$ and **(b)** $\zeta_{TP} \geq 0$.

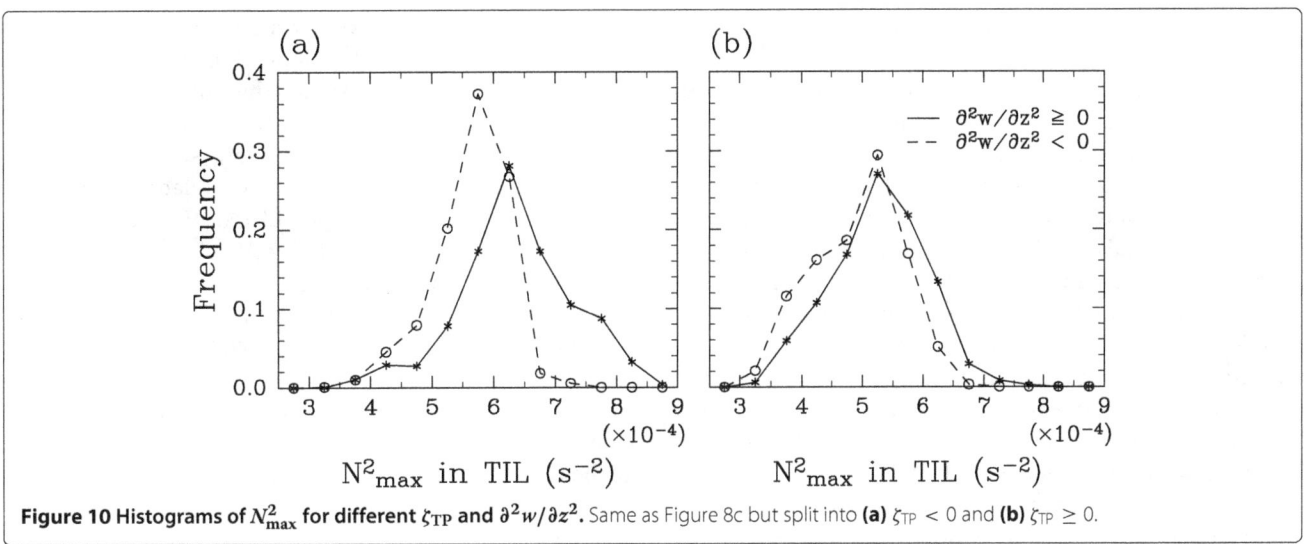

Figure 10 Histograms of N^2_{max} for different ζ_{TP} and $\partial^2 w/\partial z^2$. Same as Figure 8c but split into **(a)** $\zeta_{\mathrm{TP}} < 0$ and **(b)** $\zeta_{\mathrm{TP}} \geq 0$.

and 10 are more difficult to construct, and will not be shown in this paper.

Discussion

In our experiment, the TIL is formed as the cyclone develops. During the development and mature stages of the cyclone, the negative correlation between ζ_{TP} and N^2_{max} becomes clearer (Figure 5). In particular, N^2_{max} for negative ζ_{TP} increases, whereas N^2_{max} for positive ζ_{TP} decreases. This is consistent with the results of idealized simulations by Wirth and Szabo (2007) and Erler and Wirth (2011), in which the TIL becomes sharper as baroclinically unstable disturbances develop, and the TIL decays as the cyclone decays. Our experiment confirms that the TIL shows a similar time evolution in a realistic life cycle of an extratropical cyclone.

From Equations 6 to 10, it is expected that a steady disturbance of $\partial w/\partial z$ produces a pattern of N^2 that is in phase with $-\partial w/\partial z$, whereas a downward-propagating wave produces a pattern of N^2 that is in phase with $\partial^2 w/\partial z^2$. As shown in Figure 8, N^2_{max} has a negative correlation with $\partial w/\partial z$, whereas N^2_{max} has a positive correlation with $\partial^2 w/\partial z^2$. From these two facts, it is concluded that the TIL in our model result is strongly controlled by the dynamical processes. In particular, inertia-gravity waves, which have not been investigated in depth in previous studies on the TIL, play a significant role. Note that the western half of the analysis region is dominated by nearly stationary patterns of $\partial w/\partial z$ by the synoptic disturbances and the mountain waves, whereas the eastern half is dominated by the effect of vertically propagating gravity waves. Although both the western and eastern sides contribute to the negative correlation between ζ_{TP} and N^2_{max}, the eastern side exhibits both the highest and the lowest

N^2_{max} values. The lowest N^2_{max} appears around the center of the cyclone and the highest N^2_{max} is produced by the arc-shaped wave packet. This fact also implies that the development of the cyclone and the radiation of the gravity waves are important for the development of the negative correlation.

Radiation of inertia-gravity waves from synoptic systems has been investigated by several groups using numerical experiments (e.g., O'Sullivan and Dunkerton 1995; Plougonven and Snyder 2007; Zhang 2004; Zülicke and Peters 2006). For example, Plougonven and Snyder (2007) simulated two types of wave packets generated around the tropopause region in a so-called LC2-type baroclinic life cycle experiment. The arc-shaped packet of the inertia-gravity waves shown in Figure 6 is similar to that in previous studies and can be considered a representative feature associated with developing baroclinic disturbances. Under the assumption that gravity waves of this type are ubiquitous around synoptic systems in the extratropics, the enhancement of the TIL by gravity waves may have an impact on the climatology of the TIL. Furthermore, gravity waves generated at the exit region of jet streaks tend to experience 'wave capture' (e.g., McIntyre 2009), staying within the same region relative to the synoptic disturbances. Thus, these gravity waves may not be just linear and transient features, but could be more persistent forcings to the tropopause region through nonlinear interactions. This may be the reason the negative correlation between ζ_{TP} and N^2_{max} is clearer during the development and mature stages of the cyclone. Thus, this relationship among gravity waves, ζ_{TP}, and N^2_{max} could be a general feature associated with extratropical cyclones. Proving the statistical significance of the effect of gravity waves radiated from synoptic disturbances is a subject we plan to investigate in the future using a regional atmospheric model.

The model-simulated TILs are validated using 136 operational radiosonde profiles (distributed by the University of Wyoming) at 42 stations in the computational domain, except for those within 20 grid points from the lateral boundaries. Figure 11a shows the stations used in the analysis. Temperature profiles are resampled every 10 m in the vertical direction, and running means over 101 points are taken to smooth the profiles so that they can be compared with those in the model. The cut-off scale is about 2 km (about ten times greater than the vertical grid spacing of the model around the tropopause) to take into account the effective resolution of the model (e.g., Skamarock 2004) and small-scale gravity waves that are not resolvable in the model. Figure 11b shows the smoothed temperature profiles at 1200 UTC 20 February, 2009 at the ten stations denoted by filled circles in Figure 11a. Tropopauses are denoted by markers. The lowest tropopause is located between 300 and 500 hPa, except for the two profiles in the south. This is consistent with the tropopause height in the model (Figure 2). Although temperature inversions with much smaller depths than the smoothing scale disappear, temperature inversions just above the local tropopauses are discernible. In particular, the second and third profiles from the right (north) show clear TILs.

In total, 99 temperature profiles having local tropopauses below 14 km are selected at 35 stations from 0000 UTC 20 to 0000 UTC 21 February, 2009, and N^2_{max} is computed for each profile. Figure 11c shows the scatter diagram between the observed (the horizontal axis) and model (the vertical axis) N^2_{max} at the same time and place. The model values are linearly interpolated to the corresponding observation sites. The correlation coefficient between the observed and simulated N^2_{max} is 0.59. The mean N^2_{max} in the model is 5.3×10^{-4} s^{-2}, whereas that in the observation is 5.9×10^{-4} s^{-2}, indicating that the model underestimates the strength of the TILs. Overall, it is concluded that the model reproduced the observed N^2_{max} reasonably well.

In summary, the difference between our results and previous studies is that the effect of gravity waves has been explicitly taken into account in our analyses. As synoptic disturbances develop, gravity wave activities also increase in a specific region relative to the synoptic systems. The inertia-gravity waves are robustly reproduced with a range of model resolutions and parameters. Both synoptic-scale wave breaking and gravity wave activity seem to contribute to the time evolution of N^2_{max}. Although we have presented a single case study on TIL formation associated with extratropical explosive cyclogenesis, it is well known that explosive cyclogenesis occurs frequently in the mid-latitude storm tracks from autumn to spring (e.g., Sanders and Gyakum 1980). Furthermore, our result is consistent with the fact that the dynamical TIL formation is considered to be dominant in winter mid-latitudes (e.g., Birner 2010; Miyazaki et al. 2010b; Randel and Wu 2010).

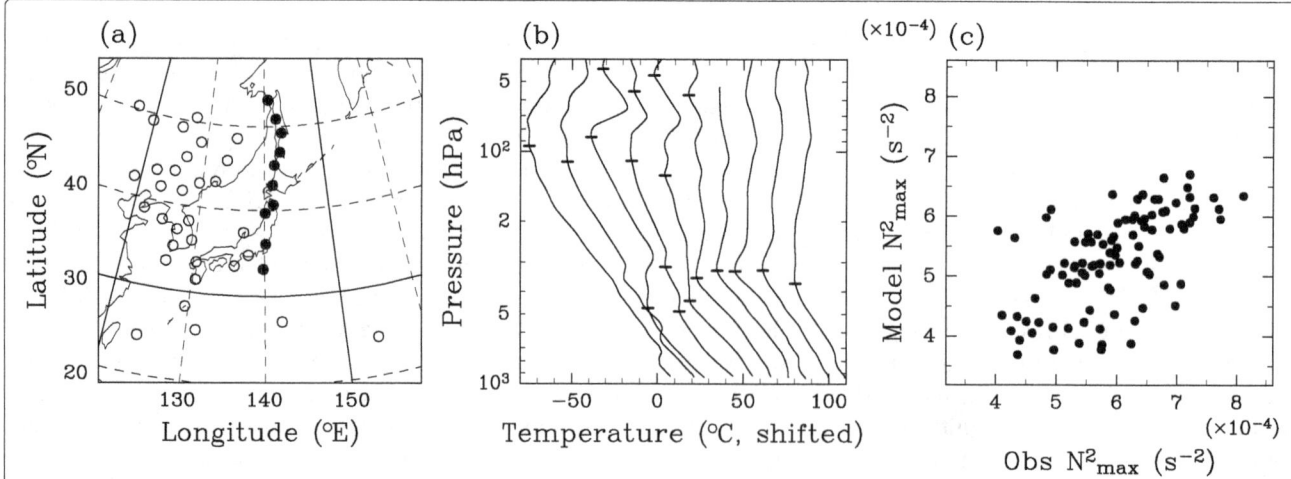

Figure 11 Radiosonde stations, profiles, and comparison with the model results. (a) Radiosonde stations used in the analysis. **(b)** Temperature profiles at the stations shown by the filled circles in **(a)** at 1200 UTC 20 February 2009. Profiles are resampled every 10 m in the vertical direction, and running means over 101 points are taken. The bottom scale is for the southmost profile, and other profiles are shifted 15°C each from south (left) to north (right). The markers show the tropopauses. **(c)** Scatter diagram of observed (the horizontal axis) and simulated (the vertical axis) N^2_{max}. All the stations in **(a)** are analyzed from 0000 UTC 20 to 0000 UTC 21 February 2009, and data with the tropopause height below 14 km are plotted. Temperature profiles are smoothed as in **(b)**.

Conclusions

We conducted a numerical experiment using a regional atmospheric model with realistic configurations to investigate the formation of the extratropical TIL associated with a realistic synoptic disturbance in mid-latitudes. The model successfully reproduced the TIL associated with an explosive cyclogenesis that occurred from 19 to 21 February, 2009 over Japan. The model also reproduced the negative correlation between the strength of the TIL (measured by the maximum buoyancy frequency squared in the TIL, N_{max}^2) and the relative vorticity near the tropopause (ζ_{TP}), which has been reported by observations (e.g., Randel et al. 2007) and previous model studies (e.g., Son and Polvani 2007). However, the negative correlation was clear only during the development and mature stages of the cyclogenesis. This suggests that the evolution of the cyclone plays an important role in the formation of the TIL.

To ascertain the dynamical formation of the TIL including the effects of gravity waves, the relationship between N_{max}^2 and the vertical divergence, $\partial w / \partial z$, and its vertical derivative, $\partial^2 w / \partial z^2$, were analyzed. Histograms of N_{max}^2 showed that regions of vertical convergence ($\partial w / \partial z < 0$) correspond to higher N_{max}^2 as well as regions of $\partial^2 w / \partial z^2 > 0$ (Figure 8). The former indicated the effect of the dynamical formation by large- to synoptic-scale motions (e.g., Wirth 2003; Birner 2010), whereas the latter indicated the effect of small-scale waves with downward phase propagation. This tendency was clearer in the regions of negative relative vorticities at the tropopause.

By taking account of the fact that the gravity wave activity associated with the cyclone and the jet streak seem to be enhanced during the development and mature stages of the cyclone, the vertical convergence by gravity waves associated with synoptic weather systems could be a key element in the formation of the negative correlation between the strength of the TIL and the local relative vorticity at the tropopause.

Competing interests
The authors declare that they have no competing interests.

Authors' contributions
SO conducted the numerical experiment, analyses, manipulation of graphics, and dynamical interpretation. MT conducted the analyses and interpretation. SY proposed the framework of the study and contributed to the analyses and interpretation. All authors read and approved the final manuscript.

Acknowledgements
The authors thank the two anonymous reviewers for their helpful comments. This work was supported by JSPS KAKENHI (S) Grant Number 24224911 and Kyoto University's Global COE Program 'Sustainability/Survivability Science for a Resilient Society Adaptable to Extreme Weather Conditions' for FY 2009–13. The figures were produced by the GFD-DENNOU Library.

Author details
[1]RIKEN Advanced Institute for Computational Science, 7-1-26 Minatojima-minami-machi, Chuo-ku, 650-0047 Kobe, Japan. [2]Weather Caster Network, 1-14-21 Ueno-Sakuragi, Taito-ku, 110-0002 Tokyo, Japan. [3]Graduate School of Science, Kyoto University, Kitashirakawa-Oiwake-cho, Sakyo-ku, 606-8502 Kyoto, Japan.

References
Andrews DG, Holton JR, Leovy CB (1987) Middle Atmosphere Dynamics. Academic Press, San Diego, p 489

Birner T, Dörnbrack A, Schumann U (2002) How sharp is the tropopause at midlatitudes?. Geophys Res Lett 29:1700. doi:10.1029/2002GL015142

Birner T, Sankey D, Shepherd TG (2006) The tropopause inversion layer in models and analyses. Geophys Res Lett 33:14804. doi:10.1029/2006GL026549

Birner T (2010) Residual circulation and tropopause structure. J Atmos Sci 67:2582–2600

Erler AR, Wirth V (2011) The static stability of the tropopause region in adiabatic baroclinic life cycle experiments. J Atmos Sci 68:1178–193

Gill AE (1982) Atmosphere-Ocean Dynamics. Academic Press, London, p 662

Japan Meteorological Agency (2009) Daily weather charts No. 85 February 2009. Tenki 56:242–243. (in Japanese)

Künsch HR (1989) The jackknife and the bootstrap for general stationary observations. Ann Statist 17:1217–1241

Müller A, Wirth V (2009) Resolution dependence of the tropopause inversion layer in an idealized model for upper-tropospheric anticyclones. J Atmos Sci 66:3491–3497

McIntyre ME (2009) Spontaneous imbalance and hybrid vortex-gravity structures. J Atmos Sci 66:1315–1326

Miyazaki K, Watanabe S, Kawatani Y, Sato K, Tomikawa Y, Takahashi M (2010a) Transport and mixing in the extratropical tropopause region in a high-vertical-resolution GCM. Part II: relative importance of large-scale and small-scale dynamics. J Atmos Sci 67:1315–1336

Miyazaki K, Watanabe S, Kawatani Y, Tomikawa Y, Takahashi M, Sato K (2010b) Transport and mixing in the extratropical tropopause region in a high-vertical-resolution GCM. Part I: potential vorticity and heat budget analysis. J Atmos Sci 67:1293–1314

Nakanishi M, Niino H (2004) An improved Mellor-Yamada Level-3 model with condensation physics: its design and verification. Boundary Layer Meteorol 112:1–31

Nakanishi M, Niino H (2006) An improved Mellor-Yamada Level-3 model: its numerical stability and application to a regional prediction of advection fog. Boundary Layer Meteorol 119:397–407

O'Sullivan D, Dunkerton TJ (1995) Generation of inertia-gravity waves in a simulated life cycle of baroclinic instability. J Atmos Sci 52:3695–3716

Plougonven R, Snyder C (2007) Inertia-gravity waves spontaneously generated by jets and fronts Part I: different baroclinic life cycles. J Atmos Sci 64:2502–2520

Randel WJ, Wu F (2010) The polar summer tropopause inversion layer. J Atmos Sci 67:2572–2581

Randel WJ, Wu F, Forster P (2007) The extratropical tropopause inversion layer: global observations with GPS data, and a radiative forcing mechanism. J Atmos Sci 64:4489–4496

Saito K, Fujita T, Yamada Y, Ishida J, Kumagai Y, Aranami K, Ohmori S, Nagasawa R, Kumagai S, Muroi C, Kato T, Eito H, Yamazaki Y (2006) The operational JMA nonhydrostatic mesoscale model. Mon Weather Rev 134:1266–1298

Saito K, Ishida J, Aranami K, Hara T, Segawa T, Narita M, Honda Y (2007) Nonhydrostatic atmospheric models and operational development at JMA. J Meteor Soc Japan 85B:271–304

Sanders F, Gyakum JR (1980) Synoptic-dynamic climatology of the "bomb". Mon Weather Rev 108:1589–1606

Skamarock WC (2004) Evaluating mesoscale NWP models using kinetic energy spectra. Mon Weather Rev 132:3019–3032

Son S-W, Polvani LM (2007) Dynamical formation of an extra-tropical tropopause inversion layer in a relatively simple general circulation model. Geophys Res Lett 34:17806. doi:10.1029/2007GL030564

Son S-W, Tandon NF, Polvani LM (2011) The fine-scale structure of the global tropopause derived from COSMIC GPS radio occultation measurements. J Geophys Res 116:20113. doi:10.1029/2011JD016030

Tomikawa Y, Nishimura Y, Yamanouchi T (2009) Characteristics of tropopause and tropopause inversion layer in the polar region. SOLA 5:141–144

Wirth V (2003) Static stability in the extratropical tropopause region. J Atmos Sci 60:1395–1409

Wirth V (2004) A dynamical mechanism for tropopause sharpening. Meteorologische Zeitschrift 13:477–484

Wirth V, Szabo T (2007) Sharpness of the extratropical tropopause in baroclinic life cycle experiments. Geophys Res Lett 34:02809. doi:10.1029/2006GL028369

Yabu S, Murai S, Kitagawa H (2005) Numerical Weather Prediction Division Report. Numerical Weather Prediction Division, Japan Meteorological Agency, Tokyo, Japan, pp 53-64. (in Japanese)

Zängl G, Wirth V (2002) Synoptic-scale variability of the polar and subpolar tropopause: data analysis and idealized PV inversions. Quart J Roy Meteor Soc 128:2301–2315

Zhang F (2004) Generation of mesoscale gravity waves in upper-tropospheric jet-front systems. J Atmos Sci 61:440–457

Zülicke C, Peters D (2006) Simulation of inertia-gravity waves in a poleward-breaking Rossby wave. J Atmos Sci 63:3253–3276

7

The seasonal variations of atmospheric ^{134,137}Cs activity and possible host particles for their resuspension in the contaminated areas of Tsushima and Yamakiya, Fukushima, Japan

Takeshi Kinase[1,2*], Kazuyuki Kita[3], Yasuhito Igarashi[4], Kouji Adachi[4], Kazuhiko Ninomiya[5], Atsushi Shinohara[5], Hiroshi Okochi[6], Hiroko Ogata[6], Masahide Ishizuka[7], Sakae Toyoda[8], Keita Yamada[8], Naohiro Yoshida[8,9], Yuji Zaizen[10], Masao Mikami[11], Hiroyuki Demizu[12] and Yuichi Onda[13]

Abstract

A large quantity of radionuclides was released by the Fukushima Daiichi Nuclear Power Plant accident in March 2011, and those deposited on ground and vegetation could return to the atmosphere through resuspension processes. Although the resuspension has been proposed to occur with wind blow, biomass burning, ecosystem activities, etc., the dominant process in contaminated areas of Fukushima is not fully understood. We have examined the resuspension process of radiocesium (^{134,137}Cs) based on long-term measurements of the atmospheric concentration of radiocesium activity (the radiocesium concentration) at four sites in the contaminated areas of Fukushima as well as the aerosol characteristic observations by scanning electron microscopy (SEM) and the measurement of the biomass burning tracer, levoglucosan.

The radiocesium concentrations at all sites showed a similar seasonal variation: low from winter to early spring and high from late spring to early autumn. In late spring, they showed positive peaks that coincided with the wind speed peaks. However, in summer and autumn, they were correlated positively with atmospheric temperature but negatively with wind speed. These results differed from previous studies based on data at urban sites. The difference of radiocesium concentrations at two sites, which are located within a 1 km range but have different degrees of surface contamination, was large from winter to late spring and small in summer and autumn, indicating that resuspension occurs locally and/or that atmospheric radiocesium was not well mixed in winter/spring, and it was opposite in summer/autumn. These results suggest that the resuspension processes and the host particles of the radiocesium resuspension changed seasonally. The SEM analyses showed that the dominant coarse particles in summer and autumn were organic ones, such as pollen, spores, and microorganisms. Biological activities in forest ecosystems can contribute considerably to the radiocesium resuspension in these seasons. During winter and spring, soil, mineral, and vegetation debris were predominant coarse particles in the atmosphere, and the radiocesium resuspension in these seasons can be attributed to the wind blow of these particles. Any proofs that biomass

(Continued on next page)

* Correspondence: t.kinase.J52@gmail.com
[1]Graduate School of Science and Engineering, Ibaraki University, 2-1-1 Bunkyo, Mito, Ibaraki 310-8512, Japan
[2]Hitachi Power Solutions Co., Ltd, 2-2-3 Saiwaicho, Hitachi, Ibaraki 317-0073, Japan
Full list of author information is available at the end of the article

(Continued from previous page)
burning had a significant impact on atmospheric radiocesium were not found in the present study.

Keywords: Radiocesium, Atmospheric radioactivity, Seasonal variation, Resuspension, Fukushima, Nuclear accident, Bioaerosol, Spore, Mineral dust, Host particle

Introduction

In March 2011, abundant and various radionuclides were released into the atmosphere (e.g., Chino et al. 2011) as a result of the nuclear accident at the Fukushima Daiichi Nuclear Power Plant (FDNPP) caused by the Great East Japan Earthquake (Holt et al. 2012), and their amounts and diffusion have been of public concern because of their health impacts (Report: Working Group on Risk Management of Low-dose Radiation Exposure 2011). Radiocesium isotopes ^{134}Cs and ^{137}Cs were two of the main radionuclides released from the FDNPP accident (Steinhauser et al. 2014). As an example, the total amount of released ^{137}Cs was estimated by Chino et al. (2011) to be 1.3×10^{16} Bq (13 PBq). The ^{134}Cs emission inventory should be almost the same as that of ^{137}Cs, i.e., 1.2×10^{16} Bq (12 PBq; Steinhauser et al. 2014) because the activity ratio of these two isotopes was almost united in the case of the FDNPP accident (e.g., Merz et al. 2013). There have been concerns that the deposited radiocesium will influence the contaminated area for an extended period, in contrast to the effects of radioiodine, because the half-lives of ^{134}Cs and ^{137}Cs are relatively long: 2.06 years and 30.17 years, respectively (Dietz and Pachucki 1973; Unterweger 2002). Figure 1 shows the contaminated deposition density map for $^{134, \, 137}$Cs (the unit of deposition density is Bq m^{-2}) measured from October to November 2011 by the Ministry of Education, Culture, Sports, Science and Technology, Japan (MEXT) (2011). The radionuclides released by the FDNPP were deposited on the ground, vegetation, houses, and other objects, causing serious contamination of widespread areas, especially areas in the northwest of the FDNPP.

Although more than 5 years have passed since the FDNPP accident, significant atmospheric radiocesium activity concentrations (hereafter called the radiocesium concentration), on the order of approximately 10^{-4} Bq m^{-3}, have still been observed in the contaminated area (e.g., Ochiai et al. 2016). These concentrations are at least two orders of magnitude higher than those observed in East Japan before the FDNPP accident (~ 1 µBq/m^3; Igarashi et al. 2015). Several studies showed that this atmospheric radiocesium could not be attributed to the direct emissions/leakage from the FDNPP site (e.g., Igarashi et al. 2015; Kajino et al. 2016) but could be supplied by secondary emissions of the deposited radiocesium, which is called resuspension to the atmosphere (e.g., Igarashi 2009; Igarashi et al. 2015; Kajino et al. 2016). Igarashi et al. (2015) concluded that the delayed primary emissions from the FDNPP could not be a major source of the current (till March 2015) radiocesium in the atmosphere based on the atmospheric observations in Tsukuba and the emission inventory data from Tokyo Electric Power Co., Inc. (TEPCO) (2012). Kajino et al. (2016) also showed that resuspension could predominantly contribute to the atmospheric radiocesium concentrations in the year 2013 using a 3D aerosol model simulation and the emission data from TEPCO.

The soluble form of primary radiocesium is carried by submicron particles (Masson et al. 2013), which would most likely be non-sea salt sulfate (nss-SO$_4^{2-}$), as suggested by Doi et al. (2013) and Kaneyasu et al. (2012). In general, water-soluble cesium ions from SO$_4^{2-}$ are considered to be adsorbed or attached to mineral particles in a soil environment (Bostic et al. 2002; Dumat and Staunton 1999; Mukai et al. 2014). These mineral particles were shown to act as resuspended host particles in an atmospheric environment by the previous investigations of the global fallout from atmospheric nuclear weapon tests (e.g., Igarashi et al. 2005; Masson et al. 2010). Therefore, a major resuspension source has been thought to be the suspension of contaminated dust particles by wind (Igarashi et al. 2016; Ishizuka et al. 2017; Sýkora et al. 2012), even for the case of the FDNPP accident contamination. In addition, a significant amount of water-soluble cesium ions has been absorbed by living organisms, such as mushrooms (Yoshida et al. 1994) and plants, and has been relocated within vegetation, from bacteria (Tomioka et al. 1992), through plant roots (Ehlken and Kirchner 2002), to pollen and then to honey (Barišić et al. 1992; Kanasashi et al. 2015). Such contaminated forest ecosystems could be another source for radiocesium re-emission into the atmosphere (also we call "resuspension" in this study). The cedar pollen in contaminated forests contains significant amounts of radiocesium. The highest total radiocesium concentration in Fukushima cedar pollen was reported to be 25.4 Bq g^{-1} (dry weight) in 2014 (Forestry Agency, Japan (FAJ) 2015), suggesting that pollen emissions could still be a candidate for the host particles of radiocesium resuspensions, although pollen emissions occur during limited durations of weeks or months each year

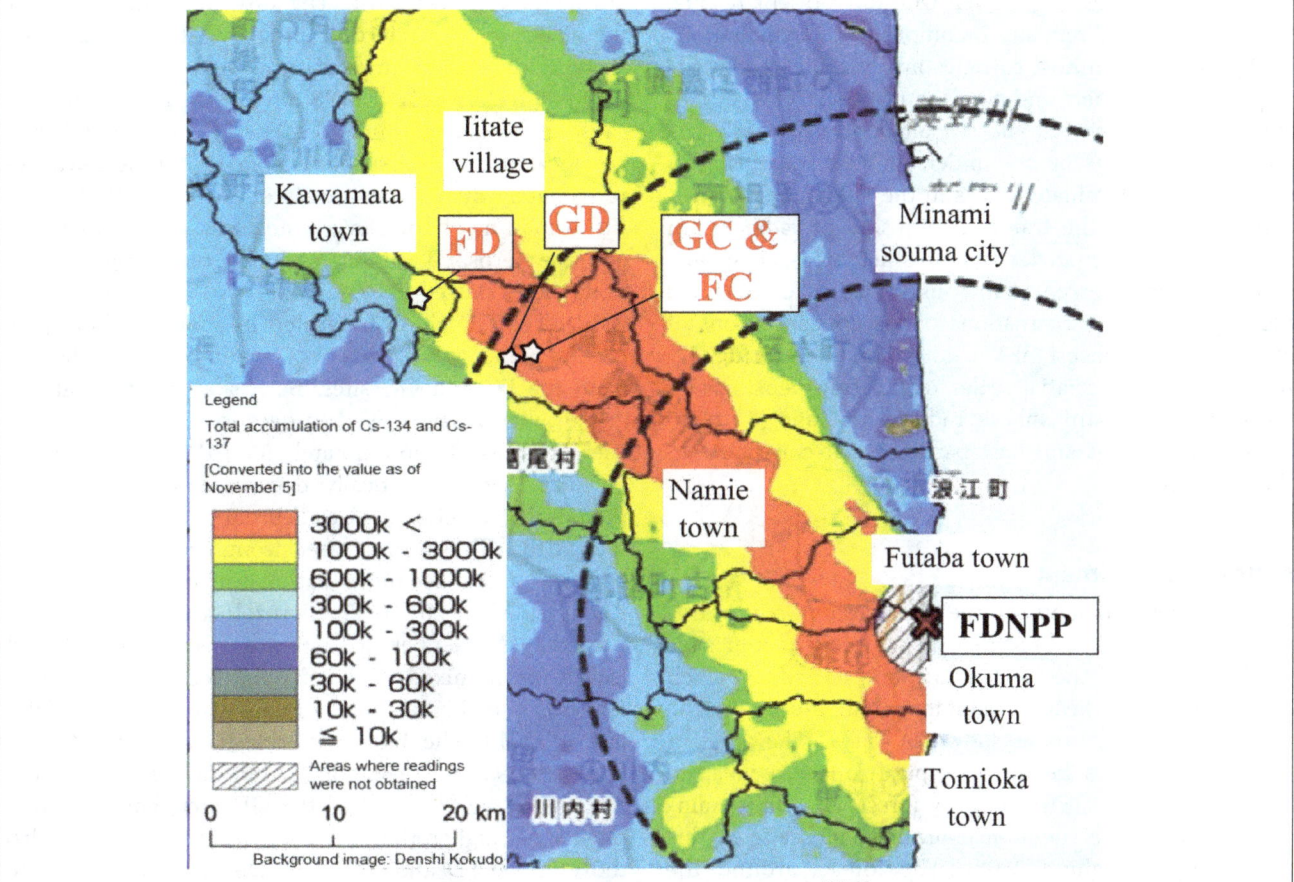

Fig. 1 The four sampling locations in this study plotted over the deposition density map for ^{134,137}Cs obtained by the Japanese government (Ministry of Education, Culture, Sports, Science and Technology, Japan (MEXT) 2011). The radius of dotted outer circle is 30 km

depending on plant species. Furthermore, open biomass burning of contaminated vegetation is also a candidate for the radiocesium resuspension process (Bourcier et al. 2010; Igarashi 2009; Lujanienė et al. 2009; Sýkora et al. 2012; Yoschenko et al. 2006).

Many studies have been carried out to address the environmental radiocesium contamination resulting from the FDNPP accident, but the processes of radiocesium resuspension have not yet been fully identified in this case. Knowledge of the host particles of such secondary radiocesium emissions and their long-term variations is limited (e.g., Igarashi et al. 2015) because of the lack of long-term and intensive studies in Fukushima. Ishizuka et al. (2017) discussed the processes contributing to the radiocesium resuspension with soil dust particles from the surface based on measurements within the contaminated area in Fukushima. Their proposed scheme can be applied to evaluate secondary emissions caused by aeolian resuspension of radioactive materials associated with mineral dust particles from the ground surface. However, their study only focused on the resuspension from winter to spring; the seasonal variations and other processes of resuspension were not discussed. A recent

work by Kajino et al. (2016) has revealed that the forest ecosystem could be another source of radiocesium during the warm seasons, in addition to that of dust suspension. These authors applied a novel scheme for radiocesium resuspension from the forest (as the green fraction) that operates according to forest activity as measured by the normalized difference vegetation index (NDVI) satellite data (Gutman and Ignatov 1998). However, Kajino et al. (2016) showed a model simulation application and did not reveal the actual host material of radiocesium. Recently, Ochiai et al. (2016) added new observational results of the monthly radiocesium concentrations (September 2012 to December 2014) at a contaminated area in western Tsushima that is near our present observation site. These authors applied a two-stage size separation (coarse and fine) in their monitoring and demonstrated that ^{137}Cs activity concentration enhancement during the warm season and the coarse fraction (> 1.1 μm) of ^{137}Cs exhibited seasonal changes along with the concentration increases. These authors interpreted this seasonal trend via the relevance of the seasonal changes in the prevailing local winds and locations of the surface contamination. Although their observational results agree well with our present results, their explanation remains incomplete; no explanations of the radiocesium host particles involved are given. It is important to understand the contributions of each resuspension process in all seasons and to break the current limitations of our understandings, especially in the heavily contaminated areas in the Fukushima prefecture, to evaluate the transport/diffusion of radiocesium from the contaminated areas to other areas. The purposes of this research were thus set as follows: (1) to identify the seasonal variations of the concentrations of atmospheric radiocesium via long-term monitoring, (2) to determine the spatial scales of the resuspension and redistribution of atmospheric radiocesium, and (3) to estimate the sources and host particles that contribute to the resuspension.

Methods/Experimental

For the measurements of the radiocesium concentrations, we have carried out aerosol sampling at four sites (labeled herein as the FD, FC, GD, and GC sites), which are located in heavily contaminated areas in the Fukushima prefecture, as shown in Fig. 1. The areas of the sampling sites belong to typical Japanese villages/towns, which are surrounded by forests in a mountainous area as is the traditional farming landscape in this region (e.g., Berglund 2008). The forests around the sampling sites are secondary forests surrounding farm villages (Satoyama) and are not dense, primeval forests. According to Berglund (2008), Satoyama is defined as "forests surrounding farm villages and managed by farmers for different needs—timber for buildings, wood for fuel and charcoal production, and leaf litter and twigs used as fertilizer for crops, particularly in the rice paddy fields situated in the lowlands." Currently, the agricultural lifestyle has been changing greatly; however, major landscape changes have not occurred. Inhabitants were evacuated from these contaminated areas by the time that this manuscript was revised (up to March 2017), so this area has been free from human activity, except for decontamination and monitoring activities.

The sites labeled with the letters F and G are the forest and open ground sites, respectively, and the sites labeled with the letters D and C are sites where decontamination was conducted or not, respectively. The FD site in the Yamakiya area, Kawamata-town, is located in a cedar forest 35 km northwest of the FDNPP, and its contamination was the lowest of the four observation sites even before decontamination because the radioactive plume was blocked by hills.

We could start the monitoring at the FD site soon after the FDNPP accident, i.e., since July 2011. The area around the FD site was decontaminated between August 2013 and October 2014. The dose rate monitoring by the Nuclear Regulation Authority showed that the average dose rate in this area had decreased from $0.90 \, \mu Sv \, h^{-1}$ in 2012 to $0.28 \, \mu Sv \, h^{-1}$ in 2014 mainly because of the decontamination. The FC, GC, and GD sites are located in the Tsushima area in Namie-town, approximately 30 km northwest of the FDNPP. The GC site was in a school playground (area is $10.7 \times 10^3 \, m^2$), and the aerosol sampling and meteorological observations here started in December 2012 (Ishizuka et al. 2017). The GD site was located in another school playground (area is $5.7 \times 10^3 \, m^2$), within a 1 km distance from the FC and GC sites. Both sites are basically flat and originally paved by sand because of their gymnastic uses. An area of approximately $5 \times 10^4 \, m^2$, including the GD site, had been locally decontaminated during the period of December 2011 to February 2012, and the dose rate at the GD site decreased from 9.60 to $2.8 \, \mu Sv \, h^{-1}$ because of the decontamination (Japan Atomic Energy Agency (JAEA) 2012). The aerosol sampling at this site started in November 2012. The FC site is in a broad-leaved forest mixed with red pine trees adjacent to the GC site, and the aerosol sampling at this site started in January 2014. The Tsushima area, including the FC and GC sites, was not decontaminated during our observation period, except around the GD area, and their dose rates were higher (by at least three to ten times) than those at the FD and GD sites. Details of the four sites are summarized in Table 1.

High-volume air samplers (120SL, Kimoto, Japan, and HV-1000R/F, Sibata, Japan) were used for aerosol sampling (a flow rate of $0.7 \, m^3 \, min^{-1}$) with quartz-fiber filters (2500QAT-UP, Pallflex, USA). Six-stage cascade impactors (TE-236, Tisch Environmental, USA) mounted on other high-volume air samplers (a flow rate of $0.556 \, m^3 \, min^{-1}$) were also used to collect the aerosols separately according to their aerodynamic diameters, with quartz-fiber filters (TE-230QZ, Tisch Environmental, USA and 2500QAT-UP, Pallflex, USA) applied as collection substrates. This impactor classifies particles as larger than $10.3 \, \mu m$ (stage #1), 4.2–$10.3 \, \mu m$ (stage #2), 2.1–$4.2 \, \mu m$ (stage #3), 1.3–$2.1 \, \mu m$ (stage #4), 0.69–$1.3 \, \mu m$ (stage #5), 0.39–$0.69 \, \mu m$ (stage #6), and smaller than

Table 1 Information about the observation locations and conditions at the four sites

Site-ID	Latitude	Longitude	Land condition	Decontaminated or contaminated
FD	37° 35′ 07″ N	140° 41′ 29″ E	Cedar forest	Decontaminated
FC	37° 33′ 41″ N	140° 46′ 06″ E	Broad leaf forest	Contaminated
GD	37° 33′ 38″ N	140° 45′ 37″ E	School ground	Decontaminated
GC	37° 33′ 44″ N	140° 46′ 07″ E	School ground	Contaminated

0.39 μm (backup filter) in diameter, in which the collection efficiency exceeds 50%.

The aerosol sampling period at the FD, FC, and GD sites was mostly between 1 week and 1 month. At the GC site, short-term sampling (1–3 days) was carried out using seven high-volume air samplers with timer-controls. The radioactivities of [134,137]Cs of the aerosols sampled in filters were measured at the Meteorological Research Institute (MRI), Osaka University, Tokyo Institute of Technology, and Ibaraki University using coaxial-type Ge semiconductor detectors coupled with computed spectrometric analyzers. The radioactivities of [134]Cs and ^{137}Cs were identified and determined from the gamma-ray peak intensities at 605 and 662 keV, respectively. Details of the detector models and the spectrometric analyzers are summarized in Table 2. The radiocesium concentration was calculated from the measured radioactivity and the total volume of the sampled air.

A scanning electron microscope (SEM; SU3500, Hitachi High-Technologies Co., Tokyo Japan) with an energy-dispersive X-ray spectroscope (EDS; X-max 50 mm, Horiba Ltd., Kyoto, Japan) was used to observe the shapes and chemical compositions of the aerosols collected in the above samplings to identify the major host particles of the resuspended radiocesium.

Results and discussion
Seasonal and long-term variations of the atmospheric concentrations of radiocesium activity

Figure 2 shows the time series of the radiocesium concentrations of [134,137]Cs obtained from the aerosol sampling at the four sites. For the case of the samples obtained using the impactor, the radiocesium concentration values were the sum of all stages. The radiocesium concentration at the FD site showed a rapid exponential decrease during the earlier period from July 2011 to October 2011. After that, the decrease became slower. Igarashi et al. (2015) showed that the direct emissions/

leaks of radiocesium from the FDNPP site significantly affected the radiocesium concentrations in 2011 and 2012 (in the early period), when spikes occurred in the radiocesium concentrations monitored at the MRI, Tsukuba (approximately 170 km south–southwest from the FDNPP), and that atmospheric radiocesium was mainly supplied by resuspension after this period. It should be noted here that radiocesium of the resuspension origin is also subjected to transportation and diffusion as is the primary emission.

The radiocesium concentrations at all of the four sites showed similar seasonal variations, being low in winter and early spring and high in late spring, summer, and early autumn, after November 2011. The annual mean, maximum, and minimum values of the ^{137}Cs concentrations measured at each site are summarized in Table 3. Similar seasonal variations at all sites indicate that resuspension occurred with identical processes/sources in both open areas and forests. This result is opposite to that of the seasonal variation (high in winter and low in summer) found in urban areas by other studies (e.g., Igarashi et al. 2015; Sýkora et al. 2012; Watanabe 2014), suggesting that the dominant resuspension processes/sources are different because of the locations/surface conditions of the study areas, specifically between the urban sites and the forest sites. The sampling sites in our study are located in low mountainous areas, where the major land uses are agricultural fields (abandoned because of the evacuation) placed along valleys and forests covering mountain slopes. Because it is a typical surface condition in heavily contaminated areas in Fukushima as mentioned previously, the present seasonal trend is probably typical of these areas. Similar seasonal variations of radiocesium concentrations were found in the contaminated area of the western part of Namie (near the GC site) by Ochiai et al. (2016).

Between spring and autumn, two maxima were found in May and from August to September, especially for the 2014 results; Fig. 3 compares the temporal variations of

Table 2 Information about the instruments used in this study

Measured at	Detector model	Production	Spectrometric analyzer	Production
MRI	GEM50	Seiko EG&G	Multiport	Oxford-Tennelec
	IGC60210	Princeton Gamma-Tech		
	EGPC40	Canberra		
	GEM-90205-P	Seiko EG&G	MCA760092x	Seiko EG&G
Osaka University	GC6020	Canberra		
	GEM40	Ortec	MCA7600	Seiko EG&G
	GEM 80	Ortec		
Tokyo Institute of Technology	GC3018	Canberra	DSA1000	Canberra
Ibaraki University	GC4020	Canberra	Linx	Canberra

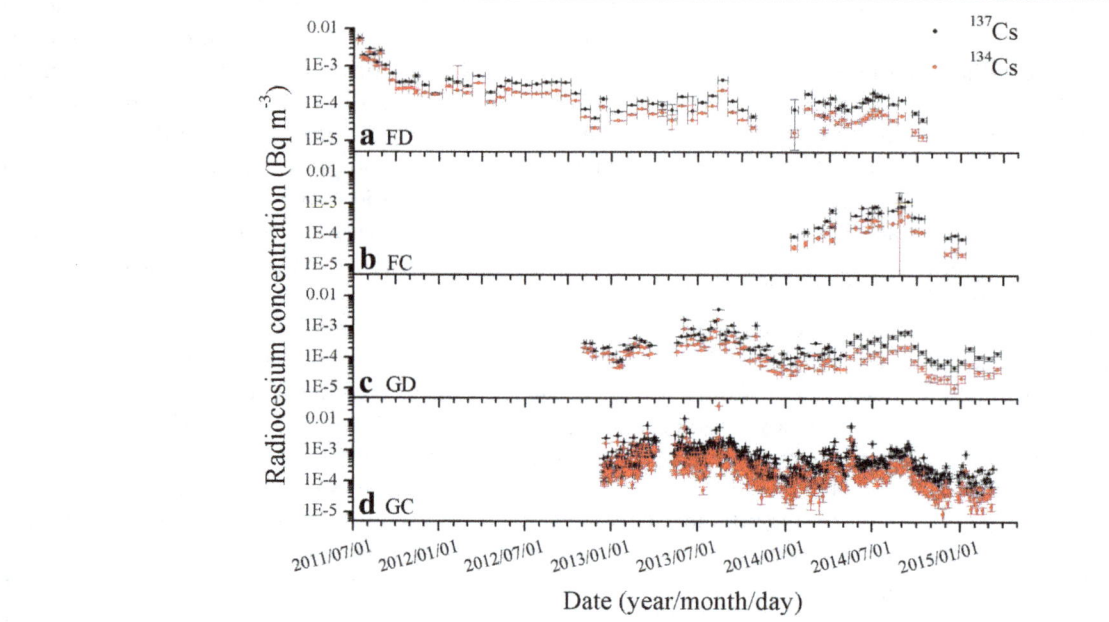

Fig. 2 Activity concentrations of ^{134,137}Cs (Bq m^{-3}) measured at the FD, FC, GD, and GC sites in the Fukushima prefecture in July 2011 to March 2015. These results show high concentrations in late spring, summer, and early autumn and low concentrations in winter and early spring at all of the presented observation locations

(a) the radiocesium concentrations at the GC site with the 30 min averages of various meteorological parameters: (b) wind speed, (c) atmospheric temperature, and (d) precipitation. At the first maximum, occurring around May, the wind speed peaks generally coincided with the radiocesium concentration peaks. During the second maximum, occurring around September, the radiocesium concentration and temperature had similar variations, but the wind speeds were negatively correlated with these variables. These results indicate that the

processes or sources of the resuspension are different in these seasons: the physical processes (suspension by wind) are significant in late spring, and another process is dominant in summer/autumn.

Seasonal variations of resuspended host particles found by impactor sampling

To indicate the fractions of ^{137}Cs captured on each stage as well as on the backup filter in the cascade impactor

Table 3 Averages, maximums, and minimums of the ^{137}Cs activity concentration (µBq m^{-3}) at each site

Location (sampling duration)	Average			Maximum				Minimum			
	Year	^{137}Cs	Error	Date (start) m/d h:m	Date (end) m/d h:m	^{137}Cs	Error	Date (start) m/d h:m	Date (end)[a] m/d h:m	^{137}Cs	Error
FD (July 2011–October 2014)	2011	1450	33.7	7/9 16:40	7/18 12:10	5600	66.4	12/9 12:47	2012/1/6 11:16	186	11.0
	2012	287	9.7	3/9 10:43	4/6 11:15	547	7.8	11/16 12:01	12/7 10:18	41.2	2.0
	2013	127	13.6	8/9 14:16	8/30 11:20	432	4.6	10/18 14:48	10/28 10:45	45.9	3.2
	2014	115	8.3	6/29 9:41	7/4 10:30	202	6.4	10/3 10:27	10/20 11:07	38.7	4.6
FC (May 2014–January 2015)	2014	537	78.6	8/22 16:13	8/29 12:52	1620	843	12/26 15:41	2015 1/11 13:24	74.3	5.7
GD (November 2012–June 2015)	2012	213	2.7	11/3 13:40	11/9 12:41	294	2.1	12/28 12:45	2013/1/11 13:46	130	1.4
	2013	521	4.1	8/9 11:55	8/16 11:16	3690	8.0	12/13 12:04	12/23 12:08	72.0	3.0
	2014	243	26.4	9/19 11:16	10/3 10:59	684	94.6	12/26 11:28	2015 1/11 12:07	47.4	8.5
	2015	123	18.7	5/4 11:26	6/4 10:52	225	31.9	4/25 10:41	5/4 11:11	56.8	9.9
GC (December 2012–November 2014)	2012	462	12.2	12/21 13:00	12/22 13:00	2550	28.2	12/19 13:00	12/20 13:00	95.2	4.7
	2013	1170	19.2	8/14 13:00	8/15 13:00	60,400	132	11/12 13:00	11/13 13:00	68.9	11.6
	2014	546	11.6	5/16 13:00	5/18 13:00	6750	31.0	1/15 13:00	1/17 11:38	48.4	7.8

[a]In some cases, the sampling duration crossed over years

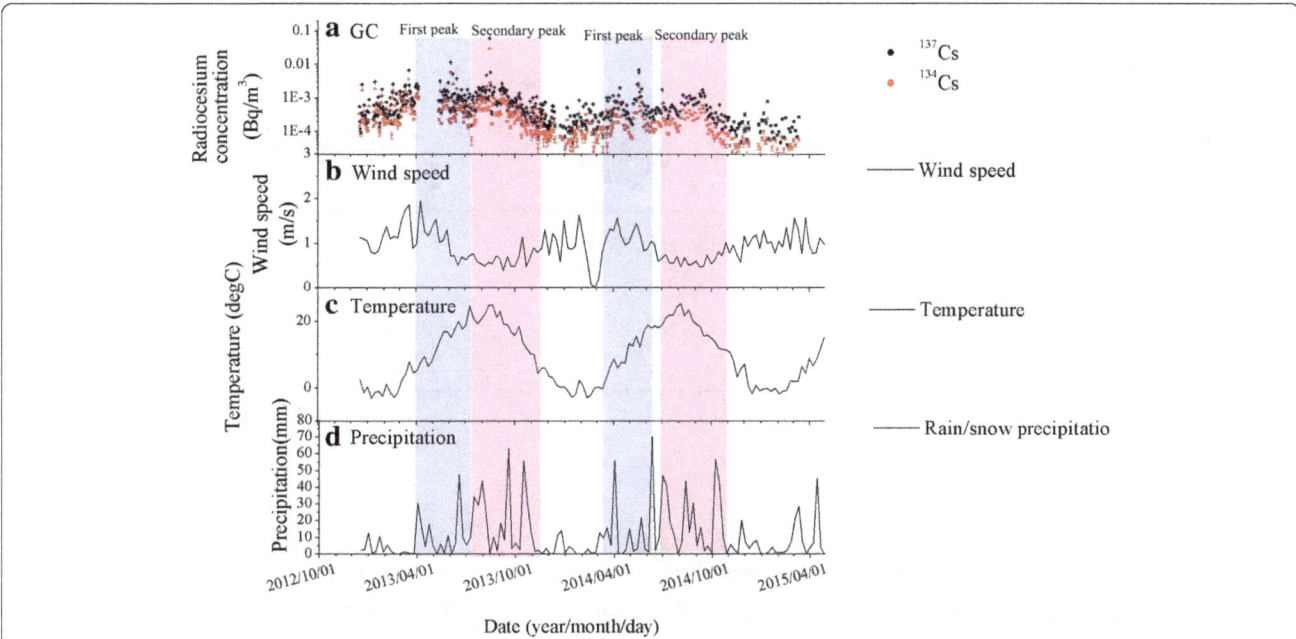

Fig. 3 Time series of (**a**) radiocesium concentrations, (**b**) wind speeds, (**c**) temperatures, and (**d**) rain/snow precipitation. At the GC site, we can see two maxima in the higher concentration seasons: in May and September. The first maxima occurred with wind speed peaks, and the second maxima occurred with atmospheric temperature peaks

sampling, the measured values of the ^{137}Cs radioactivity on the filters ^{137}Cs$_{(i)}$ were normalized as shown in Eq. 1:

$$\text{Normalized }^{137}\text{Cs}_{(i)} = \frac{^{137}\text{Cs}_{(i)}}{^{137}\text{Cs}_{(\text{Total})}}, \qquad (1)$$

where (*i*) is the stage number, ^{137}Cs$_{(i)}$ is the ^{137}Cs radioactivity of the (*i*)-th sampling stage collection substrate, and ^{137}Cs$_{(\text{Total})}$ is the total of the ^{137}Cs radioactivity of all of the sampling stage substrates, including the backup filter for each sampling period. The radioactivity of the backup filter was also normalized to the normalized ^{137}Cs$_{(\text{Total})}$ value as ^{137}Cs$_{(7)}$. Figure 4 shows the time series of the ^{137}Cs fraction of stage #2 (nominal aerosol diameter of 4.2–10.3 μm) and the backup filter (< 0.39 μm), where the highest fraction was most often found. In the fraction of stages #1–5, the normalized ^{137}Cs$_{(1-5)}$ showed similar seasonal variations among themselves, i.e., high in summer/autumn (June to November) and low in winter/ spring (December to April or May). This result is consistent with that of Ochiai et al. (2016); the coarse fraction (> 1.1 μm) of ^{137}Cs exhibited similar seasonal changes. In contrast, the fraction of the backup filter of normalized ^{137}Cs$_{(7)}$ showed the opposite trend; it was high in winter/spring and low in summer/autumn. In this discussion, we omitted the ^{137}Cs$_{(6)}$ values because they were frequently lower than the detection limit. This seasonal variation of the ^{137}Cs$_{(1-5)}$ and ^{137}Cs$_{(7)}$ fractions could be attributed to a measurement artifact, i.e., the bounce

effect, rather than the actual size variations of the host aerosols for the radiocesium resuspension. Bouncing of particles, particularly coarse ones, occurred in stages #1–6, especially during winter/spring, resulting in the considerable amount of coarse particles sampled at the backup filter, although the filter was expected to capture only finer particles. Figure 5 shows a picture of a typical backup filter sample at the GC site in spring, showing many coarse particles, such as pollen and soil/ mineral particles. These coarse particles would nominally have been trapped in stages #1 and #2, but they bounced in these earlier stages and accumulated on the backup filter. As indicated by the red rectangles, accumulations of coarse particles were found near the edge of the backup filter, presumably because of the stagnation of air flow or narrower gaps between the filter and the cover plate at the edge. This bounce effect was significant in winter and spring when the normalized ^{137}Cs$_{(7)}$ values were high. This finding enables us to discuss the seasonal differences of the host particles for the radiocesium resuspension, although this finding also makes discussing their size distributions difficult.

Figure 6a, b shows the SEM images of the stage #2 substrate samples obtained (a) on April 10–24, 2014 (NHVB-260424) and (b) on October 17–30, 2014 (NHVB-261030). The ^{137}Cs$_{(2)}$ values were 44.8 ± 7.1 and 28.3 ± 11.3 μBq m^{-3}, respectively, for the NHVB-260424 and NHVB-261030 substrate samples. As shown in Fig. 6b (denoted by the red curve), the adhesion of

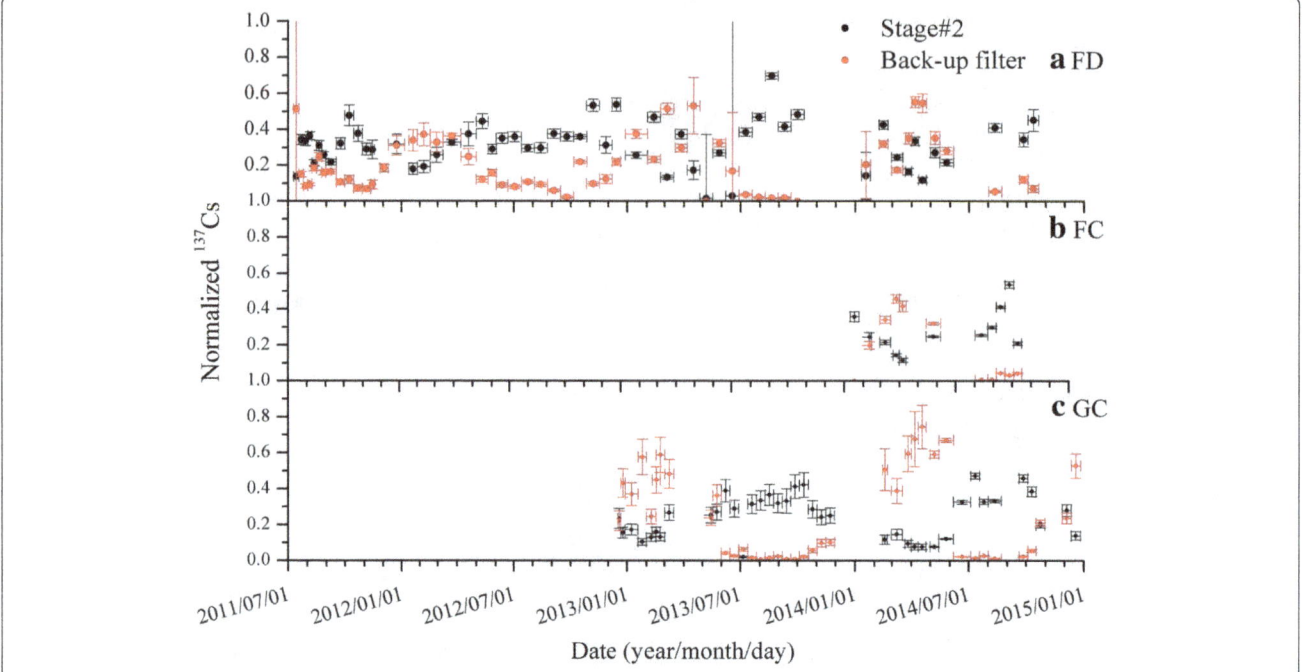

Fig. 4 Time series of the normalized $^{137}Cs_{(2)}$ and $^{137}Cs_{(7)}$ measured at the (**a**) FD, (**b**) FC, and (**c**) GC sites. These results clearly showed the seasonal variation but did not accurately reflect the size distribution of radiocesium in the atmosphere because of the bouncing particles. The amount of bouncing particles significantly changes the collection efficiency of the cascade impactor

organic matter on quartz fibers was frequently found in each stage of substrates in the summer/autumn samples. In contrast, such organic matter adhesion was not found in winter/spring. We also found that most coarse particles could not be removed from the summer/autumn sample substrates via soaking them in water, but that the coarse particles could be easily removed from the winter/spring sample filters. The adhesive organic matter on the quartz fibers probably worked as glue to fix the host particles of radiocesium to the sample substrates and the filter, or the host particles themselves could have firmly attached to the sample filter, reducing bouncing of coarse particles significantly in summer and autumn. In contrast, the host particles did not firmly adhere to the fibers in winter and spring.

The spatial extent of the resuspension

The atmospheric activity concentrations at the GC and GD sites are compared by adopting their ratio, R_{scale}, as follows:

$$R_{scale} = \frac{^{137}Cs_{\text{-GD}}}{^{137}Cs_{\text{-GC}}}, \qquad (2)$$

where $^{137}Cs_{\text{-GC}}$ and $^{137}Cs_{\text{-GD}}$ denote the ^{137}Cs concentrations observed at the GC or GD sites, respectively. There are two important points for interpreting the R_{scale} values. Although the GC and GD sites are situated very close together (about 0.75 km), the amount of radiocesium remaining over the ground surface (0–5 cm depth) at the GC site was much higher than that at the GD site because of the local decontamination (approximately 50,000 m^2) around the GD site. By sampling the surface soil (0–5 cm depth) at six points around the high-volume air samplers at each site on June 4, 2015, their radiocesium

Fig. 5 Picture of a typical backup filter sampled at the GC site in spring (April 2014). We found many coarse particles, which include pollen and coarse soil/mineral dusts, in the areas indicated by the red rectangles

Fig. 6 SEM images of filters for stage #2 sampling at the GC site in **a** spring and **b** autumn. The adhesive organic matter that sticks to the quartz fiber is surrounded by the red line. From the morphology, we suspect that the material was dense liquid when it was suspended in air. This matter probably affects the collection efficiency of the coarse particles on the filter

concentrations were measured. The dried soil samples (approximately 150 g) were put in U-8 plastic containers (AS ONE Corporation, Osaka, Japan), and their ^{137}Cs radioactivities were measured at Ibaraki University. The average soil activity concentrations at the GC and GD sites were 128 ± 3.9 and 2.17 ± 0.07 Bq g^{-1}, respectively. Considering quite similar ground surface conditions, we can assume that the resuspension of ^{137}Cs occurred similarly at both sites. In this case, if the resuspension of ^{137}Cs occurred very locally and/or the host particles were not transported, the R_{scale} values would be approximately 0.017, which is the ratio of the GD surface soil activity to that of GC. In contrast, when the resuspension occurred at larger spatial scales and/or the host particles could be distantly transported and well mixed, R_{scale} could be nearly one. Therefore, R_{scale} is expected to indicate the spatial extent of the resuspension, i.e., R_{scale} is determined by a combination of (1) the extent

of those areas wherein similar resuspensions occur and (2) the transport distance of the host particles.

Figure 7a shows that the time series of R_{scale} and its median values varied from 0.24 to 0.82 and were larger than the ratio of the ground surface activities at GD to GC (0.017), indicating that the activity concentration at the GD site was not only governed by local resuspension but also by one in the broader areas around the GD site. These observations also indicate that the host particles in the atmosphere were not completely mixed by their transport/diffusion, especially when R_{scale} was low. Because the absolute value of R_{scale} probably depends on the extent of the decontaminated area around the GD site as well as on the ratio of the surface radioactivity densities at both sites, comparisons of R_{scale} values at various places are meaningless. Therefore, the R_{scale} values indicate the relative variations of the emission scales and the magnitudes of the transport/mixing at a specific pair of sites.

Note that R_{scale} showed clear seasonal variations, such that smaller values occurred in winter/spring and larger values occurred in summer/autumn. In addition, these observed variations were synchronized with those of the fractions of activity of the aerosol particles trapped at the stage #2 substrate and the backup filter in the impactor, as shown in Fig. 4. Figure 7b compares the time series of the normalized ^{137}Cs$_{(7)}$ (the red line) with that of the R_{scale} values, and Fig. 7c shows the correlation between R_{scale} and the normalized ^{137}Cs$_{(7)}$. The green lines in Fig. 7a show the median values (since the data quantities are not sufficient, we used median values instead of average ones) of R_{scale} during summer/autumn, which were $0.60 + 0.07/- 0.05$ (central 50% range) in 2013 and $0.61 + 0.21/- 0.03$ in 2014, and the blue lines show those during winter/spring, which were $0.33 + 0.1/- 0.09$ in 2012–2013 and $0.43 + 0.11/- 0.09$ in 2013–2014. Comparing Fig. 7a, b, the R_{scale} and normalized ^{137}Cs$_{(7)}$ values show nearly synchronized seasonal variations with each other. Fig. 7c shows an evident correlation between them. These results can be interpreted both by the transport of host particles and by the extent of the resuspension area. The transport/mixing of resuspended host particles depends on the wind speed and particle deposition speed. Considering that the average wind speed was lower in summer/autumn than in winter/spring, as shown in Fig. 3b, the R_{scale} decrease in summer/autumn indicates that the deposition speed should decrease considerably as the properties of the particle change (i.e., changes in the size and mass density or both of the host particle). In addition, if the resuspension activity in the areas surrounding the GD sites was similar to that around the GC site in summer/autumn and was lower in winter/spring, the observed seasonal

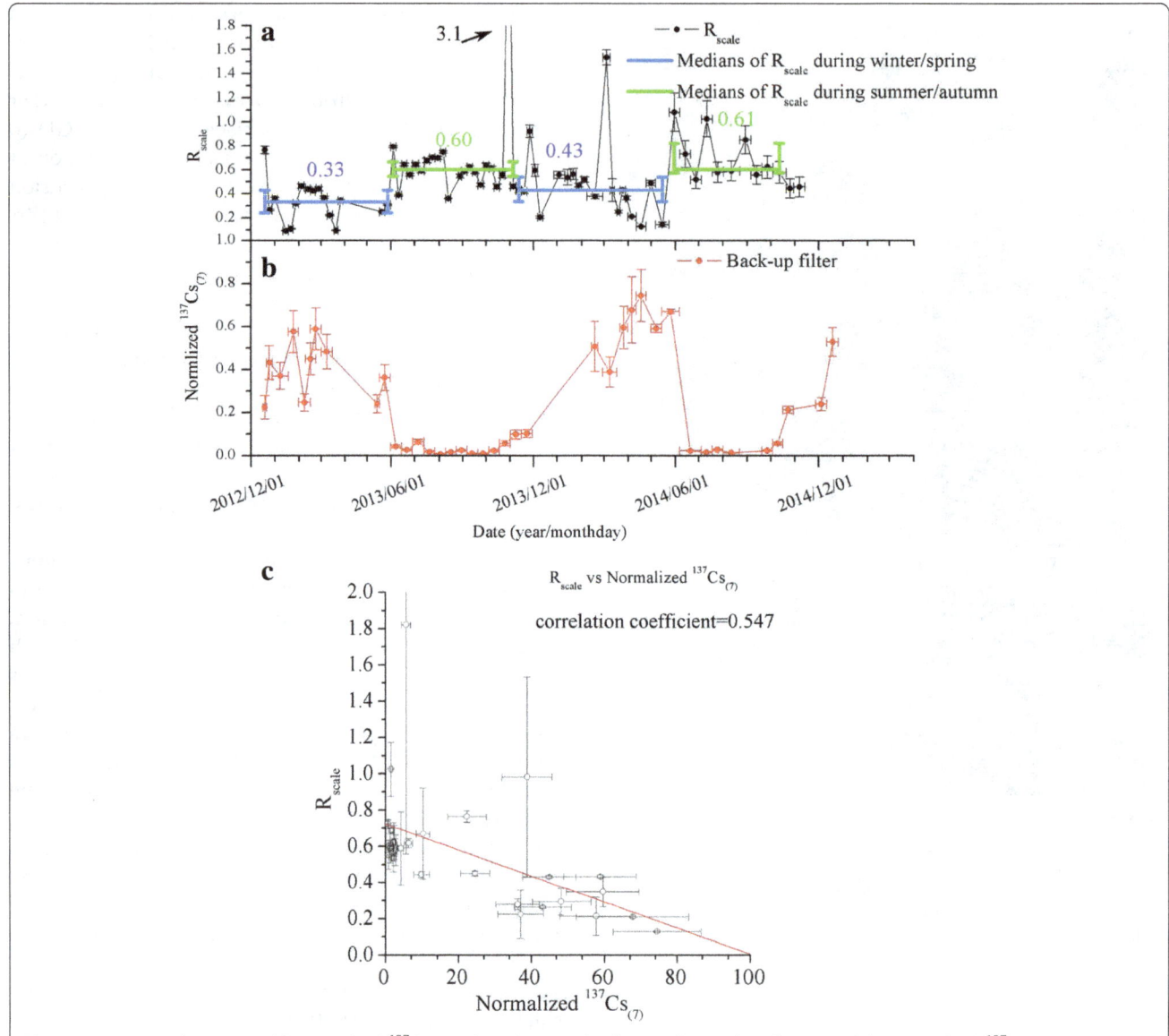

Fig. 7 Time series of **a** R_{scale} and **b** normalized $^{137}Cs_{(7)}$ at the GC site and **c** the correlation plot of R_{scale} and the normalized $^{137}Cs_{(7)}$. Green and blue curves in **a** are the medians of R_{scale} in summer/autumn and winter/spring, respectively. The error bars for each curve in **a** and **b** show the range of central 50%

variations could have occurred. However, because the seasonal variations of the R_{scale} values were synchronized with the normalized $^{137}Cs_{(7)}$, this seasonal variation of the resuspension scale can (at least partly) be attributed to the seasonal changes of the host particle materials.

Possible host particles for radiocesium resuspension

Radiocesium was primarily emitted in the form of submicron water-soluble aerosols during the FDNPP accident (e.g., Kaneyasu et al. 2012). Another type of primary radiocesium emission has been reported to

consist of glassy, water-insoluble particles with diameters of approximately 2 μm (Adachi et al. 2013; Abe et al. 2014; Yamaguchi et al. 2016) and with high specific activity (Igarashi et al. 2014). These particles are stable at least for a few years in the environment and have recently been found at the ground surface (Satou et al. 2016). If such particles are scattered by wind, they would contribute to atmospheric radiocesium. However, in the present observations, we could not find serious concentration increases due to such insoluble radiocesium-bearing particles.

Studies performed after the Chernobyl accident showed that biomass burning was a plausible resuspension process of radiocesium in the contaminated area (Bourcier et al. 2010; Igarashi 2009; Lujaniene et al. 2009; Yoschenko et al. 2006). Thus, this study tested the hypothesis of resuspension due to biomass burning. Levoglucosan is known to be an organic tracer of biomass burning (Simoneit et al. 1999), and the positive correlation of radiocesium with levoglucosan indicates the contribution of biomass burning to radiocesium re-emission (Bourcier et al. 2010). In March and May, biomass burning usually increases in the Fukushima prefecture (Local government of Fukushima prefecture 2014). A forest fire occurred on March 17, 2013, at Ootaki, approximately 28 km away from the observation sites (Real time disaster information of Fukushima). We compared the concentration of ^{137}Cs activity with that of levoglucosan in the air. Figure 8 shows the time series of (a) the concentration of the ^{137}Cs activity and (b) that of levoglucosan at GC. In addition, we also compared them with water-soluble organic carbon (WSOC), which is not a tracer of biomass burning but is emitted from various organic aerosol sources. Although the levoglucosan and WSOC showed similar trends (except for short-term spike changes), there was little correlation between the concentrations of ^{137}Cs activity and the levoglucosan and WSOC concentrations in the present work. We focused on the events of the biomass burning on March 17, 2013, and the singular increasing data on March 25. The forest fire on March 17 did not affect the levoglucosan and WSOC concentrations at this observation site. Figure 9a, b illustrates the maps showing the air mass forward and backward trajectories calculated using the NOAA HYSPLIT model (Stein et al. 2015) during the forest fire event periods. Figure 9a shows the forward trajectories from the fire site on March 17 at 16:00 (LT) to March 18 at 13:00.

The air mass did not pass over the observational sites; instead, it went north. Therefore, the increased concentration of ^{137}Cs activity on March 17 was not attributable to the biomass burning event. For the event on March 25, with an increased levoglucosan concentration, the backward trajectories from the measurement site GC show two origin areas (Fig. 9b, which is calculated from March 24 at 13:00 (LT) to March 25 at 13:00 (LT)): the Sea of Japan and the mountainous areas to the northwest which is the local contaminated areas. In these areas, there was no official information about the fire disasters at that time, although the possibility of small fires is not fully neglected. Overall, the present results indicate that biomass burning was not a dominant source of the resuspension during our observational period at the sites.

The elements in the aerosol samples taken at the GC and GD sites were analyzed quantitatively using SEM-EDS, showing that the major components were different in winter/spring and summer/autumn. Figure 10a shows a SEM image of the sample filter obtained at the GD site in July 2015 (summer), and Fig. 10b–e are the element map images of C, Fe, Al, and Na, respectively, in the same viewing field with Fig. 10a. Numerous spots of C indicate that carbonaceous aerosols were composed of major coarse particles during summer/autumn, and the Al, Fe, and Na spots probably represent that soil/mineral dust, metal, and (sea) salt particles were rare. We tried to identify these carbonaceous particles as some bioaerosol particles on the basis of their sizes and shapes on the SEM image (Carrera et al. 2007; Chaturvedi et al. 1998; Healy et al. 2012; Laucks et al. 2000). Figure 11a shows an example image of these organic particles in a wide field of view with close-up images of pollen (Fig. 11b), a spore (Fig. 11c), and microorganisms (indicated by the

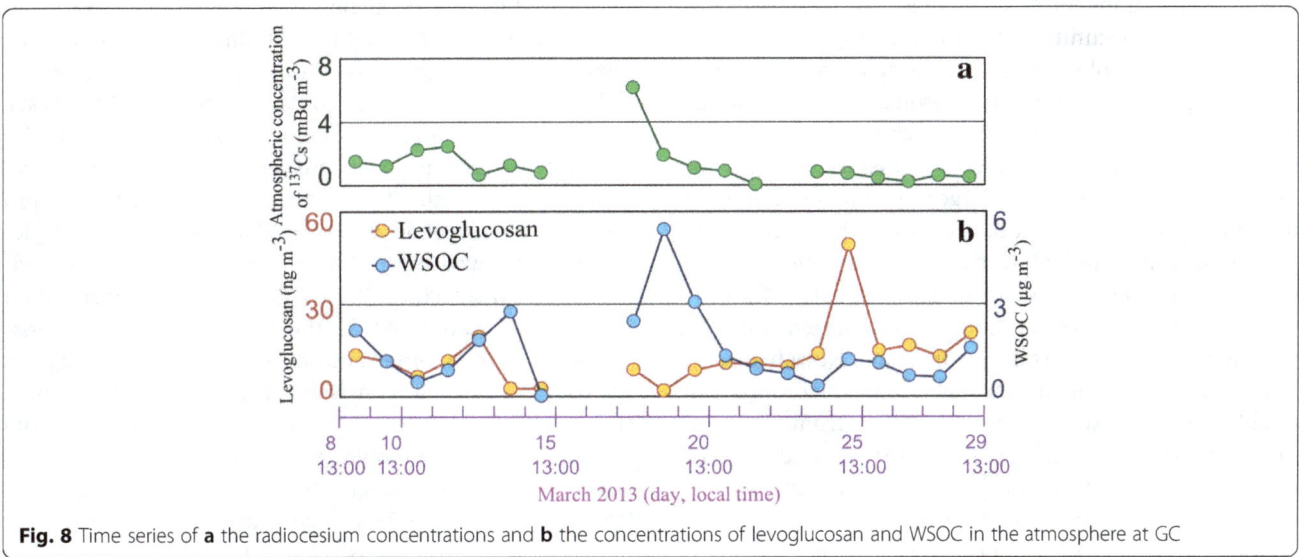

Fig. 8 Time series of **a** the radiocesium concentrations and **b** the concentrations of levoglucosan and WSOC in the atmosphere at GC

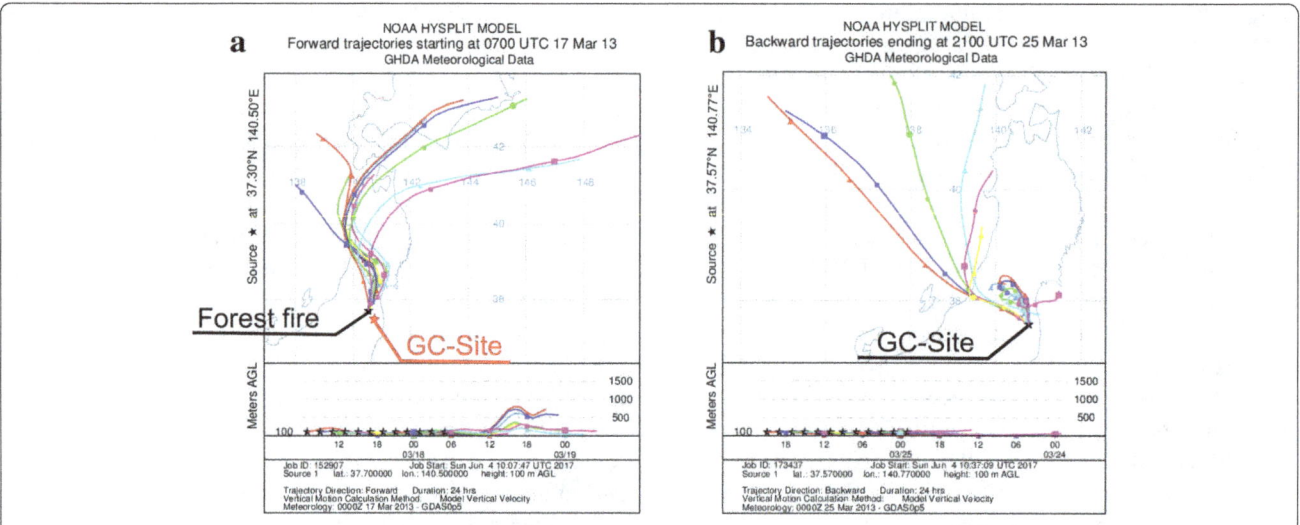

Fig. 9 Maps show the forward and backward trajectories calculated with NOAA HYSPLIT model (Stein et al. 2015) during the forest fire event. **a** The forward trajectories from Ootaki (forest fire area), which are calculated from March 17 at 16:00 (LT) to March 18 at 13:00 (LT). **b** Backward trajectories from the measurement site (GC), which are calculated from March 24 at 13:00 (LT) to March 25 at 13:00 (LT)

red circle in Fig. 11d) at the GD site. However, it was difficult to distinguish spores and microorganism particles only from the SEM images in the present work. We separately counted pollen and microorganism/spores in the SEM images at × 500 and × 1000 magnifications, respectively, of the four sample filters obtained at the GD site in summer (July and August 2014) and winter (December 2014 and January 2015). To compare the magnitudes of these particle counts, the count numbers of the four filter samples were normalized by the sampling air volume and were scaled by the values for the January sample. Figure 12 shows the scaled particle counts of pollen and microorganism/spores with their Poisson standard deviations and compares them with the concentrations of ^{137}Cs activity. Because so many particles had accumulated and overlapped with each other on the sample filters, it was difficult to count these particles precisely. Thus, the absolute values were not highly accurate, and the counts showed only relative amounts. The number of microorganisms/spores as well as that of pollen was much larger in summer than those in winter. This result indicates that these bioaerosol emissions can significantly contribute to the resuspension process. Several studies suggested that ^{137}Cs processed by the ecosystem can be re-emitted into the atmosphere as bioaerosols. For instance, Kuwahara et al. (2005) showed that mushrooms and microorganisms could absorb and accumulate ^{137}Cs from the soil. Furthermore, Kanasashi et al. (2015) showed that the acropetal translocation of ^{137}Cs occurred in Japanese cedars and that their pollen contained ^{137}Cs. Several studies have described the processes of bioaerosol

emissions and the seasonal variations in the bioaerosol concentrations and residence time. For instance, Jones and Harrison (2004) showed that bacteria are emitted from nearly all surfaces, such as the ground, vegetation, and so on. Burrows et al. (2009a, 2009b) and Schumacher et al. (2013) described an increase in the bioaerosol concentrations and an extension of the residence time during the warm season because of the greater turbulence and vertical mixing. The results in these studies are consistent with our results in that bioaerosols, such as microorganism, spores, and pollen, which are actively emitted from nearly all surfaces in the contaminated areas in summer/ autumn, would increase ^{137}Cs in the atmosphere. The R_{scale} values approached unity, as shown in Fig. 7, in these seasons, which can be attributed to longer bioaerosol residence times, which would result in well-mixed atmospheric ^{137}Cs. In this study, the sampling sites are located in low mountainous areas and are surrounded by forests, and thus, it is natural that the increases of bioaerosols significantly contribute to resuspension, unlike in urban areas. Kajino et al. (2016) simulated resuspension from the soil (Ishizuka et al. 2017) and that from vegetation (details were showed in the "Introduction" section), succeeding in reconstructing the ^{137}Cs activity concentration levels that we observed. In their results, most of atmospheric ^{137}Cs in summer/autumn was supplied by biogenic emissions, which is consistent with our results. These studies show that the enhancement of ecosystem activity in summer/autumn could allow increased emissions of larger amounts of pollen, spores, and microorganisms from contaminated forests than those in winter, resulting in a more active resuspension of ^{137}Cs.

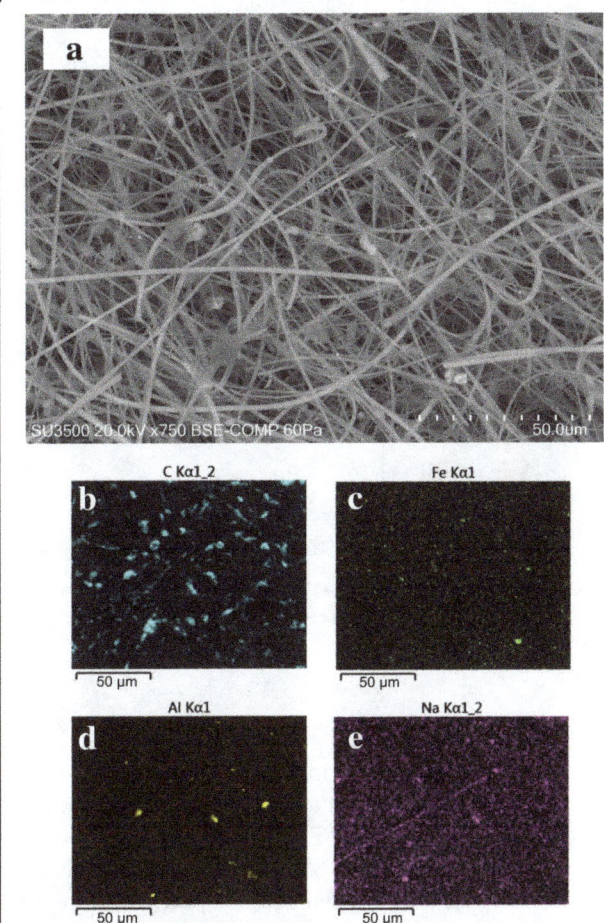

Fig. 10 The SEM image and elemental mapping images of the sample filters obtained at the GD site in July 2015. **a** The SEM image, (**b**) the elemental map of C, (**c**) the elemental map of Fe, (**d**) the elemental map of Al, and (**e**) the elemental map of Na. The characteristic organic particles commonly exist in the filter samples

Fig. 11 Examples of organic particles in the samples obtained in September 2014 at the GD site. **a** A SEM image of the wide visible range. **b** Pollen. **c** Spore. **d** Microorganisms (surrounded by red circles)

Figure 13a–e shows the SEM and element map images, which are similar to Fig. 10a–e, of the sample filters obtained at the GD site in January 2015. In contrast to Fig. 10, Fig. 13 shows the soil/mineral dust particles indicated by the numerous Al spots that were the dominant coarse aerosols observed in winter/spring. Although some C spots were detected, they were mostly debris from vegetation rather than from living bioaerosols because these particles have different shapes and sizes than pollen, spores, and microorganisms. The smaller values of R_{scale} in Fig. 7 suggest that resuspension would occur from local surfaces, such as the playground surface, and that the deposition speeds of the host particles were high in winter/spring. Because there were no significant sources of metal and salt particles around the observational sites, these particles were probably emitted by sources far from the observational sites (Igarashi 2009) and were unlikely to be major host

particles. On the basis of these results and considerations, the soil and mineral dust particles from nearby ground surfaces as well as the debris from the vegetation were the most likely host particles of the ^{137}Cs resuspension in winter/spring.

Many studies have indicated that ^{137}Cs is firmly adsorbed to clay mineral particles (e.g., Dumat and Staunton 1999; Igarashi et al. 2005; Masson et al. 2010; Mukai et al. 2014; Ishizuka et al. 2017), and the transport of mineral dust particles by wind is considered to be a significant resuspension process. As shown in Fig. 3, we can see a positive correlation between the radiocesium concentrations and wind speeds in spring, indicating that the soil/mineral dust, pollen, forest debris, etc., that were suspended by wind blow were plausible host particles of the resuspension. Ishizuka et al. (2017) used the winter/spring samples obtained with the cascade impactor at the GC site, which was also used in this study, and evaluated the relation between the soil/mineral dust particles and activity concentrations of ^{137}Cs during winter/spring; then, they compared these relationships with those estimated using a size-resolved one-dimensional resuspension scheme, and their conclusion is consistent with our results.

Conclusions

To understand the resuspension process of radiocesium in areas heavily contaminated by the FDNPP accident,

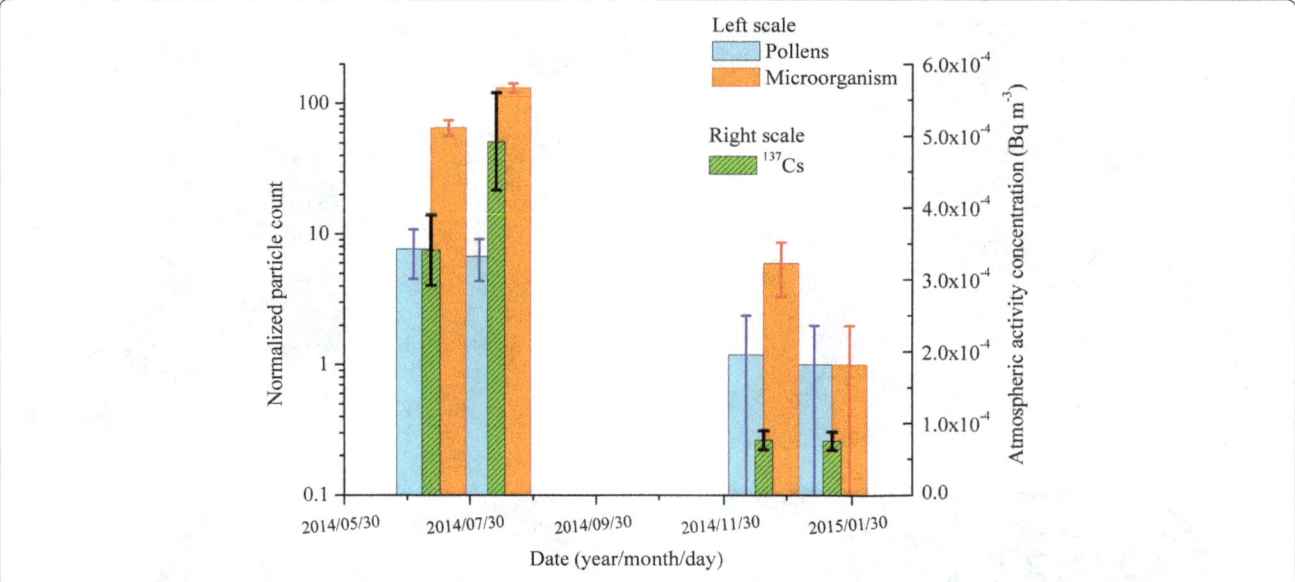

Fig. 12 The normalized particle count of the bioaerosols using SEM (in relative values, the error bar shows the Poisson standard deviation) and atmospheric concentrations of ^{137}Cs activity sampled at the GD site during summer (July and August 2014) and winter (December 2014 and January 2015). The counts of microorganisms include those of spores (see the text)

we conducted long-term measurements of atmospheric radiocesium with high-volume air samplers and impactors in the Tsushima and Yamakiya areas of the Fukushima prefecture. We analyzed the temporal variations of the radiocesium concentrations and examined the particles on the sample substrates of the impactor and its filters using SEM-EDS. We also introduced R_{scale}, which is the ratio of the ^{137}Cs air concentrations from two close sites with similar land uses but different contamination levels, in order to estimate the spatial extent of the resuspension. The main finding in this study is that the seasonal differences of both the radiocesium resuspension processes and the host particle material can be described as shown below:

1. In summer and autumn, the atmospheric radiocesium activity concentration was higher, and both the higher normalized ^{137}Cs$_{(1-5)}$ values and SEM-EDS analysis results indicated that coarse organic particles, such as spores, pollen, and microorganisms, with adhesive organic matter were the dominant host particles of radiocesium resuspension. The larger values of R_{scale} in these seasons are consistent with this result, considering that the deposition velocity of the light biogenic organic particles was probably low and that the resuspended radiocesium on them should be well mixed. These organic particles could be emitted from forest ecosystems.

2. In winter and early spring, the atmospheric radiocesium activity concentrations were lower, and they increased in May. The higher normalized ^{137}Cs$_{(7)}$ values and the SEM-EDS analysis results showed that the dominant coarse particles were dust (soil and mineral) and vegetation debris and that these were the dominant host particles for radiocesium resuspension. The temporal variations of the radiocesium concentrations showed similar variations to those of wind speed, suggesting that wind suspension of contaminated particles from the ground surface is considered as the main resuspension process in these seasons. The smaller R_{scale} values in these seasons are still consistent with this result, considering that the deposition velocity of the dust particles with higher density was probably high and that the resuspended radiocesium on them should be less mixed.

3. No significant correlation between the concentrations of levoglucosan and WSOC and that of radiocesium was found in spring, indicating that biomass burning was not a significant resuspension process of radiocesium.

4. Although many studies have suggested that soil/mineral dusts and bioaerosols could be major host particles of $^{134, 137}$Cs, we first identified the seasonal variations of the dominant resuspension processes/host particles. Further research on bioaerosols, such as working to identify specific bioaerosols and evaluating the quantitative relationship between

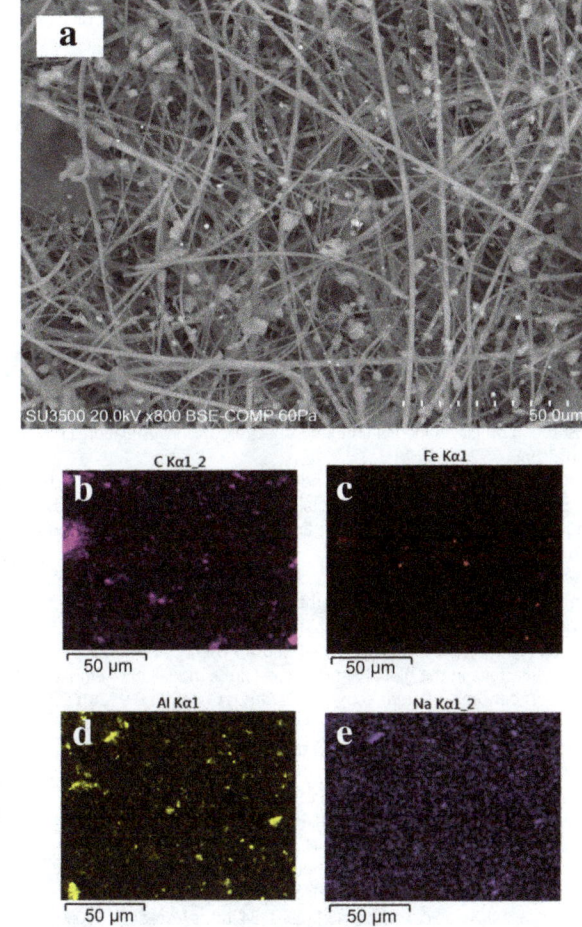

Fig. 13 The SEM and elemental mapping images of the sample filter obtained at the GD site in January 2015. **a** The SEM image, (**b**) the elemental map of C, (**c**) the elemental map of Fe, (**d**) the elemental map of Al, and (**e**) the elemental map of Na. Major coarse particles include Al (clay mineral) and Na (salt or mineral)

bioaerosols and radiocesium, is required. In the near future, we will include more observations of bioaerosols to reveal more detailed radiocesium resuspension processes. A flux measurement of these particles would be necessary to quantitatively evaluate the radiocesium resuspension rate.

Abbreviations

EDS: Energy-dispersive X-ray spectroscopy; FAJ: Forestry Agency Japan; FDNPP: Fukushima Daiichi Nuclear Power Plant; HYSPLIT: The Hybrid Single Particle Lagrangian Integrated Trajectory Model; JAEA: Japan Atomic Energy Agency; MEXT: Ministry of Education, Culture, Sports, Science and Technology, Japan; MRI: Meteorological Research Institute; NDVI: Normalized Difference Vegetation Index; NOAA: National Oceanic and Atmospheric Administration; SEM: Scanning electron microscopy; TEPCO: Tokyo Electric Power Company; WSOC: Water-soluble organic carbon

Acknowledgements

We acknowledge T. Kimura, K. Inukai, K. Kamioka (Atox Co. Ltd.), S. Kakitani (Osaka University), and N. Hayashi (Ibaraki University) for their sample preparation and measurements of radioactivity; T. Kanari (Green Blue Co. Ltd.) for the periodical maintenance of the measurement stations in the field; members of Ibaraki University and the College of Engineering of Ibaraki University for their maintenance of the tools after sampling and work the setting field instruments; Drs. T. Sekiyama (MRI), M. Kajino (MRI), Y. Hatano (University of Tsukuba), and T. Maki (Kanazawa University) for their discussions; and the many people who cooperated this study. We also express our gratitude to the local governments of Kawamata and Namie as well as to Fukushima prefectural government, who kindly offered this monitoring opportunity.

Funding

This work was financially supported by MEXT KAKENHI (a Grant-in-Aid for Scientific Research on Innovative Areas under the A01–1, A01–2 and A04–8 research teams in the "Interdisciplinary Study on Environmental Transfer of Radionuclides from the Fukushima Daiichi NPP Accident"; the grant numbers are 24110002, 24110003, and 24110009, respectively) and the MEXT Japanese Radioactivity Survey at the MRI. Additionally, this study was partly supported by a grant from the Asahi group foundation at Ibaraki University. Thus, this study was discussed and supported by the workshop of ISET-R for writing articles.

Authors' contributions

TK collected and analyzed the samples and summarized these data in this article. KK planned and carried out the measurements as well as the samplings. YI is the leader of our project team, helped in writing the manuscript, and provided important discussions for this article. KA and YZ supported the identification of the aerosols by SEM-EDS. KN, AS, ST, KY, and NY analyzed the radioactivity of filter samples. HOk and HOg analyzed the levoglucosan and WSOC of filter samples. MI and HD supported the maintenance of measurements in the field. MM and YO provided important suggestions for this article. All authors read and approved the final manuscript.

Authors' information

TK studied atmospheric environments, especially focusing on black carbon (BC) aerosols, and received his PhD in atmospheric chemistry from Ibaraki University through the study of Antarctic aerosol science. He has been working at Hitachi Power Solutions Co., Ltd. since 2016 and wrote this article while working. In his studies, he participated in the project of the workshop of Interdisciplinary Study on Environmental Transfer of Radionuclides from the Fukushima Daiichi NPP Accident (ISET-R) to shed light on the environmental transfer of radionuclides and to contribute to the revival of the affected areas in the Fukushima prefecture.
KK received his PhD from Tokyo University in 1990 and worked at Tokyo University until 2002 as a research assistant. He moved to Ibaraki University in 2002. His studies focus on the atmospheric O_3 and NO_2, the effects of BC on global warming, and the resuspension processes of radiocesium.
YI received his PhD from the University of Tsukuba in 1987. From 1987 to 1991, he worked at the former National Institute of Radiological Sciences and moved to the MRI in 1991. His current interests are atmospheric aerosols; their precursors, including Asian dust and $PM_{2.5}$; their possible influences on climate and general environmental changes; and other related phenomena.
KA received his PhD from Kobe University in 2005, worked at Arizona State University between 2005 and 2011 as a postdoctoral/faculty research associate and is currently studying atmospheric aerosols at the MRI.
KN received his PhD from Osaka University in 2007 and worked at Osaka University in 2008. He moved to JAEA and worked as a research scientist between 2009 and 2012; then, he returned to Osaka University. His main research field is nuclear- and radiochemistry.
AS received his PhD from Osaka University in 1985 and then worked at the Japan Society for the Promotion of Science. He moved to Nagoya University, where he worked between 1987 and 1998. Then, he moved to Kyoto University in 1998, before moving to Osaka University in 1999. His main research field is nuclear- and radiochemistry.
HOk worked at Kanagawa University between 1991 and 1999 and received his PhD from Tokyo University in 1997 while he was working. From 1999 to 2003, he worked at the University of East Anglia, UK. He moved to the Tokyo Metropolitan Institute of Technology, where he worked from 2003 to 2005. Then, he moved to Waseda University. His studies are focused on the environmental science topics of the atmosphere, hydrosphere and ecosystems of the forest, and on the development of a technique for environmental remediation via soft-chemical methods.

HOg received her PhD from the Prefectural University of Kumamoto in 2011. She worked at Waseda University between 2011 and 2016, and then, she moved to Sibata Scientific Technology Ltd. Her main research field is atmospheric aerosols.

MI received his PhD from Osaka University in 2000 and then worked at the Faculty of Systems Engineering, Wakayama University from 2000 to 2005. He moved to the Faculty of Engineering, Kagawa University. His studies are centered on the resuspension processes of radionuclides with dust particles as forced by wind, the processes of the suspension of Asian dust, and researching the hydrological environment and its processes.

ST received his PhD from the University of Tokyo in 2007 and started his study on the development of analytical methods for isotopocules, i.e., isotopically substituted molecules, of atmospheric nitrous oxide (N_2O) at Nagoya University as a postdoctoral fellow of the Japan Science and Technology Corporation. After moving to the Tokyo Institute of Technology in 2008, he pursued an investigation of the geochemical cycle of N_2O and is currently expanding his research interests into other trace gasses and aerosols as an associate professor.

KY received his PhD from Tokyo Metropolitan University in 1997. He has worked at the Tokyo Institute of Technology since 2000. He studies the relationship between radionuclides and organics in aerosols.

NY received his PhD from the Tokyo Institute of Technology in 1984. He was with the Mitsubishi-Kagaku Institute of Life Sciences, Toyama Univ., Nagoya Univ., and has been a full professor in charge of two Environmental Departments of the Tokyo Institute of Technology since 1998 and is a Principal Investigator at the Earth-Life Science Institute. He studies "isotopomers: isotopic substituted molecules" as powerful tracers that can help to reduce uncertainties in the biological, chemical, physical, and anthropogenic source and sink processes of environmental molecules, especially for global warming-related species.

YZ graduated the Meteorological College in 1984 and worked at the Nagasaki Marine Observatory, Japan Meteorological Agency from 1984 to 1989. He has been studying atmospheric aerosols at the MRI.

MM received his PhD from Tohoku University in 1996 and has worked at the MRI from 1985 to 2015. He held the position of Senior Researcher for Research Affairs at the Institute. Currently, he is working at the Japan Meteorological Business Support Center.

HD graduated the Faculty of Maritime Sciences of Kobe University in 2005 and worked at a general company of the manufacturing industry. In 2010, he moved to the Technical Services Division of the College of Engineering, Ibaraki University. He works as an engineer providing the technical support at Ibaraki University and helped with our research.

YO graduated and received his PhD from the University of Tsukuba in 1990, working at Nagoya University from 1992 to 1999. Currently, he is working at the University of Tsukuba, where he researches hydrology and the environmental dynamics of radionuclides as well as being the PI of the whole ISET-R project.

Competing interests

The authors declare that they have no competing interests.

Author details

[1]Graduate School of Science and Engineering, Ibaraki University, 2-1-1 Bunkyo, Mito, Ibaraki 310-8512, Japan. [2]Hitachi Power Solutions Co., Ltd, 2-2-3 Saiwaicho, Hitachi, Ibaraki 317-0073, Japan. [3]College of Science, Ibaraki University, 2-1-1 Bunkyo, Mito, Ibaraki 310-8512, Japan. [4]Atmospheric Environment and Applied Meteorology Research Department, Meteorological Research Institute, 1-1 Nagamine, Tsukuba, Ibaraki 305-0052, Japan. [5]Graduate School of Science, Osaka University, 1-1, Machikaneyama, Toyonaka, Osaka 560-0043, Japan. [6]Faculty of Science and Engineering, Waseda University, 3-4-1 Okubo, Shinjuku, Tokyo 169-8555, Japan. [7]Faculty of Engineering, Kagawa University, 2217-20 Hayashi-cho, Takamatsu, Kagawa 761-0396, Japan. [8]Department of Chemical Science and Engineering, School of Materials and Chemical Technology, Tokyo Institute of Technology, 4259 Nagatsuta, Yokohama, Kanagawa 226-8502, Japan. [9]Earth-Life Science Institute, Tokyo Institute of Technology, 2-12-1 Ookayama, Meguro, Tokyo 152-8551, Japan. [10]Forecast Research Department, Meteorological Research Institute, Japan Meteorological Agency, 1-1 Nagamine, Tsukuba, Ibaraki 305-0052, Japan. [11]Japan Meteorological Business Support Center, 3-17 Nishikicho, Kanda, Chiyoda, Tokyo 101-0054, Japan. [12]College of Engineering, Ibaraki University, 4-12-1 Narusawa, Hitachi, Ibaraki 316-8511, Japan. [13]Center for Research in Isotopes and Environmental Dynamics, University of Tsukuba, 1-1-1 Tennodai, Tsukuba, Ibaraki 305-8577, Japan.

References

Abe Y, Iizawa Y, Terada Y, Adachi K, Igarashi Y, Nakai I (2014) Detection of uranium and chemical state analysis of individual radioactive microparticles emitted from the Fukushima nuclear accident using multiple synchrotron radiation X-ray analyses. Anal Chem 86:8521–8525

Adachi K, Kajino M, Zaizen Y, Igarashi Y (2013) Emission of spherical cesium-bearing particles from an early stage of the Fukushima nuclear accident. Nature Sci Rep 3:2554. doi:https://doi.org/10.1038/srep02554.

Barišić D, Lulić S, Kezić N, Vertačnik A (1992) ^{137}Cs in flowers, pollen and honey from the Republic of Croatia four years after the Chernobyl accident. Apidologie 23:71–78

Berglund BE (2008) Satoyama, traditional farming landscape in Japan, compared to Scandinavia. Japan Review 20:53–68

Bostic BC, Vairavamurthy MA, Karthikeyan KG, Chorover J (2002) Cesium adsorption on clay minerals: an EXAFS spectroscopic investigation. Environ Sci Technol 36:2670–2676

Bourcier L, Sellegri K, Masson O, Zangrando R, Barbante C, Gambaro A, Pichon J-M, Boulon J, Laj P (2010) Experimental evidence of biomass burning as a source of atmospheric [137]Cs. Atmos Environ 44:2280–2286. https://doi.org/10.1016/j.atmosenv.2010.04.017

Burrows SM, Elbert W, Lawrence MG, Pöschl U (2009a) Bacteria in the global atmosphere–Part 1: review and synthesis of literature data for different ecosystems. Atmos Chem Phys 9:9263–9280

Burrows SM, Butler T, Jöckel P, Tost H, Kerkweg A, Pöschl U, Lawrence MG (2009b) Bacteria in the global atmosphere–Part 2: modeling of emissions and transport between different ecosystems. Atmos Chem Phys 9:9281–9297

Carrera M, Zandomeni RO, Fitzgibbon J, Sagripanti J-L (2007) Difference between the spore sizes of Bacillus anthracis and other Bacillus species. J Appl Microbiol 102:303–312. https://doi.org/10.1111/j.1365-2672.2006.03111.x

Chaturvedi M, Datta K, Nair PKK (1998) Pollen morphology of Oryza (Poaceae). Grana 37:79–86. https://doi.org/10.1080/00173139809362647

Chino M, Nakayama H, Nagai H, Terada H, Katata G, Yamazawa H (2011) Preliminary estimation of release amounts of ^{131}I and ^{137}Cs accidentally discharged from the Fukushima Daiichi Nuclear Power Plant into the atmosphere. J Nucl Sci Technol 48:1129–1134. https://doi.org/10.1080/18811248.2011.9711799

Dietz LA, Pachucki CF (1973) ^{137}Cs and [134]Cs half-lives determined by mass spectrometry. J Inor Nucl Chem 35:1769–1776

Doi T, Masumoto K, Toyoda A, Tanaka A, Sibata Y, Hirose K (2013) Anthropogenic radionuclides in the atmosphere observed at Tsukuba: characteristics of the radionuclides derived from Fukushima. J Environ Radioactiv 122:55–62. https://doi.org/10.1016/j.jenvrad.2013.02.001

Dumat C, Staunton S (1999) Reduced adsorption of caesium on clay minerals caused by various humic substances. J Environ Radioactiv 46:187–200

Ehlken S, Kirchner G (2002) Environmental processes affecting plant root uptake of radioactive trace elements and variability of transfer factor data: a review. J Environ Radioactiv 58:97–112

Forestry Agency, Japan (FAJ) (2015) Heisei 26 nendo Sugi obana ni fukumareru houshasei sesiumu noudo no chousa kekka ni tsuite (About FY H26 investigation results on radiocesium concentrations contained in cedar pollen). http://www.rinya.maff.go.jp/j/press/kaihatu/150130.html. in Japanese. Accessed 19 Feb 2018

Gutman G, Ignatov A (1998) The derivation of the green vegetation fraction from NOAA/AVHRR data for use in numerical weather prediction models. Int J Remote Sensing 19:1533–1543

Healy DA, O'Connor DJ, Burke AM, Sodeau JR (2012) A laboratory assessment of the Waveband Integrated Bioaerosol Sensor (WIBS-4) using individual samples of pollen and fungal spore material. Atmos Environ 60:534–543. https://doi.org/10.1016/j.atmosenv.2012.06.052

Holt M, Campbell RJ, Nikitin MB (2012) Fukushima nuclear disaster, congressional research service

Igarashi Y (2009) Anthropogenic radioactivity in aerosol-a review focusing on studies during the 2000s. Jpn J Health Phys 44:313–323

Igarashi Y, Aoyama M, Hirose K, Povinec P, Yabuki S (2005) What anthropogenic radionuclides (^{90}Sr and ^{137}Cs) in atmospheric deposition, surface soils and aeolian dusts suggest for dust transport over Japan. Water Air Soil Poll 5:51–69

Igarashi Y, Adachi K, Kajino M, Zaizen Y (2014) Characteristic of spherical Cs-bearing particles collected during the early stage of FDNPP accident. IAEA meeting https://www-pub.iaea.org/iaeameetings/cn224p/Session3/Igarashi.pdf. Accessed 19 Feb 2018

Igarashi Y, Kajino M, Zaizen Y, Adachi K, Mikami M (2015) Atmospheric radioactivity over Tsukuba, Japan: a summary of three years of observations after the FDNPP accident. Prog Earth Planet Sci 2:44

Ishizuka M, Mikami M, Tanaka T, Igarashi Y, Kita K, Ymada Y, Yoshida N, Toyoda S, Satou Y, Kinase T, Ninomiya K, Shinohara A (2017) Use of a size-resolved 1-D resuspension scheme to evaluate resuspended radioactive material associated with mineral dust particles from the ground surface. J Environ Radioactiv 166:436–448. https://doi.org/10.1016/j.jenvrad.2015.12.023

Japan Atomic Energy Agency (JAEA) (2012) Fukushima daiichi gennshiryokuhatudennsho ni kakawaru hinannkuikutou ni okeru josennjisshougyoumu houkokusho gaiyoubann. Decontamination trial summary report for evacuation areas with the Fukushima daiichi nuclear power plant accident, JAEA https://fukushima.jaea.go.jp/initiatives/cat01/pdf/summary.pdf. Acessed 19 Feb 2018

Jones AM, Harrison M (2004) The effects of meteorological factors on atmospheric bioaerosol concentrations—a review. Sci Total Environ 326:151–180

Kajino M, Ishizuka M, Igarashi Y, Kita K, Yoshikawa C, Inatsu M (2016) Long-term assessment of airborne radiocesium after the Fukushima nuclear accident: re-suspension from bare soil and forest ecosystems. Atmos Chem Phys 16:13149–13172. https://www.atmos-chem-phys.net/16/13149/2016/. https://doi.org/10.5194/acp-16-13149-2016

Kanasashi T, Sugiyama Y, Takenaka C, Hijii N, Umemura M (2015) Radiocesium distribution in sugi (Cryptomeria japonica) in Eastern Japan: translocation from needles to pollen. J Environ Radioactiv 139:398–406

Kaneyasu N, Ohashi H, Suzuki F, Okuda T, Ikemori F (2012) Sulfate aerosol as a potential transport medium of radiocesium from the Fukushima Nuclear Accident. Environ Sci Technol 46:5720–5726. https://doi.org/10.1021/es204667h

Kuwahara C, Fukumoto A, Ohsone A, Furutya N, Sibata H, Sugiyama H, Kato F (2005) Accumulation of radiocesium in wild mushrooms collected from a Japanese forest and cesium uptake by microorganisms isolated from the mushroom-growing soils. Sci Total Environ 345:165–172

Laucks ML, Roll G, Shcweigner G, Davis EJ (2000) Physical and chemical (raman) characterization of bioaerosols–pollens. J Aerosol Sci 31:307–319

Local government of Fukushima prefecture (2014) Shoubou bousai nennpou (H24 nenn ban) (About the monthly number of forest fire at Fukushima prefecture during 2013). https://www.pref.fukushima.lg.jp/uploaded/attachment/37264.pdf in Japanese. Accessed 19 Feb 2018

Lujanienė G, Aninkevičius V, Lujanas V (2009) Artificial radionuclides in the atmosphere over Lithuania. J Environ Radioactiv 100:108–119. https://doi.org/10.1016/j.jenvrad.2007.07.015

Masson O, Piga D, Gurriaran R, D'Amico D (2010) Impact of an exceptional Saharan dust outbreak in France: PM_{10} and artificial radionuclides concentrations in air and in dust deposit. Atmos Environ 44:2478–2486

Masson O, Ringer W, Malá H, Rulik P, Dlugosz-Lisiecka M, Eleftheriadis K, Meisenberg O, Vismes-Ott AD, Gensdarmes F (2013) Size distributions of airborne radionuclides from the Fukushima Nuclear Accident at several places in Europe. Environ Sci Technol 47:10995–11003. https://doi.org/10.1021/es401973c

Merz S, Steinhauser G, Hamada N (2013) Anthropogenic radionuclides in Japanese food: environmental and legal implications. Environ Sci Technol 47:1248–1256

Ministry of Education, Culture, Sports, Science and Technology, Japan (MEXT) (2011) Results of the fourth airborne monitoring survey by MEXT.

Mukai H, Hatta T, Kitazawa H, Yamada H, Yaita T, Kogure T (2014) Speciation of radioactive soil particles in the Fukushima contaminated area by IP autoradiography and microanalyzes. Environ Sci Technol 48:13053–13059. https://doi.org/10.1021/es502849e

Ochiai S, Hasegawa H, Kakiuchi H, Akata N, Ueda S, Tokonami S, Hisamatsu S (2016) Temporal variation of post-accident atmospheric ^{137}Cs in an evacuated area of Fukushima Prefecture: size-dependent behaviors of ^{137}Cs-bearing particle. J Environ Radioactiv:131–139. https://doi.org/10.1016/j.jenvrad.2016.09.014

Satou Y, Sueki K, Sasa K, Adachi K, Igarashi Y (2016) First successful isolation of radioactive particles from soil near the Fukushima Daiichi Nuclear Power Plant. Anthropocene 14:71–76

Schumacher CJ, Pohlker C, Aalto P, Hiltunen V, Petaja T, Kulmala M, Poschl U, Huffman JA (2013) Seasonal cycles of fluorescent biological aerosol particles in boreal and semi-arid forests of Finland and Colorado. Atmos Chem Phys 13:11987–12001

Simoneit BRT, Shcauer JJ, Nolte CG, Ortos DR, Elias VO, Fraser MP, Rogge WF, Cass GR (1999) Levoglucosan, a tracer for cellulose in biomass burning and atmospheric particles. Atmos Environ 33:173–182

Stein AF, Draxler RR, Rolph GD, Stunder BJB, Cohen M (2015) NOAA's HYSPLIT atmospheric transport and dispersion modeling system. Am Meteorol Soc 96:2059–2078

Steinhauser G, Brandl A, Johnson E (2014) Comparison of the Chernobyl and Fukushima nuclear accidents: a review of the environmental impacts. Sci Total Environ 470–471:800–817

Sýkora I, Povince PP, Brest'áková L, Florek M, Holý K, Masarik L (2012) Resuspension processes control variations of ^{137}Cs activity concentrations in the ground-level air. J Radioanal Nucl Chem 293:595–599

Tokyo Electric Power Co., Inc. (TEPCO) (2012): Genshirotateyakarano tuikatekihoushuturyou no hyoukakekka (2012 nenn 10 gatsu 22 nichi) (Estimation of the additional emission from nuclear reactor buildings (22 October 2012)), available at: http://www.meti.go.jp/earthquake/nuclear/pdf/121022/121022_01s.pdf in Japanese

Tomioka N, Uchiyama H, Yagi O (1992) Isolation and characterization of cesium-accumulating bacteria. Appl Environ Microb 58:1019–1023

Unterweger MP (2002) Half-life measurements at the National Institute of Standards and Technology. Appl Radiat Isotopes 56:125–130

Watanabe A (2014) Houshaseibusshitsu no taikichuunoudo / koukaryou nado no choukihenndou. Long time trend of atmospheric concentration/ fallout of radiocesium. J Jpn Soc. Atmos Environ 49:A89–A90. https://www.jstage.jst.go.jp/article/taiki/49/6/49_A89/_pdf

Working Group on Risk Management of Low-dose Radiation Exposure (2011) Report:working group on risk management of low-dose radiation exposure. http://www.cas.go.jp/jp/genpatsujiko/info/twg/Working_Group_Report.pdf

Yoschenko VI, Kashparov VA, Protsak VP, Lundin SM, Levchuk SE, Kadygrib AM, Zvarich SM, Khomutinin YV, Maloshtan IM, Lanshin VP, Kovtun MV, Tschiersch L (2006) Resuspension and redistribution of radionuclides during grassland and forest fires in the Chernobyl exclusion zone: part I. Fire experiments J Environ Radioactiv 86:143–163. https://doi.org/10.1016/k.jenvrad.2005.08.003

Yoshida S, Muramatsu Y, Ogawa M (1994) Radiocesium concentrations in mushrooms collected in Japan. J Environ Radioactiv 22:141–154

8

Retrieval of radiative and microphysical properties of clouds from multispectral infrared measurements

Hironobu Iwabuchi[1]*[iD], Masanori Saito[1], Yuka Tokoro[1], Nurfiena Sagita Putri[1] and Miho Sekiguchi[2]

Abstract

Satellite remote sensing of the macroscopic, microphysical, and optical properties of clouds are useful for studying spatial and temporal variations of clouds at various scales and constraining cloud physical processes in climate and weather prediction models. Instead of using separate independent algorithms for different cloud properties, a unified, optimal estimation-based cloud retrieval algorithm is developed and applied to moderate resolution imaging spectroradiometer (MODIS) observations using ten thermal infrared bands. The model considers sensor configurations, background surface and atmospheric profile, and microphysical and optical models of ice and liquid cloud particles and radiative transfer in a plane-parallel, multilayered atmosphere. Measurement and model errors are thoroughly quantified from direct comparisons of clear-sky observations over the ocean with model calculations. Performance tests by retrieval simulations show that ice cloud properties are retrieved with high accuracy when cloud optical thickness (COT) is between 0.1 and 10. Cloud-top pressure is inferred with uncertainty lower than 10 % when COT is larger than 0.3. Applying the method to a tropical cloud system and comparing the results with the MODIS Collection 6 cloud product shows good agreement for ice cloud optical thickness when COT is less than about 5. Cloud-top height agrees well with estimates obtained by the CO_2 slicing method used in the MODIS product. The present algorithm can detect optically thin parts at the edges of high clouds well in comparison with the MODIS product, in which these parts are recognized as low clouds by the infrared window method. The cloud thermodynamic phase in the present algorithm is constrained by cloud-top temperature, which tends not to produce results with an ice cloud that is too warm and liquid cloud that is too cold.

Keywords: Cloud optical thickness, Cloud-top height, Effective particle radius, Ice cloud, Optimal estimation method, Satellite remote sensing

Introduction

Clouds play a vital role in regulating the Earth's radiation budget, through shortwave cooling and longwave warming effects (Ramanathan et al. 1989). The cloud radiative effects depend on the type of cloud, and thus, the radiation budget is controlled by the occurrence of various types of clouds (Hartmann et al. 1992), which complicates our understanding of cloud roles in the climate system. In particular, the radiative effects of ice clouds are not well understood, partly because the optical properties of ice clouds are not well quantified (Baran 2009), which is a major source of uncertainty in ice cloud representations in global climate models. There are discrepancies in satellite observation climatology of ice clouds, and improvement of ice cloud processes is still a challenge (e.g., Waliser et al. 2009). Climatology and spatial and temporal variations of clouds on various scales are also important to understand cloud response and feedback in climate systems. Satellite remote sensing can provide constraints for global cloud properties that are useful for developing cloud parameterizations. Macroscopic, microphysical, and optical properties are generally used in satellite remote sensing of clouds. There are specialized methods for each property, including cloud fraction, cloud-top properties (temperature/

* Correspondence: hiroiwa@m.tohoku.ac.jp
[1]Center for Atmospheric and Oceanic Studies, Graduate School of Science, Tohoku University, 6-3 Aoba, Aramakiaza, Aoba-ku, Sendai, Miyagi 980-8578, Japan
Full list of author information is available at the end of the article

pressure/height), cloud thermodynamic phase, cloud optical thickness (COT), and cloud-particle effective radius (CER).

There are two passive remote sensing methods that are commonly used for cloud optical and microphysical properties: infrared (IR) window (split-window) (Inoue 1985; Parol et al. 1991; Giraud et al. 1997) and visible/shortwave IR (VIS/SWIR) bispectral (Nakajima and King 1990) approaches. IR window cloud retrieval is suitable for optically thin high clouds with COT of 0.1–5 (e.g., Garnier et al. 2012), whereas the VIS/SWIR method is suitable for optically thick clouds with COT greater than 1 (Nakajima and King 1990; Platnick et al. 2003). We have developed an IR method to retrieve COT and CER by using the 8.5, 11, and 12 μm bands of the moderate resolution imaging spectroradiometer (MODIS) onboard the Aqua satellite (Iwabuchi et al. 2014). In this method, inversion was based on the optimal estimation method (Rodgers 2000), which simultaneously fits the physics model to measurements and diagnoses rigorous uncertainties and retrieval quality. The optimal estimation method has been used widely for cloud remote sensing (Cooper et al. 2003; Heidinger and Pavolonis 2009; Watts et al. 2011; Walther and Heidinger 2012; Poulsen et al. 2012; Sourdeval et al. 2013; 2015; Wang et al. 2016). In a previous work (Iwabuchi et al. 2014), cloud retrieval was applied only to the ice phase cloud, and the a priori cloud-top temperature (CTT) was independently estimated by the CO_2 slicing technique (Menzel et al. 2008) in the MODIS operational product. Thus, the retrieval was strongly constrained by cloud-top prior information and affected by the CTT accuracy in the MODIS product. Because the CTT retrieval itself can depend on COT and microphysical properties, the overall retrieval performance can be obtained if the cloud-top height (CTH), COT, and effective radius are retrieved simultaneously from the window and absorption bands.

In addition, the cloud thermodynamic phase is important because liquid and ice clouds play different roles in regulating the Earth's radiation budget and hydrological cycle. Although cloud retrieval using passive sensors usually assumes single-layer ice or liquid clouds, it leads to substantial errors in estimated cloud optical and microphysical properties if there is a multilayer cloud system or if the assumed cloud phase is wrong (Davis et al. 2009). Recent studies using active remote sensing from CloudSat and Cloud-Aerosol Lidar and Infrared Pathfinder Satellite Observation (CALIPSO) satellites have obtained a globally averaged multilayered cloud occurrence of 25–28 % (Li et al. 2015). A cloud analysis algorithm should include methods for detection and property retrieval of multilayered cloud systems and determination of the cloud phase.

In this paper, an optimal estimation-based cloud retrieval algorithm is presented, where COT, CER, cloud-top pressure (CTP), and surface temperatures are simultaneously retrieved from measurements in ten thermal IR (TIR) bands of MODIS including the window and CO_2 and water vapor absorption bands. Combined use of TIR bands enables the cloud thermodynamic phase to be distinguished and allows the method to be used for multilayer clouds, as previous pioneering studies suggest. The cloud retrieval algorithm is developed as part of the Integrated Cloud Analysis System (ICAS), which we develop in this study. This paper is organized as follows. The "Methods" section describes the source data used for cloud analysis, the cloud retrieval algorithm, the forward model, and the measurement and model errors, which are thoroughly quantified by model-to-model and model-to-observation comparisons. In the "Results and Discussion" section, retrieval errors are evaluated based on retrieval simulations in idealized cases, to understand the advantages and limitations of the algorithm. The algorithm is applied to a MODIS granule, and the retrieved cloud properties are compared with the MODIS Collection 6 (C6) operational product. The conclusion is given in the "Conclusions" section.

Methods/Experimental

Source data

The measurement data used in this study are from the level 1B product of MODIS onboard the Aqua satellite. MODIS has a swath of 2330 km, and a granule every 5 min covers an area of 2330 × 2030 km. TIR bands have a 1 km resolution, and the ten TIR bands are summarized in Table 1. In addition to bands 29, 31, and 32 in the atmospheric window, the bands used include ozone absorption band 30 (9.6 μm), water vapor absorption bands 27 and 28, and carbon dioxide absorption bands 32–36. The spectral radiance of each band is converted to the brightness temperature (BT). The band mean Planck function for temperature T is defined as

$$\overline{B}(T) = \frac{\int_0^\infty B(\lambda, T)\phi(\lambda)d\lambda}{\int_0^\infty \phi(\lambda)d\lambda}, \qquad (1)$$

where B is the Planck function, λ is the wavelength, and ϕ is the response function of each MODIS band. The band mean Planck function is precalculated for different temperatures, and the Akima interpolation (Akima 1970) is used to calculate the function or its inverse function, the BT, from the look-up table.

The source data used in ICAS are summarized in Table 2. Meteorological field data, including temperature

Table 1 Characteristics of MODIS bands used in this study

Band	Center wavelength (μm)	Band range (μm)	Absorbers[a]
27	6.78683	6.535–6.895	H_2O
28	7.34963	7.175–7.475	H_2O, CH_4
29	8.55511	8.400–8.700	H_2O, N_2O, CH_4
30	9.72374	9.580–9.880	H_2O, CO_2, O_3
31	11.026	10.780–11.280	H_2O, CO_2
32	12.0423	11.770–12.270	H_2O, CO_2
33	13.3648	13.185–13.485	H_2O, CO_2, O_3
34	13.686	13.485–13.785	H_2O, CO_2, O_3
35	13.9252	13.785–14.085	H_2O, CO_2, O_3
36	14.2153	14.085–14.385	H_2O, CO_2, O_3

[a]The major absorbing gas is shown in italics

and ozone and water vapor mixing ratios, are obtained by interpolation in the space-time domain from the Modern-Era Retrospective analysis for Research and Applications (MERRA) product, IAU 3D assimilated state on pressure, which has a horizontal resolution of 1.25 ° × 1.25 °, 42 pressure levels, and a time interval of 3 h. The MERRA product is the atmosphere re-analysis product of the National Aeronautics and Space Administration, Goddard Earth Observing System Model, Version 5 (GEOS-5) with its atmospheric data assimilation system (Rienecker et al. 2011). Concentrations of carbon dioxide (CO_2), methane (CH_4), and nitrous oxide (N_2O) are global monthly mean values provided from the World Data Center for Greenhouse Gases of the World Meteorological Organization Global Atmosphere Watch program (Tsutsumi et al. 2009).

Sea surface temperature data are from the MODIS 8 day mean level 3 product that is based on the TIR split window method (Brown et al. 1999). The root-mean-square error (RMSE) of SST by the split window method is evaluated as 0.35 K. Sea surface emissivity is determined by using the Fresnel equations for a flat sea surface based on the complex refractive index and the satellite zenith angle. The effects of a rough surface, including the effects of multiple reflection and wind direction, are sufficiently small for our purposes when the satellite zenith angle is 60 ° or less (Masuda 2012).

Table 2 Summary of MODIS operational product data used in the retrieval algorithm

Quantity	Source	Spatial resolution	Temporal resolution
Atmospheric profile	MERRA	1.25 °, 42 levels	3 h
Sea surface temperature	MODIS L3	0.4167 °	8 day mean
Land surface temperature	MODIS L3	0.05 °	8 day mean
Land surface emissivity	BFED	0.05 °	Monthly
Trace gas concentration	Climatology	Global	Monthly

The complex refractive index of seawater is synthesized from that of pure water based on Downing and Williams (1975) with a correction for the salinity effect based on Friedman (1969).

The land surface temperature is from the MODIS land 8 day mean level 3 product (MYD11C2), which is based on the day–night algorithm (Wan and Li 1997). For each day and night satellite overpass, the 8 day mean values are available in the product. In the present study, the land surface temperature is temporarily interpolated by considering the diurnal variation. The RMSE of the land surface temperature is less than 1 K (Wan et al. 2004; Wang et al. 2008). The land surface emissivity is from the baseline-fit emissivity database (BFED) monthly mean product (Seemann et al. 2008). Spectral interpolation is used to infer land surface emissivity in the MODIS bands, assuming that the emissivity is linear to the wavelength as recommended by the BFED documentation. The RMSE of land surface emissivity in the BFED is 0.01 or less in the IR window region and about 0.015–0.025 in the other TIR bands.

Forward model

A physics-based forward model is developed and used in the cloud retrieval algorithm. The forward model takes auxiliary data for the atmospheric profile and background surface properties mentioned above, and it computes the BTs and their partial derivatives with respect to several atmospheric and surface variables. The radiative transfer is calculated by using the correlated k-distribution (CKD) method with six quadrature points for each band. The optimization method of Sekiguchi and Nakajima (2008) is used to determine the CKD coefficients from line-by-line radiative transfer calculations with the HITRAN2012 database (Rothman et al. 2013) and the continuum absorption model (Mlawer et al. 2012). Modeled gas species include water vapor, carbon dioxide, ozone, nitrous oxide, carbon monoxide, and methane.

The bulk optical properties of clouds are precalculated and tabulated for ice and liquid clouds with different particle size distributions and ice crystal habit distributions considering the spectral response function of MODIS spectral bands. In the forward model calculation, the optical properties are interpolated with respect to the CER from the look-up table by using the Akima interpolation. The optical properties of water droplets are computed by the Lorenz-Mie theory. The optical properties of ice particles are obtained from a database published by Yang et al. (2013), who used a combination of the discrete dipole approximation and the improved geometrical optics method for randomly oriented ice crystals of various shapes. Several models of particle habit distribution are incorporated into the model, including solid column, plate, column aggregate, the general habit mixture (Baum et al. 2011; Cole et al. 2013), and the two-habit model (Liu et al. 2014), with different degrees of surface roughness. In the present study, the column aggregate model with very rough surfaces is used because it is assumed in obtaining the MODIS C6 cloud product. TIR measurements are not strongly sensitive to the ice habit assumptions (Cooper et al. 2006).

Radiative transfer in a plane-parallel multilayered atmosphere is solved by the two-stream approximation (Nakajima et al. 2000) with the delta-M method (Wiscombe 1977). Solutions of the two-stream approximation are upward and downward irradiances at layer boundaries, from which the radiances at the top of the atmosphere in arbitrary directions can be calculated. The radiative transfer equation for a single homogeneous layer is written as

$$\mu \frac{dI(\tau,\mu)}{d\tau} = I(\tau,\mu) - \left[\frac{\varpi}{2\pi} \left\{ F^{\downarrow}\left(1 - \frac{3}{2}g\mu\right) + F^{\uparrow}\left(1 + \frac{3}{2}g\mu\right) \right\} + (1-\varpi)B(\tau) \right]$$

(2)

where $I(\tau,\mu)$ is radiance at optical depth τ from the top of the layer in a direction with $\mu = \cos\theta$ for view zenith angle θ, ϖ is the single-scattering albedo, g is the asymmetry factor, and $F^{\uparrow}$ and $F^{\downarrow}$ are upward and downward irradiances, respectively. The second term on the right-hand side of (2) is the radiative source function, $J(\tau,\mu)$, which is here approximated to be linear to τ, as

$$J(\tau,\mu) = a(\mu)\tau + b(\mu),$$

(3)

where a and b are coefficients determined by the source function values at the layer boundaries. Thus, upward radiance emergent from this layer in the μ direction at the top of layer with optical depth $\Delta\tau$ is analytically solved as

$$I_{\text{top}}(\mu) = I_{\text{bot}}(\mu)e^{-\frac{\Delta\tau}{\mu}} - (a(\mu)\Delta\tau + a(\mu)\mu + b(\mu))e^{-\frac{\Delta\tau}{\mu}} + a(\mu)\mu + b(\mu)$$

(4)

Total radiance at the top of atmosphere is computed by the sum of components emergent from all atmospheric layers and the background surface. Band mean radiance calculated by integration over the CKD terms is converted to the BT.

The error of this approximate radiative transfer model is evaluated by comparing the model with an accurate model based on the discrete ordinate method for a variety of atmosphere and cloud states. Correction formulae based on a cubic polynomial for BT bias are developed for each band. After the bias correction, the RMSE reaches a maximum of 0.3 K in band 29, where the

scattering effect is strong compared with other TIR bands. The two-stream approximation enables fast calculations, whereas the errors from the radiative transfer approximation are sufficiently small. For cloud retrieval, uncertainties in atmospheric profile and background surface properties are a major source of errors in the forward model.

Figure 1 shows the BT and BT differences (BTDs) at the split window band, calculated for liquid and ice clouds with a CTT of 247 K in a tropical atmosphere with a sea surface temperature of 300 K. As suggested by prior studies, a combination of multiple bands in the window region of 8–13 µm allows the CER to be inferred. Measurements in these bands are sensitive to clouds with a COT of 0.05–20 and an effective radius of 3–100 µm for ice clouds. The COT is defined at a wavelength of 550 nm throughout this paper. The water phase (liquid/ice) is moderately important to the spectral differences in BTs in the split window. Absorption by ice and liquid particles becomes stronger at wavelengths longer than 11 µm, although the ice and liquid phases have different spectral dependences of absorption, which means that the cloud thermodynamic phase can be determined from these bands.

Retrieval algorithm

The optimal estimation method (Rodgers 2000) is used to solve an inverse problem. The method fits the forward model to the measurement under constraints by an a priori probability distribution of the state vector in the forward model. Defining state vector $\mathbf{x}$, measurement vector $\mathbf{y}$ with the BTs in the MODIS TIR bands as elements, and the model parameter vector $\mathbf{b}$, the problem to be solved is written as

$$\mathbf{y} = \mathbf{F}(\mathbf{x},\mathbf{b}) + \mathbf{e},$$

(5)

where $\mathbf{F}$ is the forward model, and $\mathbf{e} = \mathbf{y} - \mathbf{F}(\mathbf{x},\mathbf{b})$ is a measurement–model error vector. A cost function is given by

$$J(\mathbf{x}) = [\mathbf{y} - \mathbf{F}(\mathbf{x},\mathbf{b})]^{\text{T}} S_{\text{e}}^{-1} [\mathbf{y} - \mathbf{F}(\mathbf{x},\mathbf{b})] + [\mathbf{x} - \mathbf{x}_{\text{a}}]^{\text{T}} S_{\text{a}}^{-1} [\mathbf{x} - \mathbf{x}_{\text{a}}],$$

(6)

where S_{a} is an error covariance matrix of the a priori $\mathbf{x}_{\text{a}}$, and S_{e} is a measurement–model error covariance matrix. The Levenberg-Marquardt method is used to obtain a minimized J, at which the solution converges. The final value of J is the retrieval cost, which represents the degree of fit between the model and measurement. The criterion that J is sufficiently small with an optimal solution is set as $J < 2m$, where the m is a number of the observation vector elements. A feature of the optimal estimation is that the uncertainty of the solution can be diagnosed quantitatively with an

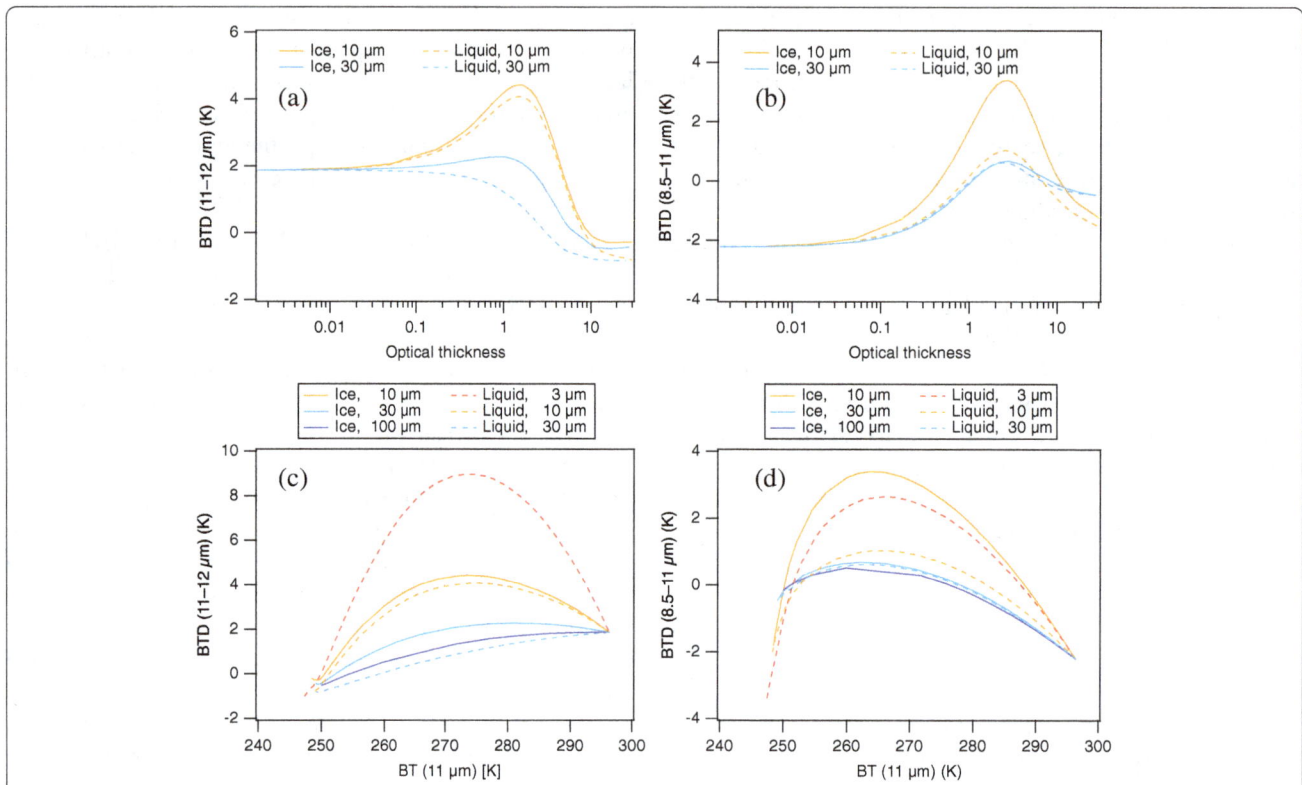

Fig. 1 Sensitivities of the split window bands to COT, effective particle radius, and cloud thermodynamic phase. Theoretical relationships of BTDs with COT (**a**, **b**) and BT (**c**, **d**) in the 11 μm band. Calculations are shown for different effective particle radii for ice and liquid clouds with the same CTTs of 247 K in a typical tropical atmosphere with sea surface temperature of 300 K. The two-habit model from Liu et al. (2014) is assumed for ice particles

error covariance matrix. In addition, diagnostics of the estimation quality, such as the degree of freedom for signal (DOFS) and the information content, are obtained.

The cloud inversion is tried first with a single-layer cloud. If an optimal solution is obtained, the single-layer assumption is accepted. Otherwise, an inversion with a double-layer assumption is tried. A double-layer cloud solution is accepted if J is smaller than that of the single-layer assumption and the COT of the upper cloud is less than 8. This is because TIR measurements lose sensitivity to the lower cloud under the double-layer cloud assumption if the upper COT is more than about 8. The state vector includes cloud properties such as cloud water path (CWP), CER, CTP, and background surface temperature in single-layer cloud cases. With nonlinearity in mind, logarithms of CTP, CWP, and CER are elements of the state vector. The top-pressure of the lower cloud is inferred in double-layer clouds, instead of background surface temperature, as a similar double-layer cloud retrieval is proposed by Watts et al. (2011).

The cloud layer is assumed to be composed of either liquid droplets or ice particles. The ice (liquid) phase is

accepted if an optimal solution is obtained with CTT below a threshold temperature, $T_f = -38$ °C (above $T_m = 0$ °C). If the CTT is between T_f and T_m, or an optimal solution is not obtained, then inversion with the other cloud phase assumptions are tried. Final judgment of cloud phase is to select a lower cloud phase cost function, R, of

$$R = \frac{P_1}{V_1} + \frac{P_2}{V_2} + \frac{P_3}{V_3} + \frac{J}{J_{opt}}, \qquad (7)$$

where weighting coefficients are set as $V_1 = 8$ K^2, $V_2 = 30^2$ K^2, $V_3 = 0.3^2$, and $J_{opt} = 2m$. As shown in Fig. 1, the split window bands are sensitive to the cloud phase. Thus, P_1 is the sum of squares of BTD between the observation and model at the split window bands at wavelengths of 8.5, 11, and 12 μm

$$P_1 = \left(\mathrm{BTD}_{8.5-11,obs} - \mathrm{BTD}_{8.5-11,mdl}\right)^2 \\ + \left(\mathrm{BTD}_{11-12,obs} - \mathrm{BTD}_{11-12,mdl}\right)^2, \qquad (8)$$

where subscripts "obs" and "mdl" denote the observation and model, respectively. Because the CTT is a main factor that prescribes cloud phase, P_2 increases with increasing deviation of CTT (T_{top}) from a critical temperature, $T_{c,ice} = -15$ °C for ice and $T_{c,liq} = -23$ °C for liquid, as

$$P_2 = \begin{cases} \left[\min\left(0, T_{top} - T_{c,liq}\right)\right]^2 & \text{for liquid} \\ \left[\max\left(0, T_{top} - T_{c,ice}\right)\right]^2 & \text{for ice} \end{cases}.$$

(9)

However, CTT is uncertain when CTP retrieval is uncertain. Thus, P_3 in Eq. (7) is the error variance of the logarithm of CTP. Finally, an index of the cloud phase, Q, is calculated as

$$Q = 1 + \frac{R_{liq}^2}{R_{liq}^2 + R_{ice}^2}, \quad (10)$$

which has a value between 1 and 2: Q is 1–1.5 for liquid and 1.5–2 for ice. If the cloud phase costs for liquid and ice phase assumptions have similar values, then Q is nearly 1.5, which means that cloud phase determination is ambiguous.

Assumptions and prior information

As shown by Cooper et al. (2003) and Garrett et al. (2009), explicit representation of cloud top and base boundaries is important in the TIR cloud retrieval. In this study, the cloud base pressure is parameterized using an empirical formula. Sassen and Comstock (2001) and Sassen et al. (2008) showed that the geometrical thickness of the cirrus with COT less than 3 is 1–3 km with a global mean of 2 km based on ground and space-based lidar observations. Sassen and Comstock (2001) and Veglio and Maestri (2011) showed that an optically thicker cirrus cloud becomes geometrically thicker. In the present algorithm, the geometric thickness, H (m), is represented by a function of CWP, W (kg m^{-2}),

$$H = \begin{cases} B_{liq} + A_{liq}\sqrt{W/W_{liq}} & \text{for liquid cloud} \\ B_{ice} + A_{ice}\sqrt{W/W_{ice}} & \text{for ice cloud} \end{cases},$$

(11)

where
$B_{liq} = 20$ m, $B_{ice} = 20$ m,
$A_{liq} = 400$ m, $A_{ice} = 2000$ m,
$W_{liq} = 0.06$ kg m^{-2}, $W_{ice} = 0.02$ kg m^{-2}.
The cloud base pressure is determined from H and atmospheric temperature and pressure profiles.

The prior information about the cloud properties and background surface temperature is given in Table 3, where T'_{sfc} is the sea or land surface temperature obtained from the MODIS level 3 product, W is CWP, and r_e is CER. It is assumed that the surface temperature RMSEs are set to include the uncertainty from daily diurnal variations. The a priori probability distributions of CWP have large dispersions to under-constrain the

Table 3 A priori information and prescribed ranges of the elements of the solution vector

Variable	A priori	Standard deviation	Min.	Max.
COT, liquid clouds	–	–	0.25	30 (8)[a]
COT, ice clouds	–	–	0.04	30 (8)
T_{sfc} (K), ocean	T'_{sfc}	0.7	$T'_{sfc} - 2.1$	$T'_{sfc} + 2.1$
T_{sfc} (K), land	T'_{sfc}	3	$T'_{sfc} - 9$	$T'_{sfc} + 9$
Liquid cloud properties				
$\ln[W$ (kg/m^2)]	$\ln 0.04$	4	–	–
$\ln[r_e$ (μm)]	$\ln 10$	1.0	$\ln 2$	$\ln 30$
Ice cloud properties				
$\ln[W$ (kg/m^2)]	$\ln 0.02$	4	–	–
$\ln[r_e$ (μm)]	$\ln 25$	1.0	$\ln 3$	$\ln 100$

[a]The values for double-layer clouds are in parentheses

CWP retrieval. In contrast, the a priori CER has a relatively small standard deviation because CERs obtained from passive remote sensing are in limited ranges (Hong et al. 2007; Wang et al. 2016). S_a is an almost diagonal matrix, with weak correlation between the CWP and CER with a correlation coefficient of 0.25. In the inversion, cloud properties are limited to prescribed ranges of realistic values. Lower and upper limits of COT are set because the TIR measurements lose sensitivity in cases with very small and large COTs.

Because the temperature ranges at which liquid and ice clouds exist are known a priori, the a priori and variance of CTP are determined by considering the vertical distribution of the air temperature profile. It is assumed that a liquid (ice) cloud is not present with CTT lower than −40 °C (higher than 2 °C). In addition, the a priori CTT (T_a) and CTP range are assumed as follows.

- For liquid clouds, $P_{flz} < P_{top} < 0.96 P_{sfc}$, $T_a = 5$ °C
- For ice clouds, $0.9 P_{trp} < P_{top} < P_{mlt}$, $T_a = -55$ °C

P_{frz} and P_{mlt} are pressures at an air temperature of −40 and 2 °C, respectively, and P_{tpp} is a tropopause pressure. The standard deviation of the a priori CTP on a logarithmic scale, $\sigma_{\ln P}$ is determined to cover the CTP ranges as

$$\sigma_{\ln P} = 0.7 \times \max[|\ln(P_a/P_{min})|, |\ln(P_a/P_{max})|],$$

(12)

where P_a is a priori CTP, and P_{min} and P_{max} are the lower and upper limits of CTP, respectively, determined as previously mentioned.

Measurement and model errors

Observations and models may have bias and noise-like error components arising from various error sources. Model errors include (1) error due to radiative transfer approximations, (2) errors from the representation of atmosphere with a finite number of atmospheric layers, (3) errors in the sea or land surface emissivity, (4) uncertainty in atmospheric temperature and gas concentration profiles, (5) error from assuming the cloud base pressure, (6) uncertainty of the ice habit model and particle size distribution, and (7) error from the vertical and horizontal heterogeneity of the clouds. Each error component may depend on the state of the atmosphere and the background surface, which make it complicated to quantify the error covariance matrix appropriately.

Simple assumptions can be made about several error components. The RMSE of sea surface temperature is assumed as 0.7 K in the inversion by considering daily and diurnal variations in sea surface temperature and possible differences between clear-sky and cloudy cases. According to the observations of Newman et al. (2005), the RMSE of sea surface emissivity due to the uncertainty of seawater optical constants is estimated to be approximately 0.001 at satellite zenith angles of less than 60 °. Over land, the surface temperature and emissivity in cloudy cases are likely to differ significantly from those in clear-sky cases, although the magnitude is uncertain. The RMSE of land surface temperature is assumed as 3 K in this study, although precise quantification is required in the future. BFED land surface emissivity product (Seemann et al. 2008) is created by using the MODIS land surface emissivity product. The error covariance matrix of the land surface emissivity is constructed considering the MODIS product error and the BFED modeling error documented in the literature.

The measurement–model error covariance matrix is divided into two components, as

$$S_e = K_b S_b K_b^T + S_{e,off}. \tag{13}$$

The first and second terms on the right-hand side are online and offline calculation terms, respectively. The online term is calculated from the error covariance matrix, S_b, and the Jacobian matrix, K_b, for the model parameters. K_b is calculated in the forward model. Model parameters included in the online calculation term are the background surface emissivity for each band and cloud base pressure. For the cloud base pressure error, the standard deviation of the logarithm of the cloud base pressure is assumed to be approximately proportional to the geometrical thickness of the cloud, which is approximately proportional to $\ln(P_{bas}/P_{top})$, as

$$\sigma = S_{bas} \ln\left(\frac{P_{bas}}{P_{top}}\right), \tag{14}$$

where factor S_{bas} is 0.2. The offline calculation term is divided into three components as

$$S_{e,off} = S_{e,RTM} + S_{e,atm} + S_{e,noise}. \tag{15}$$

The right-hand side contains the errors from radiative transfer approximation, $S_{e,RTM}$, atmospheric profile uncertainty, $S_{e,atm}$, and measurement noise, $S_{e,noise}$. $S_{e,RTM}$ is small, as previously described.

To estimate several error components in the measurements and the model, BTs from daytime clear-sky pixel measurements over the ocean are compared with those from the model. Because the RMSE of sea surface temperature is as small as 0.7 K, errors in atmospheric data and model approximations and assumptions are evaluated by the comparison. The clear-sky area is determined based on the cloud mask data in the MODIS Product (Ackerman et al. 1998). If an area about 30 km^2 is composed of only the "confidently clear" pixels, then the center area of about 20 km^2 is used for the model-measurement comparison. The covariance matrix of measurement noise, $S_{e,noise}$, is estimated from the variance and covariance of measurement–model differences within a 20 km^2 area. The maximum noise is 0.4 K at MODIS band 27, and the noise is less than 0.25 K in the window bands. Figure 2 shows comparisons of the clear-sky BTs. Each data point represents mean values over a 20 km^2 segment. Water vapor absorption bands tend to exhibit larger differences between model and measurement as uncertainties in temperature and water vapor amount increase model errors. From this comparison, the systematic difference (bias) between the observations and model calculations is evaluated for each MODIS band, and then the biases are removed from the forward model used in the subsequent analyses.

The atmospheric profile error is estimated from the covariance matrix of measurement–model differences obtained in the oceanic clear-sky comparison. The vertical distribution of error patterns of temperature and water vapor, and ozone mixing ratios are determined by fitting the simulated error covariance matrix to the observations. The best estimate of the error covariance matrix is shown in Fig. 3 along with that obtained from observation. By using the estimated atmospheric profile errors, $S_{e,atm}$ for cloudy cases is evaluated by model simulations under a variety of atmospheric conditions. The error in cloudy cases strongly depends on CTT, because a major source of atmospheric profile error in cloudy cases is the error in the amount of water vapor above the cloud top. For lower CTTs, the amount of water vapor above the cloud top generally tends to be smaller in various atmospheric profiles. Thus, $S_{e,atm}$ for cloudy cases is tabulated in five CTT ranges from 200 to 300 K with an interval of 20 K. The results for high and low CTTs are shown in Fig. 3. The error covariance matrix with high CTT is close to that for clear-sky cases. The

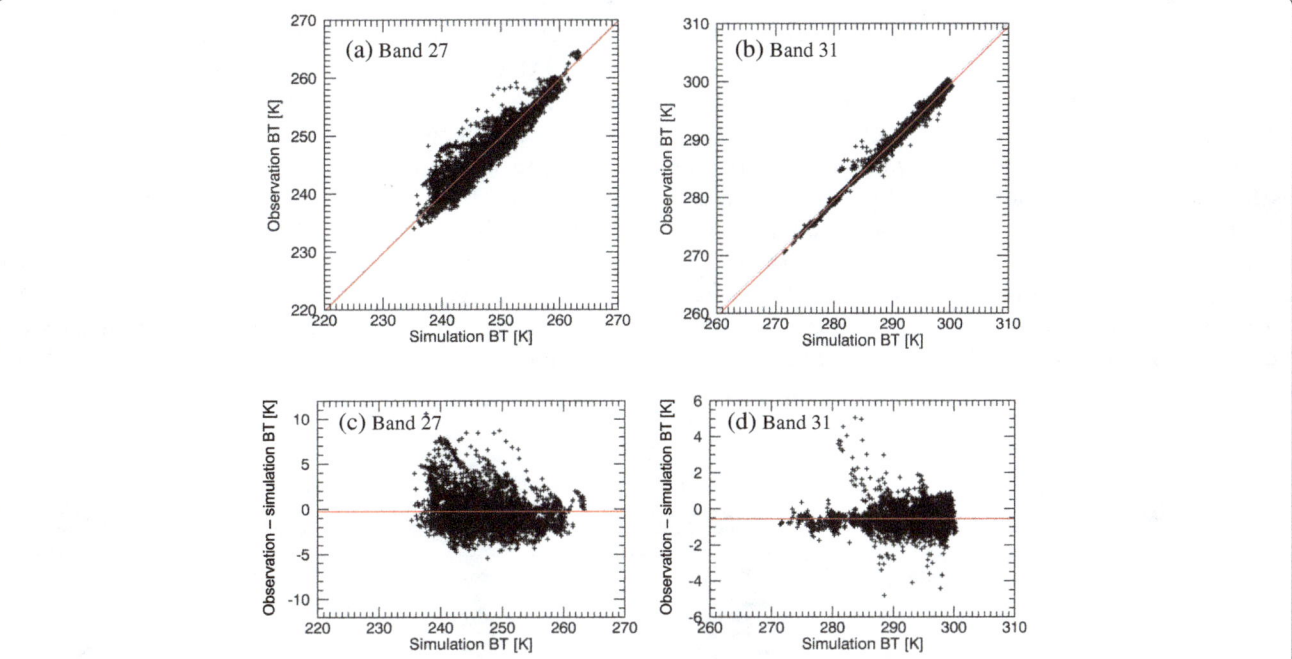

Fig. 2 Comparison of observed and simulated BTs for clear-sky pixels over the ocean. Results are shown for MODIS band 27 (water vapor absorption band) and band 31 (window band). Scatterplots of observed BTs (**a**, **b**) and differences in the observation from the simulation (**c**, **d**). *Red lines* in **a** and **b** denote fit lines with a slope constrained to be 1. *Red lines* in **c** and **d** denote biases. Each data point represents mean values over a 20 km² segment

errors decrease at the water vapor absorption bands for lower CTT.

Because not all error sources are included in Eq. (13), initial tests show that the model does not fit the measurements well if Eq. (13) is used directly in the cloud property inversion. The uncertainty due to the horizontal and vertical heterogeneity in clouds and the uncertainty in the optical properties of ice particles from the ice habit model are not included in Eq. (13). These uncertainties are difficult to quantify; however, based on by trial and error, we artificially set the diagonal elements of the error covariance matrix obtained from Eq. (13) as 20 % larger.

Results and discussion

Retrieval error evaluation by simulations

The errors and performance of cloud retrieval are tested by retrieval simulations. Measurement signals with errors are simulated by the forward model calculations for perturbed atmospheric and surface states with random noise that obey the error covariance matrices assumed above. Retrieval errors are evaluated by comparing the retrieved cloud properties from the noise-superimposed measurement signals with the initial values. This methodology is identical to that used by Iwabuchi et al. (2014). For each state, a series of 1000 retrieval simulations are performed to evaluate the mean bias error and the RMSE. The satellite viewing

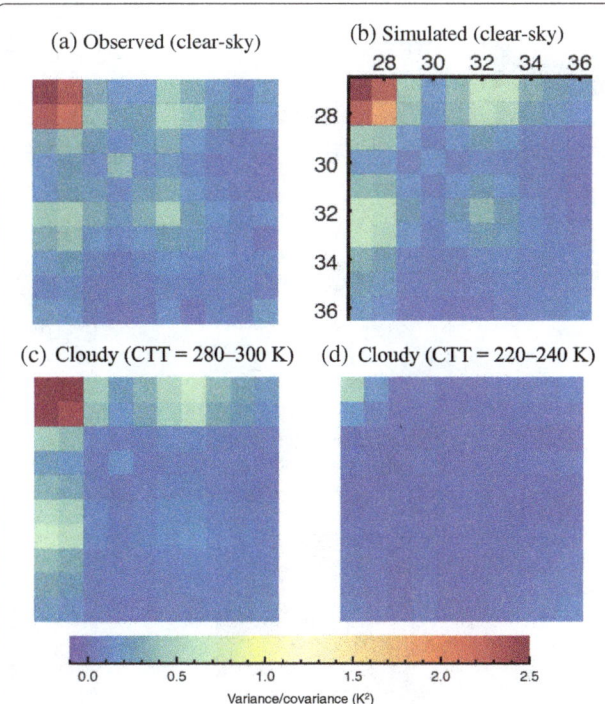

Fig. 3 Model error covariance matrices for atmospheric profile errors. **a** Estimation from comparison of clear-sky observations over the ocean to model calculations. **b**, **c**, **d** Estimations from Monte Carlo simulation of estimated errors in atmospheric profiles, for **b** clear sky, **c** cloudy sky with CTT of 280–300 K, and **d** cloudy sky with CTT of 280–300 K. Each panel shows an image of error covariance matrix. The row and columns denote ten MODIS bands

zenith angle ranges from 0 ° to 60 °, and a tropical atmosphere is assumed with a sea surface temperature of 300 K.

Figure 4 shows results for ice clouds with CTT = 221 K. The retrieval bias and RMSE, correct cloud phase discrimination rate, and DOFS are shown by color scales as functions of initially assumed values (truth values in this test) of COT and CER. The rate of the optimal cloud retrievals for all trials is also computed for each cloud state. The retrieval errors presented are evaluated only for optimal retrievals with correct cloud phase identification. The estimation of CTP is accurate with a small bias of within ±10 % for almost all cloud states tested here. CTP RMSE is less than 20 % for optically thick clouds with COT of 0.3 or more. When the initial COT is 0.1–10 and CER is 3–60 μm, the CER and COT retrieval biases are generally less than 15 %, RMSE is less than 30 %, and DOFS is close to 3. For very thin and very thick clouds, the sensitivity of the forward model to CER is low, which explains the CER retrievals close to the a priori values and the low DOFS. Optimal cloud retrievals are obtained for 100 % of clouds with COT of 0.1. Optically thinner clouds are not retrieved because they are excluded in the retrieval algorithm as clear sky. With the

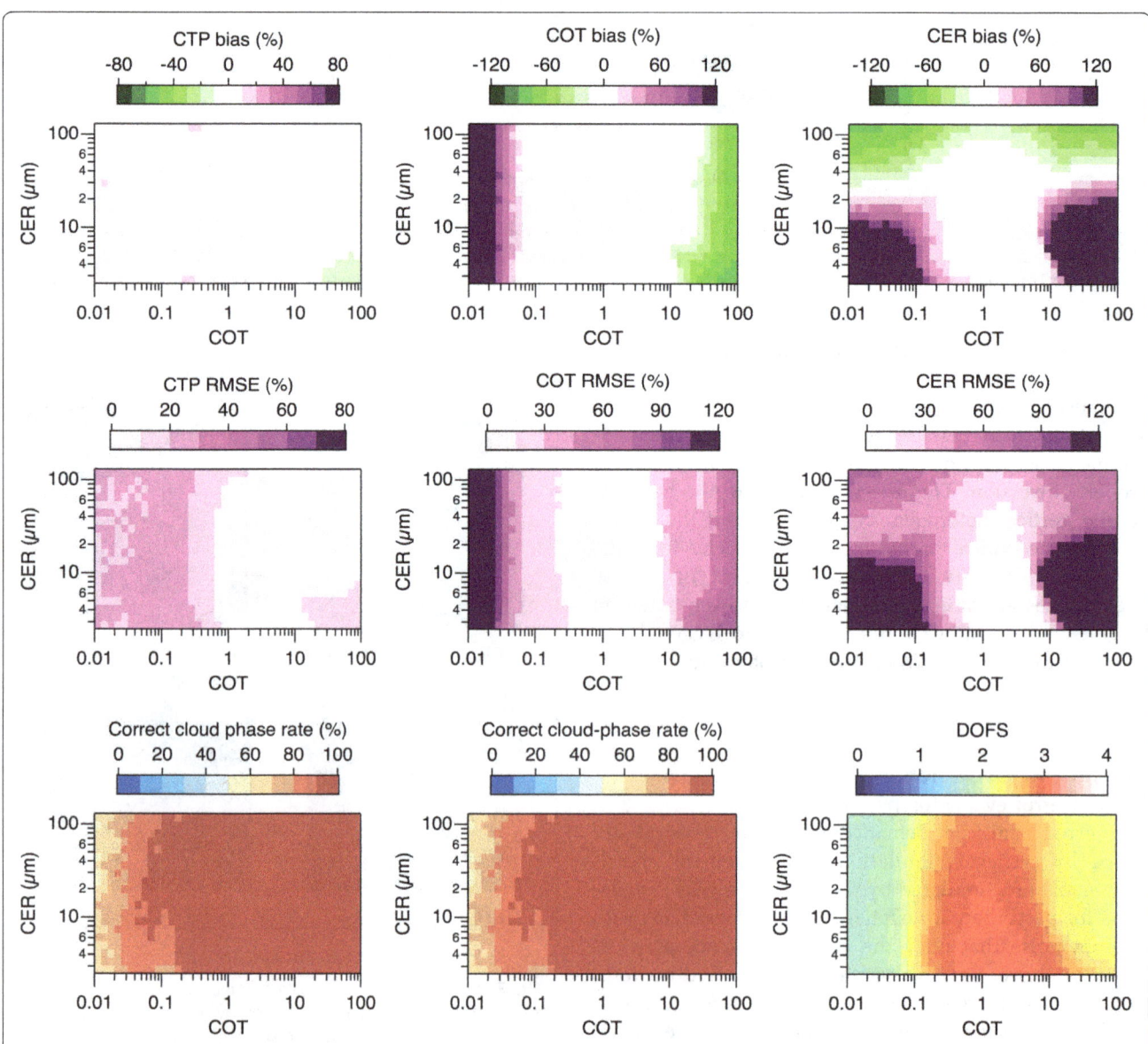

Fig. 4 Retrieval test results by retrieval simulations for ice clouds over the ocean in a tropical atmosphere. Results are shown for CTT of 221 K and different COT and CER. MBE, RMSEs, and DOFS are evaluated from retrievals with optimal solutions and correct cloud phase determination. Optimal cloud retrieval rate and correct cloud phase retrieval rate is also evaluated. For each cloud state, 1000 retrievals from noise-superimposed simulated measurements are attempted

exception of cases with very optically thin clouds, cloud phase determination is correct with a correct identification rate of about 100 %.

Figure 5 shows the results for liquid clouds with CTT = 277 K. Because the CTT is closer to the background surface temperature than for ice clouds, the retrieval performance for liquid cloud is worse than for ice clouds, resulting in lower DOFS. The CTP bias and RMSE are very small over a wide range of COT and CER, which is better than for ice clouds. This is mainly because a liquid cloud is geometrically thinner than an ice cloud. Use of a weakly absorbing CO_2 band (MODIS band 33) might improve the estimation of the top of low

clouds. Both CER and COT bias are within ±30 % when the COT is 0.3–20 and CER is 4–20 μm. The percentage of correct cloud phase identification is about 100 % in most cases shown here. When both the COT and CER are small, cloud phase identification is not correct with a score of about 50 %. This is because COT is limited to be more than 0.25 for liquid clouds.

Retrieval performance is tested at various CTTs. The initial assumptions, as the truth in this test, about cloud phase are as follows. There is exclusively an ice cloud for CTT ≤ −38 °C, exclusively a liquid cloud for CTT ≥ 0 °C, and the ice and liquid cloud fractions vary linearly with CTT between −38 and 0 °C. CERs are assumed to

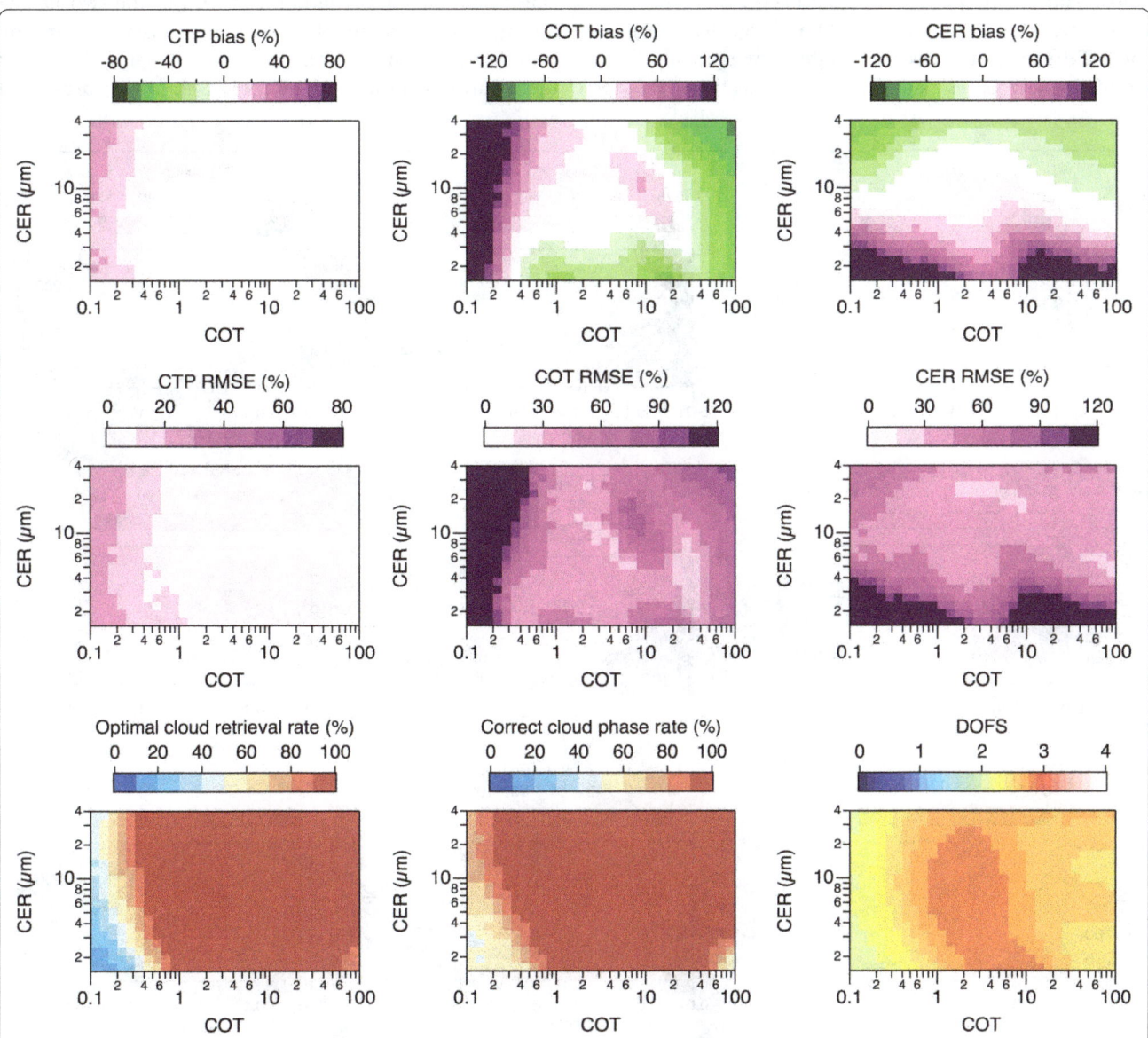

Fig. 5 Retrieval test results by retrieval simulations for ice clouds over the ocean in a tropical atmosphere. Results are shown for CTT of 277 K and different COT and CER. MBE, RMSEs, and DOFS are evaluated from retrievals with optimal solutions and correct cloud phase determination. Optimal cloud retrieval rate and correct cloud phase retrieval rate is also evaluated. For each cloud state, 1000 retrievals from noise-superimposed simulated measurements are attempted

be typical values of 13 μm for liquid clouds and 32 μm for ice clouds. Figure 6 shows the test results. The optimal cloud retrieval rate is approximately 100 % for clouds that are optically thick enough. Very optically thin clouds are not retrieved because they are identified as clear sky in the retrieval algorithm. Cloud phase discrimination is accurate for CTT lower than −38 °C or higher than 0 °C, except for very optically thin cases. Cloud phase determination is difficult when CTT is between the two critical temperatures, particularly for COT of less than 0.5. CTP of optically thin (COT <0.3) clouds also has a problem with biases. Low sensitivity to CTP means that CTP retrievals tend to be close to the a priori values. The poor estimation accuracy of the CTP is observed for CTT around 200 and 265 K. However, the CTP can be retrieved accurately for optically thick clouds for any CTTs. The CTP bias is less than 10 % for

COT >0.3, and the CTP RMSE is less than about 10 % for COT >0.5. For the retrieval error of COT, the bias is less than 15 %, and the RMSE is less than 30 % for high clouds with COT = 0.1–10. The COT retrieval error generally increases with increasing CTT. However, CER retrieval shows good performance over wide ranges of CTT and COT in this test, where CER truths are assumed to be close to the a priori values.

It would be desirable to clarify the benefits of using ten TIR bands to retrieve cloud macrophysical and microphysical properties simultaneously. In the CO_2 slicing technique for CTP retrieval, cloud effective temperature is estimated assuming that the cloud is isothermal and the cloud emissivities in the two neighboring bands are identical. Cloud emissivities in the split window bands can be obtained by using this cloud temperature estimate. Heidinger et al. (2015) presented

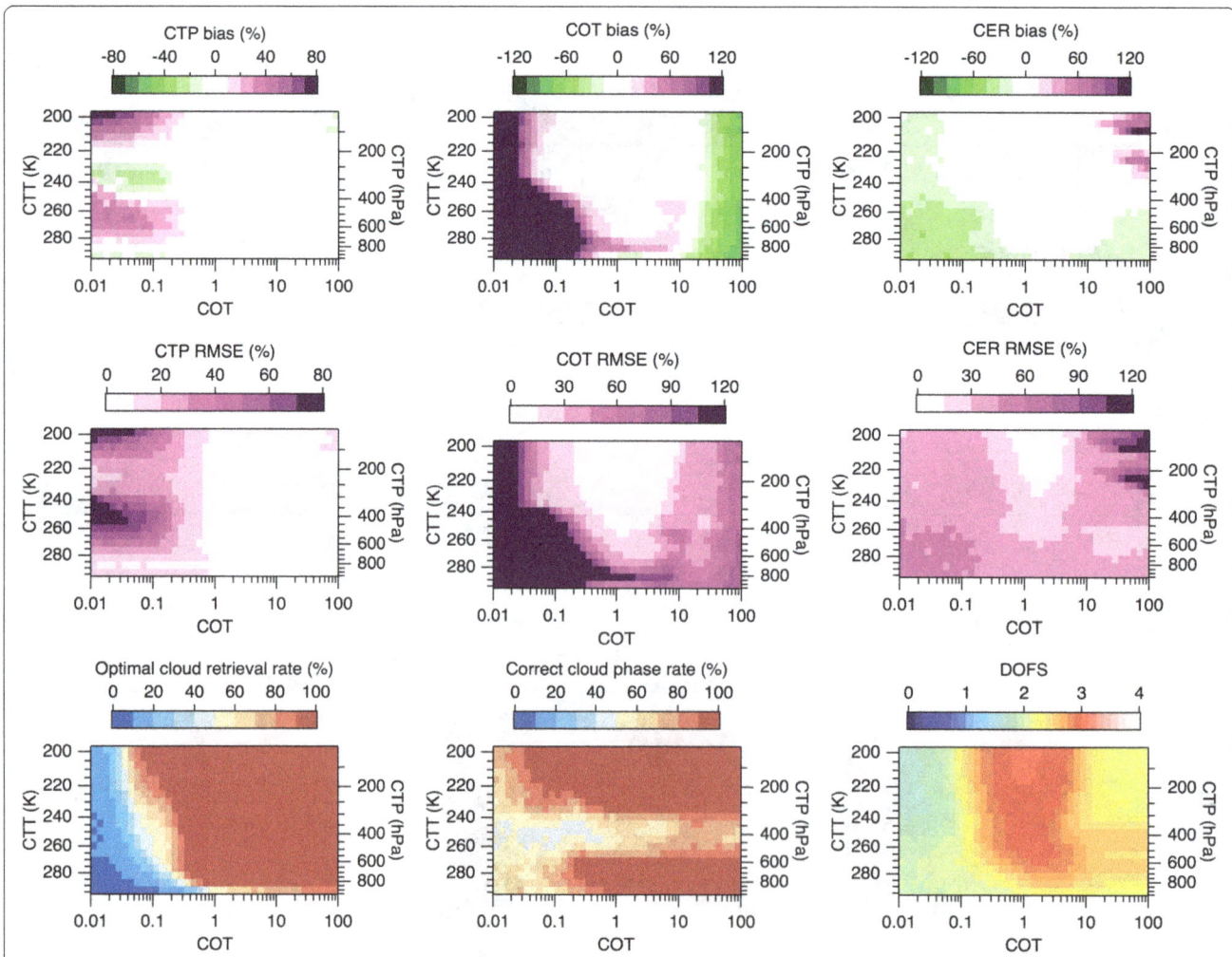

Fig. 6 Retrieval test results by retrieval simulations for clouds with different COT and CTT. It is assumed that occurrence of the ice phase increases with decreasing CTT from 0 to −40 °C. CER is assumed to be 13 μm for liquid clouds and 32 μm for ice clouds. MBE, RMSEs, and DOFS are evaluated from retrievals with optimal solutions and correct cloud phase determination. For each cloud state, 1000 retrievals from noise-superimposed simulated measurements are attempted

a simple, efficient two-stage method for retrieving COT and CER from the cloud emissivities in the split window. First, the macrophysical properties (CTP/CTH/CTT) are retrieved, followed by the optical and microphysical properties. Iwabuchi et al. (2014) also used the CTT obtained from the CO_2 slicing technique to help estimate the optical and microphysical properties from the split window bands of MODIS. In the present study, we test a surrogate method for CTP retrieval, instead of the CO_2 slicing technique itself. MODIS bands 35 and 36 are used to retrieve CTP in the optimal estimation framework, with CER strictly constrained at the a priori value for ice clouds. The two bands are sensitive to the upper troposphere. In the MODIS cloud product, CTPs of most high clouds in the tropics are retrieved by using those two bands. The fixed CER assumption is reflected in the fixed spectral dependence of the optical properties of the ice particles, which is a better assumption than identical cloud emissivities at the two bands. Figure 7 shows the CTP retrieval errors for ice clouds with a CER of the a priori value (25 μm). The two-band retrieval is compared with the ten-band retrieval with the same assumptions about the measurement and model errors. The simultaneous retrievals from using all ten bands outperform those based on using only the

two CO_2 bands, and the ten-band retrieval produces more certain estimates of CTP. If the assumed cloud temperature is highly uncertain in the second stage of retrieval of COT and CER, it is likely that the COT and CER estimates will be more erroneous. The simultaneous retrieval of macrophysical and microphysical properties can provide better consistency between the physics model and measurements for all bands.

Application to tropical cloud systems

As an initial test, the results obtained from the present retrieval algorithm in ICAS are compared with the MODIS C6 cloud products with a 1 km resolution. The MODIS granule in this case is acquired on April 1, 2007, at 3:55 UTC, which is the same as the case in Iwabuchi et al. (2014). The region is located in the ocean to the north of New Guinea. An area of about 200 × 1200 km is selected to illustrate the comparison results, covering the CloudSat/CALIPSO ground track (Fig. 8). The CTH obtained from ICAS is compared with the CloudSat/CALIPSO cloud mask product developed by Kyushu University (Hagihara et al. 2010). Because CloudSat/CALIPSO observes the Earth in a nadir view, a parallax correction is applied to the ICAS retrieval to compare the datasets coherently. Furthermore, the parallax-corrected ICAS retrieval is regridded into the CloudSat/CALIPSO track by using nearest neighbor interpolation. The evaluated pixels are only pixels with optimal retrieval. In the region of interest, there are about 1868 collocated pixels between ICAS (with optimal retrieval) and the radar–lidar product. More than 90 % of the pixels are detected as cloudy pixels by both products, whereas about 2 % are detected as clear-sky pixels. ICAS incorrectly assigns about 1.5 % of pixels as cloudy and misses about 0.5 % of cloudy pixels in the radar–lidar product, although collocation is not certain owing to the parallax.

Figure 8 shows that the radar–lidar product detects many pixels with a high cloud top. The upper clouds cover wide areas, and parts of the left and right sides of the figures are covered with thick clouds. Middle-level clouds are present in the middle of the figures. The MERRA atmospheric profile shows that the CTT of the middle cloud is about −20 °C. The upper part of the high clouds probably consists of small ice particles and it can be detected only by lidar (green). The CTH obtained from ICAS tends to be lower than that of the radar–lidar product and close to the cloud top detected by cloud radar (blue and yellow). The top height of ICAS varies from 5 to 17 km, whereas the radar–lidar product has a more uniform top height with an altitude of around 15 km. Similar to the ICAS cloud top, the cloud top detected by radar is more variable than the cloud top detected by lidar. This is an expected limitation of ICAS

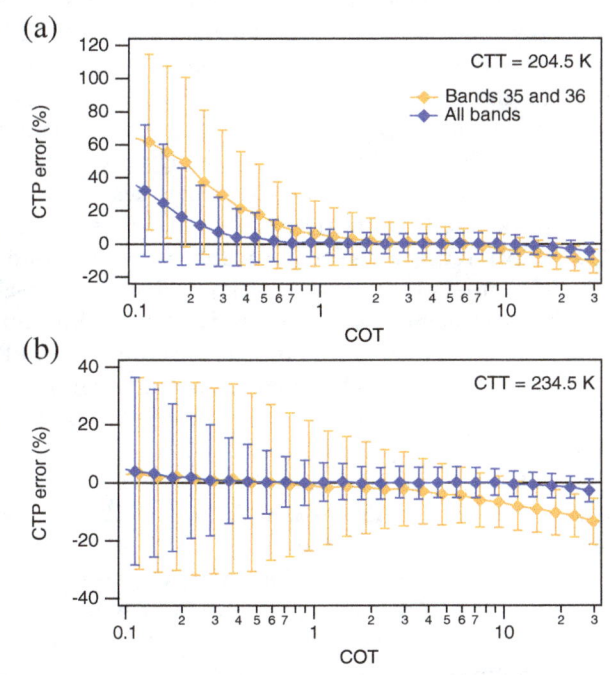

Fig. 7 CTP retrieval errors estimated by retrieval simulation for ice clouds with different CTTs. The CTTs are **a** 204.5 K and **b** 234.5 K. In this simulation, CTP are retrieved from MODIS bands 35 and 36 (*yellow*) or all bands (*blue*), with CER strictly constrained at the a priori value for ice clouds. Mean bias error and standard deviation of error are denoted by *diamond markers* and *error bars*, respectively. For visibility, results for bands 35 and 36 are shifted slightly along the horizontal axis

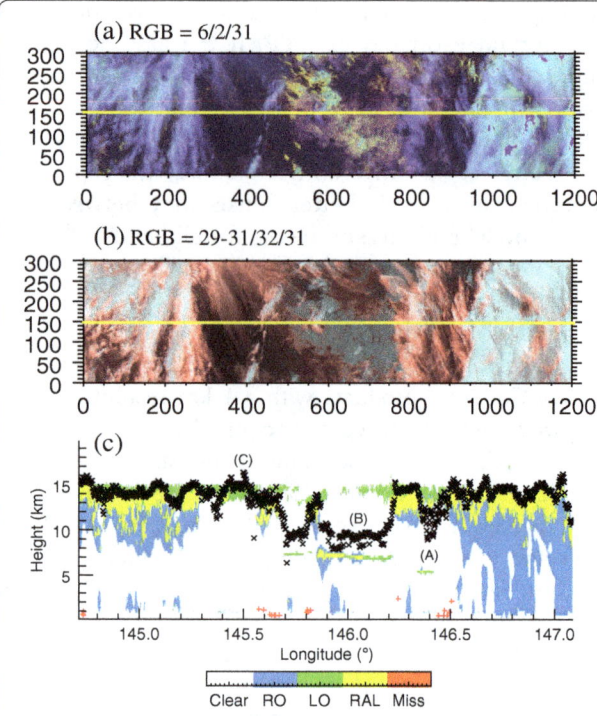

Fig. 8 A test case of tropical cloud systems over the ocean north of New Guinea. The MODIS granule is taken at 3:55–4:00 UTC on April 1, 2007. **a**, **b** False color image with band combinations of MODIS bands. **a** Reflectances in bands 6, 2, and 31 and **b** BTD between bands 29 and 31 and BTs of 32 and 31 for the *red*, *green*, and *blue* channels. *Yellow lines* in **a** and **b** denote the ground track of CloudSat and CALIPSO satellite. **c** Comparison of retrieved CTHs with cloud mask data obtained from CloudSat radar and CALIPSO lidar measurements. *Blue* denotes clouds detected by radar only (RO), *green* by lidar only (LO), *yellow* by both radar and lidar (RAL), and *orange* denotes missing data (Miss). The scattered crosses show the location of the ICAS cloud top at the respective longitude, where *black crosses* represent cloud top of the first layer and *red crosses* represent cloud top of the second layer

because previous studies have shown that IR measurements are not sensitive to very optically thin clouds, which can be sensed only by lidar and not by radar (Watts et al. 2011). ICAS detects more pixels with a cloud top above 15 km compared with radar–lidar products. Some pixels correspond to CTHs that are higher than the lidar measurements (around region C). This is probably an erroneous retrieval because the ICAS results are influenced primarily by the uncertainty of the atmospheric profile, and possibly by the ice habit assumption and vertical inhomogeneity within the cloud, yielding an ICAS cloud top near the tropopause that is too high compared with the lidar measurement. Red crosses in Fig. 8c denote the cloud top of the lower cloud of the double-layer cloud retrieval in ICAS. Several parts of the second layer top in ICAS match the cloud top of the third cloud layer in the radar–lidar profile, although there are parts that deviate greatly. The cloud tops of the upper cloud layer in double-

layer cases are well estimated, similar to single-layer cloud cases. ICAS misses most multilayer clouds at longitudes of 146.35 °–146.45 ° (region A), where the upper first and second cloud layers are detected only by lidar. ICAS wrongly identified these pixels as single-layer cloud and retrieved CTHs between the first and second cloud layers. A similar problem occurs at longitudes of 146.1 °–146.2 ° (region B). In these cases, the uppermost cloud is too optically thin, and ICAS cannot identify the upper cloud in a multilayer cloud system, probably because the retrieval with the single-layer cloud assumption has an optimal solution with clouds at the wrong height. These results suggest that the algorithm requires re-examination and improvement.

Figure 9 shows a comparison of the cloud properties in ICAS and C6. In C6, the VIS/SWIR method is used for cloud optical and microphysical properties (COT and CER), and the CO_2 slicing method and the IR window method are used to estimate the cloud top in C6. Two methods are used for cloud thermodynamic phase determination in the MODIS 1 km products. One is a method based on the BT and BTD in the TIR bands supplemented by using the cloud emissivity ratios (TIR method; Baum et al. 2012). The other is a method used in the retrieval of daytime optical properties of clouds (shortwave algorithm; Marchant et al. 2016), in which cloud phase is determined from "votes" from several tests based on the BTs of the TIR window bands, CER values determined from different combinations of multiple VIS and SWIR bands, CTT, and the water vapor absorption band at 1.38 μm. There is a high correlation between COTs in ICAS and C6, although the COT in ICAS tends to be smaller and limited to less than 20. ICAS works well for retrieval of thin clouds. In C6, COT (CER) retrieval is not available for thin clouds with COT of less than about 0.5 (Eq. (1)). The retrieval in ICAS is available for many pixels of optically thin, high clouds. As shown in the previous subsection, CER of optically thick clouds has a large uncertainty in ICAS because of low sensitivity. In the optically thin parts at the edges of the upper clouds, the CTH in C6 is low, whereas ICAS CTH has high values. In the CO_2 slicing method, C6 and ICAS agree well. In the IR window method, C6 shows low clouds with CTH of 0–3 km, whereas the CTH is 10–15 km in ICAS, resulting in significant differences as large as 10 km or more. ICAS retrieval has more ice cloud pixels (Fig. 9i–k), particularly at optically thin edges of high clouds. Over the central area of the images, thin upper layer clouds overlap over the middle-level cloud, as seen in the radar–lidar profile. In these areas, many pixels are determined as ice phases in ICAS, whereas CTHs are in the middle of the thin upper cloud and middle cloud. The shortwave algorithm in C6 shows a liquid phase for these

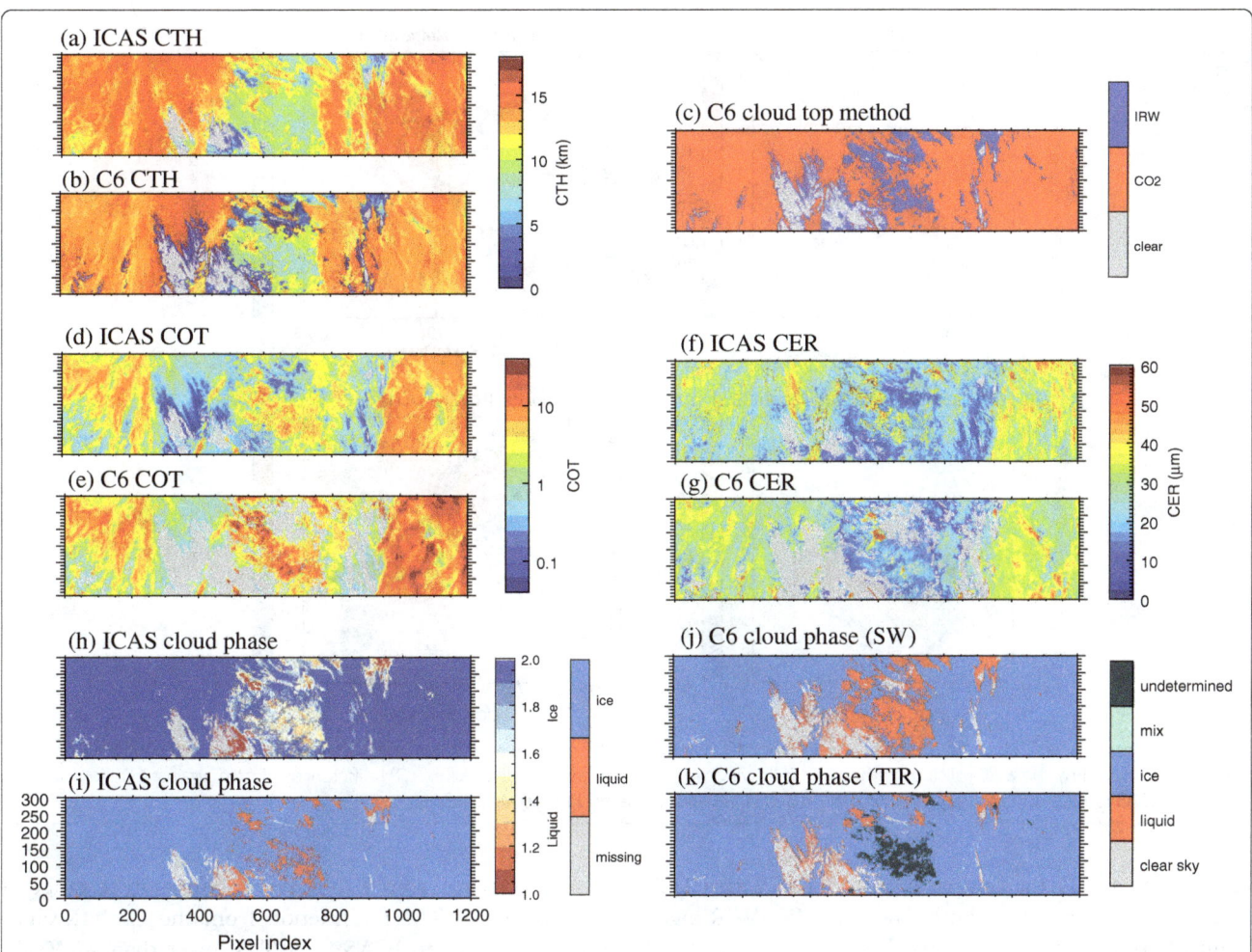

Fig. 9 Comparison of retrieved cloud properties between ICAS and MODIS products. CTH in **a** ICAS and **b** MODIS. **c** CTH method used in the MODIS product, showing pixels for the IR window and CO_2 slicing methods and clear-sky pixels. COT in **d** ICAS and **e** MODIS. CER in **f** ICAS and **g** MODIS. Cloud thermodynamic phase estimated in ICAS. **h** The phase index defined in Eq. (10), and **i** binary results for the cloud phase. Ice/liquid binary results are based on the original cloud phase index data that has continuous values from 1 to 2. Cloud thermodynamic phase estimated in the MODIS product: **j** shortwave (SW) algorithm and **k** TIR algorithm. Gray pixels in **k** represent the mixed phase

pixels, and the phase is undetermined in the TIR method in C6, indicating difficulty in determining unique cloud phases in multilayer cloud systems.

Statistical comparisons for the collocated pixels are made for a full granule that covers a 2000 km² region. Figure 10 shows the CTH comparison between ICAS and C6. In C6, CO_2 slicing is applied to the high clouds, and the CTH retrieval agrees well with the CTH in ICAS. In the optically thin ice clouds, CTH in C6 tends to be higher than the ICAS estimates (Fig. 10c). If the IR window method is used in C6, the pixels are covered by upper clouds in ICAS, whereas they are covered by low clouds in C6. When the comparison is limited to ice cloud retrieval in ICAS and IR window retrieval in C6, the ICAS–C6 difference in CTH tends to increase with decreasing COT. As seen in the spatial distribution, the optically thin parts at the edges of upper clouds are

treated as lower clouds in C6. These results suggest that ICAS can capture the CTH of optically thin upper layer clouds well.

Figure 11 shows results for the COT and CER for ice clouds. COT shows good agreement when the COT is less than about 6. For thick clouds, TIR measurements lose their sensitivity to the lower part of cloud systems, whereas the visible reflectance is sensitive to the total column COT. Many pixels with multilayer clouds are probably included in the results presented here. The COT from the TIR measurements should be considered as the COT of upper clouds in multilayer cloud systems. This explains why the C6 COT is often larger than ICAS COT, particularly when COT is larger than about 6. In contrast, CERs in ICAS and C6 exhibit significant dispersion, showing a weak correlation with a correlation coefficient of 0.24. The correlation is higher with a

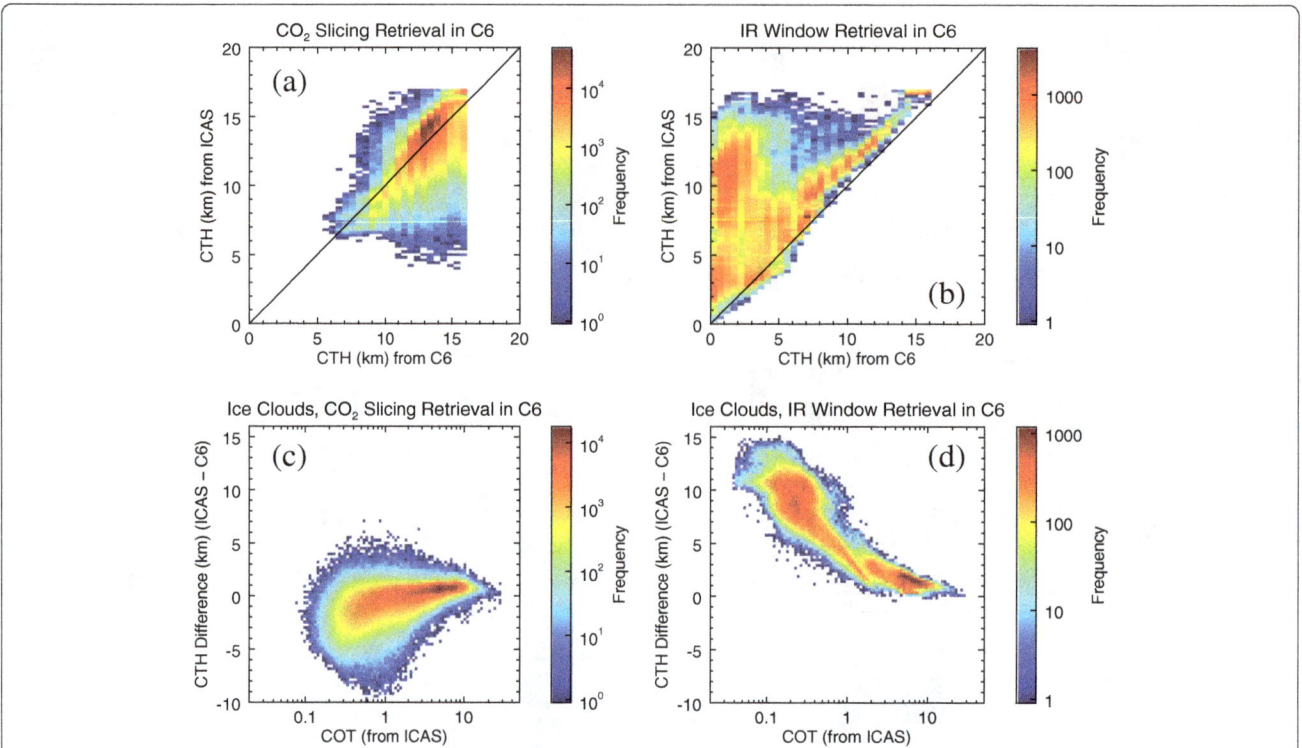

Fig. 10 Comparison of CTHs obtained from ICAS and the MODIS product. The occurrence frequency distributions are shown. Results are shown for pixels retrieved by the **a** CO_2 slicing method and **b** IR window method in the MODIS product. CTH differences plotted against COT for ice clouds in ICAS retrieval for the **c** CO_2 slicing method and **d** IR window method in the MODIS product

coefficient of 0.31 for comparing ice clouds with COTs in a range of 1–6, where both TIR and VIS/SWIR algorithms have a high sensitivity to CER.

A simple comparison of the cloud thermodynamic phase on a pixel-by-pixel basis does not make sense because different products retrieve or assume different CTTs in the cloud phase determination. We also compare the temperature dependence of the ice/liquid phase occurrence fraction. Figure 12 shows the ice cloud fraction as function of CTT. ICAS and C6 ice results are plotted against CTT for the products. The ice fraction in

ICAS decreases from 1 to 0 over a temperature range of –40 to –0 °C. The ice fraction from the C6 TIR algorithm is similar to ICAS for CTTs lower than –5 °C. In the C6 TIR algorithm, there are ice phase pixels even with CTT above 0 °C. However, it is important to check that the absolute number of such pixels is small because most pixels are covered by ice clouds in the results from all the algorithms. In contrast, for the shortwave algorithm (dotted line), the ice fraction is much lower than those in the ICAS and C6 TIR algorithms. The cloud phase in the shortwave algorithm is determined to

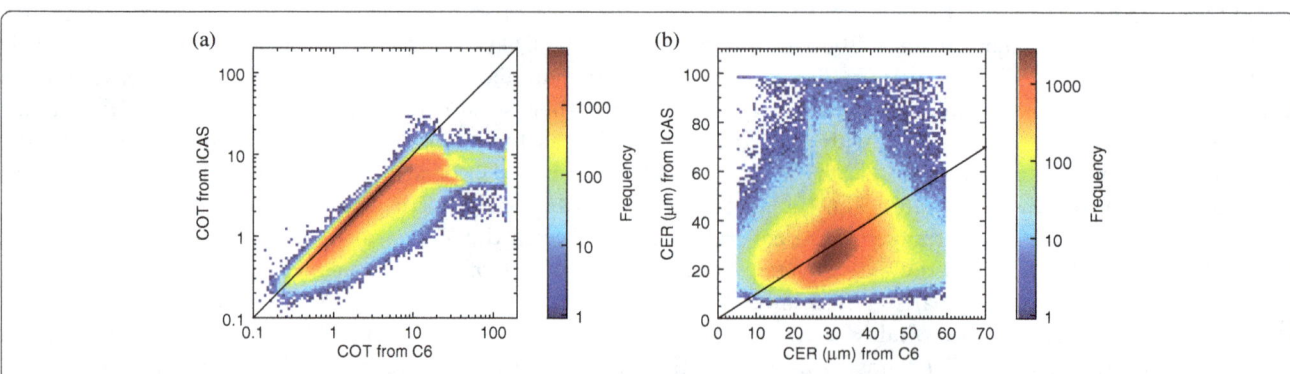

Fig. 11 Comparison of COTs and CERs obtained from ICAS and the MODIS product. Results for **a** COT and **b** CER of ice clouds, shown as the occurrence frequency distribution

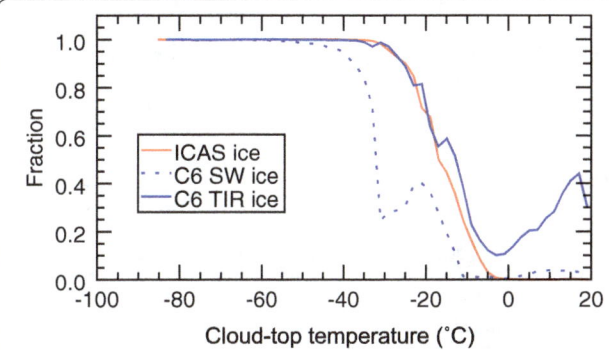

Fig. 12 Ice cloud occurrence fraction for ICAS and MODIS C6 SW and TIR algorithms. The ice fraction is shown as a function of CTT. The MODIS results are calculated from the numbers of pixels of ice, liquid, and mixed phase clouds, not including undetermined phase pixels

explain the VIS/SWIR reflectance measurements well. In multilayer cloud systems with optically thin ice clouds over optically thick liquid clouds, shortwave reflectances could be sensitive to the lower cloud. Thus, the cloud phase determination can differ from that from TIR measurement, which is sensitive to the upper cloud. In the C6 shortwave algorithm result, the ice fraction changes abruptly at a CTT of about −33 °C (~240 K). According to Marchant et al. (2016), the CTT test uses a threshold at 240 K to "vote" for the ice clouds. This could result in the discontinuity of the ice fraction. As shown previously, cloud phase inference in ICAS is based on the consistency between the model and measurements under the constraints given by a priori knowledge about the relationship between CTT and the phase, relying on BTDs in the split window with large weights. Thus, the cloud phase determination in ICAS is constrained strongly by the cloud phase dependence on the CTTs, which guarantees that ice clouds that are too warm and liquid clouds that are too cold are not retrieved. However, the strong constraint on the cloud phase is a limitation of our algorithm because a small amount of information about the cloud phase comes from measurements. Baum et al. (2012) showed that in their algorithm refinement for MODIS C6, using cloud emissivity ratios between the split window bands substantially improves the inference of the ice cloud phase, especially for optically thin ice clouds. In future work, cloud phase inference could be improved by using this index.

Conclusions

An optimal estimation-based cloud retrieval algorithm has been developed to estimate the optical and physical properties of clouds simultaneously from measurements of several TIR bands. A major source of modeling errors is uncertainties in atmospheric profiles, which are usually difficult to quantify. In this study, they are assessed by direct comparison of the clear-sky observations over the ocean with the model calculations. This type of model-measurement comparison is important for developing cloud retrieval algorithms. A feature of the present algorithm is that the COT and the CTH is retrieved well for optically thin clouds by simultaneously fitting the model to the measurement in multiple TIR bands. Although the cloud top inferred from TIR measurement fluctuates with the cloud top from the cloud radar profile, the topmost parts of clouds seen only in the lidar profile are not detected well by TIR measurements. Compared with MODIS C6 operational products, COT of less than 5 agrees well, although CER deviates greatly. CTH estimates agree well for optically thick clouds when the MODIS product is based on the CO_2 slicing method, whereas there is significant disagreement in CTHs between the present study and C6 products for optically thin clouds at the cloud edges. In the present algorithm, the determination of the cloud thermodynamic phase is strongly constrained by a priori knowledge about cloud phase dependence on the CTTs. It guarantees that ice clouds that are too warm and water clouds that are too cold are not retrieved; however, more statistical verification of the temperature dependence should be performed by increasing the number of cases.

The present algorithm will be used in studies with observations from the Himawari-8 satellite, a Japanese next-generation geostationary meteorological satellite, which has been operated by the Japan Meteorological Agency since 7 July 2015 and carries a visible-to-IR imager with greatly improved radiometric, spectral, spatial, and temporal resolution (Bessho et al. 2016). The development strategy used in this study will be used to create an algorithm for Himawari-8. The algorithm will have several modifications to accommodate the different spectral bands and will have improvements to the multilayer cloud retrieval and cloud phase discrimination. CALIPSO lidar measurements are suitable for retrieving optically thin cloud and reliable cloud phase discrimination (Hu et al. 2009). Using depolarization lidar comparison on a global scale, would help determine the performance of the TIR-based algorithm. In the future, further comparison of collocated data from different cloud products will be performed to characterize respective strengths and limitations of different methods.

Abbreviations
BFED: Baseline-fit emissivity database; BT: Brightness temperature; BTD: Brightness temperature difference; C6: Collection 6; CALIPSO: Cloud-Aerosol Lidar and Infrared Pathfinder Satellite Observation; CER: Cloud-particle effective radius; CKD: Correlated k-distribution; COT: Cloud optical thickness; CTH: Cloud-top height; CTP: Cloud-top pressure; CTT: Cloud-top temperature; DOFS: Degree of freedom for signal; ICAS: Integrated Cloud Analysis System; MBE: Mean bias error; MODIS: Moderate resolution imaging spectroradiometer; RMSE: Root-mean-square error; SWIR: Shortwave infrared; TIR: Thermal infrared; VIS: Visible

Acknowledgements

The authors are grateful to Prof. Hajime Okamoto of Kyushu University, Japan, for providing the cloud mask data made from CloudSat/CALIPSO data and Dr. Shuichiro Katagiri of Kyushu University, Japan, for the valuable comments during this study.
The MODIS data were obtained from the NASA websites.

Funding

This work was promoted and supported by Japan Society for the Promotion of Science (JSPS) KAKENHI Grant Number 25287117.

Authors' contributions

HI proposed the topic, conceived and designed the study, and conducted major parts of the study. M. Saito collaborated with the corresponding author in the development of the inversion module and carried out the evaluations of forward model and retrieval errors. YT collaborated with the corresponding author in the development and evaluation of the forward model and carried out the analysis using the MODIS data. NSP carried out the comparison analysis using CloudSat/CALIPSO product data. M. Sekiguchi developed the CKD model. All authors read and approved the final manuscript.

Competing interests

The authors declare that they have no competing interests.

Author details

[1]Center for Atmospheric and Oceanic Studies, Graduate School of Science, Tohoku University, 6-3 Aoba, Aramakiaza, Aoba-ku, Sendai, Miyagi 980-8578, Japan. [2]The Graduate School of Marine Science and Technology, Tokyo University of Marine Science and Technology, 2-1-6 Etchujima, Koto-ku, Tokyo 135-8533, Japan.

References

Ackerman SA, Strabala KI, Menzel WP, Frey RA, Moeller CC, Gumley LE (1998) Discriminating clear sky from clouds with MODIS. J Geophys Res 103:32–141. doi:10.1029/1998JD200032

Akima H (1970) A new method of interpolation and smooth curve fitting based on local procedures. J ACM 17:589–602

Baran AJ (2009) A review of the light scattering properties of cirrus. J Quant Spectrosc Radiat Transfer 110:1239–1260

Baum BA, Yang P, Heymsfield AJ, Schmitt CG, Xie Y, Bansemer A, Hu YX, Zhang Z (2011) Improvements in shortwave bulk scattering and absorption models for the remote sensing of ice clouds. J Appl Meteorol Climatol 50:1037–1056

Baum BA, Menzel WP, Frey RA, Tobin DC, Holz RE, Ackerman SA, Heidinger AK, Yang P (2012) MODIS cloud-top property refinements for Collection 6. J Appl Meteorol Climatol 51:1145–1163. doi:10.1175/JAMC-D-11-0203.1

Bessho K, Date K, Hayashi M, Ikeda A, Imai T, Inoue H, Kumagai Y, Miyakawa T, Murata H, Ohno T, Okuyama A, Oyama R, Sasaki Y, Shimizu Y, Shimoji K, Sumida Y, Suzuki M, Taniguchi H, Tsuchiyama H, Uesawa D, Yokota H, Yoshida R (2016) An introduction to Himawari-8/9—Japan's new-generation geostationary meteorological satellites. J Meteorol Soc Japan 94:151–183. doi:10.2151/jmsj.2016-009

Brown OB, Minnet PJ, Evans R, Kearns E, Kipatrick K, Kumar A, Sikorski R, Zavody A (1999) MODIS infrared sea surface temperature algorithm. Algorithm theoretical basis document Version 2.0. Miami University. http://modis.gsfc.nasa.gov/data/atbd/atbd_mod25.pdf. Accessed 28 May 2016.

Cole BH, Yang P, Baum BA, Riedi J, Labonnote LC, Thieuleux F, Platnick S (2013) Comparison of PARASOL observations with polarized reflectances simulated using different ice habit mixtures. J Appl Meteorol Climatol 52:186–196

Cooper SJ, L'Ecuyer TS, Stephens GL (2003) The impact of explicit cloud boundary information on ice cloud microphysical property retrievals from infrared radiances. J Geophys Res 108:4107. doi:10.1029/2002JD002611

Cooper SJ, L'Ecuyer TS, Gabriel PK, Baran A, Stephens GL (2006) Objective assessment of the information content of visible and infrared radiance measurements for cloud microphysical property retrievals over the global oceans. Part 2: Ice clouds. J Appl Meteorol 45:42–62

Davis SM, Avallone ML, Kahn BH, Meyer KG, Baumgardner D (2009) Comparison of airborne in situ measurements and Moderate Resolution Imaging Spectroradiometer (MODIS) retrievals of cirrus cloud optical and microphysical properties during the Midlatitude Cirrus Experiment (MidCiX). J Geophys Res 114:D02203. doi:10.1029/2008JD010284

Downing H, Williams D (1975) Optical constants of water in the infrared. J Geophys Res 80:1656–1661

Friedman D (1969) Infrared characteristics of ocean water. Appl Opt 8:2073–2078

Garnier A, Pelon J, Dubuisson P, Faivre M, Chomette O, Pascal N, Kratz DP (2012) Retrieval of cloud properties using CALIPSO imaging infrared radiometer. Part I: Effective emissivity and optical depth J Appl Meteorol Climatol 51: 1407–1425. doi:10.1175/JAMC-D-11-0220.1

Garrett KJ, Yang P, Nasiri SL, Yost CR, Baum BA (2009) Influence of cloud-top height and geometric thickness on a MODIS infrared-based ice cloud retrieval. J Appl Meteorol Climatol 48:818–832

Giraud V, Buriez JC, Fouquart Y, Parol F, Seze G (1997) Large-scale analysis of cirrus clouds from AVHRR data: assessment of both a microphysical index and the cloud-top temperature. J Appl Meteorol 36:664–675

Hagihara Y, Okamoto H, Yoshida R (2010) Development of combined CloudSat/CALIPSO cloud mask to show global cloud distribution. J Geophys Res 115: D00H33. doi:10.1029/2009JD012344

Hartmann DL, Ockert-Bell ME, Michelsen ML (1992) The effect of cloud type on earth's energy balance: global analysis. J Climate 5:1281–1304

Heidinger AK, Pavolonis MJ (2009) Gazing at cirrus clouds for 25 years through a split window. Part I: methodology. J Appl Meteorol Climatol 48:1100–1116

Heidinger AK, Li Y, Baum BA, Holz RE, Yang P (2015) Retrieval of cirrus cloud optical depth under day and night conditions from MODIS Collection 6 cloud property data. Remote Sens 7:7257–7271. doi:10.3390/rs70607257

Hong G, Yang P, Gao BC, Baum BA, Hu YX, King MD, Platnick S (2007) High cloud properties from three years of MODIS Terra and Aqua collection-4 data over the tropics. J Appl Meteorol Climatol 46:1840–1856

Hu Y, Winker D, Vaughan M, Lin B, Omar A, Trepte C, Flittner D, Yang P, Nasiri SL, Baum B, Sun W, Liu Z, Wang Z, Young S, Stamnes K, Huang J, Kuehn R, Holz R (2009) CALIPSO/CALIOP cloud phase discrimination algorithm. J Atmos Ocean Technol 26:2293–2309

Inoue T (1985) On the temperature and effective emissivity determination of semi-transparent cirrus clouds by bi-spectral measurements in the 10 μm window region. J Meteorol Soc Jpn 63:88–89

Iwabuchi H, Yamada S, Katagiri S, Yang P, Okamoto H (2014) Radiative and microphysical properties of cirrus cloud inferred from the infrared measurements made by the moderate resolution imaging spectroradiometer (MODIS). Part I: retrieval method. J Appl Meteorol Climatol 53:1297–1316. doi:10.1175/JAMC-D-13-0215.1

Li J, Huang J, Stamnes K, Wang T, Lv Q, Jin H (2015) A global survey of cloud overlap based on CALIPSO and CloudSat measurements. Atmos Chem Phys 15:519–536. doi:10.5194/acp-15-519-2015

Liu C, Yang P, Minnis P, Loeb NG, Kato S, Heymsfield AJ, Schmitt CG (2014) A two-habit model for the microphysical and optical properties of ice clouds. Atmos Chem Phys 14:13719–13737. doi:10.5194/acp-14-13719-2014

Marchant B, Platnick S, Meyer K, Arnold GT, Riedi J (2016) MODIS Collection 6 shortwave-derived cloud phase classification algorithm and comparisons with CALIOP. Atmos Meas Tech 9:1587–1599. doi:10.5194/amt-9-1587-2016

Masuda K (2012) Influence of wind direction on the infrared sea surface emissivity model including multiple reflection effect. Pap Meteorol Geophys 63:1–13. doi:10.2467/mripapers.63.1

Menzel WP, Frey RA, Zhang H, Wylie DP, Moeller C, Holz RE, Maddux B, Baum BA, Strabala KI, Gumley LE (2008) MODIS global cloud-top pressure and amount estimation: algorithm description and results. J Appl Meteorol Climatol 47: 1175–1198. doi:10.1175/2007JAMC1705.1

Mlawer EJ, Payne VH, Moncet JL, Delamere JS, Alvarado MJ, Tobin DD (2012) Development and recent evaluation of the MT_CKD model of continuum absorption. Phil Trans R Soc A 370:1–37. doi:10.1098/rsta.2011.0295

Nakajima T, King MD (1990) Determination of the optical thickness and effective particle radius of clouds from reflected solar radiation measurements. Part I: theory. J Atmos Sci 47:1878–1893

Nakajima T, Tsukamoto M, Tsushima Y, Numaguti A, Kimura T (2000) Modeling of the radiative process in an atmospheric general circulation model. Appl Opt 39:4869–4878

Newman SM, Smith JA, Glew MD, Rogers SM, Taylor JP (2005) Temperature and salinity dependence of sea surface emissivity in the thermal infrared. Q J R Meteorol Soc 610:2539–2557. doi:10.1256/qj.04.150

Parol F, Buriez JC, Brogniez G, Fouquart Y (1991) Information content of AVHRR channels 4 and 5 with respect to the effective radius of cirrus cloud particles. J Appl Meteorol 30:973–984

Platnick S, King MD, Ackerman SA, Menzel WP, Baum BA, Riedi JC, Frey RA (2003) The MODIS cloud products: algorithms and examples from Terra. IEEE Trans Geosci Remote Sens 41:459–473. doi:10.1109/TGRS.2002.808301

Poulsen CA, Siddans R, Thomas GE, Sayer AM, Grainger RG, Campmany E, Dean SM, Arnold C, Watts PD (2012) Cloud retrievals from satellite data using optimal estimation: evaluation and application to ATSR. Atmos Meas Tech 5:1889–1910

Ramanathan V, Cess RD, Harrison EF, Minnis P, Barkstrom BR, Ahmad E, Hartmann DL (1989) Cloud-radiative forcing and climate: results from the earth radiation budget experiment. Science 243:57–63

Rienecker MM, Suarez MJ, Gelaro R, Toding R, Bacmeister J, Liu E, Bosilovich MG, Schubert SD, Takacs L, Kim GK, Bloom S, Chen J, Collins D, Conaty A, Silva A, Gu W, Joiner J, Koster RD, Lucchesi R, Molod A, Owens T, Pawson S, Pegion P, Redder CR, Reichle R, Robertson FR, Ruddick AG, Sienkiewicz M, Woollen J (2011) MERRA: NASA's modern-era retrospective analysis for research and applications. J Climate 24:3624–3648. doi:10.1175/JCLI-D-11-00015.1

Rodgers CD (2000) Inverse methods for atmospheric sounding. World Scientific, Singapore

Rothman LS, Gordon IE, Babikov Y, Barbe A, Benner DC, Bernath PF, Birk M, Bizzocchi L, Boudon V, Brown LR, Campargue A, Chance K, Cohen EA, Coudert LH, Devi VM, Drouin BJ, Fayt A, Flaud JM, Gamache RR, Harrison JJ, Hartmann JM, Hill C, Hodges JT, Jacquemart D, Jolly A, Lamouroux J, Le Roy RJ, Li G, Long DA, Lyulin OM et al (2013) The HITRAN2012 molecular spectroscopic database. J Quant Spectrosc Radiat Transfer 130:4–50

Sassen K, Comstock JM (2001) A midlatitude cirrus cloud climatology from the facility for atmospheric remote sensing. Part III: radiative properties. J Atmos Sci 58:2113–2127

Sassen K, Wang Z, Liu D (2008) Global distribution of cirrus clouds from CloudSat/Cloud-Aerosol Lidar and Infrared Pathfinder Satellite Observations (CALIPSO) measurements. J Geophys Res 113:D00A12. doi:10.1029/2008JD009972

Seemann SW, Borbas EE, Knuteson RO, Stephenson GR, Huang HL (2008) Development of a global infrared land surface emissivity database for application to clear-sky sounding retrievals from multispectral satellite radiance measurements. J Appl Meteorol Climatol 47:108–123

Sekiguchi M, Nakajima T (2008) A k-distribution-based radiation code and its computational optimization for an atmospheric general circulation model. J Quant Spectrosc Radiat Transfer 109:2779–2793

Sourdeval O, Labonnote LC, Brogniez G, Jourdan O, Pelon J, Garnier A (2013) A variational approach for retrieving ice cloud properties from infrared measurements: application in the context of two IIR validation campaigns. Atmos Chem Phys 13:8229–8244. doi:10.5194/acp-13-8229-2013

Sourdeval O, Labonnote LC, Baran AJ, Brogniez G (2015) A methodology for simultaneous retrieval of ice and liquid water cloud properties. Part I: information content and case study. Q J R Meteorol Soc 141:870–882. doi:10.1002/qj.2405

Tsutsumi Y, Mori K, Hirahara T, Ikegami M, Conway TJ (2009) Technical report of global analysis method for major greenhouse gases by the World Data Center for Greenhouse Gases. GAW Report No. 184 (WMO/TD 1473), Geneva, pp 29

Veglio P, Maestri T (2011) Statistics of vertical backscatter profiles of cirrus clouds. Atmos Chem Phys 11:12925–12943. doi:10.5194/acp-11-12925-2011

Waliser DE, Li JLF, Woods CP, Austin RT, Bacmeister J, Chern J, Del Genio A, Jiang JH, Kuang Z, Meng H, Minnis P, Platnick S, Rossow WB, Stephens GL, Sun-Mack S, Tao W, Tompkins AM, Vane DG, Walker C, Wu D (2009) Cloud ice: a climate model challenge with signs and expectations of progress. J Geophys Res 114:D00A21. doi:10.1029/2008JD010015

Walther A, Heidinger AK (2012) Implementation of the daytime cloud optical and microphysical properties algorithm (DCOMP) in PATMOS-x. J Appl Meteorol Climatol 51:1371–1390

Wan Z, Li ZL (1997) A physics-based algorithm for retrieving land-surface emissivity and temperature from EOS/MODIS data. IEEE Trans Geosci Remote Sens 35:980–996. doi:10.1109/36.602541

Wan Z, Zhang Y, Zhang Q, Li ZL (2004) Quality assessment and validation of the MODIS global land surface temperature. Int J Remote Sens 25:261–274

Wang W, Liang S, Meyers T (2008) Validating MODIS land surface temperature products using long-term nighttime ground measurements. Remote Sens Environ 112:623–635

Wang C, Platnick S, Zhang Z, Meyer K, Yang P (2016) Retrieval of ice cloud properties using an optimal estimation algorithm and MODIS infrared observations: 1. Forward model, error analysis, and information content. J Geophys Res. doi:10.1002/2015JD024526

Watts PD, Bennartz R, Fell F (2011) Retrieval of two-layer cloud properties from multispectral observations using optimal estimation. J Geophys Res 116: D16203. doi:10.1029/2011JD015883

Wiscombe W (1977) The delta-M method: rapid yet accurate radiative flux calculations for strongly asymmetric phase functions. J Atmos Sci 34:1408–1422

Yang P, Bi L, Baum BA, Liou KN, Kattawar GW, Mishchenko MI, Cole B (2013) Spectrally consistent scattering, absorption, and polarization properties of atmospheric ice crystals at wavelengths from 0.2 to 100 μm. J Atmos Sci 70: 330–347. doi:10.1175/JAS-D-12-039.1

9

Meridional march of diurnal rainfall over Jakarta, Indonesia, observed with a C-band Doppler radar: an overview of the HARIMAU2010 campaign

Shuichi Mori[1][*] ⓘ, Jun-Ichi Hamada[1,2], Miki Hattori[1], Pei-Ming Wu[1], Masaki Katsumata[1], Nobuhiko Endo[1,3], Kimpei Ichiyanagi[1,4], Hiroyuki Hashiguchi[5], Ardhi A. Arbain[6,7], Reni Sulistyowati[6], Sopia Lestari[6,8], Fadli Syamsudin[6], Timbul Manik[9] and Manabu D. Yamanaka[10,11]

Abstract

This paper presents an overview of the HARIMAU2010 campaign focusing on convective activity with the diurnal rainfall meridional march (DRMM) over Jakarta, which is located on the northern coast of Jawa Island of the Indonesian maritime continent (IMC), based on 1-month intensive observations by a C-band Doppler radar and multi-point atmospheric sounding array conducted during 16 January–14 February 2010. The campaign period corresponded to a phase after large-scale Madden–Julian oscillation (MJO) active convections passed over Jakarta (MJO inactive phase). The cross-equatorial northerly surge (CENS) intruded into the Jawa Sea with a cold tongue (CT) of sea surface temperature (SST) in the beginning of the period (CENS active period: 16–26 January), and then, it started to retreat (transition period: 27 January–05 February); afterward, only a few signs of it were apparent (CENS inactive period: 06–14 February). The observational results showed that (1) rainfall over Jakarta has the nature of DRMM during the MJO inactive phase at least, (2) the DRMM is likely driven primarily by "land-breeze"-like local meridional circulation, and (3) the meridional spatiotemporal variation of rainfall over Jakarta is thus controlled by activities of both the CENS and CT over the Jawa Sea.

Keywords: Indonesian maritime continent (IMC), Land–sea breeze circulation, Diurnal cycle, Cross-equatorial northerly surge (CENS), Madden–Julian Oscillation (MJO), Cold tongue (CT)

Introduction

The Indonesian maritime continent (IMC) lies along the equator and has a zonal width of more than 5000 km. It is composed of various-sized islands with massive orography and complicated coastlines, which are surrounded by oceans with high sea surface temperature (SST). Consequently, a substantial amount of rainfall is produced over the IMC by various kinds of convective activities both in time and space. These convective activities play a major role as an energy source that drives the Walker and Hadley circulations through released heat and moisture

transport (Ramage 1968). Satellite observations have shown that the spatial distribution of tropical rainfall is not homogeneous, and it tends to be well concentrated in "coastal regions," which can be defined as regions within a distance of 300 km from the coastlines (Ogino et al. 2016, 2017). For example, a coastal heavy rainband (CHeR) formed by diurnally developed convections along the western coastline of Sumatera Island, Indonesia, was identified clearly by satellite and radar observations (Mori et al. 2011). Thus, the IMC with its long coastlines receives more substantial rainfall amounts than those over other regions in the tropical lands and open oceans.

Diurnally developed convections and local circulations over the IMC have been studied widely, in particular over equatorial western Sumatera Island (e.g., Wu et al.

* Correspondence: morishu@jamstec.go.jp
[1]Japan Agency for Marine-Earth Science and Technology (JAMSTEC), 2-15 Natsushima-cho, Yokosuka 237-006, Japan
Full list of author information is available at the end of the article

2003, 2009a; Mori et al. 2004, 2006, 2011; Sakurai et al. 2005, 2009, 2011; Shibagaki et al. 2006a; Hamada et al. 2008; Kawashima et al. 2011) and in relation to intraseasonal variations (ISVs) including the Madden–Julian oscillation (MJO, Madden and Julian 1972) (e.g., Shibagaki et al. 2006b; Fujita et al. 2011; Kamimera et al. 2012). These studies were based on a number of campaign-related observations that were collected by using X-band Doppler radars, intensive atmospheric soundings, rain gauge networks, disdrometers, and surface weather stations, as well as satellite observations. Results from those previous studies have shown the nature of diurnal land–sea migrations of precipitating systems, their variations with the MJO, and seasonal migration patterns; furthermore, such works resulted in the climatological identification of the CHeR along the southwestern coastline of Sumatera. Recently, Yokoi et al. (2017) examined the mechanism of a diurnal migrating precipitating system along the southwestern coastline of Sumatera by using mainly intensive atmospheric soundings and both shipboard and ground-based C-band radars during November–December 2015 for a preliminary campaign of the Years of the Maritime Continent (YMC); the results of their research implied that there is a crucial role of offshore preconditioning by gravity waves. See Yamanaka (2016) and Yamanaka et al. (2018) for more comprehensive reviews of field observational studies over the IMC.

Jakarta, the capital megacity of Indonesia located in the northwestern "coastal region" of Jawa Island, Indonesia (Fig. 1), has frequently suffered from serious floods in the downtown areas and landslides in the mountain foothill regions caused by torrential rainfall events, which occurred mainly during the boreal winter season from December to February (Wu et al. 2007, 2013; Trilaksono et al. 2011, 2012; Sulistyowati et al. 2014); this timeframe corresponds to the primary rainy season around Jakarta (e.g., Hamada et al. 2002, 2012). Therefore, it is crucial to clarify the environmental conditions associated with torrential rainfall events and obtain a better understanding of the generation and development mechanisms to ensure the socioeconomic security of the coastal megacity Jakarta, as well as to contribute to the current meteorological and climatological points of view. Wu et al. (2007) conducted case studies of torrential rainfall that occurred in January–February 2007 around Jakarta based on the Quick Scatterometer (QuikSCAT) sea surface winds and both operational sounding and radar observation data. They found that these events were generated by the interaction between cross-equatorial northerly surges (CENSs) intruding into the Jawa Sea from the Northern Hemisphere (Hattori et al. 2011) and local circulations over Jakarta, which were closely related to diurnally developed convections over mountain foothill regions in southern Jakarta. In addition, convective

activities over the IMC are known to be significantly affected by eastward traveling synoptic disturbances of the MJO or ISV. A case study of the extreme rainfall that occurred during January 2013 over Jakarta showed that an active phase of the MJO approaching the IMC played an essential role in the development of long-lasting convections, which were generated by interactions between the CENS in the Jawa Sea and diurnal local circulations over Jawa Island (Wu et al. 2013). Furthermore, Trilaksono et al. (2011, 2012) examined the same case as Wu et al. (2007) by using the Japan Meteorological Agency (JMA)/ Meteorological Research Institute (MRI)–nonhydrostatic model (JMA/MRI–NHM: Saito et al. 2006) and suggested the importance of a cold anomaly in the lower troposphere associated with the CENS, which may be essential for generating the extreme rainfall that results in serious flooding in Jakarta; notably, there was fine horizontal resolution in the model simulations, which allowed them to resolve the complicated topography and diurnally developed local circulations.

Early in the observation history, van Bemmelen (1922) first examined diurnal variations of local winds over Batavia (the old name for the capital Jakarta in the Dutch colonial era) scientifically based on pilot balloon observations in 1905–1915 and found a pronounced land–sea circulation in the lower troposphere that showed a predominant northerly sea breeze in daytime below a height of 1 km with a counter southerly flow above and a comparatively weak southerly land breeze in reverse during the nighttime to the early morning period. Since then, the local winds, namely, land–sea circulations, and their relation to convective activity over Jakarta have been widely studied by Hashiguchi et al. (1995a, 1995b, 1995c), Hadi et al. (2000, 2002), Renggono et al. (2001), and Araki et al. (2006). These works were conducted by using an L-band wind profiling radar (WPR, also named as a boundary layer radar (BLR)), which was located at Serpong (Fig. 1b) in the southern part of the "greater Jakarta" region (Jakarta city and surrounding urban agglomerations of Jakarta, Bogor, Depok, Tangerang, and Bekasi (JABODETABEK); hereafter, just referred to as Jakarta) shown as the area surrounded by a solid blue line in Fig. 1b. These researchers thoroughly examined the nature of these circulations and possible mechanisms, although the data obtained were basically limited to only vertical measurements at one place rather than spatiotemporal images from the Geostationary Meteorological Satellite (GMS). Recently, Realini et al. (2014) and Oigawa et al. (2017) examined the spatiotemporal variation of precipitable water vapor (PWV) and its relation to diurnally developed convections by using a multi-point global satellite navigation system (GNSS) over Jakarta and Bandung, as well as C-band and X-band radars, respectively. In addition, numerical

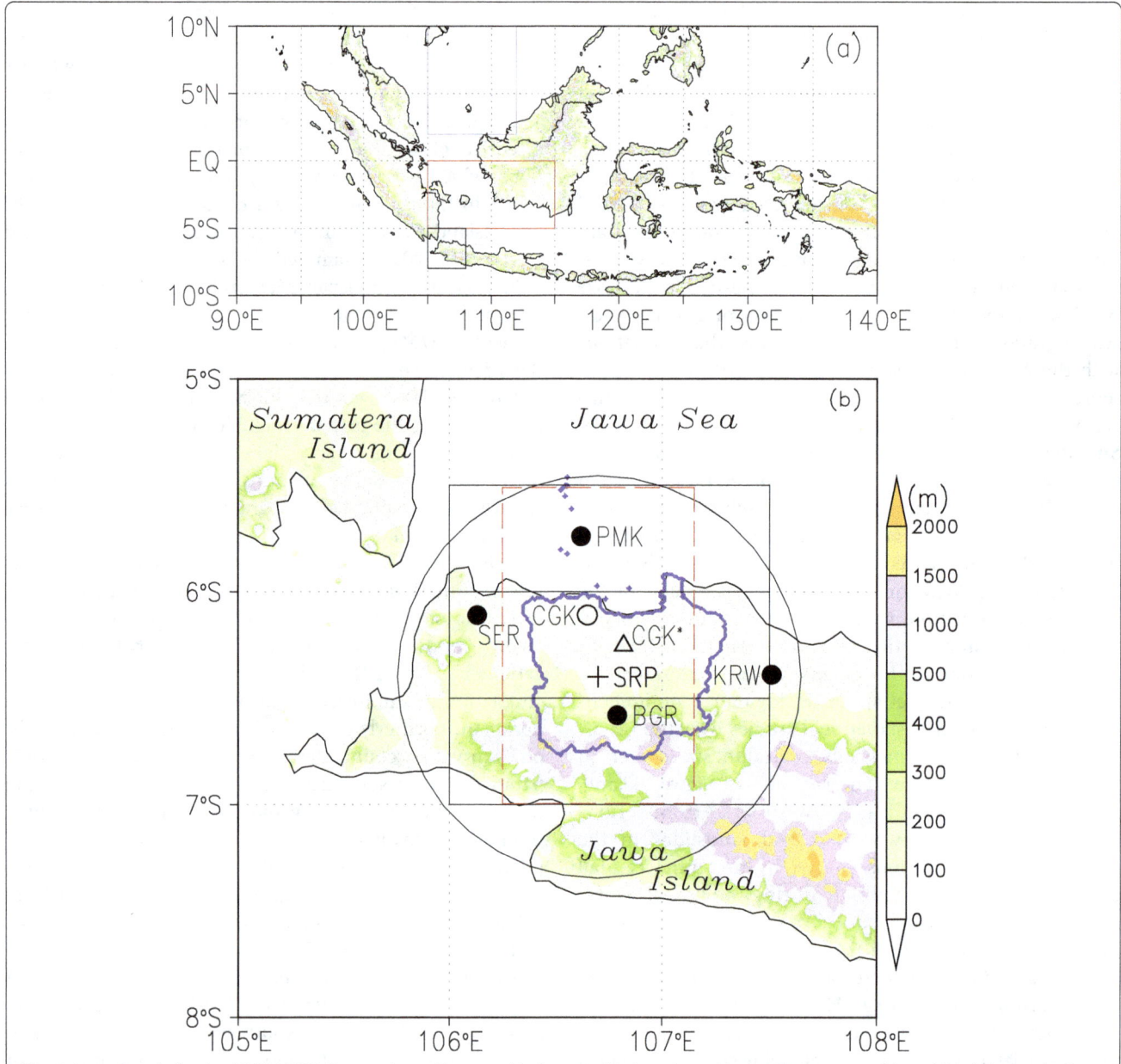

Fig. 1 Area of study for the HARIMAU2010 campaign. **a** Topographical maps of Indonesia and **b** the western part of Jawa Island. Rectangles in black, red, and blue solid lines in panel **a** indicate a location in panel **b**, definition areas for CENS (5.0° S–EQ, 105° E–115° E), and definition areas for CT (2° N–10° N, 105° E–112° E), respectively. The four closed black circles indicate the sounding stations used in the present study: PMK, SER, KRW, and BGR stand for Pramuka, Serang, Karawang, and Bogor, respectively. A small open circle shows the location of the Cengkareng (CGK) BMKG meteorological station for the Jakarta Soekarno–Hatta International Airport. The middle point between Serang and Karawang is indicated as Cengkareng* (CGK*) with an open triangle, and it was used for the analyses in Figs. 7 and 9. SRP with a plus sign stands for Serpong, where both the C-band Doppler radar (CDR) and wind profiling radar (WPR) are installed; the large open circle indicates the observational range (105 km in radius) of the CDR volume scan mode. Large three rectangles depicted by solid black lines over Jakarta (106.0° E–107.5° E) indicate the areas for GSMaP rain analyses in Fig. 3, which included coastal sea (6.0° S–5.5° S), land (6.5° S–6.0° S), and mountain foothill (7.0° S–6.5° S) regions. The rectangle depicted by the dashed red lines indicates the area (7.0° S–5.5° S, 106.25° E–107.15° E) for CDR data analyses in Figs. 6, 7, 8, and 9. The area surrounded by a solid blue line shows the "greater Jakarta" region (Jakarta city and surrounding urban agglomerations of Jakarta, Bogor, Depok, Tangerang, and Bekasi (JABODETABEK); hereafter, just referred to as Jakarta)

model simulations have been utilized to diagnose interactions between local winds, namely, land–sea circulations and diurnally developed convections over Jakarta, and the results implied a vital role of the CENS or cold tongue (CT) of SST over the South China Sea (SCS) in generating active convections at the land-breeze front, in particular for nocturnal heavy precipitation around the coastal region (Trilaksono et al. 2011, 2012; Koseki et al. 2012); however, no spatiotemporal radar or in situ observational data have been examined yet.

As the final campaign of the 5-year Japan Earth Observation System (EOS) Promotion Program (JEPP)–Hydrometeorological Array for ISV-Monsoon Automonitoring (HARIMAU) project (2005–2010, Yamanaka et al. 2008), we have carried out an integrated observation study, named HARIMAU2010 from 00 UTC (coordinated universal time) on 16 January 2010 to 24 UTC on 14 February 2010 (30 days) during the major rainy season over Jakarta. As shown in Fig. 1b, we deployed a C-band Doppler radar (CDR), intensive atmospheric sounding network, and surface weather stations during the campaign period so as to investigate the nature and spatiotemporal variation of convective activities that can generate torrential rainfall over Jakarta as well as those interactions among the MJO, CENS, and CT. This paper provides an overview of the HARIMAU2010 campaign and examines the spatiotemporal variation of diurnally developed convections over Jakarta by focusing on their relation to local circulations as the first case study based on CDR observations with a multi-point sounding array.

Methods/Experimental

Figure 1 shows the topography of the western part of Jawa Island along with the locations of observational sites for this study, which are described in detail in the following subsections. Jakarta is located at the most northern tip of a plain extending between the mountain range and Jawa Sea to the south and north, respectively. Averaged precipitation over the region is approximately 2000 mm year^{-1} and 400 mm month^{-1} in the rainy season (Hamada et al. 2012). Local time (LT) used in this study represents West Indonesian standard time (WIB) over the western part of Indonesia (WIB meridian: 105° E) including Jakarta (LT = UTC + 7 h). Therefore, the sunrise (sunset) time over Jakarta is approximately 06 (18) LT throughout the year.

C-band Doppler and wind profiling radars

We employed the CDR and WPR installed at Serpong (6.40° S, 106.70° E, 46 m above mean sea level (AMSL)) to observe the behaviors of convections over Jakarta and wind profiles, respectively, with high temporal and spatial resolutions. Major specifications of the CDR and WPR are summarized in Table 1. The CDR obtains the three-dimensional reflectivity and Doppler velocities every 6 min through a series of conical scans with antenna elevations ranging from 0.6° to 50° (volume scan mode) and surveillance observations at one elevation angle (0.6°). Observation radii of the CDR were 105 km (Fig. 1b) and 175 km for the volume scan mode and surveillance observations, respectively, and only the former data were examined in the present study.

Table 1 Major specifications of Serpong CDR and WPR, and their operating condition during the HARIMAU2010 campaign

Parameter		Value	
Radar	C-band Doppler radar (CDR)	Wind profiling radar (WPR)	
Location		Serpong (6.40° S, 106.70° E, 46 m AMSL)	
Manufacturer and model	Toshiba Corporation JMA-237B	Sumitomo Electric Industries (SEI) WPR LQ-0 (with 3-Parabolic Antennas)	
Frequency	5.320 GHz	1357.5 MHz	
Peak power	140 kW	400 W	
Pulse width	1.0 μs	0.67 μs	
Pulse repetition frequency (PRF)	1800 Hz (surveillance) 1360 Hz (volume scan)	10 kHz	
Beam width	0.98°	7.6°	
Signal processor	Sigmet RVP8	SEI LQ	
Maximum range	175 km (surveillance) 105 km (volume scan)	0.3–10 km (vertically)	
Sampling resolution	200 m	100 m	
Antenna rotation speed	30° s^{-1}	–	
Elevation angles	0.6° (surveillance) 0.6°–50.0° (18 elev./volume scan)	Beam directions (Az, Ze) (0°,0°), (0°,15°), (90°,15°)	
Scan interval	6 min	1 min	

Data obtained by the CDR were first processed by the Sigmet "interactive radar information system (IRIS)" (Vaisala 2008), and the following three signal filters were applied: "log receiver signal-to-noise ratio (LOG)," "clutter-to-signal ratio (CSR)," and "signal quality index (SQI)." Range bins with LOG < 1.0 dB, CSR < 23 dB, and SQI < 0.4 were omitted as "no data," which is where reflectivity was below the noise level and Doppler velocity was not available. Then, the IRIS "speckle filter" was applied to remove isolated data bins that must not have been weather targets, for example, towers and aircrafts. Atmospheric gaseous attenuation was corrected by 0.0016 dB km^{-1} as the IRIS default setting, whereas rain attenuation was not applied to the obtained data because only qualitative rain distributions were analyzed in the present study and no quantitative discussions were presented,[1] for example, comparisons with ground-based rain gauge data or satellite-derived rainfall data. The area between 6.83° S and 6.72° S was disregarded from the analyses because of the serious ground clutter below

2 km AMSL caused by the mountain range in the southern part of Jawa Island, as shown in Fig. 1.

The reflectivity and radial Doppler velocity were interpolated over the Cartesian coordinates with grid intervals of 0.5 km (421 grids × 421 grids in a 105 km × 105 km square) in the horizontal and 0.5 km in the vertical (40 levels from 0.5 to 20 km AMSL) plane for this study. Reflectivity fields were partitioned into convective and stratiform regions by using a technique proposed by Steiner et al. (1995). We calculated the hourly rainfall rates from the reflectivity field by using Z–R relations for each rainfall type, where Z and R are the radar reflectivity (dBZ) and rainfall rate (mm h^{-1}), respectively. In this study, we employed the Z–R relations as first-guess values for rainfall rate estimations in the Tropical Rainfall Measuring Mission (TRMM) 2A25 version 5 algorithm, i.e., $Z = 148 \, R^{1.55}$ for convective rainfall, $Z = 276 \, R^{1.49}$ for stratiform rainfall (Schumacher and Houze Jr., 2003), to facilitate comparisons of the rainfall characteristics among various regions. Then, the rainfall rates were averaged over the maximum volume-scan coverage area to estimate radar echo coverage (%) in the following analyses. Radar echo coverage was defined as the area of radar echo (km^2) with a radar reflectivity of more than 10 dBZ at a specific height divided by the area of radar detection (km^2) at the same height. The CDR data used in the present study started from 09 LT on 17 January 2010 because of electrical troubles and ended at 07 LT on 15 February 2010.

The WPR obtained three components of the wind velocity with high spatiotemporal resolutions (see Table 1) during the campaign period successfully. The WPR data were valuable for analyzing the fine structures of the wind distribution in and out of mesoscale convections

around the Serpong WPR site as a number of previous studies have showed, e.g., Hashiguchi et al. (1995a, 1995b, 1995c), Hadi et al. (2000), Renggono et al. (2001), and Araki et al. (2006). However, the WPR data were not analyzed in this paper because these data were not suitable for direct comparisons with spatiotemporally averaged CDR and atmospheric sounding array data over the "greater Jakarta" region, which are mentioned later in detail.

Intensive atmospheric soundings

We made intensive soundings from 00 UTC on 16 January to 24 UTC on 14 February 2010 (30 days) at 6-h intervals and additional 3-hourly soundings from 00 UTC on 24 January to 24 UTC on 06 February 2010 (14 days) at Pramuka (05.74° S, 106.62° E, 1 m AMSL), Serang (06.11° S, 106.13° E, 71 m AMSL), Karawang (06.39° S, 107.51° E, 53 m AMSL), and Bogor (06.58° S, 106.79° E, 248 m AMSL) to obtain the atmospheric structure at sub-diurnal temporal scales. In addition, we employed operational (00 and 12 UTC) and extra (06 and 18 UTC) sounding data obtained at Cengkareng (06.11° S, 106.65° E, 9 m AMSL), where there is a meteorological station for the Jakarta Soekarno–Hatta International Airport managed by the Agency for Meteorology, Climatology and Geophysics, Indonesia (BMKG). Locations of sounding stations are depicted in Fig. 1b, and specifications for the sounding time and types of transmitters, receivers, and balloons are summarized in Table 2. All soundings provided vertical profiles of pressure, temperature, relative humidity, and horizontal wind every 2 s, which corresponded to a height resolution of approximately 10 m, and the data were averaged vertically into 100-m intervals. In addition,

Table 2 Specifications of intensive sounding network during the HARIMAU2010 campaign. (00 UTC 16 January–24 UTC 14 February 2010)

Site name	Bogor	Pramuka	Serang	Karawang	Cengkareng
Location	06.58° S 106.79° E	05.74° S 106.62° E	06.11° S 106.13° E	06.39° S 107.51° E	06.11° S 106.65° E
Height (AMSL)	248 m	1 m	71 m	53 m	9 m
Sounding time (UTC)		00, 03, 06, 09, 12, 15, 18 (24 January–06 February) 00, 06, 12, 18 (other period)			00, 06 12, 18
Missing data period	–	00 UTC 16–06 UTC 17 January	–	00 UTC 17–12 UTC 20 January	–
Transmitter			Vaisala RS92-SGPD		Meisei RS-06G
Receiver	Vaisala MW15		Vaisala MW31	Vaisala MW21 (MW15 used until 18 UTC 16 January)	–
AWS		Vaisala MAWS201		Davis Weather Monitor II (All data missing due to troubles)	–
Disdrometer	Persivel M300	–	–	–	–
Rainwater sampling	X	X	–	–	–

3-hourly interpolated data were created during the 6-hourly sounding period at each station for the following analyses.

We adopted the sounding stations at Pramuka (PMK), Serang (SER), Cengkareng (CGK), Karawang (KRW),[2] and Bogor (BGR) for representative locations of the "coastal sea" over the Jawa Sea, western, central, and eastern "coastal land," and "mountain foothills" of southern Jakarta, respectively, as shown in Fig. 1b. However, we found serious biases in the operational sounding data collected at Cengkareng, in particular for the relative humidity in the lower troposphere as well as the temperature and wind data. Therefore, we made a composite dataset as a proxy of the soundings at Cengkareng by using those data obtained at both Serang and Karawang, and then, we used it in the following analyses designated as "Cengkareng* (CGK*)" at the middle point (06.25° S, 106.82° E, 62 m AMSL) between Serang and Karawang (see Fig. 1b), the representative location of "central coastal land." In addition, portions of sounding data in the beginning of the campaign were lacking because of receiver troubles at Pramuka and Karawang (see Table 2 for details).

Surface observations

Automatic weather stations (AWSs) were deployed at Bogor, Pramuka, Serang, Serpong (Vaisala MAWS201), and Karawang (Davis Weather Monitor II). They provided surface pressure, temperature, relative humidity, solar radiation, wind direction and speed, and rainfall data. The minimum sensitivity for the rainfall amount and the temporal resolution of the recorded data were 0.2 mm and 1 min, respectively. However, the whole AWS dataset obtained at Karawang was lost during a mechanical malfunction. Raindrop size distribution data were obtained every 1 min with a Parsivel M300 optical disdrometer at Bogor. Moreover, rainwaters were sampled at Pramuka, Bogor, and the Pondok Betung BMKG station (06.25° S, 107.61° E, 26 m AMSL) every 3 or 6 h for stable isotope analyses (e.g., Fudeyasu et al. 2011; Belgaman et al. 2016), though these data are not examined in this paper.

Other data

Optimum interpolation sea surface temperature (OISST; Banzon et al. 2016) and blended sea winds (BSW; Zhang et al. 2006) data provided by the U.S. National Oceanic and Atmospheric Administration (NOAA) and objective reanalysis data (Kalnay et al. 1996) supplied by the U.S. National Centers for Environmental Prediction (NCEP) were used to analyze large-scale environmental conditions over the Jawa Sea and Jawa Island during the campaign period. The spatiotemporal resolutions of the OISST and BSW datasets were 0.25° × 0.25° and 1 day, and those of the NCEP reanalysis were 2.5° × 2.5° and 1 day. An all-season real-time multivariate MJO (RMM)

index (Wheeler and Hendon 2004) provided by the Bureau of the Meteorology Research Centre (BMRC) of Australia was also used to confirm the phases of MJO activity during this period. A global satellite mapping of precipitation (GSMaP; Okamoto et al. 2005) near-real-time (NRT) dataset provided by the Japan Aerospace Exploration Agency (JAXA) was used to examine the temporal and spatial distribution of rainfall over the whole IMC region. In addition, the GTOPO30 (U.S. Geological Survey 1993) global digital elevation model (DEM) dataset with a spatial resolution of 30-min arcseconds was used to confirm the area of radar detection and depict the topography around the study region. Moreover, we made numerical model simulations for experimental near real-time rainfall forecasts over Jakarta with the JMA/MRI-NHM, though those results are not presented in this paper because their reproducibility has not been validated sufficiently for analyses yet.

Results

Synoptic view during the campaign period

A Hovmöller diagram of the GSMaP rain rate and NCEP wind barbs at 925 hPa during January–February 2010 is shown in Fig. 2 along with a RMM diagram for the same period. Although a large-scale disturbance recognized as a MJO passed over the IMC region in early January, only small-scale and isolated rain areas were identified over western Jawa (106°–107° E) during the campaign period of 16 January–15 February 2010. Westerly wind prevailed over the IMC in the beginning of the campaign period, which blew into the convection center of the MJO in the east. As a result, the northwestern part of Jawa including Jakarta city was regarded as being in the MJO inactive phase and free from effects caused by such large-scale equatorial disturbances. Indeed, the RMM index (Fig. 2b) showed that the center of the MJO activity passed through the IMC (regions 4–5) by 20 January 2010, and then it reached into the western Pacific Ocean (regions 6–7) in February.

Each panel in Fig. 3 shows (a) temporal variations of the NCEP horizontal wind vectors and meridional wind speed (shade) at 925 hPa, (b) the CENS index defined as the area averaged meridional sea surface wind over 105° E–115° E, 5° S–EQ (Hattori et al. 2011) based on BSW data, (c) the CT index defined as the area averaged SST at 2° N–10° N and 105° E–112° E based on OISST data (Koseki et al. 2012), (d) the area averaged GSMaP daily rainfall over the western part of Jawa including Jakarta shown in the rectangular areas in Fig. 1b, and (e) the CDR echo coverage at a height of 2 km with the convective echo fraction during the campaign period. As seen in Fig. 3a, CENS originated from the far north (> 20° N) Siberia–Mongolia area and ran through the East and South China Seas, and then it finally intruded into the northern coastline of Jawa Island (6° S), with the

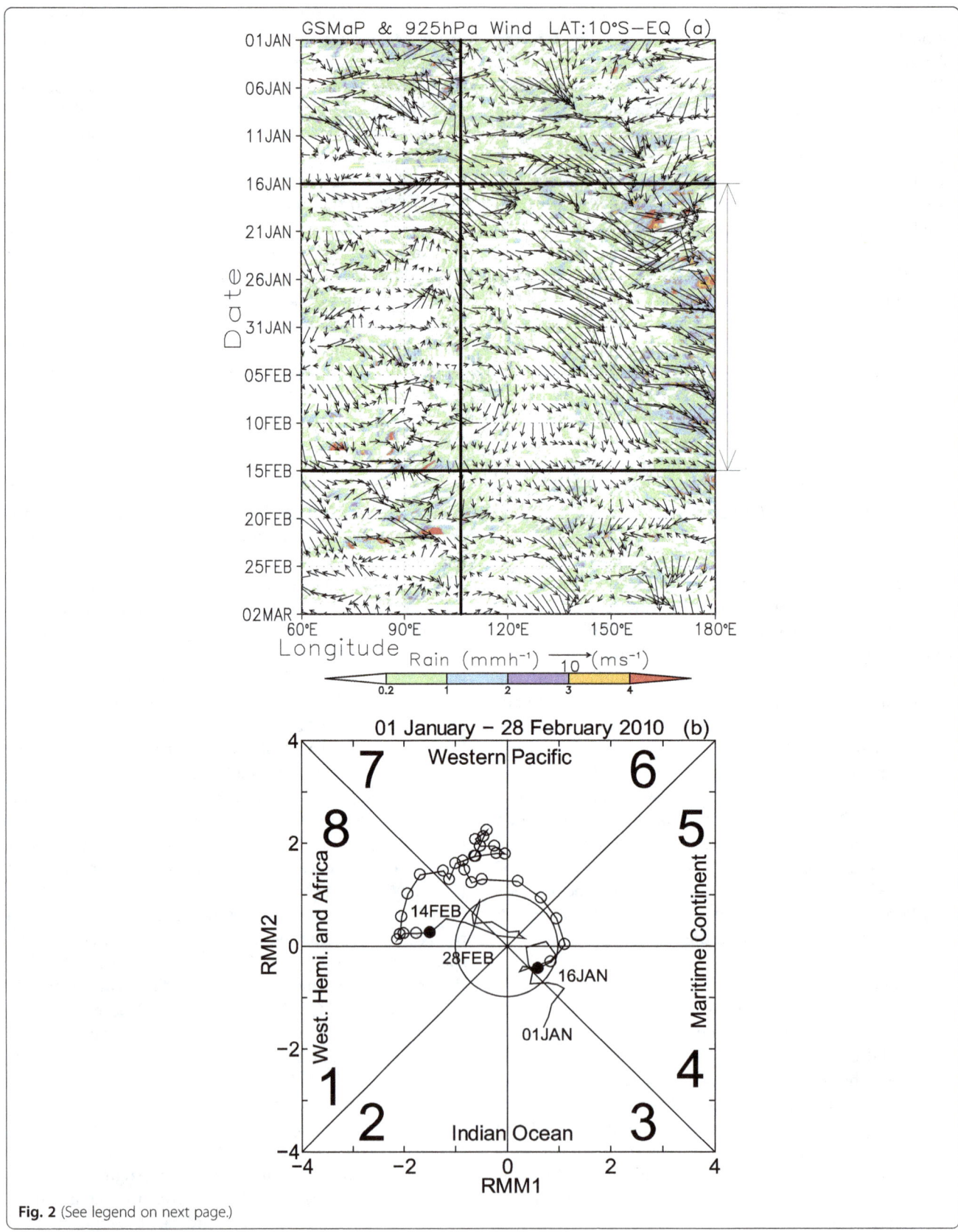

Fig. 2 (See legend on next page.)

(See figure on previous page.)

Fig. 2 Synoptic disturbances during the campaign period. **a** Hovmöller diagram of the GSMaP rain intensity with NCEP wind vectors at 925 hPa and **b** the RMM index based on data distributed from the BMRC during January and February 2010. The vertical line in panel **a** indicates the longitudinal location Serpong, Jakarta (106.7° E). The vertical arrow between the two horizontal lines shows the period of the HARIMAU2010 campaign (16 January–14 February 2010). Open and closed circles in panel **b** show daily positions of the MJO center during the campaign period and starting/ending dates of the campaign, respectively

largest negative CENS index at < -5 m s^{-1} (Fig. 3b; we used the definition of a "CENS event" by Hattori et al. 2011) at the beginning of the campaign (16 January). This was followed by a short break period (19–24 January), and then its strength decreased gradually with time until the end of the campaign (14 February). As a result, moderate northerly wind up to 3–4 m s^{-1} prevailed over Jakarta in the first half of the campaign followed by weak southerly wind less than 2 m s^{-1} in the latter half of the period, though northerly wind was still observed over the Jawa Sea. Simultaneously, the CT index (Fig. 3c) reached its minimum on 17 January and then increased gradually with intermittent minimal peaks. Consequently, the campaign period was regarded as a withdrawal or weakening phase of both the CENS and CT from the synoptic point of view.

Figure 3d shows the area averaged daily rainfall over the western part of the Jawa Island, which is divided into the coastal sea (red), land (green), and mountain foothill (gray) regions as shown in Fig. 1b. The daily rainfall over the coastal sea and land regions before the campaign started showed maximums of approximately 60 mm and 90 mm, respectively, while that over the mountain foothills was below 20 mm. Conversely, the values were comparatively small (i.e., mostly below 20 mm) in the former 20 days (16 January–04 February) except on 26 and 29 January over the coastal sea and land regions. On the other hand, the daily rainfall over the mountain foothill region amounted to more than 50 mm in the latter 10 days (05–14 February) while that over the coastal sea and land regions stayed mostly below 30 mm and 10 mm, respectively. The total CDR echo coverage (Fig. 3e) in the beginning of the campaign showed diurnal variations with peaks of around 25–40% until 23 January. The convective echo fraction during the same period had similar variations with peaks below 20% just in advance of each peak in the total echo coverage. Conversely, diurnal peaks of convective echo fractions after 23 January increased gradually from approximately 20 to 30% until 13 February when the next northerly wind approached Jawa Island, though they decreased a little over a short period during 28 January and 02 February.

Time–height cross sections of sounding data during the campaign period obtained at Bogor, where the most southern sounding station is located alongside the mountain foothills, are presented in Fig. 4. We first examined the data obtained at Bogor because this was the only station that had no missing data period during the campaign, including for 3-hourly soundings (Table 1), and it was one of the fundamental stations used for further analyses in the following subsections. Although meridional wind in the lower troposphere (Fig. 4d) showed a weak northerly component in the beginning of the period, which was consistent with the sign of the CENS in Fig. 3a, it was not sustainable and was followed by a frequent southerly component by the end of the period. Meanwhile, zonal wind (Fig. 4c) in the lower troposphere showed a westerly velocity of more than 10 m s^{-1} in the beginning of the campaign period, which blew into the MJO convection center located in the western Pacific Ocean as mentioned above. On the other hand, that in the upper troposphere was strong easterly wind, which formed stable vertical wind shear, and this wind was strong in the former half of the period. The westerly wind in the lower troposphere decreased by the end of January and then changed into weak easterly wind in the latter half of the period. This zonal wind regime, i.e., strong westerly, westerly, and easterly winds, also characterized the three sub-periods of the campaign.

Although equivalent potential temperature (Fig. 4a) showed vital diurnal variations throughout the analysis period, it displayed a cold anomaly in the former half of the period followed by a warm anomaly, which occurred 5–10 days in advance of the zonal wind shift in the lower troposphere. The specific humidity (Fig. 4b) showed massive dry layers in the middle troposphere around a height of 5 km in the beginning (16–20 January) and the end (10–14 February) of the analysis periods as well as diurnal variations in the lower troposphere. Sounding data in the middle to upper troposphere, i.e., above a height of 5 km, which were observed at all five stations, had quite similar synoptic characteristics to those data collected at Bogor (figures not shown) because they were obtained within a horizontal range of 100 km at maximum (see Fig. 1b). Those in the lower troposphere, however, showed pronouncedly different features from each other in particular from the point of their diurnal variations, which are described in detail in the following subsections.

Based on the results of the synoptic views above, we divided the campaign period into three 10-day sub-periods as follows: period-I (16–25 January 2010), period-II (26 January–04 February 2010), and period-III (05–14 February 2010), hereafter referred to as the CENS active, transition, and inactive periods,

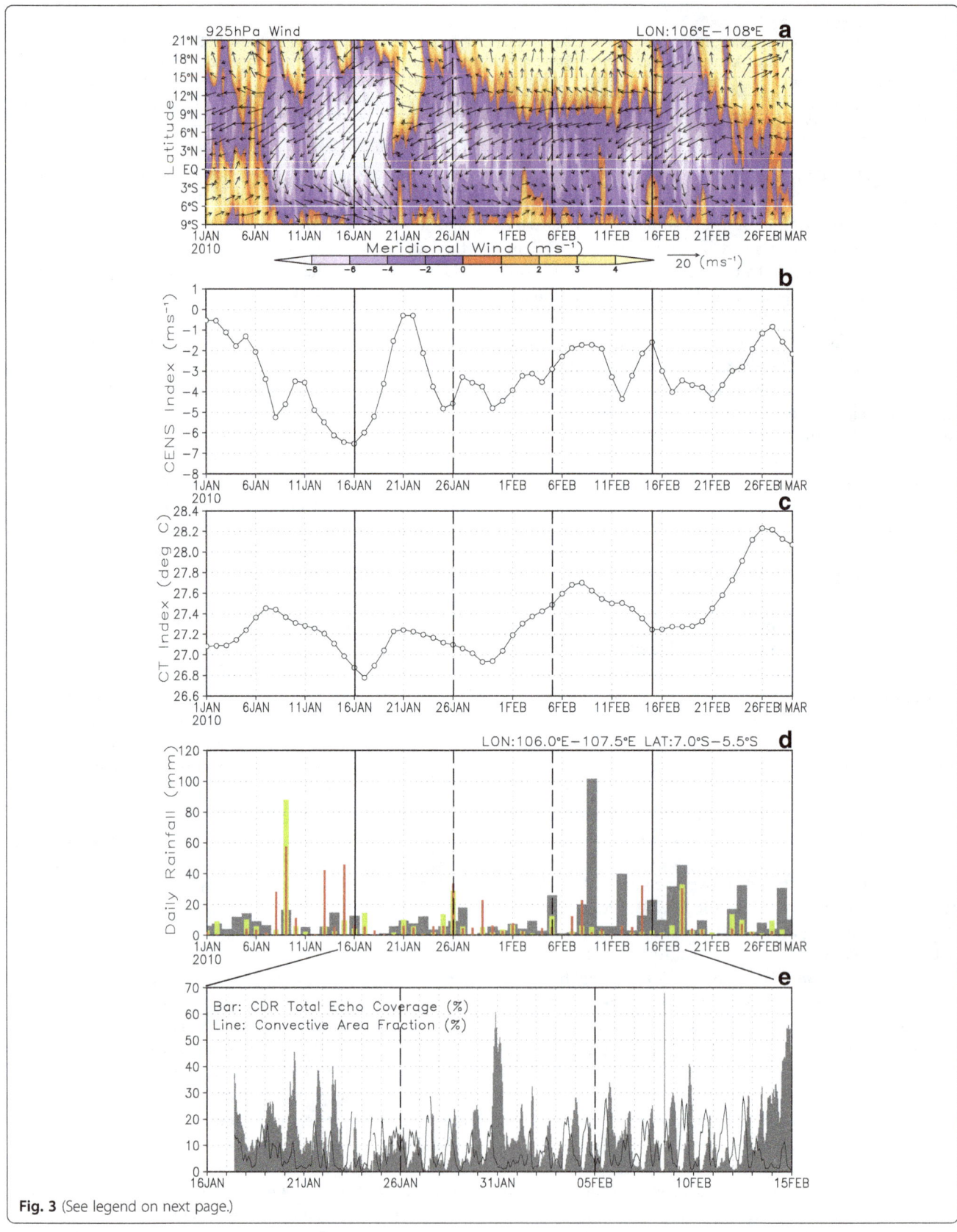

Fig. 3 (See legend on next page.)

(See figure on previous page.)
Fig. 3 Temporal variations of the CENS, CT, and rainfall during the campaign period. **a** Temporal variations of NCEP horizontal wind vectors and meridional wind speed (shade) at 925 hPa averaged between 106° S–108° E, **b** CENS and **c** CT indices, **d** the GSMaP daily rainfall amount over Jakarta, and **e** CDR echo coverages. **d** The GSMaP rain rate is divided into three regions as designated in Fig. 1b: coastal sea (red), plain (green), and mountain foothills (gray). **e** Bars (gray) and lines (black) in the CDR echo coverages indicate the total echo coverage (%) and convective rain area fraction (%)

respectively, for convenience in the analyses. The CENS and CT indices and brief descriptions of the lower troposphere in each sub-period are summarized in Table 3.

Meridional march of diurnal rainfall over Jakarta and local circulations

Figure 5 shows the daily and diurnal variations of surface rainfall during the campaign period observed by AWSs at Pramuka (a), Serang (b), Serpong (c), and Bogor (d), which correspond to the north (coastal sea), west (coastal land), central (plain), and south (mountain foothills) areas of the sounding array, respectively, as shown in Fig. 1b. Note that the AWS data at Serang were employed as "coastal land" because there was no AWS installed at Cengkareng and that at Karawang was completely in a no data status during the whole campaign period (see Table 2). Diurnal rainfall over the "coastal sea" observed at Pramuka increased rapidly after sunset around 18 LT; its primary peak occurred at 21 LT. Then, it decreased gradually until 06 LT with a secondary peak at 03 LT. This diurnal variation seems to be a combination of typical variations over the land with evening showers and over the sea with nocturnal rain. The Pramuka daily rainfall was comparatively large in the middle of the campaign period (25 January–5 February), but not much occurred at both the beginning and ending of the period. The rainfall over the "coastal land" observed at Serang (Fig. 5b) did not show any specific diurnal peaks or daily variations, though there was a slight tendency for more rainfall in the early morning (late evening) in the former (latter) half of the period, whereas that over the "plain" area observed at Serpong (Fig. 5c) showed weaker and gentle diurnal peaks in the afternoon at 12 LT and 16 LT according to period averaged characteristics throughout the whole period without specific daily variations. Conversely, that over the "foothills" observed at Bogor (Fig. 5d) showed a pronounced diurnal variation with a striking peak at around 17 LT as an averaged appearance (right panel), though the peak time varied much day by day during the period. In addition, the daily rainfall amount increased gradually from the beginning to the end of the period with a phase delay of the diurnal rainfall peak from around noon (11 LT) to the evening (18 LT). Similar phase delay features were also seen at the other AWS stations but around differential local times, e.g., at Pramuka (Fig. 5a) from 18 LT to

03 LT in the beginning to middle of the period, though this was not so clear in comparison with that in Bogor (Fig. 5d). These findings indicate that diurnal and daily rainfall variations over Jakarta have quite different characteristics along the meridional direction perpendicular to the coastline.

We then proceeded to examine the meridional variation of diurnally developed convections during the campaign period, which have been partially reported on by Sulistyowati et al. (2014) as well. Figure 6a shows the meridional variation of the CDR echo distribution at a height of 2 km over Jakarta averaged over a 100-km width (106.25° E–107.15° E) in the east–west direction from the CDR site as shown by a rectangle with dashed red lines in Fig. 1b. Note that the width decreases only from 5.6° S (100 km) to 5.5° S (70 km) according to our analysis, and this condition also applies to Figs. 6, 7, 8, and 9. The weak and broad extent of the radar echo was evident in the first quarter of the period, which corresponded to the intrusion of the active CENS (Fig. 3a), and similar echoes were seen in the last few days of the period. A major part of the echo distribution showed pronounced diurnal variation with strong reflectivity in the southern mountainous region in the afternoon and northern coastal region in the nighttime to next early morning. Both major reflectivity regions were connected apparently in some cases from south (north) to north (south) in the late evening (early morning); however, this was not always obvious for the whole period. Therefore, the diurnal variation (48 h for convenience) of the radar echo distribution averaged over the whole campaign period is presented in Fig. 6b, and it was composed of the following three major parts: (1) a large extent of the echo area developed in the daytime to late evening around the mountains on the southern side of Jakarta, (2) northward traveling echoes moved from the mountains to the coastline in the evening (approximately 15 LT) at a speed of approximately 5 m s^{-1}, and then they stayed and redeveloped around the coastal region through the night, and (3) some parts of the echoes started to return back southward from the coastal to mountain foothill regions in the morning (approximately 06 LT) with a speed of approximately 2 m s^{-1}, although this was not more obvious than the northward propagation. Note that, however, these are period averaged characteristics created by the superposition of diurnal cycles, which varied much from day to day. Indeed, these

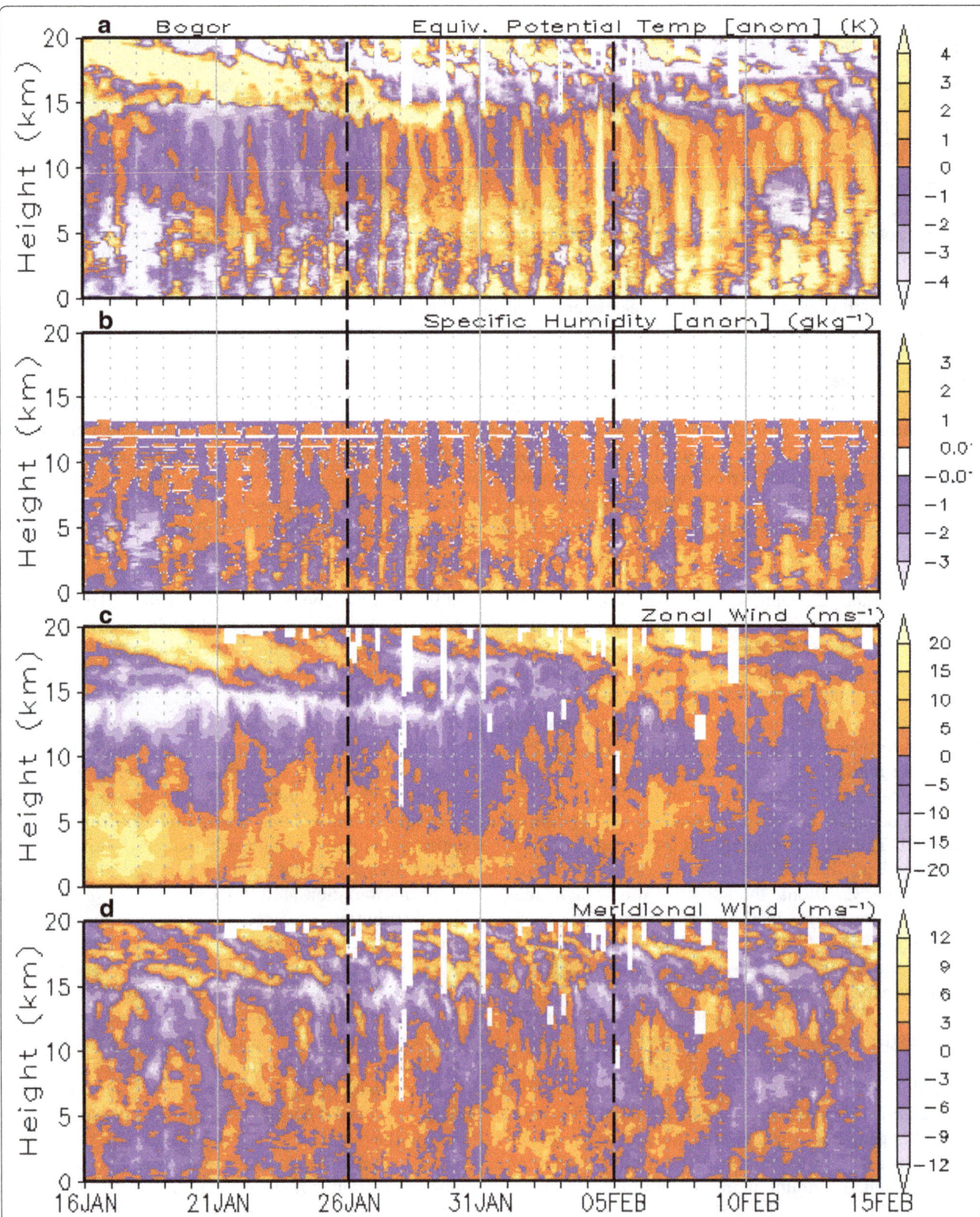

Fig. 4 Temporal variations of the sounding data obtained at Bogor. Time–height cross-sectional views of the sounding data obtained at Bogor during the campaign period. **a** Equivalent potential temperature (anomaly), **b** specific humidity (anomaly), and **c** zonal and **d** meridional wind speeds. Anomalies were calculated from the period averaged data of Bogor at each height

Table 3 Indices and brief descriptions of lower troposphere in each sub-period

Sub-periods	Period-I (CENS active)	Period-II (transition)	Period-III (CENS inactive)
Date	16–25 January	26 January–04 February	05–14 February
CENS Index [avg] (ms^{-1})	−3.42	−3.83	−2.25
CENS Index [min] (ms^{-1})	*−6.53*	−4.80	−4.30
CT Index [avg] (deg.C)	27.07	27.14	27.54
CT Index [min] (deg.C)	26.78	26.93	27.35
Temperature	Cooler	Getting warmer	Warmer
Zonal wind	Strong westerly	Westerly	Easterly
Meridional wind	Most northerly	Southerly	Little northerly
Remarks	CENS reached its maximum in the beginning of the sub-period with the lowest CT, cool atmosphere, and strong westerly wind in the lower troposphere followed by an intermittent break.	CENS started to retreat accompanied by both warmer CT and atmospheric temperature in the lower troposphere with decreasing easterly wind.	CENS mostly retreated with the warmest CT and atmospheric temperature in the lower troposphere followed by easterly wind.

Italicized letter in the CENS Index [min] shows above the criteria of "CENS" by Hattori et al. (2011). See text in detail

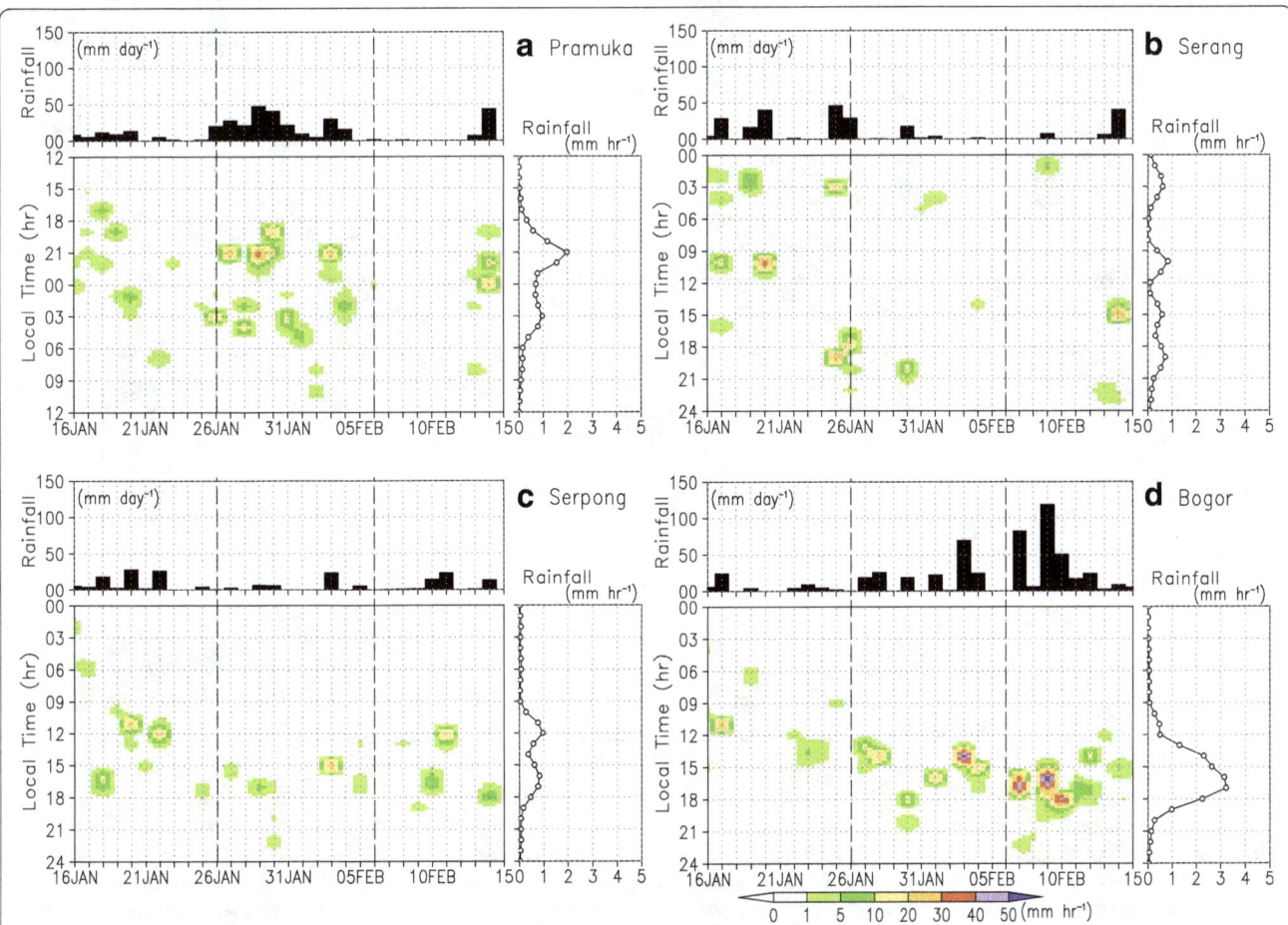

Fig. 5 Surface rainfall data obtained by AWSs during the campaign period. Daily and diurnal variations of the surface rainfall observed by AWSs during the campaign period: **a** Pramuka, **b** Serang, **c** Serpong, and **d** Bogor. Top and side panels for each site indicate the daily rainfall amount and period averaged rainfall diurnal variation, respectively. Note that the time in the bottom panel only at Pramuka (**a**) runs downward from 12 LT to 12 LT on the following day

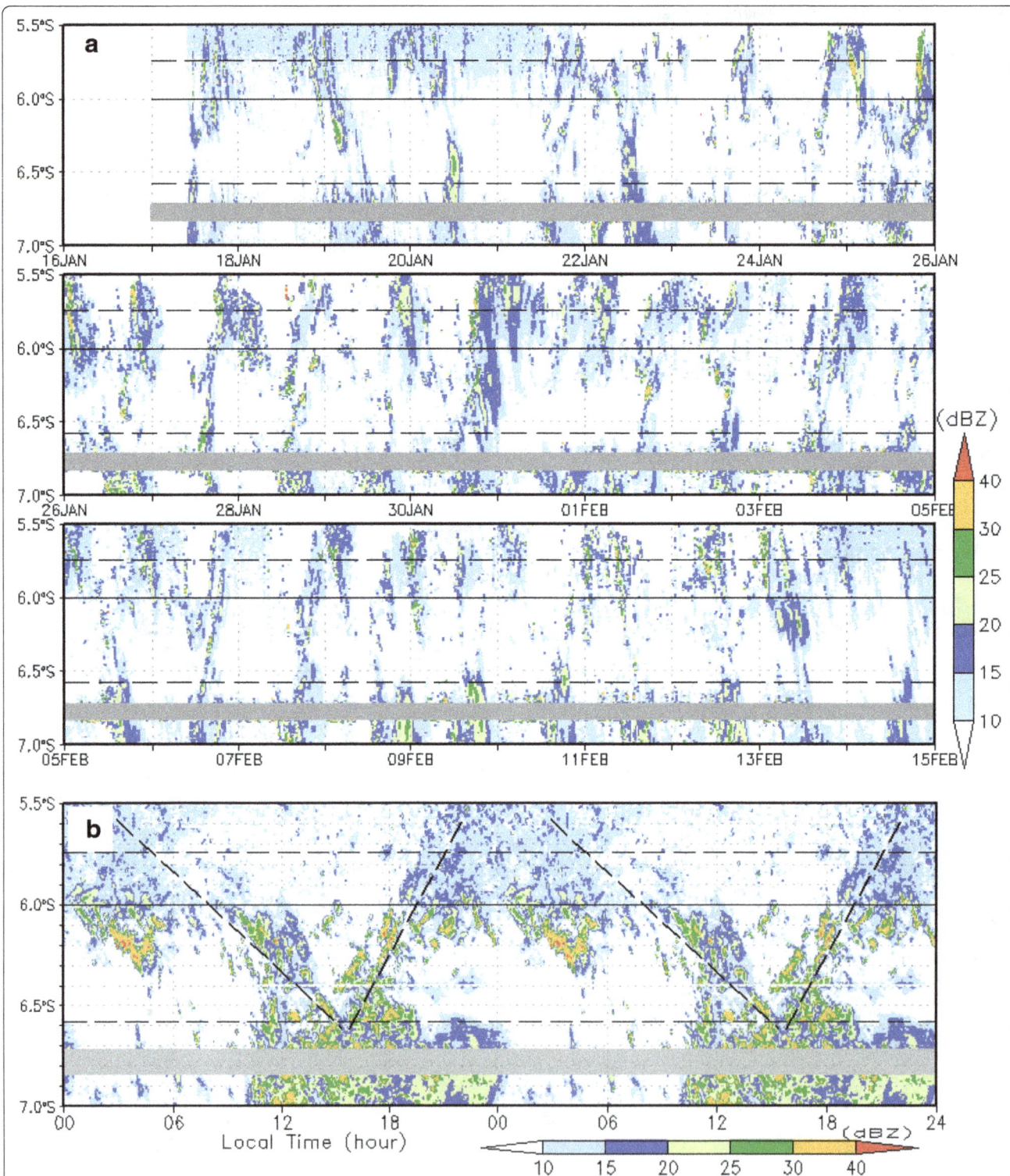

Fig. 6 Daily and diurnal meridional variations of the CDR echo. **a** Meridional variation of the CDR echo distribution at a height of 2 km during the campaign period averaged over a 100-km width (106.25° E–107.15° E) in the east–west direction from the Serpong CDR site as shown by the rectangle with dashed red lines in Fig. 1b and **b** its mean diurnal cycle (duplicated twice). Data in gray-shaded areas (6.83° S–6.72° S) were disregarded from the analyses because of serious ground clutter. Thin black solid horizontal lines indicate the averaged latitudinal location of the coastline (6.0° S) of the western part of Jawa Island as shown in Fig. 1b. Latitudinal locations of Pramuka (5.74° S) and Bogor (6.58° S) are also indicated by thin dashed horizontal black lines. Thick dashed lines in panel **b** indicate the estimated speed of DRMMs at approximately 2 m s^{-1} (5 m s^{-1}) southward (northward) during the early morning (late evening) for reference. Note that, however, these DRMM dashed lines show period averaged characteristics created by superposition of diurnal cycles that varied much from day to day

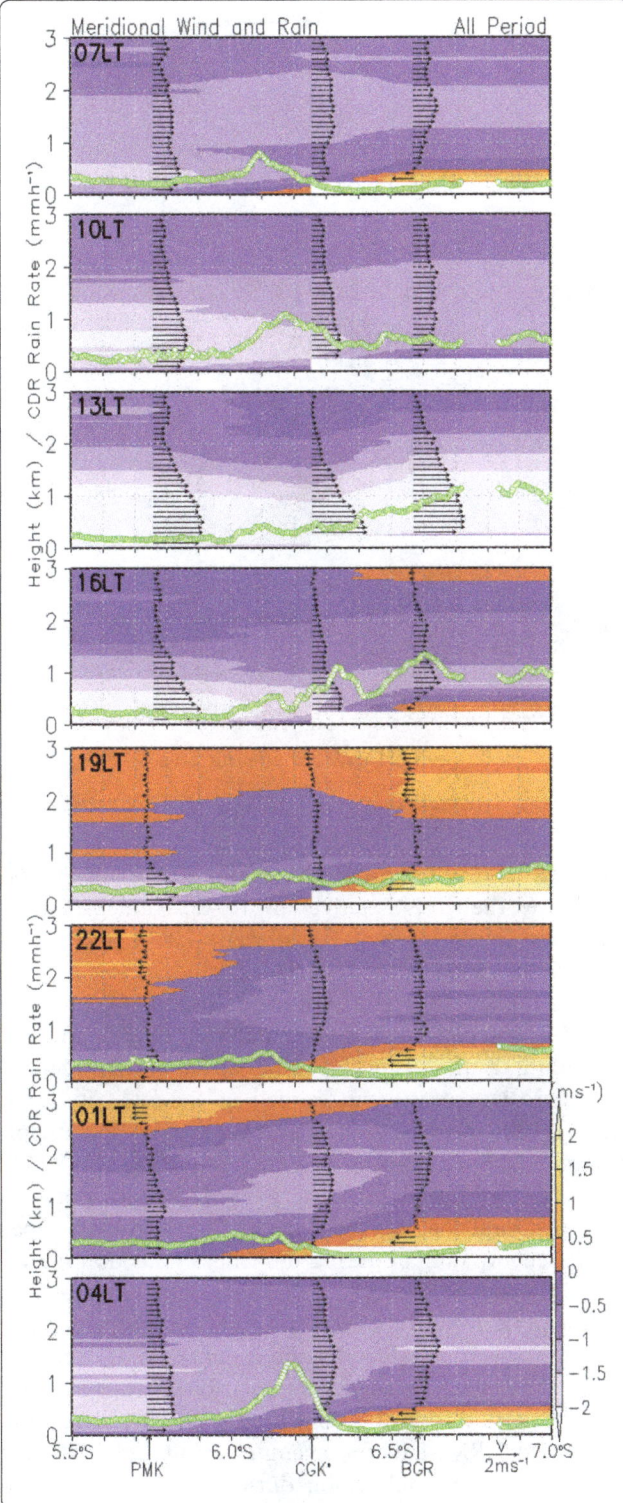

Fig. 7 Diurnal meridional circulations over Jakarta. Cross-sectional views of diurnal meridional circulation over Jakarta obtained by the multi-point sounding array averaged for the campaign period (color shades and wind vectors) with the meridional distribution of the CDR estimated rain rate at a height of 2 km (green lines with small circles) averaged over a 100-km width in the east–west direction from Serpong. Pramuka, Cengkareng, and Bogor indicate the sounding stations at the coastal sea over the Jawa Sea (north), the coastline (center), and the mountain foothills (south), respectively

characteristics are apparently different among the CENS active, transition, and inactive phases, as shown in Fig. 8, and they should be examined in detail individually as we will do in our next study.

Finally, in this subsection, we examine the diurnal variation of the meridional local circulation over Jakarta and its relation to the rainfall spatial variation, namely, the diurnal rainfall meridional march (hereafter, DRMM). Figure 7 shows cross-sectional views of meridional wind circulation in the lower troposphere perpendicular to the coastline from the "coastal sea (Pramuka)" to the "mountain foothills (Bogor)" through the "coastal land (Cengkareng*)" every 3 h from 07 LT to 04 LT on the following day averaged for the campaign period as well as the area averaged rainfall intensity calculated from the CDR reflectivity observations. Meridional wind at 13 LT showed substantial sea-breeze-like northerly wind in the whole lower troposphere below a height of 3 km from the Jawa Sea to mountain foothills with a maximum close to the surface layer. Rainfall intensity increased gradually from the coastline to foothill region at the same time and reached its daily maximum of approximately 1 mm h^{-1} over the foothills around Bogor. Afterward, a land-breeze-like southerly wind appeared first at 16 LT close to the surface over the foothills and the rainfall intensity over the foothill (coastal land) region started to decrease (increase) simultaneously. The southerly wind near the surface over the land increased its depth and meridional extent after sunset at 19 LT, and thus, a frontline formed at the location of the surface northerly wind over the coastal land (Cengkareng*). Most of the rainfall over the foothill and plain regions disappeared at the same time, and another weak and broad rainfall area started to appear instead, which stretched from the coastline to the coastal sea region. Finally, the southerly wind close to the surface reached over the Jawa Sea (Pramuka) at 22 LT at its maximum extent in a day, and then it started to retreat from the coastal sea southward at 01 LT and was balanced with the predominant synoptic northerly wind close to the sea surface. A rainfall peak was generated around the coastline at 04 LT ahead of a front that formed between the intrusion of northerly wind from the Jawa Sea to the coastal land and the retreating land-breeze-like southerly wind. The sea-breeze-like northerly wind developed much after the sunrise during 07–10 LT and covered over the whole study

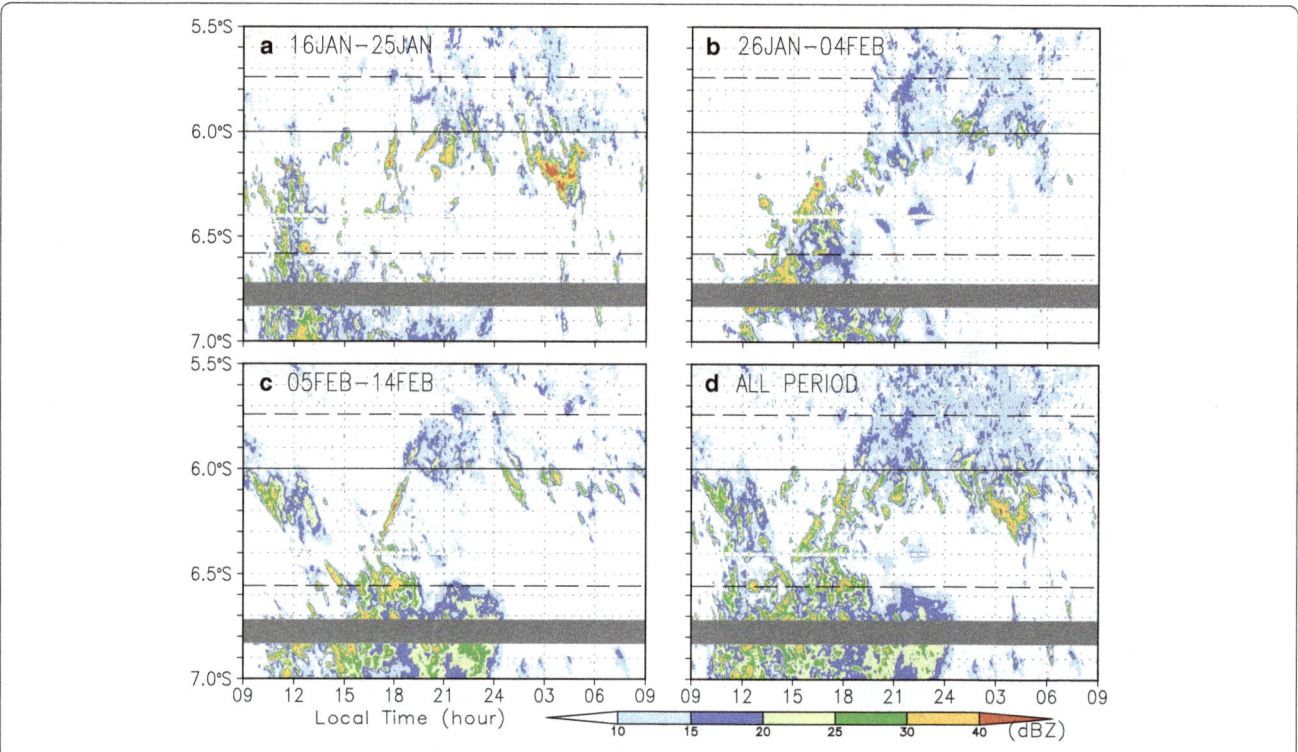

Fig. 8 Diurnal meridional variations of the CDR echo in each period. Same as Fig. 6b but 24-h views of DRMM averaged for the **a** CENS active, **b** transition, and **c** inactive periods, as well as the **d** whole periods

area quite rapidly, whereas a moderate rainfall area remained over the coastline.

The aforementioned results can be summarized as follows. (1) Most rainfall was generated originally over the mountain foothill region in the area south of Jakarta in the afternoon when the sea breeze was predominant over the whole study region, and then, it decreased with time after reaching a maximum in the evening. (2) Another rainfall event started to form over the "front" of land-breeze-like southerly wind after sunset and traveled toward the north until it reached the diurnal peak over the coastline region in the early morning of the following day. (3) The peak of coastal rainfall decreased with time after sunrise and seemed to be traveling southward as daytime convections developed gradually over the plain and then mountain foothill region by afternoon.

Role of CENS and CT in the DRMM over Jakarta

Based on the results presented in the last subsection, we found that the diurnal cycle of local meridional circulation along the coastline of the Jawa Sea played an essential role in the formation of DRMM over Jakarta. We next examine how the DRMM over Jakarta varies in response to the CENS with CT intrusion into or retreat from Jawa Island, which is a fundamental environment for the local meridional circulation. Figure 8 shows the averaged DRMM over Jakarta observed by

the CDR, and the data are the same as those in Fig. 6b but show the CENS active (a: 16–25 January 2010), transition (b: 26 January–04 February 2010), and CENS inactive (c: 05–14 February 2010) periods.

First, rainfall over the mountain foothill region (6.7° S–6.4° S) showed a narrow but not very developed peak at around 12 LT during the CENS active period (Fig. 8a) and there was mostly no rainfall in the late afternoon (15–21 LT), whereas that over the coastal land region around 6.3° S–6.0° S was quite active during 18–06 LT. Both northward and southward DRMMs were not well identified, and the two major rainfall areas looked separated. Second, rainfall during the CENS transition period (Fig. 8b) over the foothill region increased in its intensity with a peak around 15 LT, and then it weakened gradually but was maintained by the evening (18 LT); meanwhile, that over the coastal region showed a weak and broad extent during 19–06 LT after the arrival of the northward DRMM in the evening. Finally, rainfall over the mountain foothill region during the CENS inactive period (Fig. 8c) increased markedly in its intensity with a maximum at around 18 LT and was maintained until the middle of the night (24 LT); thus, there was mostly no rainfall in the morning (06–12 LT). Meanwhile, that over the coastal region showed only weak and scattered echoes. Both northward and southward DRMMs were well identified. These characteristics observed by the

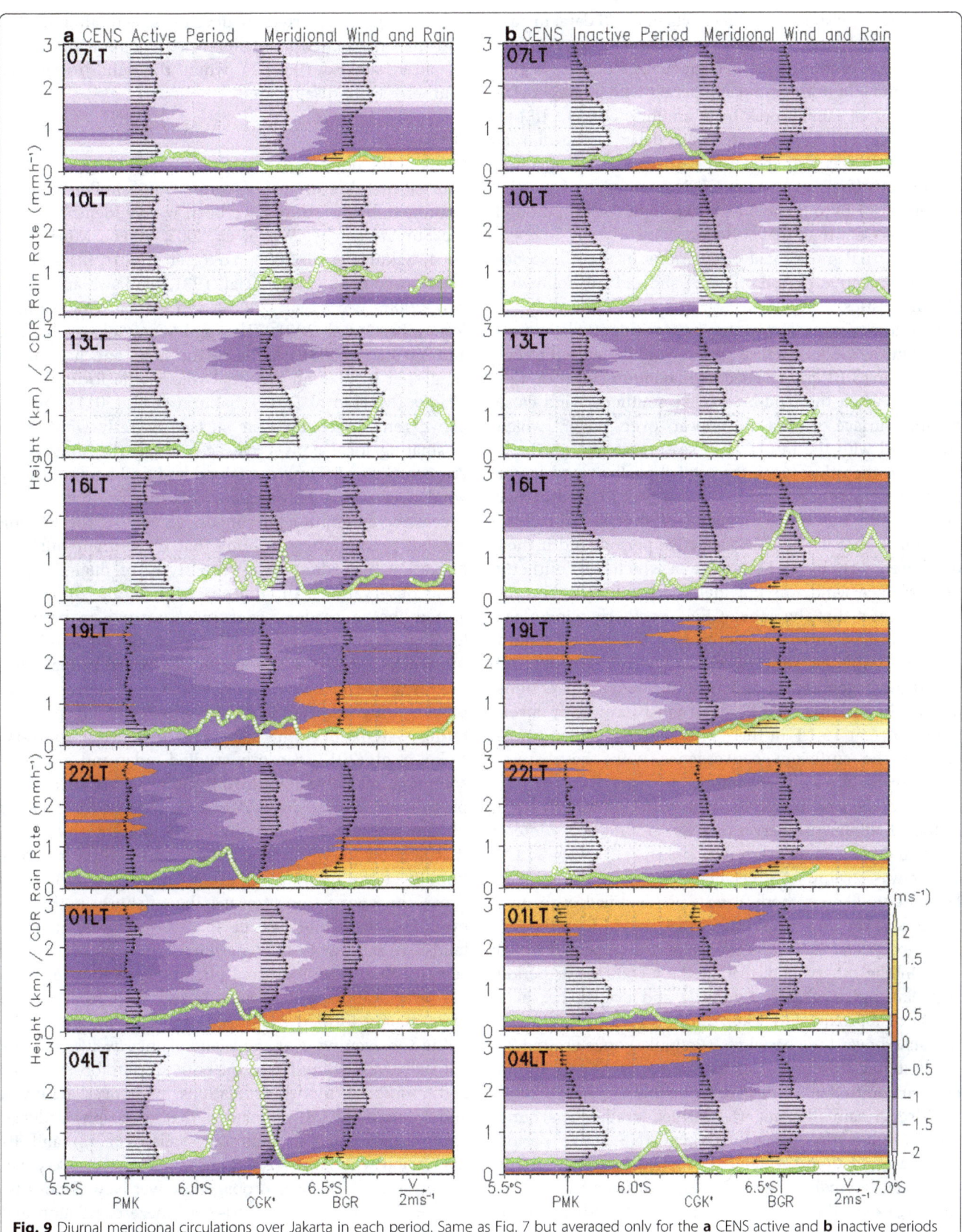

Fig. 9 Diurnal meridional circulations over Jakarta in each period. Same as Fig. 7 but averaged only for the **a** CENS active and **b** inactive periods

CDR were consistent with the results of AWS data, in particular for the results obtained at Bogor (Fig. 5d), i.e., the daily rainfall amount increased gradually from the beginning to the end of the campaign period with a phase delay of the diurnal rainfall peak from around noon (11 LT) to the evening (18 LT). Similar phase delay in the diurnal rainfall peak was also shown in both AWS data obtained at Pramuka (Fig. 5a) and the CDR observations from the evening (18 LT) to the middle of the night (24 LT), though it was less clear than that at Bogor.

The diurnal meridional circulations in the CENS active and inactive periods (Fig. 9) showed the following characteristics.

First, a rainfall peak over the mountain foothills around Bogor in the CENS inactive period (Fig. 9b) started to develop after 13 LT and reached its maximum (> 2 mm h^{-1}) at 16 LT when the land-breeze-like southerly wind close to the surface started to appear over Bogor, which converged with the intrusion of sea-breeze-like northerly wind. On the other hand, the rainfall peak around Bogor in the CENS active period (Fig. 9a) was formed earlier (13 LT) and was much smaller (~ 1 mm h^{-1}) in comparison to that in the CENS inactive period, and it then decreased early by 16 LT, a time in which the southerly near surface wind was not identified well, yet. The results suggest that the active CENS northerly wind close to the surface intruded deeply into the foothills until evening and then suppressed (or weakened) both the land-breeze generation and resulting local convections around Bogor after 13 LT, which must have been developed more at 16 LT in the CENS inactive period as result of the sufficient convergence with the distinct land-breeze-like southerly wind.

Second, the land-breeze-like near surface southerly wind in the CENS inactive period (Fig. 9b) around Bogor intensified and reached its maximum (> 3 m s^{-1}) at 19–22 LT, and it was shallow at a height of less than 500 m. The land-breeze-like southerly wind was identified even over the coastal sea (Pramuka) from 22 LT to 04 LT on the following day, and then it weakened and disappeared by 10 LT. On the other hand, that in the CENS active period (Fig. 9a) was weaker (< 2 m s^{-1}) and deeper, i.e., at a height of approximately 1 km, during 19–22 LT. Although it reached over the coastal sea (Pramuka) once around 22 LT, it started to retreat back to land sooner (by 01 LT) in comparison to that in the inactive period.

Third, a rainfall peak was formed gradually around the coastline after 16 LT in the CENS active period (Fig. 9a), and then the rainfall reached its maximum at 04 LT (> 3 mm h^{-1}) with a weak but broad in extent rain area over the coastal sea when the land-breeze-like southerly wind was retreating across the coastline after reaching its maximum extent at 22 LT. On the other hand, rainfall around the coastline in the CENS inactive

period (Fig. 9b) started to develop slowly after 01 LT, and then it reached a smaller peak (< 2 mm h^{-1}) at 10 LT after sunrise (06 LT) when the land-breeze-like southerly wind disappeared.

Discussion
Cause of the northward DRMM in the nighttime

Diurnal marches of coastal convections similar to those examined in this study have been widely identified over the Sumatera (e.g., Wu et al. 2003, 2009a; Mori et al. 2004; Sakurai et al. 2005; Yokoi et al. 2017), Kalimantan/Borneo (e.g., Houze et al. 1981; Ichikawa and Yasunari 2006; Wu et al. 2008), and Papua/New Guinea (e.g., Liberti et al. 2001; Zhou and Wang 2006; Ichikawa and Yasunari 2008) islands of the IMC and have been suggested to play an essential role in the formation of CHeRs (Mori et al. 2011). Sakurai et al. (2011) showed convergence between near surface easterly wind from dissipating local convective cells along the southwestern coastline of Sumatera Island and ambient southerly wind over the Indian Ocean, and this resulted in the successive generation of convective cells at the leading edge of precipitating systems that drove their offshore migration according to a case study of dual-Doppler radar observations. In addition, Mori et al. (2011) suggested that a seeder-feeder mechanism acts between leading-edges of westward spreading anvil clouds formed by local convections and eastward propagating large-scale MJO convections, which contributes to the offshore migration and re-development of the precipitating systems over the same region. On the contrary, Yokoi et al. (2017) recently studied radar and intensive sounding data obtained during the pre-YMC2015 campaign conducted around Bengkulu (3.86° S, 102.33° E) on the southwestern coastline of Sumatera and found that shallow gravity waves generated by initial convections over the mountain foothill region in the daytime drove the convections toward the offshore in the late evening. Meanwhile, Wu et al. (2008) showed that the land-breeze (gravity current) circulation played an essential role in the nocturnal offshore migration of rainfall over Kalimantan/Borneo Island, and such events were mainly formed by evaporation cooling after rainfall in the evening over the island plain as well as nocturnal radiative cooling based on both satellite observations and numerical model simulations. Moreover, Ichikawa and Yasunari (2008) concluded that orography induced local circulation (i.e., mountain–valley breeze) and its interaction with large-scale monsoon winds were key factors for rainfall offshore propagation over New Guinea Island in addition to both land–sea breeze circulation and gravity waves.

Based on the observational results presented above, the northward (southward) DRMM is suggested to be driven by convections formed over converged frontlines ahead of land-breeze (sea breeze)-like southerly (northerly) wind close to the surface. Characteristics of southward DRMM in this study are consistent with the results reported by Hadi et al. (2002), which showed a southward traveling sea breeze in the daytime that formed along the northern coastline of Jawa Island, as identified by line-shaped clouds based on a series of GMS visible images displaying remarkable increases of surface relative humidity. Although their data were obtained in August, which corresponds to the dry season over Jakarta, the typical onset time of sea breeze over the coastline was 10 LT in their study, and it traveled southward with a speed of approximately 10 km h^{-1} (2.8 m s^{-1}), which roughly matches with our results shown in Fig. 6b. The northward DRMM has been previously reported on observationally by Wu et al. (2007) during the CENS active phase in February and both Sulistyowati et al. (2014) and (Katsumata M, Mori S, Hamada JI, Hattori M, Syamsudin F, and Yamanaka MD (2018) Diurnal cycle over a coastal area of the maritime continent as derived by special networked soundings over Jakarta during HARIMAU2010. Unpublished) during the same campaign period of this study. In addition, Koseki et al. (2012) showed a northward DRMM in their control runs with a CT during November and March (major rainy season) with a traveling speed of approximately 4 m s^{-1} (it traveled 140 km in 9 h, see their Fig. 9a), which was also roughly consistent with our results though they did not mention it explicitly. In addition, they concluded that nocturnal rainfall over the Jawa Sea formed by the convergence between the northerly monsoon with a CT and the nocturnal land breeze originating from the southern side of Jakarta (i.e., foothill and plain regions); this rainfall amount was much reduced when the CT was suppressed because of the weakened land-breeze "front" resulting in decreased northerly monsoon flow. Observational results in this study, however, as shown in Figs. 8 and 9, are not consistent with their model-simulated conclusions. A possible reason why there was an inconsistency is that they discussed the rainfall over the Jawa Sea apart from the coastline at a distance of approximately 30–100 km, whereas that of our study was the area just over the coastline region (Figs. 7 and 9), which might involve other mechanisms as discussed briefly below.

In regard to the rainfall peaks that formed just over the coastline region as shown at 22–04 LT (04–10 LT) in Fig. 9a (Fig. 9b) when the land (sea) breeze was retreating (developing) across the coastline, specific processes may have worked to enhance them, e.g., land surface roughness decreased the northerly wind speed intruding from the Jawa Sea and this produced the coastal convergence (e.g., McPherson 1970; Pielke 1974;

Alestalo and Savijärvi 1985). In addition to this, other mechanisms that may have contributed as drivers of the DRMM over Jakarta, e.g., mountain–valley breeze circulations (e.g., Qian 2008), gravity waves (e.g., Wu et al. 2009b; Yokoi et al. 2017), self-sustaining and/or cumulus merger processes in mesoscale convective systems (e.g., Qian 2008; Sakurai et al. 2009, 2011), and the heat and moisture budgets studied by (Katsumata M, Mori S, Hamada JI, Hattori M, Syamsudin F, and Yamanaka MD (2018) Diurnal cycle over a coastal area of the maritime continent as derived by special networked soundings over Jakarta during HARIMAU2010. Unpublished) and their local transportation, should be examined in detail based on more case studies and/or by using numerical model simulations in future studies.

Large-scale northerly wind influence on the southward DRMM in the daytime

Hadi et al. (2000, 2002) showed that there are intrusions of sea breezes around 14 LT over Serpong, typically during the dry season, based on WPR observed data in 2000, and the intrusions were identified well by the abrupt increase of relative humidity and decrease of both temperature and solar radiation as well as the arrival of northerly wind below a height of 1 km. They also found, however, that northerly wind intrusions could be observed in the rainy season over Serpong, but these occurred much earlier at around 12 LT than those in the dry season. In addition, the sea-breeze intrusions occurred earlier on cloudy days even in the same dry season, though the intensity of sea-breeze circulation weakened accordingly. This seasonal variation was also identified climatologically by Araki et al. (2006) based on the same WPR observations, and the results were consistent with the time of surface maximum temperature at Serpong, i.e., 12 LT (14 LT) with a maximum temperature 30 °C (34 °C) in the rainy (dry) season. Hadi et al. (2002) suggested that the advanced arrival of northerly wind in both the rainy season and cloudy days in the dry season was not directly driven by typical sea-breeze circulations but closely related to cloud development over the plain region, i.e., inflow toward active convections that had already formed in the southern mountain foothill region, although Hadi et al. (2000) explained that the Kelvin–Helmholtz (KH) instability in the sea breeze delayed its intrusion speed during clear days.

Results of sounding and CDR analyses in this study were consistent with those in previous studies, in particular those in the rainy season. In addition, contrasts between the CENS active and inactive periods, e.g., advanced (delayed) times of less (more) rainfall peaks over the land in the CENS active (inactive) period, were similar to those between both rainy and dry seasons and cloudy and clear sky days during the dry season. Predominant northwesterly wind in the CENS active period resulted after the eastward passage of the MJO convention center (Fig. 2), which masked over the local sea-breeze circulation in the daytime, and it is suggested here that this suppresses or makes it difficult to maintain local convections over the land, which develop mainly until the evening in the CENS inactive period. The assumption is that enhancement of the northerly wind in the daytime of the CENS active period is accompanied by thermally driven local circulation that is suppressed soon after noon because of the much greater extent in cloudiness (Araki et al. 2006). This idea is also consistent with the results of Qian et al. (2010) in that weakening of northwesterly monsoon wind over the IMC amplifies the local diurnal cycle of land–sea breezes and mountain–valley winds over the western part of Jawa Island, which produces more rainfall over the mountainous region. Another possibility is that abundant rainfall over the coastal region in the night to early morning period might consume water vapor much in advance so that it would not be transported to the plain and foothill regions, which could make the atmosphere unstable in the daytime. Further studies that examine the spatiotemporal variation of moisture and local atmospheric stability characteristics during the CENS active and inactive periods will be required to settle these questions. For example, zonal variations of convective activity, which were predominant in the CENS active period with strong northwesterly synoptic winds, namely, eastward moving rain echoes mostly originating from Sumatera Island (figure not shown), might play another important role in the generation of nocturnal rainfall over the coastal area of the Jawa Sea; these provide a good contrast to the DRMMs under the remarkable local meridional circulations during the CENS inactive period.

Conclusions

This paper presented an overview of the HARI-MAU2010 campaign focusing on convective activity with the DRMM over Jakarta, Jawa Island, in the IMC based on 1-month intensive observations by a CDR and data from a multi-point sounding array that were first collected in this area during 00 UTC on 16 January to 24 UTC on 14 February 2010. The campaign period corresponded to a phase after large-scale MJO active convections passed over Jakarta (MJO inactive phase). The CENS intruded into the Jawa Sea with a CT in the beginning of the period (CENS active period: 16–25 January), and then it started to retreat (transition period: 26 January–04 February); finally, few signs of it could be detected (CENS inactive period: 05–14 February). The results of this study imply that (1) rainfall over Jakarta has the nature of DRMM during the MJO inactive phase

at least, (2) the DRMM is suggested to be driven primarily by "land-breeze"-like local meridional circulation, and (3) the meridional spatiotemporal variation of rainfall over Jakarta is controlled by activities of both the CENS and CT over the Jawa Sea, consequently.

Daily and diurnal variations of surface rainfall observed by AWSs at all sounding and CDR/WPR sites showed distinctly different characteristics, in particular for those in the meridional directions among the northern (Pramuka: coastal sea), central (Serpong: plain), and southern (Bogor: mountain foothills) sides of Jakarta. Rainfall over the coastal sea had a gentle peak in the night followed by a weak one until the next morning, whereas that over the mountain foothills had a pronounced peak in the evening and was delayed gradually from the beginning to the end of the period with increases in its intensity. The CDR confirmed those characteristics formed by the meridional marches of diurnal rainfall (DRMMs), which traveled northward (southward) in the nighttime (daytime) over Jakarta with speed of approximately 5 m s^{-1} (2 m s^{-1}). In addition, spatiotemporal variation of meridional winds over Jakarta based on the intensive soundings showed a land–sea-breeze-like circulation close to the surface, namely, the development and expansion of land-breeze-like southerly wind from the mountain foothill region after sunset (19 LT), and then an increase of its intensity and intrusion into the Jawa Sea occurred across the coastline in the nighttime (22 LT) followed by its retreat back to the land region early the next morning (04 LT). Both results suggest that nocturnal rainfall formed over the front of land-breeze-like southerly wind close to the surface and traveled from the mountain foothill region with evening showers toward the coastal region through the night coincident with the meridional advance and retreat of the front. Furthermore, the observational results suggested that the northward (southward) DRMM was driven primarily by convections formed over the converged frontline ahead of land-breeze (sea-breeze)-like southerly (northerly) wind close to the surface. Although the nature of the DRMM was fundamentally consistent with the results in a previous numerical simulation study (Koseki et al. 2012), other mechanisms that may play a contributing role as drivers of the DRMM over Jakarta, e.g., mountain–valley breeze circulations, gravity waves, self-sustaining processes in mesoscale convective systems, and coastal convergence effects, as well as heat and moisture budgets and their local transportation, should be examined in detail in future studies.

The DRMM as well as the local meridional circulation showed distinct behaviors in the CENS active and inactive periods, which can be summarized as follows. (1) Rainfall over the mountain foothill region peaked weaker (stronger) and earlier (later) around 12–13 LT (16–18 LT) in the CENS active (inactive) period. On the other hand, (2) that over the Jawa Sea peaked quite stronger (weaker) and earlier (later) around 04 LT (10 LT) in the CENS active period. (3) Land-breeze-like southerly wind close to the surface was weaker (stronger) and stayed over the Jawa Sea for a shorter (longer) time until around 04 LT (07 LT) after crossing the coastline in the CENS active (inactive) period. Consequently, these results explained the unique characteristics in daily and diurnal variations of rainfall at the mountain hill region (Bogor) well (Fig. 5d). Predominant northerly wind in the CENS active period, which masked over the local sea-breeze circulation in the daytime, may suppress or make it difficult to maintain local convections over the land, which develop mainly until the evening in the CENS inactive period. Assumptions speculated from previous studies, e.g., suppression of afternoon rain showers by greater extents of cloudiness and lesser supplies of precipitable water vapor from the coastal to foothill regions due to the consumption by abundant nocturnal rainfall in the CENS active period also, should be studied by examining the spatiotemporal variation of moisture and local atmospheric stability characteristics during the CENS active and inactive periods. This would help to settle these questions in the next phase of research. In addition, WPR and another surface observation dataset, which were not analyzed in this paper, should be utilized for case studies in the next step to focus more on the mesoscale structure of diurnally developed convections together with other data, for example, spatiotemporal variations of echo top heights obtained by the CDR (Katsumata M, Mori S, Hamada JI, Hattori M, Syamsudin F, and Yamanaka MD (2018) Diurnal cycle over a coastal area of the maritime continent as derived by special networked soundings over Jakarta during HARIMAU2010. Unpublished) and intensive atmospheric soundings during the campaign to compare with the results of previous studies based on single WPR observations. Furthermore, quantitative analyses of rainfall derived from CDR will be needed for hydrometeorological studies over Jakarta involving comparisons with ground-based rain gauge data as well as GSMaP satellite observations.

Note that the HARIMAU2010 campaign was the first to have conducted comprehensive meteorological research observations over the "greater Jakarta" region with an intensive atmospheric sounding network. The CDR used in the campaign continued its observations until the end of May 2013. After that, BMKG started operational C-band Doppler radar observations at Tangerang (6.13° S, 106.66° E), which covered the "greater Jakarta" region, and this effort continues at present.

Endnotes

[1]We experimentally applied a typical rain attenuation correction of 0.0018 $R^{1.05}$ (dB km^{-1}) for C-band (wavelength = 5 cm) radars (Doviak and Zrnić 1993) to the obtained data and found that there is no serious difference between the original and corrected data. For example in Fig. 3e, total echo coverage (convective rain area fraction) of the original and corrected data during the campaign period were 13.0% (7.7%) and 13.2% (7.5%), respectively.

[2]Although Karawang is located at a higher latitude than Serang, distances from the nearest coastline to Karawang are approximately 20 km. It is not so far from these locations to Cengkareng and Serang (approximately 5 km and 10 km, respectively), the "coastal land" stations, in comparison with the distances to Serpong and Bogor (approximately 40 km and 50 km, respectively), the "plain" and "mountain foothill" stations, respectively. In addition, Karawang is located over a large open field facing the northern coastline and there are only broad paddy fields and ponds for cultivations between Karawang and the coastline, i.e., there are no specific obstacles that would disturb meridional winds in the lower troposphere. It is also a quite different environment than Serpong (plain) and/or Bogor (mountain foothills) because there is the large downtown area of Jakarta with its numerous skyscrapers between the coastline and these stations. Consequently, we deal with Karwang as the representative location of the eastern "coastal land" in the present study.

Abbreviations

AMSL: Above mean sea level; AWS: Automatic weather station; Az: Azimuth angle; BLR: Boundary layer radar; BMKG: Agency for Meteorology, Climatology and Geophysics, Indonesia; BMRC: Bureau of Meteorology Research Centre; BPPT: Agency for the Assessment and Application of Technology, Indonesia; BSW: Blended sea winds; CDR: C-band Doppler radar; CENS: Cross-equatorial northerly surge; CGK: A three-letter code for the Soekarno–Hatta International Airport, Indonesia; CHeR: Coastal heavy rainband; CSR: Clutter-to-signal ratio; CT: Cold tongue; DEM: Digital elevation model; DRMM: Diurnal rainfall meridional march; EOS: Earth Observation System; GMS: Geostationary Meteorological Satellite; GNSS: Global navigation satellite system; GSMaP: Global satellite mapping of precipitation; GTOPO30: Global 30 arcsecond elevation; HARIMAU: Hydrometeorological Array for ISV-Monsoon Automonitoring (name of a project); IMC: Indonesian maritime continent; IRIS: Interactive radar information system; ISV: Intraseasonal variation; JABODETABEK: Jakarta, Bogor, Depok, Tangerang, and Bekasi; JAXA: Japanese Aerospace Exploration Agency; JEPP: Japanese EOS Promotion Plan (name of a project/fund); JFY: Japanese fiscal year; JICA: Japan International Cooperation Agency; JMA: Japan Meteorological Agency; JSPS: Japan Society for the Promotion of Science (name of a fund); JST: Japan Science and Technology Agency; KAKENHI: Grants-in-Aid for Scientific Research (name of a fund); KH: Kelvin–Helmholtz; LAPAN: National Institute of Aeronautics and Space; LOG: Log receiver signal-to-noise ratio; LT: Local time; MAHASRI: Monsoon Asian Hydro-atmosphere Scientific Research and Prediction Initiative (name of a project); MCCOE: Maritime Continent Center of Excellence; MJO: Madden–Julian oscillation; MRI: Meteorological Research Institute, Japan; NCEP: National Centers for Environmental Prediction, USA; NHM: Nonhydrostatic model; NOAA: National Oceanic and Atmospheric Administration, USA; NRT: Near real-time; OISST: Optimum Interpolation Sea Surface Temperature; PI: Principal investigator; PMM: Precipitation Measuring Mission (name of a project/fund); PWV: Precipitable water vapor; QuikSCAT: Quick Scatterometer; RMM: Real-time multivariate MJO; SATREPS: Science and Technology Research Partnership for Sustainable Development (name of a project /fund); SCS: South China Sea; SQI: Signal quality index; SST: Sea surface temperature; TRMM: Tropical Rainfall Measuring Mission; U.S.: United States of America; UTC: Coordinated universal time; WIB: West Indonesian standard time; WPR: Wind profiling radar; YMC: Years of the Maritime Continent (name of a project); Ze: Zenith angle

Acknowledgements
The authors express their thanks to all who were engaged in the HARIMAU2010 campaign observations, in particular, the young scientists and engineers from BPPT, BMKG, LAPAN, and Kyoto University. Thanks are also extended to Professor Jun Matsumoto of Tokyo Metropolitan University who encouraged the JEPP–HARIMAU and SATREPS–MCCOE projects as a part of Monsoon Asian Hydro-atmosphere Scientific Research and Prediction Initiative (MAHASRI) activities in Indonesia and offered valuable comments on the manuscript.

Funding
This work, including the HARIMAU2010 field campaign and analyses of the data obtained, was supported mainly by the Japanese Earth Observation System (EOS) Promotion Program (JEPP) Hydrometeorological Array for ISV–Monsoon Automonitoring (HARIMAU) project (JFY2005–2009, PI: Manabu D. Yamanaka) and the Science Technology Research Partnership for Sustainable Development (SATREPS) Maritime Continent Center of Excellence (MCCOE) project (JFY2009–2013, PI: Manabu D. Yamanaka) of the Japan Science and Technology Agency (JST)/Japan International Cooperation Agency (JICA). In addition, this work was partially supported by JSPS (Japan Society for the Promotion of Science) KAKENHI Grant Numbers JP22310115 (JFY2010–2012, PI: Shuichi Mori), JP25350515 (JFY2013–2015, PI: Shuichi Mori), JP 23340142 (JFY2011–2013, PI: Hiroyuki Hashiguchi), and JP15K01167 (JFY2015–2017, PI: Hamada Jun-Ichi), and the Japan Aerospace Exploration Agency (JAXA) Precipitation Measuring Mission (PMM) 6th and 7th Research Announcements (JFY2010–2012, JFY2013–2015, PI: Shuichi Mori).

Authors' contributions
SM proposed the topic of this study, including both the types of observations and analyses. SM and FS led the onsite observations by the Japanese and Indonesian teams, respectively. All authors contributed to the design, preparation, and execution of the HARIMAU2010 field campaign as well as discussions of the results presented in the final manuscript. MDY was the principal investigator of the JEPP–HARIMAU and SATREPS–MCCOE projects. All authors read and approved the final manuscript.

Competing interests
The authors declare that they have no competing interests.

Author details
[1]Japan Agency for Marine-Earth Science and Technology (JAMSTEC), 2-15 Natsushima-cho, Yokosuka 237-006, Japan. [2]Research Center for Climatology and Department of Geography, Tokyo Metropolitan University, 1-1 Minami-Osawa, Hachioji 192-0397, Japan. [3]Present Address: National Institute for Agro-Environmental Sciences (NIAES), National Agriculture and Food Research Organization (NARO), 3-1-3 Kan-non-dai, Tsukuba 305-8604, Japan. [4]Graduate School of Advanced Science and Technology, Kumamoto University, 2-39-1 Kurokami, Chuo-ku, Kumamoto 860-8555, Japan. [5]Research Institute for Sustainable Humanosphere (RISH), Kyoto University, Gokasho, Uji 611-0011, Japan. [6]Center for Regional Resources Development Technology (PTPSW), Agency for the Assessment and Application of Technology (BPPT), Jl. Raya Puspiptek, South Tangerang 15314, Indonesia. [7]Atmosphere and Ocean Research Institute (AORI), The University of Tokyo, 5-1-5, Kashiwanoha, Kashiwa-shi, Chiba 277-8564, Japan. [8]School of Earth Sciences, The University of Melbourne, Victoria 3010, Australia. [9]Space Science Center, National Institute of Aeronautics and Space (LAPAN), Jl. Dr. Junjunan, Bandung 40173, Indonesia. [10]Research Institute for Humanity and Nature (RIHN), Kamigamo-motoyama, Kita-ku, Kyoto 603-8047, Japan. [11]Kobe University, Rokkodai-cho, Nada-ku, Kobe 657-8501, Japan.

References
Alestalo M, Savijärvi H (1985) Mesoscale circulations in a hydrostatic model: coastal convergence and orographic lifting. Tellus 37A:156–162 https://doi.org/10.1111/j.1600-0870.1985.tb00277.x

Araki R, Yamanaka MD, Murata F, Hashiguchi H, Oku Y, Sribimawati T, Kudsy M, Renggono F (2006) Seasonal and interannual variations of diurnal cycles of wind and cloud activity observed at Serpong, west Jawa, Indonesia. J Meteor Soc Japan 84A:171–194 https://doi.org/10.2151/jmsj.84A.171

Banzon V, Smith TM, Chin TM, Liu C, Hankins W (2016) A long-term record of blended satellite and in situ sea-surface temperature for climate monitoring, modeling and environmental studies. Earth Syst Sci Data 8:165–176 https://doi.org/10.5194/essd-8-165-2016

Belgaman HA, Ichiyanagi K, Tanoue M, Suwarman R (2016) Observational research on stable isotopes in precipitation over Indonesian maritime continent. J Japanese Assoc Hydrol Sci 46:7–28 https://doi.org/10.4145/jahs.46.7

Doviak RJ, Zrnić DS (1993) Doppler radar and weather observations, 2nd edn. Academic Press, San Diego, p 562 https://doi.org/10.1016/C2009-0-22358-0

Fudeyasu H, Ichiyanagi K, Yoshimura K, Mori S, Sakurai N, Hamada JI, Yamanaka MD, Matsumoto J, Syamsudin F (2011) Effects of large-scale moisture transport and mesoscale processes on precipitation isotope ratios observed at Sumatera, Indonesia. J Meteor Soc Japan 89A:49–59 https://doi.org/10.2151/jmsj.2011-A03

Fujita M, Yoneyama K, Mori S, Nasuno T, Satoh M (2011) Diurnal convection peaks over the eastern Indian Ocean off Sumatra during different MJO phases. J Meteor Soc Japan 89A:317–330 https://doi.org/10.2151/jmsj.2011-A22

Hadi TW, Horinouchi T, Tsuda T, Hashiguchi H, Fukao S (2002) Sea-breeze circulation over Jakarta, Indonesia: a climatology based on boundary layer radar observations. Mon Wea Rev 130:2153–2166 https://doi.org/10.1175/1520-0493(2002)130<2153:SBCOJI>2.0.CO;2

Hadi TW, Tsuda T, Hashiguchi H, Fukao S (2000) Tropical sea-breeze circulation and related atmospheric phenomena observed with L-band boundary layer radar in Indonesia. J Meteor Soc Japan 78:123–140 https://doi.org/10.2151/jmsj1965.78.2_123

Hamada JI, Mori S, Yamanaka MD, Haryoko U, Lestari S, Sulistyowati R, Syamsudin F (2012) Interannual rainfall variability over northwestern Jawa and its relation to the Indian Ocean dipole and El Nino southern-oscillation events. SOLA 8:69–72 https://doi.org/10.2151/sola.2012-018

Hamada JI, Yamanaka MD, Matsumoto J, Fukao S, Winarso PA, Sribimawati T (2002) Spatial and temporal variations of the rainy season over Indonesia and their link to ENSO. J Meteor Soc Japan 80:285–310 https://doi.org/10.2151/jmsj.80.285

Hamada JI, Yamanaka MD, Mori S, Tauhid YI, Sribimawati T (2008) Differences of rainfall characteristics between coastal and mountainous areas of Sumatra, Indonesia. J Meteor Soc Japan 86:593–611 https://doi.org/10.2151/jmsj.86.593

Hashiguchi H, Fukao S, Tsuda T, Yamanaka MD, Tobing DL, Sribimawati T, Harijono SWB, Wiryosumarto H (1995c) Observations of the planetary boundary layer over equatorial Indonesia with an L-band clear-air Doppler radar: initial results. Radio Sci 30:1043–1054 https://doi.org/10.1029/95RS00653

Hashiguchi H, Tsuda T, Fukao S, Yamanaka MD, Harijono SWB, Wiryosumarto H (1995a) Boundary layer radar observations of the passage of the convection center over Serpong, Indonesia (6°S, 107°E) during the TOGA-COARE intensive observation period. J Meteor Soc Japan 73:535–548 https://doi.org/10.2151/jmsj1965.73.2B_535

Hashiguchi H, Yamanaka MD, Tsuda T, Yamamoto M, Nakamura T, Adachi T, Fukao S, Sato T, Tobing DL (1995b) Diurnal variations of the planetary boundary layer observed with an L-band clear-air Doppler radar. Boundary-Layer Meteor 74:419–424 https://doi.org/10.1007/BF00712381

Hattori M, Mori S, Matsumoto J (2011) The cross-equatorial northerly surge over the maritime continent and its relationship to precipitation patterns. J Meteor Soc Japan 89A:27–47 https://doi.org/10.2151/jmsj.2011-A02

Houze RA, Geotis SG, Marks FD, West AK (1981) Winter monsoon convection in the vicinity of North Borneo. Part I: structure and time variation of the clouds and precipitation. Mon Wea Rev 109:1595–1614 https://doi.org/10.1175/1520-0493(1981)109<1595:WMCITV>2.0.CO;2

Ichikawa H, Yasunari T (2006) Time-space characteristics of diurnal rainfall over Borneo and surrounding oceans as observed by TRMM-PR. J Clim 19:1238–1260 https://doi.org/10.1175/JCLI3714.1

Ichikawa H, Yasunari T (2008) Intraseasonal variability in diurnal rainfall over New Guinea and the surrounding oceans during austral summer. J Clim 21:2852–2868 https://doi.org/10.1175/2007JCLI1784.1

Kalnay E, Kanamitsu M, Kistler R, Collins W, Deaven D, Gandin L, Iredell M, Saha S, White G, Woollen J, Zhu Y, Leetmaa A, Reynolds R, Chelliah M, Ebisuzaki W, Higgins W, Janowiak J, Mo KC, Ropelewski C, Wang J, Jenne R, Joseph D (1996) The NCEP/NCAR 40-year reanalysis project. Bull Amer Meteor Soc 77:437–471 https://doi.org/10.1175/1520-0477(1996)077<0437:TNYRP>2.0.CO;2

Kamimera H, Mori S, Yamanaka MD, Syamsudin F (2012) Modulation of diurnal rainfall cycle by the Madden–Julian oscillation based on one-year continuous observations with a meteorological radar in West Sumatera. SOLA 8:111–114 https://doi.org/10.2151/sola.2012-028

Kawashima M, Fujiyoshi Y, Ohi M, Mori S, Sakurai N, Abe Y, Harjupa W, Syamsudin F, Yamanaka MD (2011) Case study of an intense wind event associated with a mesoscale convective system in west Sumatera during the HARIMAU2006 campaign. J Meteor Soc Japan 89A:239–257 https://doi.org/10.2151/jmsj.2011-A15

Koseki S, Koh TY, Teo CK (2012) Effects of the cold tongue in the South China Sea on the monsoon, diurnal cycle and rainfall in the maritime continent. Q J R Meteorol Soc 139(675):1566–1582 https://doi.org/10.1002/qj.2052

Liberti GL, Chéruy F, Desbois M (2001) Land effect on the diurnal cycle of clouds over the TOGA COARE area, as observed from GMS IR data. Mon Wea Rev 129:1500–1517 https://doi.org/10.1175/1520-0493(2001)129<1500:LEOTDC>2.0.CO;2

Madden RA, Julian PR (1972) Description of global-scale circulation cells in the tropics with a 40–50 day period. J Atmos Sci 29:1109–1123 https://doi.org/10.1175/1520-0469(1972)029<1109:DOGSCC>2.0.CO;2

McPherson RD (1970) A numerical study of the effect of a coastal irregularity on the sea breeze. J Appl Meteorol 9:767–777 https://doi.org/10.1175/1520-0450(1970)009<0767:ANSOTE>2.0.CO;2

Mori S, Hamada JI, Sakurai N, Fudeyasu H, Kawashima M, Hashiguchi H, Syamsudin F, Arbain AA, Sulistyowati R, Matsumoto J, Yamanaka MD (2011) Convective systems developed along the coastline of Sumatera Island, Indonesia, observed with an X-band Doppler radar during the HARIMAU2006 campaign. J Meteor Soc Japan 89A:61–81 https://doi.org/10.2151/jmsj.2011-A04

Mori S, Hamada JI, Tauhid YI, Yamanaka MD, Okamoto N, Murata F, Sakurai N, Hashiguchi H, Sribimawati T (2004) Diurnal land-sea rainfall peak migration over Sumatera Island, Indonesian maritime continent, observed by TRMM satellite and intensive rawinsondes soundings. Mon Wea Rev 132:2021–2039 https://doi.org/10.1175/1520-0493(2004)132<2021:DLRPMO>2.0.CO;2

Mori S, Hamada JI, Yamanaka MD, Kodama YM, Kawashima M, Shimomai T, Shibagaki Y, Hashiguchi H, Sribimawati T (2006) Vertical wind characteristics in precipitating cloud systems over west Sumatera, Indonesia, observed with equatorial atmosphere radar: case study of 23-24 April 2004 during the first CPEA campaign period. J Meteor Soc Japan 84A:113–131 https://doi.org/10.2151/jmsj.84A.113

Ogino SY, Yamanaka MD, Mori S, Matsumoto J (2016) How much is the precipitation amount over the tropical coastal region? J Clim 29:1231–1236 https://doi.org/10.1175/JCLI-D-15-0484.1

Ogino SY, Yamanaka MD, Mori S, Matsumoto J (2017) Tropical coastal dehydrator in global atmospheric water circulation. Geophys Res Lett 44:11636–11643 https://doi.org/10.1002/2017GL075760

Oigawa M, Matsuda T, Tsuda T, Noersomadi (2017) Coordinated observation and numerical study on a diurnal cycle of tropical convection over a complex topography in West Java, Indonesia. J Meteor Soc Japan 95:261–281 https://doi.org/10.2151/jmsj.2017-015

Okamoto K, Iguchi T, Takahashi N, Iwanami K, Ushio T (2005) The global satellite mapping of precipitation (GSMaP) project, 25th IGARSS Proceedings, pp 3414–3416

Pielke RA (1974) A three-dimensional numerical model of the sea breezes over South Florida. Mon Wea Rev 102:115–139 https://doi.org/10.1175/1520-0493(1974)102<0115:ATDNMO>2.0.CO;2

Qian J (2008) Why precipitation is mostly concentrated over islands in the maritime continent. J Atmos Sci 65:1428–1441 https://doi.org/10.1175/2007JAS2422.1

Qian J, Robertson AW, Moron V (2010) Interactions among ENSO, the monsoon, and diurnal cycle in rainfall variability over Java, Indonesia. J Atmos Sci 67:3509–3524 https://doi.org/10.1175/2010JAS3348.1

Ramage CS (1968) Role of a tropical "maritime continent" in the atmospheric circulation. Mon Wea Rev 96:365–370 https://doi.org/10.1175/1520-0493(1968)096<0365:ROATMC>2.0.CO;2

Realini E, Sato K, Tsuda T, Susilo, Manik T (2014) An observation campaign of precipitable water vapor with multiple GPS receivers in western Java, Indonesia. Prog Earth Planet Sci 1:17 https://doi.org/10.1186/2197-4284-1-17

Renggono F, Hashiguchi H, Fukao S, Yamanaka MD, Ogino SY, Okamoto N, Murata F, Harijono SWB, Kudsy M, Kartasasmita M, Ibrahim G (2001) Precipitating clouds observed by 1.3-GHz L-band boundary layer radars in equatorial Indonesia. Ann Geophys 19:889–897 https://doi.org/10.5194/angeo-19-889-2001

Saito K, Fujita T, Yamada Y, Ishida J, Kumagai Y, Aranami K, Ohmori S, Nagasawa R, Kumagai S, Muroi C, Kato T, Eito H, Yamazaki Y (2006) The operational JMA nonhydrostatic mesoscale model. Mon Wea Rev 134:1266–1298 https://doi.org/10.1175/MWR3120.1

Sakurai N, Kawashima M, Fujiyoshi Y, Hashiguchi H, Shimomai T, Mori S, Hamada JI, Murata F, Yamanaka MD, Tauhid YI, Sribimawati T, Suhardi B (2009) Internal structure of migratory cloud system with diurnal cycle over Sumatera Island during CPEA-I campaign. J Meteor Soc Japan 87:157–170 https://doi.org/10.2151/jmsj.87.157

Sakurai N, Mori S, Kawashima M, Fujiyoshi Y, Hamada JI, Shimizu S, Fudeyasu H, Tabata Y, Harjupa W, Hashiguchi H, Yamanaka MD, Matsumoto J, Emrizal, Syamsudin F (2011) Migration process and 3D wind field of precipitation systems associated with a diurnal cycle in west Sumatera: dual Doppler radar analyses during the HARIMAU2006 campaign. J Meteor Soc Japan 89:341–361 https://doi.org/10.2151/jmsj.2011-404

Sakurai N, Murata F, Yamanaka MD, Mori S, Hamada JI, Hashiguchi H, Tauhid YI, Sribimawati T, Suhardi B (2005) Diurnal cycle of cloud system migration over Sumatera Island. J Meteor Soc Japan 83:835–850 https://doi.org/10.2151/jmsj.83.835

Schumacher C, Houze RA Jr (2003) Stratiform rain in the tropics as seen by the TRMM precipitation radar. J Clim 16:1739–1756 https://doi.org/10.1175/1520-0442(2003)016<1739:SRITTA>2.0.CO;2

Shibagaki Y, Kozu T, Shimomai T, Mori S, Murata F, Fujiyoshi Y, Hashiguchi H, Fukao S (2006a) Evolution of a super cloud cluster and the associated wind fields observed over the Indonesian maritime continent during the first CPEA campaign. J Meteor Soc Japan 84A:19–31 https://doi.org/10.2151/jmsj.84A.19

Shibagaki Y, Shimomai T, Kozu T, Mori S, Fujiyoshi Y, Hashiguchi H, Yamamoto MK, Fukao S, Yamanaka MD (2006b) Multi-scale convective systems associated with an intraseasonal oscillation over the Indonesian maritime continent. Mon Wea Rev 134:1682–1696 https://doi.org/10.1175/MWR3152.1

Steiner M, Houze RA Jr, Yuter SE (1995) Climatological characterization of three-dimensional storm structure from operational radar and rain gauge data. J Appl Meteorol 34:1978–2007 https://doi.org/10.1175/1520-0450(1995)034<1978:CCOTDS>2.0.CO;2

Sulistyowati R, Hapari RI, Syamsudin F, Mori S, Oishi ST, Yamanaka MD (2014) Rainfall-driven diurnal variations of water level in Ciliwung River, West Jawa, Indonesia. SOLA 10:141–144 https://doi.org/10.2151/sola.2014-029

Trilaksono NJ, Otsuka S, Yoden S (2012) A time-lagged ensemble simulation on the modulation of precipitation over west Java in January–February 2007. Mon Wea Rev 140:601–616 https://doi.org/10.1175/MWR-D-11-00094.1

Trilaksono NJ, Otsuka S, Yoden S, Saito K, Hayashi S (2011) Dependence of model-simulated heavy rainfall on the horizontal resolution during the Jakarta flood event in January-February 2007. SOLA 7:193–196 https://doi.org/10.2151/sola.2011-049

U.S. Geological Survey (1993) Digital Elevation Models, Data User Guide 5, Reston, 53 pp http://pubs.er.usgs.gov/publication/70038376

Vaisala (2008) RVP8 Digital IF Receiver/Doppler Signal Processor User's Manual. Vaisala Inc, p 437 ftp://ftp.sigmet.com/outgoing/releases/8.12.1/RHEL4/iris/man.tgz

van Bemmelen W (1922) Land- und Seebrise in Batavia. Beitr Phys Frei Atmos 10:169–177 (in German)

Wheeler MC, Hendon HH (2004) An all-season real-time multivariate MJO index: development of an index for monitoring and prediction. Mon Wea Rev 132:1917–1932 https://doi.org/10.1175/1520-0493(2004)132<1917:AARMMI>2.0.CO;2

Wu PM, Arbain AA, Mori S, Hamada JI, Hattori M, Syamsudin F, Yamanaka MD (2013) The effects of an active phase of the Madden-Julian oscillation on the extreme precipitation event over western Jawa Island in January 2013. SOLA 9:79–83 https://doi.org/10.2151/sola.2013-018

Wu PM, Hamada JI, Mori S, Tauhid YI, Yamanaka MD, Kimura F (2003) Diurnal variation of precipitable water over a mountainous area of Sumatra Island. J Appl Meteorol 42:1107–1115 https://doi.org/10.1175/1520-0450(2003)042<1107:DVOPWO>2.0.CO;2

Wu PM, Hamada JI, Yamanaka MD, Matsumoto J, Hara M (2009b) The impact of orographically-induced gravity wave on the diurnal cycle of rainfall over Southeast Kalimantan Island. Atmos Ocean Sci Lett 2:35–39 https://doi.org/10.1080/16742834.2009.11446773

Wu PM, Hara M, Fudeyasu H, Yamanaka MD, Matsumoto J, Syamsudin F, Sulistyowati R, Djajadihardja YS (2007) The impact of trans-equatorial monsoon flow on the formation of repeated torrential rains over Java Island. SOLA 3:93–96 https://doi.org/10.2151/sola.2007-024

Wu PM, Hara M, Hamada JI, Yamanaka MD, Kimura F (2009a) Why a large amount of rain falls over the sea in the vicinity of western Sumatra Island during nighttime. J Appl Meteor Climatol 48:1345–1361 https://doi.org/10.1175/2009JAMC2052.1

Wu PM, Yamanaka MD, Matsumoto J (2008) The formation of nocturnal rainfall offshore from convection over western Kalimantan (Borneo) Island. J Meteor Soc Japan 86A:187–203 https://doi.org/10.2151/jmsj.86A.187

Yamanaka MD (2016) Physical climatology of Indonesian maritime continent: an outline to comprehend observational studies. Atmos Res 178–179:231–259 https://doi.org/10.1016/j.atmosres.2016.03.017

Yamanaka MD, Mori S, Wu PM, Hamada JI, Sakurai N, Hashiguchi H, Yamamoto MK, Shibagaki Y, Kawashima M, Fujiyoshi Y, Shimomai T, Manik T, Erlansyah SW, Tejasukmana B, Syamsudin F, Djajadihardia YS, Anggadiredja JT (2008) HARIMAU radar-profiler network over Indonesian maritime continent: a GEOSS early achievement for hydrological cycle and disaster prevention. J Disaster Res 3:78–88 https://doi.org/10.20965/jdr.2008.p0078

Yamanaka MD, Ogino SY, Wu PM, Hamada JI, Mori S, Matsumoto J, Syamsudin F (2018) Maritime continent coastlines controlling Earth's climate. Prog Earth Planet Sci 5:21 https://doi.org/10.1186/s40645-018-0174-9

Yokoi S, Mori S, Katsumata M, Geng B, Yasunaga K, Syamsudin F, Nurhayati, Yoneyama K (2017) Diurnal cycle of precipitation observed in the western coastal area of Sumatra Island: offshore preconditioning by gravity waves. Mon Wea Rev 145:3745–3761 https://doi.org/10.1175/MWR-D-16-0468.1

Zhang HM, Bates JJ, Reynolds RW (2006) Assessment of composite global sampling: sea surface wind speed. Geophys Res Lett 33:L17714 https://doi.org/10.1029/2006GL027086

Zhou L, Wang Y (2006) Tropical rainfall measuring mission observation and regional model study of precipitation diurnal cycle in the New Guinean region. J Geophys Res 111:D17104 https://doi.org/10.1029/2006JD007243

10

Overview of the development of the Aerosol Loading Interface for Cloud microphysics in Simulation (ALICIS)

Takamichi Iguchi[1,2]*, In-Jin Choi[3], Yousuke Sato[4], Kentaroh Suzuki[5] and Teruyuki Nakajima[6]

Abstract

This review summarizes the scientific background of and past/prospective update strategies for the development of the Aerosol Loading Interface for Cloud microphysics In Simulation (ALICIS). ALICIS provides a novel approach for coupling downscaled mesoscale cloud-resolving simulations to large-scale aerosol-transport simulations. Realistic aerosol loading, including spatio-temporal aerosol variations and particle-size spectra, is implemented in the cloud-resolving simulations. Prior studies employing ALICIS have demonstrated how the interface introduction significantly improves the reproducibility of the simulated microphysical cloud structure through better representation of aerosol effects on cloud.

Keywords: Cloud microphysics, Cloud and aerosol, Regional modeling, Dynamical downscaling

Review

Introduction

Aerosol effects on cloud in atmospheric models

The coexistence of the three phases of hydrogen monoxide is, within the solar system, unique to the Earth's atmosphere. The liquid and solid phases in the atmosphere (i.e., cloud and precipitation) play critical roles in determining the Earth's radiation balance and hydrological cycle by driving considerable changes in the atmospheric hydrometeor distribution and characteristics. The hydrometeor behavior is controlled by various factors, not limited to phase changes affected by temperature, dry-air and vapor pressure, advection by background winds, and gravitational falling.

Particulate matter that is dispersed colloidally with a variety of the particle sizes, shapes, and chemical components in the Earth's atmosphere is referred to as aerosol. Aerosol particles are closely related to the liquid and solid particles making up clouds and precipitation; they are not only necessary for the nucleation process of hydrometeor particles but are also a source of impurities in these particles. To date, the aerosol influence on cloud has been investigated in a number of ways, from the perspectives of weather and climate research, as well as in the context of weather modification and geo-engineering, starting from a series of studies by Twomey and Squires who examined microphysical cloud structure under various airmass conditions using in situ aircraft measurements in the 1950s and 1960s (e.g., Squires 1956). A well-known example demonstrating the effects of aerosol is that of the systematic difference in microphysical properties between continental and oceanic clouds. Continental clouds generally have a smaller mean particle size and a higher particle number concentration than oceanic clouds. These differences are attributed to the disparity in aerosol concentration and characteristics between the regions.

Aerosol was not represented explicitly in most numerical models of the Earth's atmosphere on global or regional scales until the 1990s, despite its significance for clouds and climate. This was due in part to the fact the models were generally aimed at analyses of the atmosphere's dynamical structure and at land-surface rainfall predictions. Researchers thus considered aerosol unimportant in the basic framework of atmospheric models, because it had no direct influence on mass or energy conservation in the atmosphere and its fluid dynamics. In addition, computational resources were insufficient to explicitly include

* Correspondence: takamichi.iguchi@nasa.gov
[1]Earth System Science Interdisciplinary Center, University of Maryland, College Park, MD, USA
[2]Code 612 NASA Goddard Space Flight Center, Greenbelt, MD, USA
Full list of author information is available at the end of the article

aerosol. Aerosol effects on cloud were thus tacitly included in the models in the form of parametrizations consisting of empirical equations. For example, saturation adjustment (Soong and Ogura 1973; Tao et al. 1989) is a reasonable assumption for calculations of the nucleation of cloud droplets and their condensation growth from water vapor. Even though the presence of aerosol is a requirement for the droplets' nucleation process, the details are not important because of the wealth of aerosol particles that can be activated as cloud condensation nuclei (CCN) in the troposphere.

In the 1990s and 2000s, increasing interest in aerosol brought the issue under consideration even in the context of atmospheric modeling approaches. The most significant turning point was the recognition that aerosol might counter the anthropogenic gases that cause global warming through their greenhouse effect. Tropospheric aerosol particles have a direct impact on the Earth's radiative balance through the scattering and absorption of short- and longwave radiation (i.e., aerosol direct radiative forcing). In addition, aerosol particles have an indirect influence on the Earth's radiative balance through altering optical cloud properties by serving as CCN (i.e., aerosol indirect radiative forcing). An increase in the concentration of tropospheric aerosol particles may enhance cloud reflectance for solar irradiance as a result of the increased droplet number concentration and reduced droplet size for a fixed liquid water path (Twomey 1974, 1977; Twomey et al. 1984). On the other hand, a decrease in droplet size may inhibit droplet growth through collisional coagulation during warm-rain formation; this effect has an influence on the Earth's cloud coverage through altering cloud lifetimes owing to the suppression of rain formation (Albrecht 1989).

Including the aerosol influence proved necessary in assessments of the Earth's radiation balance and, consequently, the surface air temperature, not only in the future but also in the past. Simultaneously, researchers have become gradually more interested in changes in the precipitation amount through modification of cloud dynamical/microphysical structures by anthropogenic factors related to aerosol (Levin and Cotton 2008; Tao et al. 2012). Many numerical models have been developed competitively to represent aerosol–cloud interactions and their impacts on climate and weather phenomena on global and regional scales. In addition, remote and in situ measurements have been conducted worldwide to provide chances for modelers to justify their results. Satellite measurements have produced global projections of the distribution and characteristics of aerosol and clouds on the basis of measurements of their optical properties through spaceborne remote sensors (e.g., Nakajima et al. 2001; Rosenfeld et al. 2014). Continuous observations of aerosol by ground-based networks allow researchers to trace long-term transitions in their characteristics (e.g., Aerosol Robotic Network (AERONET) Holben et al. 1998; US Department of Energy (DOE), atmospheric radiation measurement (ARM) Stokes and Schwartz 1994). Many intensive field campaigns have also been conducted to produce comprehensive views of aerosol–cloud interaction by coupling observation results obtained from spaceborne satellites with ground-based, seaborne, and airborne instruments, as well as model simulations (e.g., INDian Ocean Experiment (INDOEX) Ramanathan et al. 2001; Asian Pacific Regional Aerosol Characterization Experiment (ACE–Asia) Huebert et al. 2003; Atmospheric Brown Clouds (ABC) Nakajima et al. 2007; Ramanathan et al. 2007; and Distributed Regional Aerosol Gridded Observation Networks (DRAGON) Eck et al. 2012).

Modeling approaches to aerosol and its activation in droplet nucleation

An essential physical process connecting aerosol with cloud is the nucleation of hydrometeor particles. All aerosol particles that can potentially be activated as nuclei of cloud droplets are referred to as condensation nuclei (CN). CCN are a subset of CN; they can be activated to initiate droplet formation at the low supersaturation observed in the troposphere. In contrast, ice-forming nuclei (IN) are defined as aerosol particles that act as the base of the deposition nucleation of cloud–ice particles or the heterogeneous freezing of supercooled droplets. At present, our knowledge of IN is less complete than that of CCN because of insufficient availability of measurements and experimental data to model the complicated ice-nucleation process (Pruppacher and Klett 1997).

A fundamental theory about CCN activation is summarized in a chart relating the relationship between the vapor saturation ratio and the equilibrium droplet size formed in CCN (Köhler 1936, Appendix 1). CN size distribution spectra, as well as ambient supersaturation and temperature, are essential factors in determining the droplet nucleation process. Several additional factors have a potential impact on the balance of the vapor saturation ratio and the equilibrium droplet size under more complex assumptions, other than the basic Köhler theory, such as changes in the amount of dissolving solutes in a droplet, temperature variations through latent heating/cooling by vapor condensation and evaporation, and the existence of insoluble and multiple soluble chemical components in impacted CN.

The question of how to numerically handle the two critical components (i.e., the size distribution spectra of aerosol particles and ambient supersaturation) is an important issue in the representation of the droplet nucleation process in cloud microphysics schemes. Existing approaches are generally categorized into the following four types, according to the temporal and spatial resolutions, and the aerosol representation in the models: (i) the use of supersaturation and aerosol size distribution spectra explicitly predicted for each spatial grid point; (ii) the use of parametrization involving predicted grid-scale supersaturation or updraft wind velocity, as well as the mass and/or number concentration of aerosol particles; (iii) the use of parametrization including the diagnosed maximum supersaturation or updraft wind velocity instead of predicted grid-scale values, as well as the mass and/or number concentration of aerosol particles; and (iv) the use of parametrization that implicitly assumes the presence of aerosol.

The first approach has been employed in a limited number of cloud-resolving models (CRMs) that aim at accurate calculation of cloud microphysics (e.g., Kogan 1991; Chen and Lamb 1994; Feingold et al. 1999; Khain et al. 2000; Geresdi and Rasmussen 2005; Lynn et al. 2005a; Kuba and Fujiyoshi 2006; Suzuki et al. 2006, 2010; Li et al. 2008; Shima et al. 2009; Misumi et al. 2010; Xue et al. 2010; Fan et al. 2012). The so-called sectional representation (e.g., Abdul-Razzak and Ghan 2002) is employed to numerically handle aerosol size distribution spectra in most of these models. Köhler theory is used to directly calculate the CCN number concentration and, consequently, the increased cloud droplet number concentration (CDNC) from the predicted aerosol size distribution spectra and supersaturation. Consumption of aerosol particles through the nucleation process is explicitly represented in the size distribution spectrum predicted. Miscellaneous aerosol types are often simplified to monotype CN to reduce the complexity. These CRMs sometimes lack a grid structure of full three-dimensional Cartesian coordinates, i.e., they may have only a two- or one-dimensional grid structure. Large computational resources are required to ensure sufficient accuracy to explicitly resolve aerosol size distribution spectra. In addition, very fine grid spacings in both time and space are required to represent minute variations in supersaturation. Such CRMs are generally employed to simulate idealized and simplified clouds to investigate the basic mechanisms of cloud and precipitation formation, and to discuss the sensitivity to various factors, including aerosol concentration. Typical size distribution forms of aerosol particles are often prescribed using reference measurement data.

The second approach is employed in high-resolution regional models that can be used for aerosol–cloud interaction studies based on idealized experiments or case studies. Generalized aerosol size distribution functions are usually assumed instead of their explicitly resolved spectra. The total or species-distinguished aerosol concentration is explicitly predicted in bulk-moment forms in some models (e.g., Hashino and Tripori 2007; Muhlbauer and Lohmann 2009; Onishi and Takahashi

2012; Saleeby and van den Heever 2013; Thompson and Eidhammer 2014), whereas the aerosol concentration is not predicted but tacitly parametrized in some models (e.g., Rasmussen et al. 2002; Seifert and Beheng 2006; Morrison and Grabowski 2007; Seiki and Nakajima 2014; Seiki et al. 2015). In most of these models, the activated CCN number concentration is parametrized on the basis of the basic Köhler theory, using the predicted supersaturation or vertical wind velocity. An empirical parametrization was suggested by Twomey (1959a, 1959b) to calculate the activated CCN number concentration from the predicted supersaturation or vertical wind velocity and prescribed coefficients representing the ambient aerosol loading. A more generalized parametrization was proposed by Abdul-Razzak et al. (1998), which combines Twomey's parametrization with a lognormal representation of the aerosol size distribution (Ghan et al. 1993); these authors also extended the parameterization to cases including multiple externally mixed aerosol types (Abdul-Razzak and Ghan 2000). On the other hand, Saleeby and Cotton (2004) proposed the use of pre-calculated lookup tables through Lagrangian parcel model calculations instead of empirical parametrizations; aerosol size distribution spectra are explicitly represented using a sectional approach in the parcel model. The lookup tables are used to calculate the proportion of aerosol particles activated for droplet nucleation by referring to the vertical wind velocity predicted in their main model framework.

The third approach is used in most global or coarse-resolution regional models for studies of aerosol–cloud interactions, even on a climatic scale (e.g., Storelvmo et al. 2006, 2008; Lohmann et al. 2007; Morrison and Gettelman 2008; Suzuki et al. 2008; Takemura et al. 2009; Wang et al. 2011; Wang et al. 2013; Park et al. 2014; Shi et al. 2014). Supersaturation and the vertical wind velocity cannot be resolved well on model grid points with coarse intervals. Saturation adjustment approaches are necessarily employed in the calculation of phase changes between vapor and cloud water/ice. The approach used to introduce aerosol-to-cloud effects is significantly different among models according to the type of bulk cloud microphysics employed. Models using double-moment bulk cloud microphysics generally employ empirical parametrizations (e.g., Abdul-Razzak et al. 1998; Abdul-Razzak and Ghan 2000) to obtain the relationship between the aerosol number concentration and CDNC. Since the grid-scale vertical velocity is not a suitable input into the parametrizations, diagnostic variables (e.g., turbulent kinetic energy) are often included in modified parametrizations to determine the activated CCN number concentration through diagnosis of the maxima of the vertical wind velocity and supersaturation (e.g., Morrison et al. 2005). In contrast, single-moment bulk cloud microphysics cannot directly incorporate the activated CCN number concentration into the microphysics calculations in the form of a CDNC change. Diagnosed CDNC can be included in the calculation of the autoconversion rate from cloud to rain to represent the ambient CN effect on clouds (e.g., Liu and Daum 2004).

The fourth approach is adopted in model simulations on any scale. Since aerosol particles are necessary for the nucleation of cloud particles, the presence of aerosol particles is always assumed in cloud microphysics in a steady form. Explicit representation of aerosol and its effect on cloud may cause excess computational costs and uncertainties in regular operational runs for short-time weather forecasting. Speedy and stable simulation is important in these runs. Even coarse parametrizations based on empirical formulas work reasonably well for such short-time simulations where aerosol loading has little direct influence and minor feedback on the prognostic variables.

Solutions to the problem related to aerosol-to-cloud modeling

The best approach to accurately simulating aerosol-to-cloud effects is obviously the first option discussed in "Modeling approaches to aerosol and its activation in droplet nucleation", i.e., using the explicitly predicted supersaturation and aerosol size distribution. However, predictions of the two components under realistic conditions are contradictory problems in a single-model framework because of scale gaps between aerosol and cloud. A small computational domain is preferable to define the grid spacing as finely as possible to represent minor variations in supersaturation. In contrast, a large computational domain is preferred to strictly predict transitions in the aerosol concentration and size distribution from sources of aerosol-particle formation. The limited capability of present computational resources places results in insufficient domain size and grid spacing in a single-model framework.

Several methods can be employed to solve the scale dilemma in aerosol-to-cloud modeling in the present computing environment. Thompson and Eidhammer (2014) used monthly aerosol climatology data as input into their Weather Research and Forecasting (WRF) model simulations (Skamarock et al. 2005). The aerosol climatology data were derived from multiannual global aerosol model simulations using the NASA Goddard Earth Observing System model coupled with the Goddard Chemistry Aerosol Radiation and Transport model (GEOS–GOCART, Ginoux et al. 2001; Chin et al. 2002; Colarco et al. 2010). The climatological data were composed of mass-mixing ratios of the five aerosol species with $0.5° × 1.25°$ horizontal grid spacing. The mass-mixing ratios were converted into two bulk number concentrations. One is the number

concentration of simplified hygroscopic aerosol particles based on combining all aerosol species except for dust and black carbon. The second is that of simplified ice-nucleating particles based on the accumulation of dust particles with diameters larger than 0.5 μm. Both number concentrations were explicitly predicted in the WRF model simulations as tracers affected by interactions with hydrometeors and emission from the land surface. The predicted aerosol number concentrations were used in calculating the activation process of aerosol particles in the bulk cloud microphysics scheme. Aerosol activation (i.e., cloud droplet nucleation) was calculated using a method employing pre-calculated lookup tables (e.g., Saleeby and Cotton 2004). Cloud-ice nucleation and heterogeneous freezing of supercooled droplets were calculated using the parametrizations of Bigg (1953), Koop et al. (2000), Phillips et al. (2008), and DeMott et al. (2010).

Shi et al. (2014) used aerosol and chemical-transport model simulations as a tool to provide a realistically diagnosed aerosol field for their cloud microphysics calculation in the main model framework with the same domain configuration. They incorporated the following two individual models off-line: the GOCART solo-module version of the WRF coupled with chemistry (WRF–Chem; Grell et al. 2005), and the National Aeronautics and Space Administration (NASA) Unified WRF (NU–WRF; Peters-Lidard et al. 2015). The activated CCN number concentration in the NU–WRF simulation was diagnosed based on the air temperature and supersaturation on the NU–WRF side, and the mass concentration of multiple aerosol species on the WRF–Chem GOCART side. The diagnosed CCN number concentration was used to modulate the autoconversion rate from cloud to rain through CDNC diagnosis in the framework of the single-moment bulk cloud microphysics in the NU–WRF simulation.

Another compromise adopted to address this problem was suggested by Iguchi et al. (2008; hereafter IG08). Their approach was based on dynamical downscaling of realistic aerosol fields from large-scale aerosol (and chemical) transport model simulations to small-scale CRM simulations. The basic numerical framework of aerosol downscaling was similar to that of dynamical downscaling for standard prognostic variables such as wind velocity, temperature, and vapor-mixing ratio. Initial and time-variant aerosol boundary conditions were prepared using four-dimensional output from the large-scale simulation. However, aerosol size distribution spectra were incompletely resolved in the large-scale aerosol transport model. This deficiency might be problematic for accurate calculation of the activated CCN number concentration on the basis of Köhler theory on the CRM side. To deal with this problem in a reasonable manner, the authors proposed the use of an additional intermediate module. The intermediate aerosol loading module complemented the information of the aerosol size distribution spectra in the dynamical downscaling framework. The aerosol size distribution spectra and their total number/mass concentrations should be as realistic as possible, such as those observed at times and locations pertaining to the simulations.

The objective of this review is to summarize our development activities and prospective update strategies for the aerosol loading interface since the publication of IG08. The interface connects small-scale CRM simulations to large-scale aerosol transport simulations in a dynamical downscaling framework and complements information about aerosol size distribution spectra. This interface is currently undergoing development in order to couple it with regional models, including a detailed cloud microphysical scheme, referred to as spectral-bin microphysics (SBM; e.g., Khain et al. 2015). Since this paper aims at reviewing the aerosol loading interface in detail, detailed descriptions of SBM are not included. Such descriptions can be found in Khain et al. (2000, 2015), among others.

Description of ALICIS
Brief overview
Our aerosol loading module is referred to as the Aerosol Loading Interface for Cloud microphysics In Simulation (ALICIS). Figure 1 provides a schematic illustration of how ALICIS works as an intermediate module between large-scale aerosol transport models and regional-scale CRMs. So far, this interface has been employed in the framework connecting global-scale simulations using the spectral radiation transport model for aerosol species (SPRINTARS; e.g.,

Takemura et al. 2000) with regional-scale simulations using the Japan Meteorological Agency Non-Hydrostatic Model (JMA–NHM; Saito et al. 2006) coupled with SBM (IG08). The SBM scheme originates from the corresponding part of the Hebrew University Cloud Model (HUCM; e.g., Khain et al. 2000). ALICIS was first published by IG08 and subsequently updated by Sato et al. (2012; hereafter SA12) and Choi et al. (2014; hereafter CH14).

CN modeling in the original HUCM
The representation of aerosol in the original HUCM is significantly simplified for idealized CRM simulations (e.g., Khain et al. 1999). The CN size distribution spectrum for a single chemical component is calculated as prognostic variables for each of the model's spatial grid points. The chemical composition of CN is assumed to consist of pure ammonium sulfate. The CN particle shape is assumed to be completely spherical. The size distribution spectrum is approximated by 33 discrete size bins, covering a particle radius ranging from 1.23×10^{-3} to 2 μm. The representative particle radius of each CN size bin is defined as follows. The CN particle mass of the maximum size bin (i.e., the 33rd bin) is assumed to be identical to the particle mass of the smallest droplet size bin; the smallest droplet radius is assumed to be 2 μm. The CN particle mass of each size bin is determined by doubling the mass bins; that is, a series of CN bins are characterized in the form of a geometric progression of particle masses with a ratio of 2 between subsequent bins. The CN particle radius of each size bin is tied to the particle mass of the bin and the CN particle density.

The initial CN size distribution spectrum in the HUCM is determined using the Köhler equation (Eq. 11) and

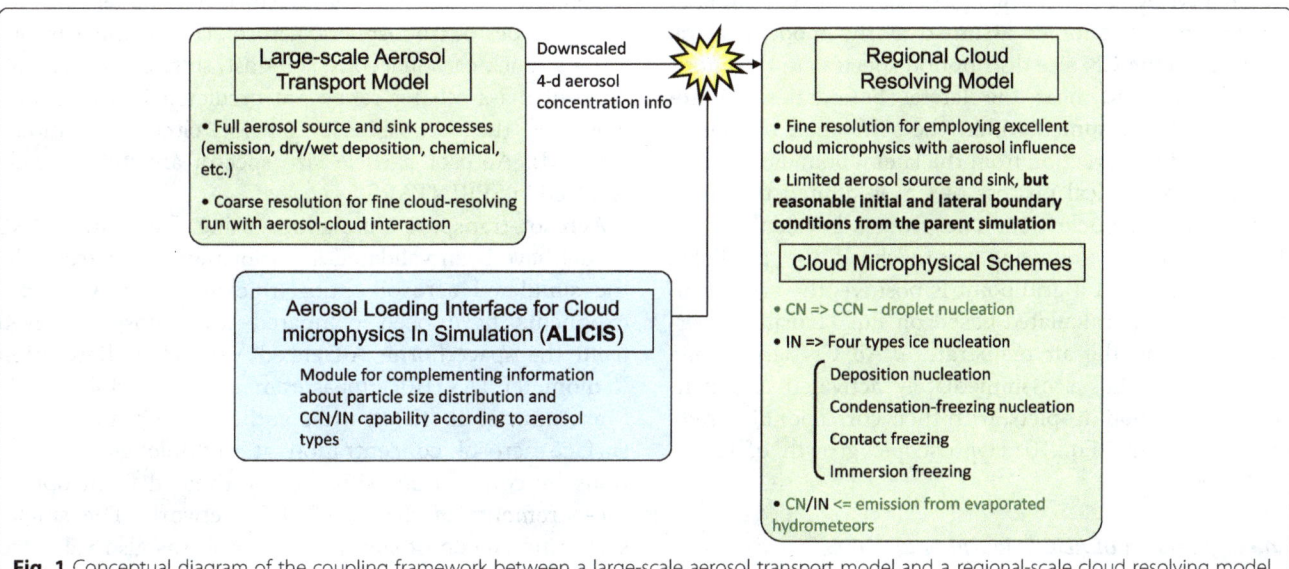

Fig. 1 Conceptual diagram of the coupling framework between a large-scale aerosol transport model and a regional-scale cloud resolving model using the ALICIS module. Gray components have not yet been implemented but will be included in future updates

Twomey's empirical parametrization described by $N_{CCN} = N_0 S^k$. Here, N_{CCN} is the number concentration of CCN activated at supersaturation S, and N_0 and k are environment-dependent coefficients (e.g., Twomey 1959a, 1959b). The following equation for the CN number density function can be obtained by coupling both equations (Khain et al. 2000):

$$\frac{dN_{CN}}{d(lnr_{CN})} = 1.5N_0 k \left(\frac{4A^3}{27Br_{CN}} \right)^{1.5k}, \tag{1}$$

where A and B are calculated from Eq. 8 for a temperature of 288.15 K, and the molecular weight, the van 't Hoff factor, and the CN particle density are identical to those of a typical ammonium sulfate aerosol particle. The spectral shape of the initial CN size distribution is common to all spatial model grid points. The initial spatial distribution is vertically inhomogeneous, since the CN number density function decreases exponentially with a scale height of 2 km above an altitude of 2 km (Khain and Sednev 1996). CN particles with radii larger than 0.6 μm are eliminated from the size distribution spectrum in standard simulations to prevent artificially rapid formation of large droplets at the initial simulation stage.

The governing equation of the number density concentration in each CN bin (Eq. 3.4 of Khain and Sednev 1996) is calculated numerically for all model grid points and each time step. Gravitational sedimentation of CN particles is neglected. CN production in the atmosphere or on the terrain surface is ignored. No direct CN source term is assumed. The CN sink is limited to consumption through the droplet nucleation process. Scattering and absorption of short- and longwave radiative fluxes by CN particles are not included, because no radiation process is calculated in the HUCM. A flat horizontal gradient of the CN number density concentration is assumed at the model's lateral boundaries; the CN size distribution spectra are thus horizontally constant near the lateral boundaries. Consequently, CN consumed within the simulation domain is replenished by advection from the lateral boundaries.

The microphysical process of CN is limited to activation of droplet nucleation. The process is calculated on the basis of the following approach. If the predicted supersaturation at a grid point is positive, the critical radius of CCN is calculated based on Eq. 11 using supersaturation and the air temperature. All CN larger than the critical radius are immediately activated and converted into cloud droplets, with their corresponding radii calculated using Eq. 10. Hygroscopic growth of CN is neglected.

An early version of ALICIS: Iguchi et al. (2008)

Recently, regional models using SBM have been widely applied, even to realistic simulations resolving clouds.

However, most studies have employed a similarly coarse approach to the CN representation (Lynn et al. 2005a, 2005b; Khain et al. 2010; Fan et al. 2012; Iguchi et al. 2012a, 2012b, 2014). Such an approach may be insufficient in realistic simulations, because in reality, the spatial distribution of aerosol particles is inhomogeneous and variable. The environment-dependent parameters in Eq. 1, N_0 and k, are determined on the basis of limited reference observations (Pruppacher and Klett 1997). Substitution of arbitrary values for those parameters causes large uncertainties in determining the CN concentration and size distribution spectra that should be consistent with their real counterparts. In particular, the uncertainties may be increased under conditions where anthropogenic pollution has a large impact on the aerosol distribution. Aerosol–cloud interaction is of great interest as a research problem under such conditions. In addition, the simplified vertical CN distribution may be problematic. The vertical CN profile assumed in the HUCM is based on a typical observed profile of the vertical aerosol distribution. However, severe aerosol contamination is sometimes observed even in the middle troposphere (Marenco et al. 2014); such highly concentrated aerosol is transported from distant intensive sources through lifting by convection and it even has a significant impact on the local cloud properties as well as on the radiation balance.

IG08 developed the aerosol loading module ALICIS to overcome the deficiencies in CN modeling associated with the original HUCM. The module provides inhomogeneous initial CN and time-variant lateral boundary conditions with a specified particle-size distribution through a dynamical downscaling approach using output from global aerosol transport simulations of SPRINTARS (e.g., Takemura et al. 2000). The SPRINTARS model includes the five types of tropospheric aerosol, i.e., organic carbonaceous, black carbonaceous, soil dust, sulfate, and sea salt. Aerosol mass-mixing ratios are predicted in the framework of the atmospheric general circulation model (AGCM). Aerosol particle size spectra are incompletely resolved in SPRINTARS.

Aerosol transport simulations using the SPRINTARS model have been validated in various ways. For example, the simulated aerosol optical thickness and Ångström exponents have been compared with those retrieved from the space-borne Advanced Very High Resolution Radiometer (AVHRR) measurements on a global scale (Takemura et al. 2000); these authors also evaluated the surface aerosol concentration at multiple global locations by comparison with those estimated from optical measurements of the AERONET network. The single-scattering albedo of simulated aerosol was also validated using AERONET data (Takemura et al. 2002). An aerosol data assimilation system has been developed on the basis of a four-dimensional variational data assimilation method (Yumimoto and Takemura 2013). The vertical aerosol distribution was validated through a comparison of the aerosol extinction with lidar measurements at several locations in Japan (Goto et al. 2015). In addition, the SPRINTARS model is included in the Aerosol Comparisons between Observations and Models (AEROCOM) project (e.g., Kinne et al. 2006). Model performance has been evaluated by comparing the global modeling outcomes of numerous aerosol modules.

IG08 determined the CN size distribution spectrum of the initial and time-variant lateral boundary conditions in their nested model simulations through the following process. Among the five types of tropospheric aerosol, organic carbonaceous, sulfate, and sea salt aerosol particles are assumed to be CN activatable as CCN. Bulk number concentrations of the three types of aerosol in SPRINTARS are calculated based on the predicted mass-mixing ratios for the grid points of the global simulation under the assumption that all aerosol particles are dry. The calculated bulk number concentrations are then temporally and spatially interpolated at the grid points of the simulation domain of the nested model. The interpolated bulk number concentrations are converted into aerosol size distribution spectra at each grid point. The conversion employs different built-in size distribution functions, according to aerosol type. Finally, accumulating the three-type aerosol size spectra yields the CN size distribution spectrum at a given grid point for the initial and boundary conditions.

The particle size distribution of organic carbonaceous aerosol is assumed to be a lognormal function with a single mode. The corresponding fraction of organic carbonaceous aerosol in the CN size distribution spectrum is given in units of the number density concentration,

$$\frac{dN_{CN(OC)}}{d(lnr_{CN})} = \frac{N_{b(OC)}}{\sqrt{2\pi}\sigma_{OC}} exp\left[-\frac{1}{2}\left\{\frac{ln\left(B_{CN}^{1/3}/B_{OC}^{1/3}\cdot r_{CN}/r_{m(OC)}\right)}{\sigma_{OC}}\right\}^2\right],$$

(2)

where N_b is the bulk aerosol number concentration of the SPRINTARS simulation, and r_m and σ are the mode radius and the standard deviation of the lognormal aerosol size distribution, respectively. The subscripts, CN and OC, distinguish the variables pertaining to CN and organic carbonaceous aerosol, respectively; $r_{m(OC)} = 0.1$ μm and $\sigma_{OC} = 1.8$ is assumed. Note that Eq. 2 applies the adjustment of the difference in the B coefficients (Eq. 8) between organic carbonaceous aerosol and CN with a certain soluble component.

Sulfate aerosol is also assumed to have a single-mode lognormal particle size distribution. The corresponding portion of the sulfate aerosol in the CN size distribution spectrum is also given by Eq. 2 in units of the number density concentration, but for parameters typical of sulfate aerosol; $r_m = 0.0695$ μm and $\sigma = 2.03$ is assumed.

The particle size distribution function of sea salt aerosol is assumed to be a power law. The corresponding fraction of sea salt aerosol in the CN number density concentration is given by

$$\frac{dN_{CN(SA)}}{d(lnr_{CN})} = 1.5N_{b(SA)}k\left(\frac{B_{SA}}{B_{CN}}\frac{r_{m(SA)}}{r_{CN}^3}\right)^k,$$

(3)

where $r_{m(SA)} = 0.1$ μm and $k = 0.6$ is assumed. The subscript SA denotes sea salt aerosol. This equation also includes adjustment of the difference in the B coefficient between CN and sea salt aerosol.

In the version of ALICIS presented here, the CN size spectrum is discretely approximated by 13 size bins with a radius ranging from 10^{-3} to 1 μm. The use of a small number of CN bins compared with the 33 bins in the original HUCM yields improved computational efficiency. The CN concentration in each bin is numerically computed every time step, using almost the same governing equations as the standard scalar tracers.

The chemical component of CN in the nested model can be arbitrarily determined by substituting a proper, corresponding value into the B coefficient of CN in the conversion equations (Eqs. 2 and 3). No microphysical processes causing a change in the CN size spectra are assumed, except for activation to droplet nucleation. Droplet nucleation is calculated using the same approach as that employed in the original HUCM; the value of the B coefficient of CN in Eqs. 2 and 3 is substituted into Eqs. 10 and 11 to determine the critical CCN radius and the corresponding radius of the generated cloud droplet. Although atmospheric radiation is calculated in the nested model, absorption and scattering of the radiative fluxes by CN are not included at present. The reason for this is that the optical properties of each aerosol species are eliminated once the multiple aerosol types have been bundled into standardized CN. In addition, the other two types of aerosol (i.e., black carbonaceous and dust aerosol) are not included in the nested model, so that this version of ALICIS has a limited capability to simulate direct aerosol effects on atmospheric radiation.

IG08 demonstrated the validity and effectiveness of the aerosol loading module in model simulations of two precipitation events observed during the 3rd Experiment of the Asian Atmospheric Particulate Environment Change Studies (APEX–E3). First, vertical profiles of the CCN number concentration in the nested model simulations were evaluated in direct comparison with those acquired by aircraft in situ measurements. The measurement methods of the CCN number concentration using a CCN counter were summarized by Ishizaka (2004) and Adhikari et al. (2005); some additional detailed information about the measurements on the event days can also be found in IG08. Figure 2 compares the results for the two individual cases. The simulations reproduced the observed CCN concentration profiles reasonably well. In the April 2 case (Fig. 2a), the observed CCN concentration ranged from approximately several tens to close to 1000/cm^3. The observed concentration decreased a little with increasing altitude. The three-type simulated CCN concentrations ranged from approximately 100 to 1000 cm^{-3} at heights of up to 5500 m; the small decrease with increasing altitude was generally reproduced. In the April 8 case (Fig. 2b), the observed CCN concentration was approximately 1000 cm^{-3} at heights below 1000 m, which decreased with increasing altitude to roughly 100 cm^{-3} at 3000 m. The vertical profiles of the simulated CCN concentrations showed a log-linear decrease with increasing altitude, similar to the observed profile.

Second, the horizontal distribution of the liquid cloud properties in the nested model simulations was compared with that estimated from data of the Modulate Resolution Imaging Spectroradiometer on the Terra satellite (Terra/MODIS) based on the satellite's retrieval algorithm (Nakajima and Nakajima 1995; Nakajima et al. 2005); plots of the horizontal distribution are shown in Fig. 6 of IG08. Figure 3 shows the relationship between the liquid cloud-top temperature and the droplets' effective radius near the cloud tops in the satellite measurements and the simulation for the April 2 case. These parameters were averaged separately in the northern and southern halves of the simulation domain to highlight spatial differences in the droplet radius distribution. The north–south difference is likely to have resulted from a horizontal bias in the CN concentration. The model simulation performed satisfactorily in reproducing the overall tendency of the north–south gradients of the observed droplet radius in warm-cloud regimes with temperatures in excess of 273 K, despite some overestimation (Fig. 3a). The observed radius monotonously increased with decreasing cloud-top temperature. The observed radii averaged in the northern half had larger values than those in the southern half; the maximum north–south difference was roughly 5 μm. The simulated radius generally increased with decreasing cloud-top temperature down to 275 K and roughly decreased or remained constant below 275 K. The simulated radii in the northern half attained larger values than those in the southern half for temperatures greater than 275 K; the north–south difference was roughly 5 μm on average. In addition, two sensitivity simulations were conducted to highlight the effects of the aerosol loading module on the simulation results (Fig. 3b, c). The

simulation assuming horizontally homogeneous initial and boundary CN conditions failed to reproduce the structure of the north–south gradients (Fig. 3b). The simulated profiles of the average droplet effective radius are almost identical for the northern and southern halves. The simulated north–south gradients in the control run were thus caused by a horizontally inhomogeneous CN distribution introduced by aerosol downscaling. This result proved that the implementation of realistic, inhomogeneous CN fields is essential for more accurate simulations of cloud microphysical properties. On the other hand, the simulation assuming a doubled CN number concentration compared with the control run had smaller effective radii (Fig. 3c). The simulated effective radius profiles were in better agreement with the observed profiles than those in the control run. This result indicates that the control run underestimated the CCN number concentration.

Update by Choi et al. (2014): Aerosol particle size distribution revision and type multiplication

The IG08 version of ALICIS still contained many limitations. Among them was the absence of two aerosol species, i.e., black carbonaceous aerosol and dust particles. In reality, they often work as CCN if soluble components are combined with the particles, either externally or internally (e.g., Sullivan et al. 2009). In addition, the modeling of the other three types of aerosol still offered room for significant improvements. For example, the application of a power-law function to the size distribution spectra of sea salt aerosol might cause an overestimation of the activated CCN number concentration because of the continuous increase in the CCN number concentration with increasing supersaturation.

The aerosol loading module was comprehensively updated by CH14. The conversion from the various types of aerosol to standardized CN, as implemented by IG08, has been removed. The same five aerosol species used in the SPRINTARS model are employed in nested models. The discretization of the size distribution spectra of each aerosol type has been extended to 17 size bins with radii ranging from 10^{-3} to 10 μm. The use of more size bins and a wider range of particle sizes compared with those in the old version allows the representation of aerosol particles with radii larger than 1 μm. Such large aerosol particles are possibly included in the sea salt and dust aerosol categories (Fig. 4).

The size distribution function of each aerosol type was assumed to be of a multi-modal lognormal form by CH14. The size distribution spectra described in this manner are consistent with those observed in reality (e.g., Brechtel et al. 1998). The number density concentration of the size distribution of each aerosol type is given by

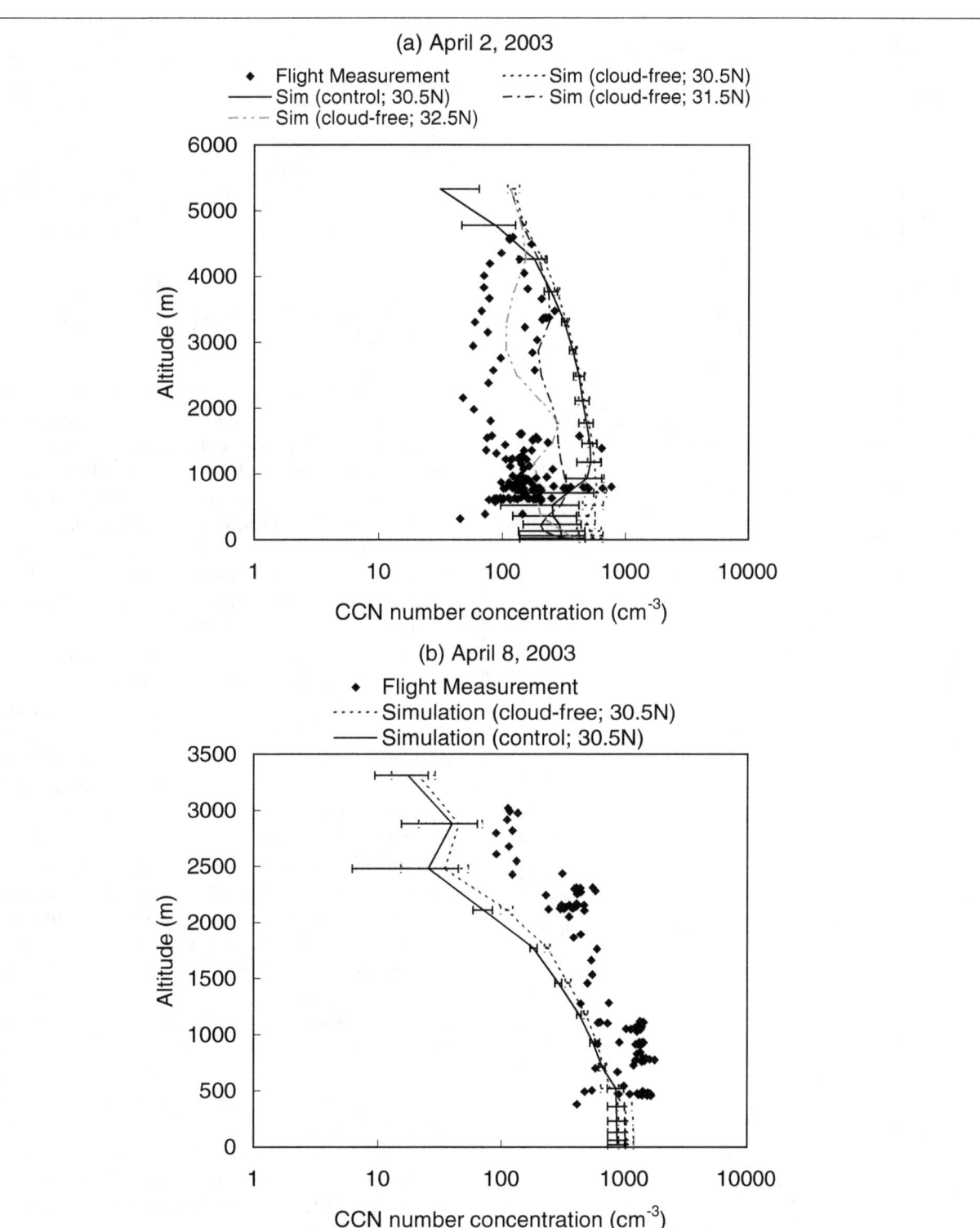

Fig. 2 Vertical distribution of the observed and simulated CCN number concentrations over the East China Sea on **a** April 2, 2003 and **b** April 8, 2003. The observed CCN number concentration in the flight measurements was obtained for liquid-phase supersaturation ranges of 0.07–0.22 % on April 2 and 0.09–0.32 % on April 8. The simulated CCN number concentration in the regional model was calculated for a liquid-phase supersaturation of 0.1 % in the April 2 case and 0.2 % in the April 8 case. Lines of simulated concentration show vertical profiles of horizontal averages in three domains of radius ±0.5° centered at 129.5° E and 30.5° N, 31.5° N, or 32.5° N. The two types of simulated concentration are plotted separately from the control and cloud-free runs (where all cloud microphysical processes were switched off to prevent CN consumption through droplet nucleation). Adapted from Iguchi et al. (2008)

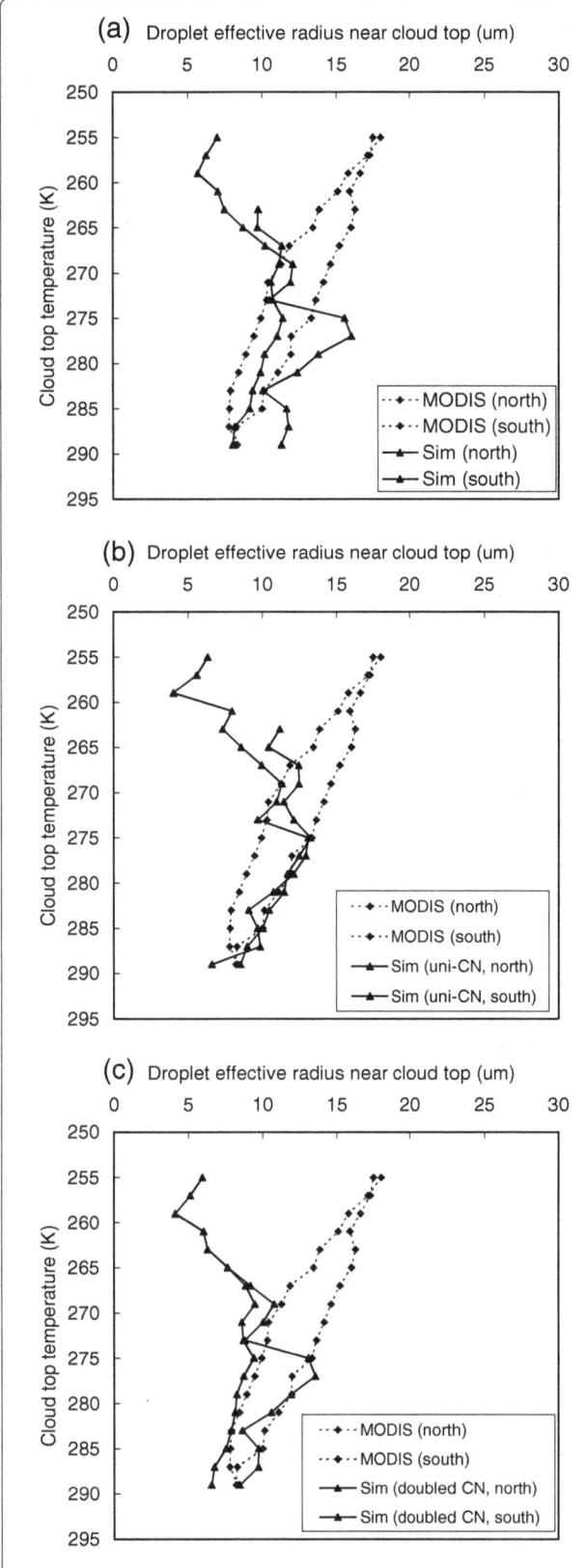

Fig. 3 Dependence of droplet effective radii near cloud tops on cloud-top temperature calculated based on simulations with three different settings and from the retrieval results derived from Terra/MODIS measurements over the East China Sea on April 2, 2003. Simulation profiles are calculated in **a** the control run, **b** the run characterized by a homogenous CN field, and **c** the run with doubled CN. Plotted radii are averaged individually in the northern and southern halves of the analysis domain. Adapted from Iguchi et al. (2008)

$$\frac{dN}{d(lnr_a)} = \sum_{i=1}^{n_m} \frac{N_{SPR}f_i}{\sqrt{2\pi}\sigma_i} exp\left(-\frac{lnr_a - lnr_{mi}}{2\sigma_i^2}\right), \qquad (4)$$

where r_a is the aerosol particle radius, i the index of a given mode, n_m the total number of modes in the size distribution, N_{SPR} the bulk number concentration of each aerosol type obtained from the SPRINTARS simulation, and f the weight factor required to normalize the multi-modal distribution. The variables n_m, f, r_m, and σ are the functions of aerosol type as listed in Table 1 of CH14; the values have been determined on the basis of observational results (d'Almeida et al. 1991; Chuang et al. 1997; Penner et al. 1998; Herzog et al. 2004).

Figure 4 compares the aerosol size distribution spectra. Two size distribution spectra were calculated for each aerosol type using the assumptions employed by CH14 and IG08. No size distribution was assumed for dust or black carbonaceous aerosol in the study by IG08. A set of observed spectra was obtained from in situ measurements during the ACE–Asia field campaign in 2001 (Huebert et al. 2003). The set of spectra was calculated from the size-segmented aerosol mass concentration measured using the Micro-Orifice Uniform Deposit Impactor (MOUDI) onboard a research vessel; the marine aerosol sampling performed during ACE–Asia 2001 is summarized by Mochida et al. (2007). Another set of observed spectra was obtained from ground-based measurements using MOUDI at the Gosan site on Jeju Island (Republic of Korea) during the Atmospheric Brown Cloud–East Asian Regional Experiment (ABC–EAREX) in 2005 (Nakajima et al. 2007).

In Fig. 4, the size distribution spectra employed by CH14 are in better agreement with those from the measurements, although the observed size spectra lack multi-modal distributions. The observed aerosol number densities monotonously decrease with increasing aerosol particle size. The observed size distributions of sulfate and organic/black carbonaceous aerosol extend to the radii of 2 µm, whereas those of dust and sea salt aerosol extend to over 5 µm. The spectra of CH14 reproduce the distinct structure of the observed spectra according to aerosol type reasonably well, especially

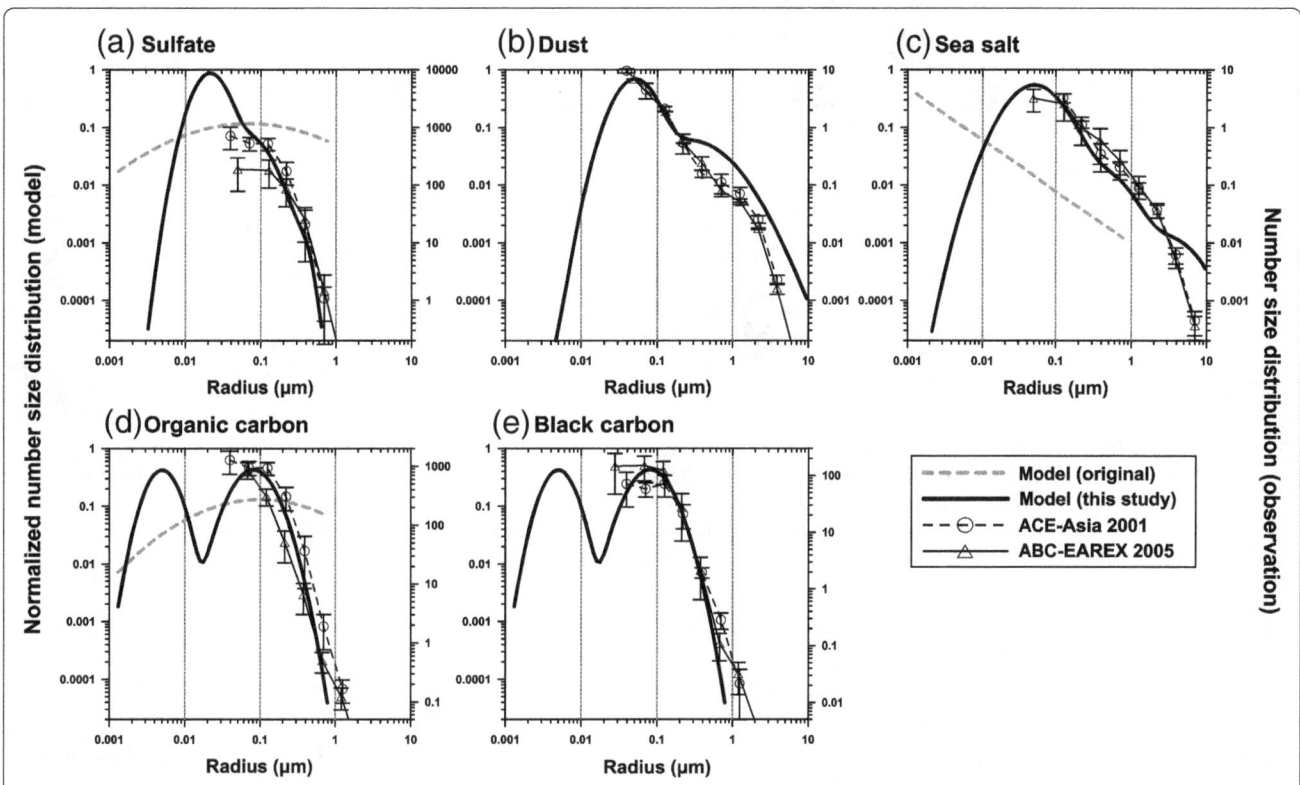

Fig. 4 Size distribution spectra of the five aerosol species, i.e., **a** sulfate, **b** dust, **c** sea salt, **d** organic carbon, and **e** black carbon, assumed in the updated and outdated ALICIS modules, and obtained from in situ measurements during ACE–Asia 2001 and ABC–EAREX 2005. Cited from Choi et al. (2014)

those of sulfate and organic/black carbonaceous aerosol. In contrast, the spectra of IG08 exhibit several severe insufficiencies. The number density concentration of sulfate aerosol is overestimated across the full particle size range shown and the sea salt aerosol concentration is significantly underestimated. The organic carbonaceous aerosol concentration is underestimated for particle radii smaller than 0.1 μm and overestimated for radii larger than 0.4 μm.

The activation process of aerosol particles to form cloud droplets was calculated for each aerosol type on the basis of basic Köhler theory by CH14, unlike in IG08. Because the five types of aerosol have different soluble capabilities, different B coefficients are employed in determining the critical radius of CCN in Eq. 11. The B coefficients are assumed to be 0.51 for sulfate, 1.16 for sea salt, 0.14 for dust, 0.14 for organic carbonaceous, and 0.05×10^{-5} for black carbonaceous aerosol (Ghan et al. 2001).

CH14 employed the same build-up process of initial and time-variant boundary conditions of binned aerosol concentrations in their nested model simulations as IG08. The time evolution of each binned concentration was numerically computed for each time step. Microphysical processes producing or losing aerosol particles were excluded, except for consumption through the nucleation of cloud droplets.

Any direct aerosol effects on atmospheric radiation were excluded, even by CH14, although the five types of aerosol were fully managed.

CH14 evaluated the performance of the updated version of the aerosol loading module in their simulation analysis. They conducted numerical simulations of two cloudy cases observed during the ABC–EAREX 2005 field campaign over an East China Sea region. Corresponding simulations using the old version of the aerosol loading module of IG08 were performed simultaneously for comparison.

First, the simulated CN and CCN number concentrations were evaluated in comparison with those obtained from ground-based measurements. The CN comparison yielded a new perspective on the validation of ALICIS, in addition to the CCN comparison also discussed by IG08. Figure 5 compares the concentrations (CCN at 0.6 % supersaturation) derived from the different assumptions made by both IG08 and CH14 with those obtained from in situ measurements at the Gosan site. The observed CN and CCN concentrations were measured using a TSI (Shoreview, MN, USA) condensation particle counter model 3010, which detected particles with diameters larger than 10 nm, and using a streamwise thermal-gradient CCN counter (Roberts and Nenes 2005); details of the measurements can be found in Yum et al. (2007). The large underestimation of the

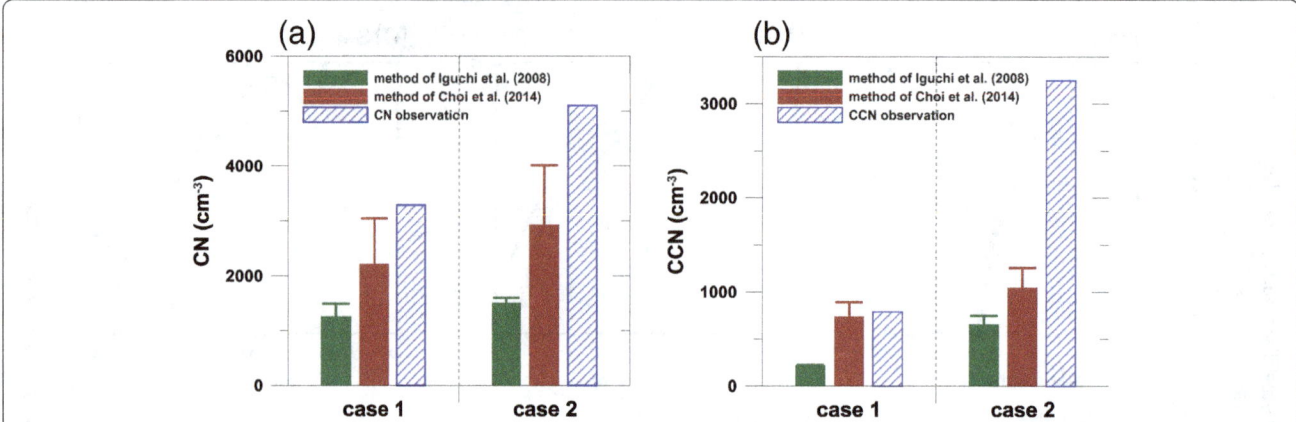

Fig. 5 Surface-level **a** CN and **b** CCN (at 0.6 % liquid-phase supersaturation) number concentrations simulated near the ABC–EAREX 2005 Gosan site and measured at the site on March 13, 2005 (case 1) and March 25, 2005 (case 2). *Green bars* show number concentrations in a regional model simulation using the ALICIS module based on Iguchi et al. (2008); the concentrations are calculated at grid points near the site (averaged within an area of 126.0° E–126.4° E and 33.1° N–33.5° N at the model bottom level) for the initial time step of the simulation. *Error bars* show standard deviations from the averages. *Red bars* show the same concentrations but using assumptions proposed by Choi et al. (2014). Observed CN and CCN were measured simultaneously using a TSI condensation particle counter and a stream-wise thermal gradient CCN counter deployed at the Gosan site, respectively

number concentrations in the outdated IG08 version has been alleviated in the updated CH14 version. The improvement is attributed to the application of more realistic distribution functions to the aerosol size spectra rather than the new inclusion of dust and black carbonaceous aerosol into the categories. However, underestimation of CN and CCN concentrations still remained in the simulations using the updated CH14 version, especially for the second case. This insufficiency is probably caused by an underestimation of the aerosol concentration in the SPRINTARS simulation over highly polluted regions affected by anthropogenic aerosol emissions.

Second, CH14 evaluated droplet effective radii near cloud tops, liquid water paths, and cloud optical thicknesses calculated in their simulations in comparison with those obtained from the Terra/MODIS satellite measurements through retrieval (see Table 2 in CH14); details of the retrieval algorithm are described by Nakajima et al. (2005). Simulations using the updated and outdated aerosol loading modules led to different values of the variables, although the cloud distribution patterns were similar. The droplet effective radii were smaller, and the liquid water paths and cloud optical thicknesses were larger in the updated simulation compared with the outdated simulation. These differences resulted from larger CDNC attributed to the larger CN and CCN number concentrations. Overall, the values in the updated simulation were in better agreement with those estimated from the satellite measurements, reflecting the better reproduction of the CN and CCN concentrations in the updated version.

Update by Sato et al. (2012): Aerosol emission from evaporated cloud droplets

SA12 introduced an aerosol particle generation process from completely evaporated droplets into the framework of the HUCM SBM, coupled with the ALICIS module. Absence of an aerosol regeneration process might cause underestimation of the CN and CCN number concentrations and, consequently, an underestimation of CDNC, particularly in long-term simulations. The process was not included in the original HUCM SBM, and was not considered by IG08 or CH14, mostly because of the following technical issue. Once the activation to droplet nucleation has been calculated in the models, the size and composition information about the aerosol particles used was eliminated. Some state-of-the-art models fully predict solute components in hydrometeor particles at the expense of increased computational cost (e.g., Chen and Lamb 1994). Such models have the advantage of enabling explicit calculations of aerosol particle emission from completely evaporated or sublimated hydrometeor particles.

The parametrization of the aerosol emission process by SA12 was based on the third approach described by Feingold et al. (1996). SA12 assumed that one aerosol particle was regenerated when one droplet had completely evaporated (Mitra et al. 1992). Solute components in the droplet were not fully predicted, so that the sizes of regenerated aerosol particles could not be calculated explicitly; alternatively, the size distribution spectrum of the regenerated aerosol was determined based on the change of the domain-averaged aerosol number concentration from the initial condition to the present time step. The additional aerosol number concentration to the k-th aerosol size bin resulting from this regeneration process (R_k) is as follows (see Eqs. 3 and 4 in Feingold et al. 1996):

$$R_k = N_R \frac{\phi_k}{\sum_{k=1}^{n_{bin}} \phi_k}; \qquad (5)$$

$$\phi_k = \frac{\bar{N}_{0,k} - \bar{N}_k}{\bar{N}_{0,k}}, \qquad (6)$$

where N_R is the total number concentration of the regenerated aerosol particles, which is equivalent to the number concentration of completely evaporated droplets; n_{bin} is the total number of aerosol size bins; and $\bar{N}_{0,k}$ and $\bar{N}_k$ are the domain-averaged aerosol number concentrations of the k-th size bin in the initial and present conditions, respectively. SA12 assumed that all CN were composed of ammonium sulfate; therefore, the chemical component of aerosol particles generated through this emission process was assumed to be ammonium sulfate.

SA12 evaluated the performance of the aerosol regeneration parametrization in their simulations by resolving stratocumulus over a region off the coast of California. Two types of simulation (i.e., with and without the regeneration parametrization) were conducted simultaneously to investigate the influence of this process. The simulation results were validated using retrieval results from the space-borne AVHRR/2 data over the region of interest (Nakajima and Nakajima 1995).

Figure 6 shows the horizontal distribution of the cloud optical thickness retrieved from the AVHRR/2 space-borne measurements (Nakajima and Nakajima 1995) and calculated in the regional model simulations excluding or including the aerosol regeneration process. Cloud optical thickness ranged from 10 to 60 over the domain of interest in the observation region. The southern half of the domain exhibited relatively thick stratocumulus. In contrast, the simulated cloud fields show a more dispersed structure. Cloud optical thicknesses simulated with the aerosol regeneration parametrization were at least in better agreement with the satellite measurements than those not based on this parameterization. Large optical thicknesses of more than 30 are patchily and widely distributed in the simulation including the regeneration process, whereas the corresponding values were less than 20 in the simulation without the regeneration process, except for areas that were close to a lateral boundary of the simulation. Implementation of the aerosol regeneration process caused an increase of approximately 20 % in the aerosol number concentration averaged over the domain. A large fraction of aerosol particles emitted from evaporated droplets was used again as CCN. The regeneration of aerosol particles caused increased CDNC and cloud optical thickness.

Sato (2012) tested another approach used by Feingold et al. (1996) for the aerosol regeneration process, because the parametrization based on Eq. 5 was not very appropriate for downscaled realistic simulations. The new approach assumed that the particle size distribution of regenerated aerosol was the same as the horizontal average at the corresponding vertical level for the initial conditions. Sato's (2012) tests using this approach showed that the reproducibility of cloud optical thickness was improved as well.

SA12 also conducted a quantitative comparison on the basis of correlation analysis between the droplet effective radii and cloud optical thicknesses (e.g., Fig. 8 of SA12). Only simulations including the aerosol regeneration parametrization reproduced a characteristic pattern in the correlation plots based on the satellite measurements. Their results demonstrate that implementation of the aerosol regeneration process is necessary to adequately simulate the cloud microphysical structure of stratocumulus over the region of interest.

Prospective ALICIS update strategy
IN modeling

The behavior of aerosol particles as IN is implicitly included in the framework of the HUCM SBM tied to the present ALICIS module. The IN effect is tacitly contained in parametrizations to calculate ice-nucleation and freezing processes without consumption of predicted aerosol particles in the model. This coarse approach is due to the present poor knowledge about complicated microphysics in ice/mixed-phase clouds. Nevertheless, introducing an explicit representation of IN activation and the resultant depletion is important for developing a more comprehensive scheme of cloud microphysics for aerosol–cloud interaction studies. Improvement of the SBM and ALICIS to manage the IN effect is a worthwhile challenge.

Solid hydrometeor particles in the atmosphere primarily form through either homogenous freezing of supercooled droplets without IN action or heterogeneous nucleation/freezing with catalysis by IN. In addition, the latter heterogeneous processes can be divided into four different categories (Vali 1985): deposition nucleation, condensation-freezing nucleation, contact freezing, and immersion freezing.

Deposition nucleation is completed when water vapor directly deposits onto the surface of an aerosol particle, provided that supersaturation with respect to ice is positive. The aerosol particle does not have to be hydroscopic. On the other hand, condensation-freezing nucleation originates from the nucleation process of a droplet under the condition that supersaturation with respect to water is positive. The aerosol particle is activated as CCN but does not have

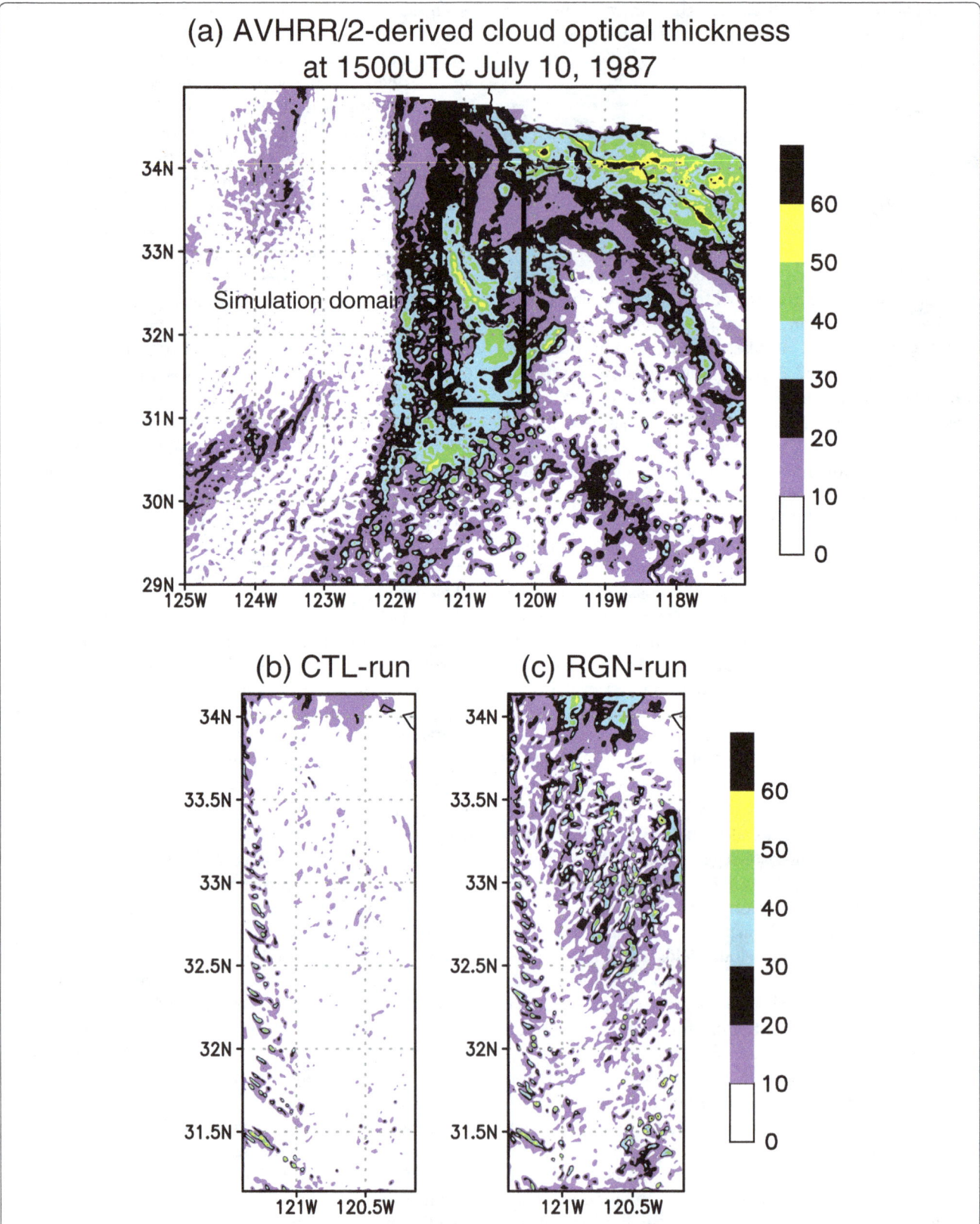

Fig. 6 Horizontal distribution of cloud optical thickness observed at 1500 UTC on July 10, 1987: **a** retrieved from the space-borne AVHRR/2 measurement results, **b** simulated by the regional model without aerosol regeneration parametrization (CTL-run) at 1400 UTC on the same day, and **c** with regeneration parameterization (RGN-run). The *red rectangle* in the top panel shows the model simulation domain

to completely dissolve in the droplet. The droplet that is formed then immediately freezes from the non-dissolved impurity. Distinguishing these two types of nucleation process is difficult in measurements where supersaturation with respect to water is positive. Supersaturation (i.e., humidity) rather than temperature is an important factor controlling the progress of these processes. Contact freezing possibly occurs when supercooled droplets collide with aerosol particles. The collision is usually caused by Brownian diffusion, thermophoresis, diffusiophoresis, electrophoresis, and/or inertial impaction. The concentration of aerosol particles that can act as contact-freezing IN at the relevant temperature is the most important factor in determining the freezing probability. Immersion freezing originates from impurities already present inside a supercooled droplet. Unlike condensation-freezing nucleation, immersion freezing does not have to start immediately upon the formation of a supercooled droplet. The freezing can occur when the temperature of the supercooled droplet decreases.

Among the four types of heterogeneous cloud glaciation processes, deposition nucleation is the simplest because the process does not proceed through a state of supercooled droplets. An increase in the number of aerosol particles that can work as deposition IN directly causes the number of ice cloud particles to increase, just like the relationship between CCN and cloud droplets. In contrast, cloud glaciation through the other three processes is more complicated. If the number concentrations of all types of aerosol increase, promotion of heterogeneous freezing through increasing IN may be in competition with depression of the freezing efficiency caused by a decrease in the particle sizes of supercooled droplets through increasing CDNC. In contrast, increased cloud height and decreased cloud temperature caused by the particle size change of supercooled droplets (e.g., Pincus and Baker 1994) may accelerate cloud glaciation through heterogeneous or homogenous freezing.

Unlike droplet nucleation summarized in Köhler theory, no comprehensive theory has been developed to formulate each ice-nucleation or freezing process. Parametrizations of these processes are constructed empirically on the basis of in situ measurement data or experimental laboratory data acquired under certain conditions. The four types of mechanisms of heterogeneous nucleation/freezing can rarely be distinguished through measurements or experiments, so that empirical parametrization may cover multiple components for each of the four types of the mechanisms. In addition, identifying the chemical components of IN is difficult for most measurements.

Appendix 2 summarizes our testbed for how to improve IN parametrization in the HUCM SBM coupled with the ALICIS module. Fan et al. (2014) improved the ice-nucleation parts of the HUCM SBM by introducing up-to-date IN parametrizations that include prognostic dust aerosol concentrations. The parametrizations were based on results by Tobo et al. (2013) and DeMott et al. (2015) for deposition/condensation-freezing nucleation and immersion freezing, and by Muhlbauer and Lohmann (2009) for contact freezing. We may modify the parametrizations into suitable forms to use the aerosol size distribution spectra predicted by models coupled with the ALICIS module. In addition, more comprehensive and generalized parametrizations of heterogeneous ice-nucleation/freezing should be implemented in future updates. Phillips et al. (2008, 2013) proposed a flexible framework where various empirical parametrizations can be bundled into a versatile form since parameters in the formulation can be easily constrained by various observational data. Note that the application of any proposed parametrizations to actual model simulations is still affected by large uncertainties. The uncertainties come from the limited applicability of findings from in situ measurement and laboratory experiment results for IN, and the coarse translation from results to developer-friendly parametrizations in a numerical scheme.

Introduction of regional variability into the aerosol size distribution and hygroscopicity parameters

The present ALICIS module includes fixed aerosol size distribution functions with constant parameters, as described in "An early version of ALICIS: Iguchi et al. (2008)". The size distribution spectra were validated for application to simulations of regions in Northeast Asia through comparisons with those obtained from on-site measurements. The spectra might thus be inappropriate if applied to simulations pertaining to other regions. For example, mineral dust aerosol particles in East Asia and North Africa (Sahara) have different physical and chemical properties (Formenti et al. 2011). The dust properties depend not only on the source region but also on the distance from the emission site.

We will introduce regional variability into the aerosol size distribution functions and the hygroscopicity parameters assumed in the ALICIS module. First, these functions and parameters should have different forms and values in accordance with the type of simulation region, e.g., continental, coastal, or maritime. For example, the parameters determining the particle size distribution of sulfate aerosol used by CH14 originate from those for maritime scenes from Chuang et al. (1997). The latter authors assumed a bimodal function with different parameter values for continental regions instead of the trimodal function employed for maritime scenes. An average of the two size spectrum forms can also be used for intermediate (i.e., coastal) scenes.

Second, introducing intercontinental variability into the ALICIS module is a valuable challenge for further improvement. Although many measurements of the aerosol size distribution have been conducted worldwide, most obtained a bulk aerosol size distribution containing all chemical components. Internal mixing of aerosol particles is highly problematic in constraining the aerosol size distribution of an individual chemical component in relation to the measurement data. Few recent studies have conducted global simulations using an aerosol transport model predicting the aerosol size distribution (e.g., Spracklen et al. 2005; Zhang et al. 2010; Jacobson 2012). These simulation results are useful for complementary provision of aerosol size distribution information, particularly over pelagic regions and in the upper troposphere where few in situ measurements are available. The results may be employed to construct climatology data of the aerosol size distribution or they may be used as direct input to simulations through dynamical downscaling approaches.

Introduction of aerosol representation to internal mixing, source/sink, and direct radiative effects

Compared with typical aerosol transport models, the present SBM coupled with the ALICIS module still lacks modeling of many aspects of aerosol behavior and its effects. Some may have a critical influence on cloud micro- and macrophysics. Their implementation is necessary to conduct more comprehensive simulations of aerosol–cloud interactions (e.g., Saleeby and van den Heever 2013). Various approaches have been proposed to introduce these aspects into models. We will include them in future updates while considering the limitations of the available computational resources.

In reality, aerosol particles are often characterized by internally mixed states with multiple chemical compositions. These aerosol particles are mostly dominant in regions at great distances from their major emission sources. The chemical and physical properties of aerosol particles may change through gas absorption/deposition or they may coagulate with other particles during migration. The question of how to numerically model an internal mixture is a delicate problem in calculating the nucleation process of cloud hydrometeor particles from aerosol particles, as well as their scattering and absorption of atmospheric radiation. Various modifications of the Köhler theory have been proposed to represent the hygroscopic growth of internally mixed particles. For example, reformulation of the Köhler equation was suggested to consider the effects of soluble gases and slightly soluble solid substances in the droplet nucleation process (Laaksonen et al. 1998). A modified Köhler theory with a single solute hygroscopicity parameter (κ-Köhler theory) has also been proposed (Petters and Kreidenweis 2007);

the new parameter represents the CCN capability of dry particles with arbitrary mixtures and can be constrained easily with observational data.

Wet scavenging (i.e., impaction of aerosol particles by hydrometeor particles) does not only cause a significant decrease in aerosol concentration but also triggers several cloud microphysical processes, such as contact or immersion freezing. The scavenging process is divided into two types: dissolution or impaction as a result of the nucleation process through condensation or deposition of water vapor on aerosol particles, and impaction through collision between aerosol particles and already extant hydrometeor particles. As highlighted in the section about contact freezing, Brownian diffusion, thermophoresis, diffusiophoresis, electrophoresis, and/or inertial impaction are major causes of collisions between aerosol and hydrometeor particles. There are significant uncertainties associated with modeling these effects, especially as regards the electromagnetic attraction between aerosol and hydrometeor particles, because of insufficient theoretical knowledge and measurement data. In addition, tracing solutions and impurity components in hydrometeor particles is technically difficult if only limited computational resources are available. At present, a few models explicitly predict solution and impurity components in hydrometeor particles in their SBM frameworks (e.g., Chen and Lamb 1994). As an alternative approach, solution and impurity components are assumed to be statistically (Poissonian) distributed according to the volume of a hydrometeor particle when the total mass of the components scavenged by hydrometeor particles is predicted (Phillips et al. 2008). Although these authors applied this approach to their parametrization for immersion freezing, the same approach is applicable to the calculation of aerosol particle emission from evaporated hydrometeor particles.

Scattering and absorption of atmospheric radiation by aerosol particles can change the radiation budget in the different atmospheric layers and on the Earth's surface; this is a well-known direct effect of aerosol on the Earth's radiation balance. Direct aerosol effects on atmospheric radiation have a little impact on cloud physics. However, strong absorption of the solar irradiance by aerosol particles potentially has a drastic influence on cloud structure. Heating through absorption may affect the surroundings of the particles on a minor scale. When such absorptive aerosol particles or impurities in hydrometeor particles exist in great volumes at layers near the cloud top, an increase in temperature through heating may promote evaporation of hydrometeor particles. This effect may cause a change in the Earth's radiation balance through cloud structure change. This kind of aerosol influence is referred to as the aerosol semi-direct radiative effect (Ackerman et al. 2000).

Aerosol optical properties are a key determinant of their influence on the Earth's radiation budget. We plan to follow standard methods employed in existing aerosol transport models (e.g., GOCART, SPRINTARS, ECHAM: Pozzoli et al. 2011; the Hadley Centre Global Environment Model HadGEM: Bellouin et al. 2011) to determine the aerosol optical properties. The single-scattering properties of spherical aerosol particles can be approximated using Mie scattering theory in radiative flux calculations. The wavelength of light, the radius of a particle, and the refractive index of its medium at the relevant wavelength are required to determine the extinction coefficient, single-scattering albedo, and asymmetry factor. The effect of water uptake on hygroscopic aerosol particles is approximated by calculating the particle radius and refractive index for a series of relative humidity values; the refractive indices are calculated on the basis of the volume-weighted mixing assumption between water and dry components (e.g., Haywood et al. 1997; Takemura et al. 2002; Eq. 3). The internal mixture effect is also approximated using the volume-weighted mixing assumption for multiple different components. Non-spherical dust particles are approximated as their corresponding spherical particles in calculating their optical properties, because relative errors introduced by this assumption are very small (Fu et al. 2009). We will generate lookup tables of extinction coefficients, single-scattering albedos, and asymmetry factors computed using a Mie theory program for the discretized series of aerosol particle radii, relative humidity values, and band wavelengths as applicable to each aerosol type. An internal mixture will be assumed only for organic and black carbonaceous aerosol particles at this stage to avoid heavy computational loads by including all mixture combinations; lookup tables for the mixed particles will also be prepared.

Conclusions

This review paper describes the scientific background and update strategies in the development of the aerosol loading module referred to as ALICIS. The module is used for coupling a downscaled CRM simulation with a large-scale aerosol transport simulation. The most important function of the module is to provide realistic temporally and spatially inhomogeneous distributions of aerosol combined with accurate aerosol size distribution information for downscaled simulations. Introduction of the module yields better simulations of cloud microphysical structure by avoiding the adoption of inadequate aerosol loading as artificially defined by users.

ALICIS has been subject to two major updates since its initiation by IG08. The assumption of the aerosol size distribution spectra was significantly improved as the

spectrum shapes better matched their observed counterparts in the first update (CH14). The aerosol emission process from evaporated cloud droplets was added in the second update (SA12). However, many points remain to be improved in future updates, including but not limited to the management of heterogeneous ice nucleation/freezing, aerosol direct/semi-direct effects on atmospheric radiation, addressing internal mixtures of different chemical components in aerosol particles, the introduction of regional variability into the aerosol size distribution, and the representation of wet scavenging.

So far, ALICIS has been employed only in the framework connecting SPRINTARS aerosol transport simulations to JMA–NHM coupled with the HUCM SBM. We are now working on implementing the aerosol loading module into the bridge between the Modern Era Retrospective analysis for Research and Applications Aerosol Reanalysis (MERRAero) or GEOS–GOCART model simulations and WRF coupled with HUCM SBM (WRF–SBM) (e.g., Iguchi et al. 2012a). This extension will readily increase the applicability of ALICIS to more diverse simulation scenes and styles. In addition, we plan to develop the coupling of ALICIS to bulk cloud microphysics instead of SBM to enhance the simulation capability.

In future updates, models coupled with the ALICIS module will include a more comprehensive representation of aerosol and related chemical species, like the WRF–Chem model. Such state-of-the-art models also enable better simulation of cloud microphysics, although they require much larger computational resources and loads. Improvement in model functionality is in competition with improvement in various resolutions such as spatial grid spacing, time-step sampling, and/or particle size bins. Keeping a balance between model performance and computational efficiency in the context of limited resources is a delicate problem in model development. Adequate sorting of the priorities between the introduction of various model functions and improvement in various resolutions is an important task assigned to researchers.

Appendix 1: Basic Köhler theory for droplet nucleation

A chart of Köhler curves (Köhler 1936) summarizes a fundamental theory about CN activation and stabilization of the formed droplets. The chart illustrates relationships between the ambient vapor saturation ratio and droplet radius under equilibrium conditions with the environment. Two competitive effects are determinants of the relationships: the Raoult effect, where dissolution depresses the surface vapor pressure on the solution; and the Kelvin effect, where the surface vapor pressure on a curved liquid surface is higher than that on a corresponding flat surface.

The supersaturation around a droplet in equilibrium with the ambient vapor is determined by the following equation (e.g., Rogers and Yau 1989), assuming that the dissolved amount of solutes in the droplet does not change:

$$S_{\mathrm{w}} = \frac{A}{r_{\mathrm{w}}} - \frac{Br_{\mathrm{CN}}^3}{r_{\mathrm{w}}^3}; \qquad (7)$$

$$A \approx \frac{3.3 \times 10^{-5}}{T}\ (\mathrm{cm}) \quad and \quad B \approx \frac{4.3v}{M_{\mathrm{CN}}}\left(\frac{4\pi\rho_{\mathrm{CN}}}{3}\right), \qquad (8)$$

where S_{w} is the equilibrium supersaturation, r_{w} the droplet radius, and T the skin temperature of the droplet; r_{CN} is the radius of the dry CN particle before dissolution; and v, M_{CN}, and ρ_{CN} are the van 't Hoff factor (e.g., Low 1969), molecular weight, and density of the CN component, respectively. The CN particle is assumed to be composed of a single soluble ingredient that has been completely dissolved in the droplet. In Eq. 7, there is a unique peak of the supersaturation $S_{\mathrm{crit(w)}}$ under the condition that $dS_w/dr_w = 0$,

$$S_{\mathrm{crit(w)}} = \frac{2A}{3r_{\mathrm{crit(w)}}}; \qquad (9)$$

$$r_{\mathrm{crit(w)}} = \sqrt{\frac{3Br_{\mathrm{CN}}^3}{A}}, \qquad (10)$$

where $r_{\mathrm{crit(w)}}$ is the critical radius of the droplet that matches $S_{\mathrm{crit(w)}}$. Once the ambient supersaturation reaches approximately $S_{\mathrm{crit(w)}}$, droplets with radii greater than $r_{\mathrm{crit(w)}}$ can exist stably. The critical radius of CN, $r_{\mathrm{crit(CN)}}$, can also be determined uniquely:

$$r_{\mathrm{crit(CN)}} = \left(\frac{4}{27}\frac{A^3}{B}\frac{1}{S_{\mathrm{crit(w)}}^2}\right)^{1/3}. \qquad (11)$$

Equation 11 represents the critical radius of CN as a function of supersaturation. If the radius of CN is larger than $r_{\mathrm{crit(CN)}}$, CN is activated as CCN and the formed droplet exists stably under the ambient supersaturation, in excess of $S_{\mathrm{crit(w)}}$.

Appendix 2: Future plans for improving IN parametrization

Deposition and condensation-freezing nucleation

A common approach to measuring ambient IN is by counting the change in the number of ice particles in a sampled airmass for a series of air temperature or supersaturation measurements. DeMott et al. (2010) provided an equation for the activated IN number concentration as a function of both temperature and the number concentration of aerosol particles with diameters larger than 0.5 μm, by combining measurement data at multiple locations from the Arctic to the Amazon Basin. This equation is represented by

$$n_{\mathrm{IN},T_k} = a(273.16 - T_k)^b (n_{a>0.5})^{(c(273.16 - T_k)+d)}, \qquad (12)$$

where $a = 0.594 \times 10^{-4}$, $b = 3.33$, $c = 0.0264$, and $d = 0.0033$; T_k is the cloud temperature (in Kelvin), $n_{a>0.5}$ is the number concentration (in cm^{-3}) of aerosol particles with diameters larger than 0.5 μm, and $n_{\mathrm{IN},Tk}$ is the active IN number concentration (std L^{-1}). This parametrization may cover all heterogeneous ice nucleation types (i.e., deposition nucleation, condensation-freezing nucleation, contact freezing, and immersion freezing) because it is based on measurement data under water saturation conditions. However, Demott et al. (2010) indicated that $n_{\mathrm{IN},Tk}$ corresponds to the maximum number concentration of ambient IN activated mostly through deposition nucleation, because deposition nucleation is the primary contributor in mixed-phase clouds at temperatures above −35 °C (Phillips et al. 2008). The measurement data were sampled at a temperature range between −9 and −35 °C.

Several issues need to be solved before implementation of the parametrization described by Eq. 12 into a model coupled with the ALICIS module. Equation 12 can be interpreted in several ways in terms of how to connect the maximum number concentration of ambient IN at a given temperature to the number growth rate of cloud ice particles in a microphysics scheme. An approach described by Khain et al. (2000) can be appropriated to deal with this problem. A possible assumption is that new IN activation occurs only if the temperature decreases at a certain grid point. The number concentration of newly activated IN per time interval is given in a differential form:

$$\frac{dn_{\mathrm{IN}}}{dt} = \begin{cases} -\dfrac{dn_{\mathrm{IN}}}{dT}\dfrac{dT}{dt}, & \text{if } dT/dt \le 0 \\[2mm] 0, & \text{if } dT/dt > 0 \end{cases}; \qquad (13)$$

$$\frac{dT}{dt} = \left(\frac{\partial T}{\partial t} + u\frac{\partial T}{\partial x} + v\frac{\partial T}{\partial y} + w\frac{\partial T}{\partial z}\right), \qquad (14)$$

where u, v, and w are the wind velocities in the framework of the three-dimensional Cartesian coordinates x, y, and z, respectively. The number of newly activated IN is directly applied to that of newly generated cloud ice particles.

Another problem is how to connect $n_{a>0.5}$ in Eq. 12 to the predicted size distribution spectra of aerosol in a model coupled with the ALICIS module. Sulfate, black carbon, organic, dust, and sea salt aerosols were assumed in the global model simulations of Demott et al. (2010); they predicted aerosol mass and number concentrations for the three specific modes in particle size for each aerosol species. The term $n_{a>0.5}$ in Eq. 12 was directly calculated from the size distribution of the four aerosol species, except for sea salt. A similar approach can be employed in a model coupled with the ALICIS module; $n_{a>0.5}$ can be calculated from the predicted size distribution spectra of the same four aerosol species. Depletion of IN is represented by reducing the number concentration of aerosol particles with diameters larger than 0.5 μm. The priority (i.e., what type and size of aerosol particles are preferentially activated) was not discussed by Demott et al. (2010). At present, larger aerosol particles are preferentially consumed; the type dependence is determined arbitrarily.

Contact freezing

Unlike deposition and condensation-freezing nucleation, contact freezing has hardly been fully parametrized or modeled in earlier atmospheric models. It is difficult to obtain appropriate measurement data to construct a parametrization scheme or an empirical formula. Lohmann and Diehl (2006) employed a parametrization based on Levkov et al. (1992) to calculate the contact-freezing probability only through Brownian diffusion of aerosol particles in their global aerosol transport model with bulk cloud microphysics. The freezing rate $Q_{frz,cnt}$ (in $m^{-3} s^{-1}$) is given by

$$Q_{frz,cnt} = m_i D_{ap} 4\pi r_w N_{a,cnt} \frac{N_w^2}{\rho q_w},$$ (15)

$$D_{ap} = \frac{kT C_c}{6\pi \eta r_{am}},$$ (16)

where m_i is the mass of the newly formed ice particle through contact freezing, D_{ap} is the Brownian aerosol diffusivity, r_w is the volume mean radius of the ambient supercooled droplets, $N_{a,cnt}$ is the number concentration of aerosol particles that can be activated as contact freezing nuclei at the relevant temperature, N_w is the number concentration of the supercooled droplets, q_w is the mass-mixing ratio of the supercooled droplets, ρ is the air density, k is the Boltzmann constant, C_c the Cunningham correction factor, η the viscosity of air, and r_{am} the mode radius of the aerosol particles.

$N_{a,cnt}$ is the most important parameter for determining the efficiency of contact freezing in Eq. 15. Lohmann and Diehl (2006) assumed that hydrophilic black carbon and accumulation-mode dust particles were able to work as contact freezing nuclei; the activation probability was dependent on the particle composition and size, and on the temperature (see their Fig. 1). Dust aerosol particles can work as contact freezing nuclei at higher temperatures than black carbonaceous aerosol particles. If mineral dust particles are composed of montmorillonite, the onset temperature of contact freezing is approximately −3 °C; all dust particles can work as contact freezing

nuclei at −8 °C. The onset temperature of contact freezing mediated by black carbonaceous particles is approximately −10 °C; the probability of freezing is largely dependent on the particle size.

This parametrization can be implemented in a model coupled with the ALICIS module without major modifications. If SBM is employed in the model, Eq. 15 is applied to all combinations between supercooled droplet bins and aerosol bins, on a one-by-one basis. On the other hand, if bulk cloud microphysics is employed, Eq. 15 directly provides the production rates of the mass-mixing ratio and the number concentration of cloud ice formed through contact freezing, as described by Lohmann and Diehl (2006).

Immersion freezing

Some prior studies already developed several empirical parametrization schemes to calculate immersion freezing rates under conditions that the particle composition and concentration of immersion freezing nuclei inside supercooled droplets are specified (e.g., Diehl and Wurzler 2004). However, the present SBM coupled with the ALICIS module does not explicitly predict the concentration or composition of impurities that potentially act as immersion freezing nuclei inside supercooled droplets. Accordingly, the present model needs to adopt an empirical parametrization that excludes the detailed characteristics of immersion freezing nuclei. As an alternative approach, Lohmann and Diehl (2006) employed the assumption that the concentration of potential IN inside supercooled droplets was identical to that in ambient air (i.e., aerosol particle concentration). Similarly, Fan et al. (2014) employed a parametrization that included a mineral dust aerosol concentration with particle diameters larger than 0.5 μm (Tobo et al. 2013; DeMott et al. 2015). These approaches are still affected by large uncertainties, even though the ambient aerosol concentration is related to the number of activatable immersion IN inside supercooled droplets through wet scavenging.

Abbreviations

ABC: Atmospheric Brown Clouds; ABC–EAREX: Atmospheric Brown Cloud–East Asian Regional Experiment; ACE–Asia: Asian Pacific Regional Aerosol Characterization Experiment; AEROCOM: Aerosol Comparisons between Observations and Models; AERONET: AErosol RObotic NETwork; AGCM: atmospheric general circulation model; ALICIS: Aerosol Loading Interface for Cloud microphysics In Simulation; APEX–E3: 3rd Experiment of the Asian Atmospheric Particulate Environment Change Studies; ARM: Atmospheric Radiation Measurement; AVHRR: (space-borne) Advanced Very High Resolution Radiometer; CCN: cloud condensation nuclei; CDNC: cloud droplet number concentration; CN: condensation nuclei; CRM: cloud resolving model; DOE: Department of Energy; DRAGON: Distributed Regional Aerosol Gridded Observation Networks; GEOS: NASA Goddard Earth Observing System model; GOCART: Goddard Chemistry Aerosol Radiation and Transport model; HadGEM: Hadley Centre Global Environment Model; HUCM: Hebrew University Cloud Model; IN: ice-forming nuclei; INDOEX: INDian Ocean EXperiment; JMA–NHM: Japan

Meteorological Agency Non-Hydrostatic Model; MERRAero: Modern Era Retrospective analysis for Research and Applications Aerosol Reanalysis; MODIS: Modulate Resolution Imaging Spectroradiometer; MOUDI: Micro-Orifice Uniform Deposit Impactor; NASA: National Aeronautics and Space Administration; NU–WRF: NASA Unified WRF; SBM: spectral bin microphysics; SPRINTARS: Spectral Radiation Transport model for Aerosol Species; WRF: Weather Research and Forecasting model; WRF–Chem: Weather Research and Forecasting model coupled with chemistry; WRF–SBM: WRF coupled with HUCM SBM.

Competing interests

The authors declare that they have no competing interests.

Authors' contributions

TI proposed and designed the study and wrote most parts of the paper. IC and YS substantially contributed to writing the subsections of "Update by Choi et al. (2014): Aerosol particle size distribution revision and type multiplication" and "Update by Sato et al. (2012): Aerosol emission from evaporated cloud droplets," respectively. KS and TN collaborated with the other authors in the context of manuscript composition. All authors read and approved submission of the final manuscript.

Acknowledgements

This research was supported by the Global Environment Research Fund B-4 of the Ministry of Environment, Japan, by project RR2002 and the Data Integration for Earth Observation project of the Japanese Ministry of Education, Sports, Science, Culture, and Technology (MEXT). YS was supported by a Grant-in-Aid for JSPS Fellows (22-7893). IC was supported by the research and development project on the development of global numerical prediction systems of the Korea Institute of Atmospheric Prediction Systems (KIAPS), funded by the Korea Meteorological Administration (KMA). The authors thank the two anonymous reviewers and the journal editor for their helpful comments in improving this paper.

Author details

[1]Earth System Science Interdisciplinary Center, University of Maryland, College Park, MD, USA. [2]Code 612 NASA Goddard Space Flight Center, Greenbelt, MD, USA. [3]Korea Institute of Atmospheric Prediction Systems, Seoul, Republic of Korea. [4]RIKEN Advanced Institute for Computational Science, Kobe, Japan. [5]Atmosphere and Ocean Research Institute, The University of Tokyo, Kashiwa, Japan. [6]Earth Observing Research Center, Japan Aerospace Exploration Agency, Tsukuba, Japan.

References

Abdul-Razzak H, Ghan SJ, Rivera-Carpio C. A parameterization of aerosol activation 1. Single aerosol type. J Geophys Res. 1998;103(D6):6123–31.

Abdul-Razzak H, Ghan SJ. A parameterization of aerosol activation 2. Multiple aerosol types. J Geophys Res. 2000;105(D5):6837–44.

Abdul-Razzak H, Ghan SJ. A parameterization of aerosol activation 3. Sectional representation. J Geophys Res. 2002;107(D3):4026. doi:10.1029/2001JD000483.

Ackerman AS, Toon OB, Stevens DE, Heymsfield AJ, Ramanathan V, Welton EJ. Reduction of tropical cloudiness by soot. Science. 2000;288:1042–7. doi:10.1126/science.288.5468.1042.

Adhikari M, Ishizaka Y, Minda H, Kazaoka R, Jensen JB, Gras JL, et al. Vertical distribution of cloud condensation nuclei concentrations and their effect on microphysical properties of clouds over the sea near the southwest islands of Japan. J Geophys Res. 2005;110:D10203. doi:10.1029/2004JD004758.

Albrecht BA. Aerosols, cloud microphysics, and fractional cloudiness. Science. 1989;245:1227–30.

Bellouin N, Rae J, Jones A, Johnson C, Haywood J, Boucher O. Aerosol forcing in the Climate Model Intercomparison Project (CMIP5) simulations by HadGEM2-ES and the role of ammonium nitrate. J Geophys Res. 2011;116: D20206. doi:10.1029/2011JD016074.

Bigg EK. The supercooling of water. Proc Phys Soc London. 1953;66B:688–94. doi:10.1088/0370-1301/66/8/309.

Brechtel FJ, Kreidenweis SM, Swan HB. Air mass characteristics, total particle concentration, and size distributions at Macquarie Island, Tasmania, during the First Aerosol Characterization Experiment (ACE 1). J Geophys Res. 1998; 103:16,351–67. doi:10.1029/97JD03014.

Chen JP, Lamb D. Simulation of cloud microphysical and chemical processes using a multicomponent framework. Part I: description of the microphysical

model. J Atmos Sci. 1994;51:2613–30. doi:10.1175/1520-0469(1994)051<2613:SOCMAC>2.0.CO;2.

Chin M, Ginoux P, Kinne S, Holben BN, Duncan BN, Martin RV, et al. Tropospheric aerosol optical thickness from the GOCART model and comparisons with satellite and sunphotometer measurements. J Atmos Sci. 2002;59:461–83.

Choi I-J, Iguchi T, Kim S-W, Nakajima T, Yoon S-C. The effect of aerosol representation on cloud microphysical properties in northeast Asia. Meteorol Atmos Phys. 2014;123:181–94. doi:10.1007/s00703-013-0288-y.

Chuang CC, Penner JE, Taylor KE, Grossman AS, Walton JJ. An assessment of the radiative effects of anthropogenic sulfate. J Geophys Res. 1997;102:3761–78.

Colarco P, da Silva A, Chin M, Diehl T. Online simulations of global aerosol distributions in the NASA GEOS-4 model and comparisons to satellite and ground-based aerosol optical depth. J Geophys Res. 2010;115:D14207. doi:10.1029/2009JD012820.

d'Almeida GA, Koepk P, Shettle EP. Atmospheric aerosols: global climatology and radiative characteristics. Hampton: A. Deepak Publishing; 1991. p. 561.

DeMott PJ, Prenni AJ, Liu X, Petters MD, Twohy CH, Richardson MS, et al. Predicting global atmospheric ice nuclei distributions and their impacts on climate. Proc Natl Acad Sci. 2010;107:11217–22.

DeMott PJ, Prenni AJ, McMeeking GR, Sullivan RC, Petters MD, Tobo Y, et al. Integrating laboratory and field data qualify the immersion freezing ice nucleation activity of mineral dust particles. Atmos Chem Phys. 2015;15:393–409. doi:10.5194/acp-15-393-2015.

Diehl K, Wurzler S. Heterogeneous drop freezing in the immersion mode: Model calculations considering soluble and insoluble particles in the drops. J Atmos Sci. 2004;61:2063–72.

Eck TF, Holben BN, Reid JS, Arola A, Ferrare RA, Hostetler CA, et al. Observations of rapid aerosol optical depth enhancements in the vicinity of polluted cumulus clouds. Atmos Chem Phys. 2012;14:11633–56. doi:10.5194/acp-14-11633-2014.

Fan J, Leung LR, Li Z, Morrison H, Chen H, Zhou Y, et al. Aerosol impacts on clouds and precipitation in eastern China: results from bin and bulk microphysics. J Geophys Res. 2012;117:D00K36. doi:10.1029/2011JD016537.

Fan J, Leung LR, DeMott PJ, Comstock JM, Singh B, Rosenfeld D, et al. Aerosol impacts on California winter clouds and precipitation during CalWater 2011: local pollution versus long-range transport dust. Atmos Chem Phys. 2014;14: 81–101. doi:10.5194/acp-14-81-2014.

Feingold G, Kreidenweis SM, Stevens B, Cotton WR. Numerical simulations of stratocumulus processing of cloud condensation nuclei through collision-coalescence. J Geophys Res. 1996;101(D16):21,391–402.

Feingold G, Cotton WR, Kreidenweis SM, Davis JT. The impact of giant condensation nuclei on drizzle formation in stratocumulus: implications for cloud radiative properties. J Atmos Sci. 1999;56:4100–17. doi:10.1175/ 1520-0469(1999)056<4100:TIOGCC>2.0.CO;2.

Formenti P, Schütz L, Balkanski Y, Desboeufs K, Ebert M, Kandler K, et al. Recent progress in understanding physical and chemical properties of African and Asian mineral dust. Atmos Chem Phys. 2011;11:8231–56. doi:10.5194/acp-11-8231-2011.

Fu Q, Thorsen TJ, Su J, Ge JM, Huang JP. Test of Mie-based single-scattering properties of non-spherical dust aerosols in radiative flux calculations. J Quant Spectrosc Radiat Transf. 2009;110:1640–53.

Geresdi I, Rasmussen R. Freezing drizzle formation in stably stratified layer clouds. Part II: the role of giant nuclei and aerosol particle size distribution and solubility. J Atmos Sci. 2005;62:2037–57.

Ghan SJ, Chang CC, Penner JE. A parameterization of cloud droplet nucleation, I, single aerosol type. Atmos Res. 1993;30:197–221.

Ghan S, Laulainen N, Easter R, Wagener R, Nemesure S, Chapman E, et al. Evaluation of aerosol direct radiative forcing in MIRAGE. J Geophys Res. 2001; 106(D6):5295–316. doi:10.1029/2000JD900502.

Ginoux P, Chin M, Tegen I, Prospero JM, Holben B, Dubovik O, et al. Sources and distributions of dust aerosols simulated with the GOCART model. J Geophys Res. 2001;106:20 255–73. doi:10.1029/2000JD000053.

Goto D, Nakajima T, Dai T, Takemura T, Kajino M, Matsui H, et al. An evaluation of simulated particulate sulfate over East Asia through global model inter-comparison. J Geophys Res. 2015;120:6247–70. doi:10.1002/2014JD021693.

Grell GA, Peckham SE, Schmitz R, McKeen SA, Frost G, Skamarock WC, et al. Fully coupled 'online' chemistry in the WRF model. Atmos Environ. 2005;39:6957–76.

Hashino T, Tripori GJ. The spectral ice habit prediction system (SHIPS). Part I: model description and simulation of the vapor deposition process. J Atmos Sci. 2007;64:2210–37.

Haywood JM, Roberts DL, Slingo A, Edwards JM, Shine KP. General circulation model calculations of the direct radiative forcing by anthropogenic sulfate and fossil-fuel soot aerosol. J Climate. 1997;10:1562–77.

Herzog M, Weisenstein DK, Penner JE. A dynamic aerosol module for global chemical transport models: model description. J Geophys Res. 2004;109: D18202. doi:10.1029/2003JD004405.

Holben BN et al. AERONET—a federated instrument network and data archive for aerosol characterization. Remote Sens Environ. 1998;66:1–16.

Huebert BJ, Bates T, Russell PB, Shi G, Kim YJ, Kawamura K, et al. An overview of ACE-Asia: strategies for quantifying the relationships between Asian aerosols and their climatic impacts. J Geophys Res. 2003;108(D23):8633. doi:10.1029/2003JD003550.

Iguchi T, Nakajima T, Khain AP, Saito K, Takemura T, Suzuki K. Modeling the influence of aerosols on cloud microphysical properties in the east Asia region using a mesoscale model coupled with a bin-based cloud microphysics scheme. J Geophys Res. 2008;113:D14215. doi:10.1029/2007JD009774.

Iguchi T, Matsui T, Shi JJ, Tao W-K, Khain AP, Hou A, et al. Numerical analysis using WRF-SBM for the cloud microphysical structures in the C3VP field campaign: impacts of supercooled droplets and resultant riming on snow microphysics. J Geophys Res. 2012a;117:D23206. doi:10.1029/2012JD018101.

Iguchi T, Matsui T, Tokay A, Kollias P, Tao W-K. Two distinct modes in one-day rainfall event during MC3E field campaign: analyses of disdrometer observations and WRF-SBM simulation. Geophys Res Let. 2012b;39:L24805. doi:10.1029/2012GL053329.

Iguchi T, Matsui T, Tao W, Khain A, Phillips V, Kidd C, et al. WRF-SBM simulations of melting layer structure in mixedphase precipitation events observed during LPVEx. J Appl Meteor Climatol. 2014;53:2710–31. doi:10.1175/JAMC-D-13-0334.1.

Ishizaka Y. Report of aircraft observations during APEX campaigns (in Japanese), report, Hydrospheric Atmos. Res. Cent. Nagoya, Japan: Nagoya University; 2004. p. 82.

Jacobson MZ. Investigating cloud absorption effects: global absorption properties of black carbon, tar balls, and soil dust in clouds and aerosols. J Geophys Res. 2012;117:D06205. doi:10.1029/2011JD017218.

Khain AP, Sednev I. Simulation of precipitation formation in the eastern Mediterranean coastal zone using a spectral microphysics cloud ensemble model. Atmos Res. 1996;43:77–110. doi:10.1016/S0169-8095(96)00005-1.

Khain AP, Pokrovsky A, Sednev I. Some effects of cloud-aerosol interaction on cloud microphysics structure and precipitation formation: numerical experiments with a spectral microphysics cloud ensemble model. Atmos Res. 1999;52:195–220. doi:10.1016/S0169-8095(99)00027-7.

Khain AP, Ovtchinnikov M, Pinsky M, Pokrovsky A, Krugliak H. Notes on the state-of-the-art numerical modeling of cloud microphysics. Atmos Res. 2000;55: 159–224. doi:10.1016/S0169-8095(00)00064-8.

Khain A, Lynn B, Dudhia J. Aerosol effects on intensity of landfalling hurricanes as seen from simulations with the WRF model with spectral bin microphysics. J Atmos Sci. 2010;67(2):365–84.

Khain AP, Beheng KD, Heymsfield A, Korolev A, Krichak SO, Levin Z, et al. Representation of microphysical processes in cloud-resolving models: spectral (bin) microphysics vs. bulk parameterization. Rev Geophys. 2015;53: 247–322. doi:10.1002/2014RG000468.

Kinne S et al. An AeroCom initial assessment—optical properties in aerosol component modules of global models. Atmos Chem Phys. 2006;6:1815–34. doi:10.5194/acp-6-1815-2006.

Kogan YL. The simulation of a convective cloud in a 3-D model with explicit microphysics. Part I: model description and sensitivity experiments. J Atmos Sci. 1991;48:1160–89. doi:10.1175/1520-0469(1991)048<1160: TSOACC>2.0.CO;2.

Köhler H. The nucleus in and the growth of hygroscopic droplets. Trans Faraday Soc. 1936;32:1152–61. doi:10.1039/tf9363201152.

Koop T, Luo BP, Tsias A, Peter T. Water activity as the determinant for homogeneous ice nucleation in aqueous solutions. Nature. 2000;406:611–4. doi:10.1038/35020537.

Kuba N, Fujiyoshi Y. Development of a cloud microphysical model and parameterizations to describe the effect of CCN on warm cloud. Atmos Chem Phy. 2006;6:2793–810.

Laaksonen A, Korhonen P, Kulmala M, Charlson RJ. Modification of the Köhler equation to include soluble trace gases and slightly soluble substances. J Atmos Sci. 1998;55:853–62.

Levin Z, Cotton WR. Aerosol pollution impact on precipitation: a scientific review. Springer Science & Business Media. 2008.

Levkov L, Rockel B, Kapitza H, Raschke E. 3D mesoscale numerical studies of cirrus and stratus clouds by their time and space evolution. Beitr Phys Atmos. 1992; 65:35–58.

Li G, Wang Y, Zhang R. Implementation of a two-moment bulk microphysics scheme to the WRF model to investigate aerosol-cloud interaction. J Geophys Res. 2008;113:D15211. doi:10.1029/2007JD009361.

Liu Y, Daum PH. Parameterization of the autoconversion process. Part I: analytical formulation of the Kessler-type parameterizations. J Atmos Sci. 2004;61:1539–48.

Lohmann U, Diehl K. Sensitivity studies of the importance of dust ice nuclei for the indirect aerosol effect on stratiform mixed-phase clouds. J Atmos Sci. 2006;63:968–82.

Lohmann U, Stier P, Hoose C, Ferrachat S, Kloster S, Roeckner E, et al. Cloud microphysics and aerosol indirect effects in the global climate model ECHAM5-HAM. Atmos Chem Phys. 2007;7:3425–46. doi:10.5194/acp-7-3425-2007.

Low RDH. A generalized equation for the solution effect in droplet growth. J Atmos Sci. 1969;26:608–11. doi:10.1175/1520-0469(1969)026<0608: AGEFTS>2.0.CO;2.

Lynn BH, Khain AP, Dudhia J, Rosenfeld D, Pokrovsky A, Seifert A. Spectral (bin) microphysics coupled with a mesoscale model (MM5). Part I: model description and first results. Mon Weather Rev. 2005a;133(1):44. doi:10.1175/MWR-2840.1.

Lynn BH, Khain AP, Dudhia J, Rosenfeld D, Pokrovsky A, Seifert A. Spectral (bin) microphysics coupled with a mesoscale model (MM5). Part II: simulation of a CaPE rain event with a squall line. Mon Weather Rev. 2005b;133(1):59. doi:10.1175/MWR-2841.1.

Marenco F, Amiridis V, Marinou E, Tsekeri A, Pelon J. Airborne verification of CALIPSO products over the Amazon: a case study of daytime observations in a complex atmospheric scene. Atmos Chem Phys. 2014;14:11871–81. doi:10.5194/acp-14-11871-2014.

Misumi R, Hashimoto A, Murakami M, Kuba N, Orikasa N, Saito A, et al. Microphysical structure of a developing convective snow cloud simulated by an improved version of the multi-dimensional bin model. Atmos Sci Let. 2010;11:186–91. doi:10.1002/asl.268.

Mitra SK, Brinkmann J, Pruppacher HR. A wind tunnel study on the drop-to-particle conversion. J Aerosol Sci. 1992;23:245–56.

Mochida M, Umemoto N, Kawamura K, Lim H-J, Turpin BJ. Bimodal size distributions of various organic acids and fatty acids in the marine atmosphere: influence of anthropogenic aerosols, Asian dusts, and sea spray off the coast of East Asia. J Geophys Res. 2007;112(15):D15209. doi:10.1029/2006JD007773.

Morrison H, Curry JA, Khvorostyanov VI. A new double-moment microphysics scheme for application in cloud and climate models. Part I: description. J Atmos Sci. 2005;62:1665–77.

Morrison H, Grabowski WW. Comparison of bulk and bin warm-rain microphysics models using a kinematic framework. J Atmos Sci. 2007;64:2839–61.

Morrison H, Gettelman A. A new two-moment bulk stratiform cloud microphysics scheme in the community atmosphere model, version 3 (CAM3). Part I: description and numerical tests. J Climate. 2008;21:3642–59. doi:10.1175/2008JCLI2105.1.

Muhlbauer A, Lohmann U. Sensitivity studies of aerosol-cloud interactions in mixed-phase orographic precipitation. J Atmos Sci. 2009;66:2517–38.

Nakajima TY, Nakajima T. Wide-area determination of cloud microphysical properties from NOAA AVHRR measurements for FIRE and ASTEX regions. J Atmos Sci. 1995;52:4043–59. doi:10.1175/1520-0469(1995)052<4043: WADOCM>2.0.CO;2.

Nakajima T, Higurashi A, Kawamoto K, Penner JE. A possible correlation between satellite-derived cloud and aerosol microphysical parameters. Geophys Res Let. 2001;28:1171–4.

Nakajima TY, Uchiyama A, Takamura T, Tsujioka N, Takemura T, Nakajima T. Comparisons of warm cloud properties obtained from satellite, ground, and aircraft measurements during APEX intensive observation period in 2000 and 2001. J Meteorol Soc Jpn. 2005;83:1085–95. doi:10.2151/jmsj.83.1085.

Nakajima T et al. Overview of the Atmospheric Brown Cloud East Asian Regional Experiment 2005 and a study of the aerosol direct radiative forcing in east Asia. J Geophys Res. 2007;112:D24S91. doi:10.1029/2007JD009009.

Onishi R, Takahashi K. A warm-bin-cold-bulk hybrid cloud microphysical model. J Atmos Sci. 2012;69:1474–97.

Park S, Bretherton CS, Rasch PJ. Integrating cloud processes in the Community Atmosphere Model, version 5. J Climate. 2014;27:6821–56. doi:10.1175/JCLI-D-14-00087.1.

Penner JE, Chuang CC, Grant K. Climate forcing by carbonaceous and sulfate aerosols. Clim Dynam. 1998;14:839–51.

Peters-Lidard CD, Kemp EM, Matsu T, Santanello Jr JA, Kumar SV, Jacob JP, et al. Integrated modeling of aerosol, cloud, precipitation and land processes at satellite-resolved scales. Environ Model Software. 2015;67:149–59. http://dx.doi.org/10.1016/j.envsoft.2015.01.007.

Petters MD, Kreidenweis SM. A single parameter representation of hygroscopic growth and cloud condensation nucleus activity. Atmos Chem Phys. 2007; 7:1961–71.

Phillips VTJ, DeMott PJ, Andronache C. An empirical parameterization of heterogeneous ice nucleation for multiple chemical species of aerosol. J Atmos Sci. 2008;65:2757–83.

Phillips VTJ, Demott PJ, Andronache C, Pratt KA, Prather KA, Subramanian R, et al. Improvements to an empirical parameterization of heterogeneous ice nucleation and its comparison with observations. J Atmos Sci. 2013;70:378–409. http://dx.doi.org/10.1175/JAS-D-12-080.1.

Pincus R, Baker MB. Effect of precipitation on the albedo susceptibility of clouds in the marine boundary layer. Nature. 1994;372:250–2. doi:10.1038/372250a0.

Pozzoli L, Janssens-Maenhout G, Diehl T, Bey I, Schultz MG, Feichter J, et al. Re-analysis of tropospheric sulfate aerosol and ozone for the period 1980–2005 using the aerosol-chemistry-climate model ECHAM5-HAMMOZ. Atmos Chem Phys. 2011;11:9563–94. doi:10.5194/acp-11-9563-2011.

Pruppacher HR, Klett JD. Microphysics of clouds and precipitation. Dordrecht, Netherlands: Kluwer Acad; 1997.

Ramanathan V et al. Indian Ocean Experiment: an integrated analysis of the climate forcing and effects of the great Indo-Asian haze. J Geophys Res. 2001;106(D22):28371–98. doi:10.1029/2001JD900133.

Ramanathan V, Ramana MV, Roberts G, Kim D, Corrigan C, Chung C, et al. Warning trends in Asia amplified by brown cloud solar absorption. Nature. 2007;448:575–8. doi:10.1038/nature06019.

Rasmussen RM, Geresdi I, Thompson G, Manning K, Karplus E. Freezing drizzle formation in stably stratified layer clouds: the role of radiative cooling of cloud droplets, cloud condensation nuclei, and ice initiation. J Atmos Sci. 2002;59:837–60.

Roberts G, Nenes A. A continuous-flow stream-wise thermal gradient CCN chamber for atmospheric measurements. Aerosol Sci Technol. 2005;39:206–21.

Rogers RR, Yau MK. A short course in cloud physics. New York: Pergamon; 1989.

Rosenfeld D et al. Global observations of aerosol-cloud-precipitation-climate interactions. Rev Geophys. 2014;52:750–808. doi:10.1002/2013RG000441.

Saito K et al. The operational JMA nonhydrostatic mesoscale model. Mon Weather Rev. 2006;134(4):1266. doi:10.1175/MWR3120.1.

Saleeby SM, Cotton WR. A large droplet mode and prognostic number concentration of cloud droplets in the Colorado State University Regional Atmospheric Modeling System (RAMS). Part I: module descriptions and supercell test simulations. J Appl Meteor. 2004;43:182–95.

Saleeby SM, van den Heever SC. Developments in the CSU-RAMS aerosol model: emissions, nucleation, regeneration, deposition, and radiation. J Appl Meteor Climatol. 2013;52:2601–22. doi:10.1175/JAMC-D-12-0312.1.

Sato Y, Suzuki K, Iguchi T, Choi I-J, Kadowaki H, Nakajima T. Characteristics of correlation statistics between droplet radius and optical thickness of warm clouds simulated by a three-dimensional regional-scale spectral bin microphysics cloud model. J Atmos Sci. 2012;69:484–503. doi:10.1175/JAS-D-11-076.1.

Sato Y. A numerical study on the microphysical properties of warm clouds off the west coast of California, CCSR report, vol. 49. Tokyo: The Univesity of Tokyo; 2012.

Seifert A, Beheng KD. A two-moment cloud microphysics parameterization for mixed-phase clouds. Part 1: model description. Meteorol Atmos Phys. 2006; 92:45–66. doi:10.1007/s00703-005-0112-4.

Seiki T, Nakajima T. Aerosol effects of the condensation process on a convective cloud simulation. J Atmos Sci. 2014;71:833–53. doi:10.1175/JAS-D-12-0195.1.

Seiki T, Kodama C, Noda AT, Satoh M. Improvement in global cloud-system-resolving simulations by using a double-moment bulk cloud microphysics scheme. J Climate. 2015;28:2405–19. doi:10.1175/JCLI-D-14-00241.1.

Shi JJ, Matsui T, Tao W-K, Tan Q, Peters-Lidard C, Chin M, et al. Implementation of an aerosol–cloud-microphysics–radiation coupling into the NASA unified WRF: simulation results for the 6–7 August 2006 AMMA special observing period. Q J Roy Meteorol Soc. 2014;140:2158–75. doi:10.1002/qj.2286.

Shima S, Kusano K, Kawano A, Sugiyama T, Kawahara S. The super-droplet method for the numerical simulation of clouds and precipitation: a particle-based and probabilistic microphysics model coupled with a non-hydrostatic model. Q J Roy Meteorol Soc. 2009;135:1307–20. doi:10.1002/qj.441.

Skamarock WC, Kemp JB, Dudhia J, Gill DO, Barker DM, Wang W, et al. A description of the advanced research WRF version 2, NCAR Tech, NCAR Tech. Note NCAR/TN-468+STR. Boulder, Colo: Natl. Cent. for Atmos. Res; 2005. p. 88.

Soong S-T, Ogura Y. A comparison between axisymmetric and slab-symmetric cumulus cloud models. J Atmos Sci. 1973;30:879–93.

Spracklen DV, Pringle KJ, Carslaw KS, Chipperfield MP, Mann GW. A global off-line model of size-resolved aerosol microphysics: I. Model development and prediction of aerosol properties. Atmos Chem Phys. 2005;5:2227–52.

Squires P. The micro-structure of cumuli in maritime and continental air. Tellus. 1956;8:443–4.

Stokes GM, Schwartz SE. The Atmospheric Radiation Measurement (ARM) Program: programmatic background and design of the cloud and radiation test bed. Bull Am Meteorol Soc. 1994;75:1201–21. doi:10.1175/1520-0477(1994)075<1201:TARMPP>2.0.CO;2.

Storelvmo T, Kristjánsson JE, Ghan SJ, Kirkevåg A, Seland Ø, Iversen T. Predicting cloud droplet number concentration in Community Atmosphere Model (CAM)-Oslo. J Geophys Res. 2006;111:D24208. doi:10.1029/2005JD006300.

Storelvmo T, Kristjánsson JE, Lohmann U. Aerosol influence on mixed-phase clouds in CAM-Oslo. J Atmos Sci. 2008;65:3214–30.

Sullivan RC, Moore MJK, Petters MD, Kreidenweis SM, Roberts GC, Prather KA. Effect of chemical mixing state on the hygroscopicity and cloud nucleation properties of calcium mineral dust particles. Atmos Chem Phys. 2009;9:3303–16.

Suzuki K, Nakajima T, Nakajima TY, Khain A. Correlation pattern between effective radius and optical thickness of water clouds simulated by a spectral bin microphysics cloud model. SOLA. 2006;2:116–9.

Suzuki K, Nakajima T, Satoh M, Tomita H, Takemura T, Nakajima TY, et al. Global cloud-system-resolving simulation of aerosol effect on warm clouds. Geophys Res Let. 2008;35:L19817. doi:10.1029/2008GL035449.

Suzuki K, Nakajima T, Nakajima TY, Khain AP. A study of microphysical mechanisms for correlation patterns between droplet radius and optical thickness of warm clouds with a spectral bin microphysics cloud model. J Atmos Sci. 2010;67:1126–41. doi:10.1175/2009JAS3283.1.

Takemura T, Okamoto H, Maruyama Y, Numaguti A, Higurashi A, Nakajima T. Global three-dimensional simulation of aerosol optical thickness distribution of various origins. J Geophys Res. 2000;105(D14):17,853–74. doi:10.1029/2000JD900265.

Takemura T, Nakajima T, Dubovik O, Holben BN, Kinne S. Single-scattering albedo and radiative forcing of various aerosol species with a global three-dimensional model. J Climate. 2002;15:333–52.

Takemura T, Egashira M, Matsuzawa K, Ichijo H, O'ishi R, Abe-Ouchi A. A simulation of the global distribution and radiative forcing of soil dust aerosols at the Last Glacial Maximum. Atmos Chem Phys. 2009;9:3061–73.

Tao W-K, Simpson J, McCumber M. An ice-water saturation adjustment. Mon Weather Rev. 1989;117:231–5.

Tao W-K, Chen J-P, Li Z, Wang C, Zhang C. Impact of aerosols on convective clouds and precipitation. Rev Geophys. 2012;50:RG2001. doi:10.1029/2011RG000369.

Thompson G, Eidhammer T. A study of aerosol impacts on clouds and precipitation development in a large winter cyclone. J Atmos Sci. 2014;71: 3636–58. doi:10.1175/JAS-D-13-0305.1.

Tobo Y, Prenni AJ, DeMott PJ, Huffman JA, McCluskey CS, Tian R, et al. Biological aerosol particles as a key determinant of ice nuclei populations in a forest ecosystem. J Geophys Res. 2013;118:10100–10. doi:10.1002/jgrd.50801.

Twomey S. The nuclei of natural cloud formation: Part II. The supersaturation in natural clouds and the variation of cloud droplet concentration. Geofis Pura Appl. 1959b;43:243–9. doi:10.1007/BF01993560.

Twomey S. Pollution and the planetary albedo. Atmos Environ. 1974;8:1251–6.

Twomey S. The influence of pollution on the shortwave albedo of clouds. J Atmos Sci. 1977;34:1149–52.

Twomey SA, Piepgrass M, Wolfe TL. An assessment of the impact of pollution on global cloud albedo. Tellus B. 1984;36:5.

Vali G. Nucleation terminology. Bull Am Meteorol Soc. 1985;66:1426–7.

Wang M, Ghan S, Ovchinnikov M, Liu X, Easter R, Kassianov E, et al. Aerosol indirect effects in a multi-scale aerosol-climate model PNNL-MMF. Atmos Chem Phys. 2011;11:5431–55. doi:10.5194/acp-11-5431-2011.

Wang Y, Fan J, Zhang R, Leung LR, Franklin C. Improving bulk microphysics parameterizations in simulations of aerosol effects. J Geophys Res Atmos. 2013;118:5361–79. doi:10.1002/jgrd.50432.

Xue L, Teller A, Rasmussen R, Geresdi I, Pan Z. Effects of aerosol solubility and regeneration on warm-phase organic clouds and precipitation simulated by a detailed bin microphysical scheme. J Atmos Sci. 2010;67:3336–54. doi:10.1175/2010JAS3511.1.

Yum SS, Roberts G, Kim JH, Song K, Kim D. Submicron aerosol size distributions and cloud condensation nuclei concentrations measured at Gosan, Korea, during the Atmospheric Brown Clouds–East Asian Regional Experiment 2005. J Geophys Res. 2007;112:D22S32. doi:10.1029/2006JD008212.

Yumimoto K, Takemura T. The SPRINTARS version 3.80/4D-Var data assimilation system: development and inversion experiments based on the observing system simulation experiment framework. Geosci Model Dev. 2013;6:2005–22. doi:10.5194/gmd-6-2005-2013.

Zhang K, Wan H, Wang B, Zhang M, Feichter J, Liu X. Tropospheric aerosol size distributions simulated by three online global aerosol models using the M7 microphysics module. Atmos Chem Phys. 2010;10:6409–34. doi:10.5194/acp-10-6409-2010.

The Non-hydrostatic Icosahedral Atmospheric Model: description and development

Masaki Satoh[1,2*], Hirofumi Tomita[3,2], Hisashi Yashiro[3], Hiroaki Miura[4,2,3], Chihiro Kodama[2], Tatsuya Seiki[2], Akira T Noda[2], Yohei Yamada[2,1], Daisuke Goto[5], Masahiro Sawada[1], Takemasa Miyoshi[3], Yosuke Niwa[6], Masayuki Hara[2], Tomoki Ohno[1], Shin-ichi Iga[3], Takashi Arakawa[7,2], Takahiro Inoue[7,2] and Hiroyasu Kubokawa[1]

Abstract

This article reviews the development of a global non-hydrostatic model, focusing on the pioneering research of the Non-hydrostatic Icosahedral Atmospheric Model (NICAM). Very high resolution global atmospheric circulation simulations with horizontal mesh spacing of approximately O (km) were conducted using recently developed supercomputers. These types of simulations were conducted with a specifically designed atmospheric global model based on a quasi-uniform grid mesh structure and a non-hydrostatic equation system. This review describes the development of each dynamical and physical component of NICAM, the assimilation strategy and its related models, and provides a scientific overview of NICAM studies conducted to date.

Keywords: Global non-hydrostatic model; Icosahedral grid; Global cloud-resolving simulations

Review

Introduction

Diabatic heating due to the release of latent heat in deep convection is the primary heat source in the atmosphere, and it is interacted with the atmospheric general circulation, especially the tropical large-scale overturning circulations such as the Hadley and Walker circulations. Individual deep convective cells are associated with meso-scale circulations that have a horizontal scale of O (10 km), and an upward convective core, along with a horizontal scale of O (km). Until recently, since the horizontal resolution of the global climate models that have been used for future climate change projections has been O (100 km), such models require the use of cumulus parameterizations in order to incorporate the effects of deep convection instead of by explicitly resolving deep convective circulations. However, it is known that cumulus parameterizations significantly affect the results of climate model simulations and that they are the most ambiguous factor used in climate models (Randall et al. 2003).

To overcome the above-mentioned cumulus parameterization issue, global non-hydrostatic models that utilize a horizontal mesh interval of O (km) for global atmospheric circulation simulations have been developed. Such models explicitly calculate deep convective circulations over the global domain without using cumulus parameterizations. At the grid-resolvable scale, water vapor is saturated into the liquid or ice phase of water in the upward flow field in order to form clouds and is eventually converted to rain and snow through cloud microphysics processes. In global non-hydrostatic models, clouds are spontaneously organized and the multi-scale structures of convective systems are reproduced over the global domain.

The Non-hydrostatic Icosahedral Atmospheric Model (NICAM) (Tomita and Satoh 2004; Satoh et al. 2008; Satoh 2013) was first designed to be run with a horizontal mesh size approximately 3.5 km over the global domain by using the Earth Simulator (http://www.jamstec.go.jp/es/en/) which was launched by the Japan Agency for Marine-Earth Science and Technology (JAMSTEC) in 2002. NICAM uses an icosahedral grid, as shown in Figure 1. Higher resolution grids are recursively subdivided from a coarser resolution grid. Hereinafter, we will refer to the grid division level as the g-level. The number of points, arcs, and triangles of the icosahedral grids with g-level l are given as follows:

* Correspondence: satoh@aori.u-tokyo.ac.jp
[1]Atmosphere and Ocean Research Institute, The University of Tokyo, 5-1-5 Kashiwanoha, Kashiwa, Chiba 277-85648, Japan
[2]Japan Agency for Marine-Earth Science and Technology, 3173-15, Showa-machi, Kanazawa-ku, Yokohama, Kanagawa 236-0001, Japan
Full list of author information is available at the end of the article

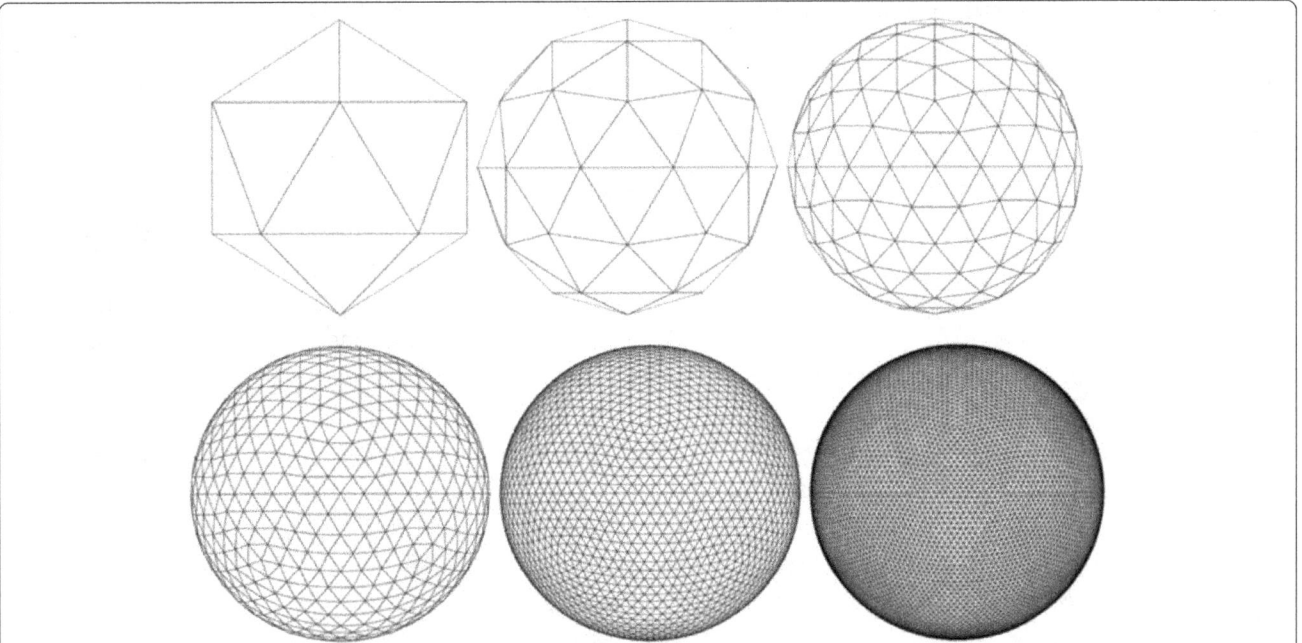

Figure 1 Icosahedral grids. Icosahedral grids for grid division levels of 0, 1, and 2 (top, from left to right), and 3, 4, and 5 (bottom, from left to right).

$$N_P = 10n^2 + 2 = 10 \times 2^{2l} + 2,$$

$$N_A = 30n^2 = 30 \times 2^{2l},$$

$$N_T = 20n^2 = 20 \times 2^{2l},$$

where $n = 2^l$. We define the average area of the triangles $\bar{A}$ and the average grid interval $\bar{\varDelta}$ as follows:

$$\bar{A} = \frac{4\pi R^2}{N_T} = \frac{\pi R^2}{5 \times 2^{2l}},$$

$$\bar{\varDelta} = \sqrt{2\bar{A}} = \sqrt{\frac{2\pi}{5}} \frac{R}{2^l},$$

where an Earth radius of $R = 6{,}371.22$ km is used. The values for each g-level are listed in Table 1.

NICAM has been shown to reproduce a realistic multi-scale cloud structure from a meso-scale to a planetary-scale cloud organization that is associated with the Madden-Julian Oscillation (MJO) (Madden and Julian 1971, 1972) at a g-level between 9 ($\bar{\varDelta} = 14$ km) and 11 ($\bar{\varDelta} = 3.5$ km) (Tomita et al. 2005; Miura et al. 2007b). By using the K computer, which is installed at the RIKEN Advanced Institute for Computational Science (AICS) in Kobe, Japan (http://www.nsc.riken.jp/index-eng.html), the resolution of NICAM has been recently increased to the subkilometer level, and it was shown that the deep convective core is more realistically resolved by using a g-level 13 ($\bar{\varDelta} = 870$ m) mesh simulation (Miyamoto et al. 2013; Figure 2). In the current study, various experiments including decadal continuous experiments and case sweep experiments (Miyakawa et al. 2014) were also conducted at g-levels between 9 and 11. Future projection studies such as those investigating changes in clouds and tropical cyclone activities are also investigated. The results are interpreted based on the more physically based cloud microphysics processes without the ambiguities of cumulus parameterizations.

This article describes the current development status, design, and concepts behind the individual components of NICAM. First, the background and an overview of studies related to the global non-hydrostatic model are reviewed. Then, the history of NICAM development and the scientific outcomes are summarized in the 'Scientific overview of NICAM' section, while the NICAM computational design is described in the 'Design, structure, development, and timeline' section. The two sections that follow describe the dynamical and physical components of NICAM, and their respective subsections describe each component of the physical processes. Next, the assimilation strategy is described. Finally, various NICAM usages that have been developed by modifying the original NICAM geometry are presented.

Review of global non-hydrostatic models

Thanks to the significant advances in high-performance computers over the last decade, global atmospheric simulations with a horizontal resolution of O (10 km) can be achieved (Ohfuchi et al. 2004; Mizuta et al. 2005; Kinter et al. 2013; Wedi 2014). At enhanced horizontal resolution of less than 10 km, traditional atmospheric

Table 1 Number of points, arcs, triangles, and average grid interval of the icosahedral grids

g-level	Points N_P	Arcs N_A	Triangles N_T	Average area $\bar{A}$ (km²)	Average interval $\bar{\Delta}$ (km)
0	12	32	20	5,100,996.991	7,142.126
1	42	122	80	1,275,249.248	3,571.063
2	162	482	320	318,812.312	1,785.532
3	642	1,922	1,280	79,703.078	892.766
4	2,562	7,682	5,120	19,925.769	446.383
5	10,242	30,722	20,480	4,981.442	223.191
6	40,962	122,882	81,920	1,245.361	111.596
7	163,842	491,522	327,680	311.340	55.798
8	655,362	1,966,082	1,310,720	77.835	27.899
9	2,621,442	7,864,322	5,242,880	19.459	13.949
10	10,485,762	31,457,282	20,971,520	4.865	6.975
11	41,943,042	125,829,122	83,886,080	1.216	3.487
12	167,772,162	503,316,482	335,544,320	0.304	1.744
13	671,088,642	2,013,265,922	1,342,177,280	0.076	0.872
14	2,684,354,562	8,053,063,682	5,368,709,120	0.019	0.436

The g-level is the grid division level. Here, $R = 6371.22$ km.

general circulation models (AGCMs) encounter fundamental difficulties in their dynamic framework formulation as well as in their computational efficiency. As their resolutions increase, AGCMs can capture flow features with comparable scales of motion in the horizontal and vertical directions (such as deep convection and fine-scale gravity waves) that can invalidate the hydrostatic approximation. Although deep convective systems in the tropics play key roles in global atmospheric circulations, they have not been directly resolved by AGCMs, and their effects have only been considered in a parameterized form

(Arakawa 2004). The effects of fine-scale gravity waves that are parameterized as gravity wave drag in AGCMs can be captured as the resolution increases, but the propagation of such gravity waves will be incorrectly calculated unless the non-hydrostatic effect is taken into consideration (Iwasaki et al. 1989).

Regarding the numerical algorithm, many of the existing AGCMs employ the spectral transform method to represent spherical fields, and it has been pointed out that spectral transforms become increasingly inefficient for high-performance computing as the horizontal resolution

Figure 2 Cloud distribution simulated by the NICAM 870 m grid spacing experiment for 6:00 UTC 25 August 2012 (Miyamoto et al. 2013).

increases (e.g., Stuhne and Peltier 1996; Taylor et al. 1997; Randall et al. 2000; Satoh et al. 2005; Cheong 2006; Tomita et al. 2008; Wedi 2014). Another problem that occurs during computation on a massively parallel computer is that the spectral transform method requires extensive data movement between computer nodes. Although the double Fourier transformation method has been proposed as an alternative (e.g., Cheong 2006), this method still requires global communication between computer nodes.

To increase the horizontal resolution beyond O (10 km) in a global atmospheric model, the governing equations and numerical algorithms must be reconsidered. More specifically, the governing equations must be non-hydrostatic, and the grid point method replaces the spectral method. As for the familiar latitude-longitude grid (lat-lon grid), however, the grid spacing near the poles becomes drastically reduced as the horizontal resolution is increased, which means that reduced grids are generally required to avoid severe time interval restrictions for the Courant-Friedrichs-Lewy (CFL) condition. In principle, the semi-Lagrangian, semi-implicit (SLSI) approach (cf. Laprise 2008; Staniforth and Wood 2008) could be employed to overcome the requirement of the CFL condition. Several authors (Semazzi et al. 1995; Cullen et al. 1997; Qian et al. 1998; Côté et al. 1998; Yeh et al. 2002; Davies et al. 2005; Wedi and Smolarkiewicz 2009; Wood et al. 2013) have used the lat-lon grid to solve a set of non-hydrostatic equations using the SLSI approach in order to acquire a larger time interval for integration. However, it is unclear how effective the elliptic solvers, which were developed for SLSI schemes, would be for ultra-high-resolution calculations, even though it has been recently proven that a multi-grid approach is an ideal solver for massively parallel computers (Heikes et al. 2013).

The pole problem can be overcome by using grid systems with quasi-homogeneous grids over the sphere. One such grid is the icosahedral grid, which is currently one of the major grid systems used for high-resolution global atmospheric modeling. Primitive (hydrostatic) equation global models using icosahedral grids have been developed at Colorado State University for climate modeling (CSU AGCM) (Ringler et al. 2000), Deutscher Wetterdienst for numerical prediction modeling (GME) (Majewski et al. 2002), the National Oceanic and Atmospheric Administration (NOAA) for the Flow-following finite-volume Icosahedral Model (FIM) (http://fim.noaa.gov/), and the Laboratoire de Météorologie Dynamique (DYNAMICO) (http://www.lmd.polytechnique.fr/~dubos/DYNAMICO/). Global non-hydrostatic models using icosahedral grids are also being developed by several international groups including the geodesic grid model at Colorado State University (UZIM; http://www.cmmap.org/research/models.html), the Icosahedral Non-hydrostatic model (ICON) at the Deutscher Wetterdienst and the Max Planck Institute for Meteorology (http://www.mpimet.mpg.de/en/science/models/icon.html; Zängl et al, 2014), the Model for Prediction Across Scales (MPAS) at the National Center for Atmospheric Research (Skamarock et al. 2012), and the Non-hydrostatic Icosahedral Model (NIM) at NOAA (http://www.esrl.noaa.gov/gsd/ab/ac/GPU_Parallelization_NIM.html). Entries for all these models can also be found at https://www.earthsystemcog.org/projects/dcmip-2012/. As for other types of grid models, cubic grids are also candidates for high-resolution atmospheric models (McGregor 1996; Lin 2004; Putman and Suarez 2011).

In Japan, the icosahedral atmospheric model has been created by using a non-hydrostatic system, i.e., NICAM (Tomita and Satoh 2004; Satoh et al. 2008; Satoh 2013; http://nicam.jp/). Development of NICAM began around 2000. Since high-resolution modeling has now entered the mainstream worldwide, NICAM has joined a number of international high-resolution numerical modeling projects, such as the Athena Project (Kinter et al. 2013) with the Integrated Forecast System (IFS) (Jung et al. 2012) and the Icosahedral-grid Models for Exascale Earth system simulations (ICOMEX) (Zängl et al. 2011) with ICON, MPAS, and DYNAMICO.

Scientific overview of NICAM

While more than a decade has passed since the development of NICAM began, the first milestone was reached in 2004 when Tomita and Satoh (2004) began finalizing the dynamical core of the model. Two unique characteristics of the NICAM dynamical core are the use of the spring dynamics to construct a modified icosahedral grid (Tomita et al. 2001, 2002) and the use of a flux form non-hydrostatic system which guarantees conservation of mass and energy (Satoh 2002, 2003). Tomita et al. (2001, 2002) constructed NICAM's icosahedral grids by improving the numerical accuracy of differential operators and using spring dynamics to improve the homogeneity of the grid system. An improved version of the spring dynamics method, which is more homogeneous and applicable even for higher resolution, is also now available (Iga and Tomita 2014). NICAM uses a fully compressible (elastic) non-hydrostatic system to obtain statistically equilibrium states by performing long-timescale simulations. For this purpose, we devised a non-hydrostatic numerical scheme that guarantees the conservation of mass and energy (Satoh 2002, 2003). Tomita and Satoh (2004) implemented this non-hydrostatic scheme in their global model using an icosahedral grid configuration, with which they developed a dynamical core for NICAM. Since the split-explicit time integration scheme (Klemp and Wilhemson 1978) is used for the horizontal propagation of fast waves and for the implicit treatment of the vertical propagation of fast waves, multi-dimensional elliptic solvers are not required. To extend the original non-hydrostatic scheme to the global domain, the set of equations provided in Satoh (2002, 2003) is reformulated for spherical geometry and modified in order to make the equations suitable for an icosahedral grid configuration. The finite volume method is used for numerical discretization, so that the total mass and energy over the domain are conserved. The resulting model is suitable for long-term climate simulations.

NICAM is used for 'cloud-resolving simulations' by explicitly resolving convective circulation. This is accomplished by drastically increasing the horizontal resolution using high-power computing systems, such as Earth Simulator and the K computer. After finalizing the development of the dynamical core (Tomita and Satoh 2004), the first 3.5 km mesh global NICAM simulation was performed on Earth Simulator by Tomita et al. (2005) using the aqua planet configuration (Neale and Hoskins 2001). The results of this simulation clearly show multi-scale convective systems propagating near the equator (Satoh et al. 2005), which is similar to the propagation of the observed cloud clusters and super cloud cluster structure (Hayashi and Sumi 1986; Nakazawa 1988; Takayabu et al. 1999).

Tomita et al. (2005) conducted aqua planet experiments at three horizontal resolutions, g-levels 9, 10, and 11 ($\bar{\Delta}$ = 14, 7, 3.5 km), using the same physical schemes without cumulus parameterization. In their results, the multi-scale convective structure along the equatorial zone and the diurnal cycle of convective precipitation were reproduced similarly in all simulations, although the propagation speed along the equator and the phase lag of the diurnal cycle depends on the resolution. This result motivated us to use a combination of different horizontal resolution experiments with the same physical schemes when studying convective properties simulated by NICAM.

The results of the NICAM aqua planet experiment were analyzed intensively by Nasuno et al. (2007, 2008), Nasuno (2008), and Nasuno and Satoh (2011a, b), which showed the roles of the equatorial convective systems embedded in the multi-scale structure. Mapes et al. (2008) also investigated the persistent structure in the tropics in terms of the predictability of disturbances. The diurnal cycle of tropical convective precipitation was clearly simulated with the dependency of the phase lag on the horizontal resolution (Tomita et al. 2005). The semi-diurnal cycle was also captured. This was further investigated by Yasunaga et al. (2013) by using the NICAM aqua planet experiment. The detailed structures of convective systems near the tropopause were analyzed by Kubokawa et al. (2010).

The aqua planet configuration is used to test cloud changes that are related to increased surface temperatures in order to imitate global warming conditions. Miura et al. (2005) compared the cloud cover responses between NICAM and the Model for Interdisciplinary Research on Climate (MIROC) (Hasumi and Emori 2004), which is an ordinary resolution climate model (denoted as CCSR/NIES/ FRCGC in Miura et al. 2005), under the aqua planet condition. The comparison results show that the cloud cover simulated by NICAM increases in high-latitude regions, while that by MIROC decreases. Such a contrasting response to cloud cover is very interesting and should be analyzed in more detail, as will be described later in this section. The change in the meridional distribution of relative humidity is shown and compared to the other model results by Sherwood et al. (2010).

The aqua planet experiments of NICAM were compared to the other model results in Blackburn et al. (2013), and additional aqua planet experiments were conducted by Yoshizaki et al. (2012a, b) in order to investigate the multiple-scale convective structure along the equator.

A NICAM simulation that includes a realistic configuration with a land and sea contrast was performed by Miura et al. (2007a). The results of this simulation show not only realistically simulated tropical cyclogenesis but also a clear dependency of convective organization on a planetary boundary layer (PBL) scheme, specifically, water vapor transport. Using the simulated dataset, the diurnal variation of the convective systems over the Tibetan Plateau was analyzed by Sato et al. (2007, 2008).

A successful simulation of a realistic MJO event was presented by Miura et al. (2007b), in which a 1 month integration with a 7 km grid spacing and a 1 week integration with a 3.5 km grid spacing were performed. A large-scale cloud organization, its eastward propagation, and the multiple-scale structure of convective systems were realistically reproduced (Nasuno et al. 2009). The similarity of the simulation to the observed MJO was further analyzed by Liu et al. (2009). In addition, multiple tropical cyclones were realistically generated from the active cloud areas of the MJO. Particularly, Fudeyasu et al. (2008, 2010a, 2010b) analyzed the evolution of a tropical storm in detail, which was generated from the MJO 2 weeks after the initial condition.

Thanks to the high-resolution numerical simulations performed by Miura et al. (2007b), simulation results can be analyzed or used in various ways. The dataset consists of 7 km mesh and 14 km mesh grid data for 1 month from 15 December 2006 and 3.5 km mesh grid data for 1 week from 25 December 2006. The diurnal cycle of tropical convective systems was intensively analyzed by Sato et al. (2009). Convective momentum transports of the MJO were analyzed by Miyakawa et al. (2012). The dataset was compared with the satellite observations of the Tropical Rainfall Measuring Mission (TRMM) and CloudSat (Masunaga et al. 2008), the geostationary satellite, the Multi-functional Transport Satellite (MTSAT-1R) (Inoue et al. 2008), and CloudSat and Cloud-Aerosol Lidar and Infrared Pathfinder Satellite Observations (CALIPSO) (Inoue et al. 2010; Satoh et al. 2010; Ham et al. 2013), and with the estimation of cloud radiative forcing (Sohn et al. 2010). The dataset was also used to develop a cloud parameterization for coarser resolution AGCMs (Watanabe et al. 2009). The convective structure near the tropopause was also analyzed by Kubokawa et al. (2012). The energy spectrum and similar frequency spectra of the high-resolution data were analyzed by Terasaki et al. (2009) and Tsuchiya et al. (2011).

Following Tomita et al. (2005) and Miura et al. (2007b), studies of intra-seasonal variability (ISV) including MJOs and boreal summer intra-seasonal oscillations (BSISO) (Fu and Wang 2004) are unique areas of NICAM studies. Various aspects of the MJO and tropical disturbances simulated by Tomita et al. (2005) and Miura et al. (2007b) are being analyzed as mentioned above. NICAM studies are also collaborated with observational field campaigns of MJOs. Miura et al. (2009) examined an onset of MJO observed at the Mirai Indian Ocean cruise for the Study of the MJO-convection Onset (MISMO) field campaign (Yoneyama et al. 2008), suggesting a role of the sea surface temperature (SST) gradient in the western Indian ocean. Also, for the Cooperative Indian Ocean Experiment on Intraseasonal Variability in the Year 2011/Dynamics of the MJO (CINDY2011/DYNAMO) campaign (Yoneyama et al. 2013), Nasuno (2013) summarized a result from quasi-real-time weather forecasting experiments that showed simulated characteristics of MJOs during the CINDY2011/DYNAMO period in the boreal winter between 2011 and 2012. The quasi-real forecasting system using the stretch NICAM (Tomita 2008a; see 'Models related to NICAM' section) was developed particularly for use in collaboration with field experiments, as introduced by Oouchi et al. (2012). More statistical results for MJO simulations are shown by Miyakawa et al. (2014), who reported a good performance of MJO predictability by the hind cast approach with NICAM. BSISO simulations were analyzed by Oouchi et al. (2009a, 2014), Taniguchi et al. (2010), Yanase et al. (2010b), and Satoh et al. (2012a). In particular, they focused on relations between ISV and tropical cyclones, as described below, since ISV modulates large-scale vorticity distributions that generally lead to preferable conditions for cyclogenesis.

Tropical cyclones can also be adequately researched by using NICAM. Tropical cyclogensis and its evolution from the MJO were analyzed for boreal winter cases by Fudeyasu et al. (2008, 2010a, 2010b). For boreal summer, tropical cyclogenesis was found to be frequently related to ISV activities (Kikuchi and Wang 2010). The cyclone Nargis, which caused severe damage to Myanmar in May 2008, was successfully simulated and its relation to the northward propagation of BSISO in the Bay of Bengal was analyzed by Taniguchi et al. (2010) and Yanase et al. (2010a, 2012a). Oouchi et al. (2009a) simulated tropical cyclogenesis in the northwestern Pacific and showed its relation to the MJO. The effects of an equatorial Kelvin wave on an abrupt development of a tropical cyclone were clearly shown by Yanase et al. (2010b). Yanase et al. (2012b) also analyzed some statistical behaviors of tropical cyclones simulated by NICAM especially over the eastern Pacific. The impacts of a tropical cyclone on the Baiu rainfall were analyzed by Yamaura et al. (2013). The preferable conditions of tropical cyclogenesis were analyzed over the Atlantic domain (Satoh et al. 2013).

Tropical cyclone climatology and possible changes in the future projection of tropical cyclone activities are actively being researched using NICAM. For example, Satoh et al. (2012a) analyzed 8 years of boreal summer experiments that were conducted in the Athena Project (Kinter et al. 2013) in order to show the simulated climatology of tropical cyclones and produced results that show good relations between tropical cyclone activities and MJO activities. Future changes in the tropical cyclone structure were analyzed by Yamada et al. (2010, 2012), Yamada and Satoh (2013), and Emanuel et al. (2010).

NICAM simulations also contribute to the studies on Asian monsoon and tropical convective systems. For example, Oouchi et al. (2009b) provided a realistic distribution of precipitation during a boreal summer and showed its variation for diurnal and intra-seasonal timescales. Additionally, the resolution dependency of the diurnal cycle of convective systems was analyzed by Sato et al. (2009), and their results were found to be generally consistent with those of the aqua planet experiment (Tomita et al. 2005); as resolution becomes coarser, the timing of precipitation peak becomes increasingly delayed. Sato et al. (2009) also analyzed the diurnal evolution of tropical convective cold pools, while Fujita et al. (2011) analyzed the behavior of the diurnal convection over maritime continental regions. Dirmeyer et al. (2012) and Noda et al. (2012) also analyzed the diurnal cycle of precipitation over the global domain. A mechanism for the diurnal variation of summer precipitation over southern China was investigated by Satoh and Kitao (2013) using the stretched version of NICAM (see 'Models related to NICAM' section).

The use of a cloud microphysics scheme in NICAM experiments without cumulus parameterization leads to realistic behavior of cloud distribution in the horizontal and vertical directions. Furthermore, as seen in the paragraph below, it might even produce a different behavior of cloud feedback from that obtained by conventional climate models that use cumulus parameterization. Iga et al. (2007a) analyzed the climatological distribution of cloud cover and compared it with the data of the International Satellite Cloud Climatology Project (ISCCP)

(Rossow and Schiffe 1999), while in a later study, Iga et al. (2011) analyzed upper tropospheric ice clouds and their sensitivity using various model parameters. Noda et al. (2010) investigated the importance of the subgrid turbulent process on cloud distributions. Kodama et al. (2012) showed the sensitivities of upper clouds to the choice of various cloud microphysics parameters. Satoh and Matsuda (2009) investigated how cloud microphysics affects upper clouds under an idealized radiative convective equilibrium condition.

As for the future simulations of cloud change by NICAM, Miura et al. (2005) showed the model's different behavior when compared to a coarse resolution AGCM (MIROC). In addition, Collins and Satoh (2009) showed that as a response that was different from the existing models, the upper cloud cover might increase, whereas the ice water path decreases under future warming conditions. Satoh et al. (2012b) analyzed their results and speculated that a future reduction in convective circulation could lead to a reduction in the ice water path under the warming condition. Tsushima et al. (2014) further investigated longwave cloud feedback and showed the possibility of a stronger positive feedback than other coarse resolution AGCMs. Yamada and Satoh (2013) analyzed the relation between cloud changes over the global domain and tropical cyclone changes. Kodama et al. (2014a) also analyzed the cloud changes associated with storm tracks simulated by the NICAM aqua planet experiments.

Recently, use of the K computer has allowed the horizontal resolution of NICAM to be increased. Miyamoto et al. (2013) performed (for the first time) a subkilometer mesh global domain simulation with grid spacing of 870 m (g-level 13) and 96 vertical layers (Figure 2). They showed that at a resolution of less than 1.7 km, convective cores could be resolved by multiple grid cells. The ability to use the K computer introduces a new area of study for NICAM. Numerous MJO case studies are currently being conducted in order to evaluate their statistical predictability (Miyakawa et al. 2014). More than 10 years of continuous simulations have and are being conducted under present and future conditions in order to investigate future changes to tropical cyclones and other extreme events. The subsequent sections describe the improvements in the simulated climatology that have been made possible by improving the physical processes.

As described above, NICAM is usually run without using cumulus parameterization. In general, the horizontal scale of meso-scale phenomena ranges from a few to several hundred kilometers. It is not readily understood why a model with horizontal mesh spacing of O (10 km) reproduces such meso-scale phenomena without subgrid convective parameterization. It is speculated that for relatively large-scale meso-scale convective systems, the statistical effects of circulation resolved by grids behave

similarly to nature. Figure 3 shows resolution dependency of upper cloud distributions in the range of $\bar{\Delta}$ = 28, 14, 7, 3.5, 1.7, 0.87 km for the NICAM simulations conducted by Miyamoto et al. (2013). Although in that study, it was analyzed that the convective core becomes resolved at $\bar{\Delta}$ = 1.7 km. Figure 3 shows that the large-scale structure of upper clouds is nearly unchanged for these resolutions. This result implies that the usefulness of a model without cumulus parameterization depends on research targets of meteorological phenomena. Large-scale organized convective structures such as MJOs are, in general, well reproduced without using cumulus parameterization, even at the coarser horizontal resolutions around O (10 km) as also indicated by Holloway et al. (2013), while the precise simulation of details of meso-scale convective structure needs much finer resolutions (Bryan et al. 2003).

Design, structure, development, and timeline

A timeline of the development of NICAM is summarized in Table 2. As was previously mentioned, the development of NICAM started in 2000 at JAMSTEC. Through the construction of icosahedral grid systems (Tomita et al. 2001, 2002), a shallow water model (SWM) (Tomita et al. 2001, 2002) was initially tested. At the same time, new non-hydrostatic numerical schemes were being developed (Satoh 2002, 2003). By combining the two approaches, a three-dimensional (3D) dynamical core was eventually developed (Tomita and Satoh 2004). The resulting 3.5 km mesh high-resolution dynamical core experiments were examined by Iga et al. (2007b). Then, full-physics experiments have been underway since 2004. Furthermore, the dynamical schemes, computational stability, and a tracer advection scheme (Miura 2007; Niwa et al. 2011a) have improved. For physical processes, we have actively introduced more sophisticated schemes, especially for cloud microphysics (Grabowski 1998; Tomita 2008b; Seiki and Nakajima 2014) and PBL schemes (Nakanishi and Niino 2006, 2009; Noda et al. 2010). An accurate semi-Lagrangian scheme for sedimentation processes was developed for cloud microphysics schemes (Xiao et al. 2003). Aerosol processes (Suzuki et al. 2008) and atmospheric chemistry processes have also been introduced. Each component of the abovementioned physical schemes will be described in later sections of this review. Several NICAM-based systems are also currently under development, including atmosphere-ocean coupling and ensemble-based data assimilation systems, and downscaling methods for regional scale experiments have been developed for more advanced uses.

Since high-resolution simulations require large computational resources, the efficiency of state-of-the art supercomputers needs to be maximized. In general, weather/climate models use a large number of physical

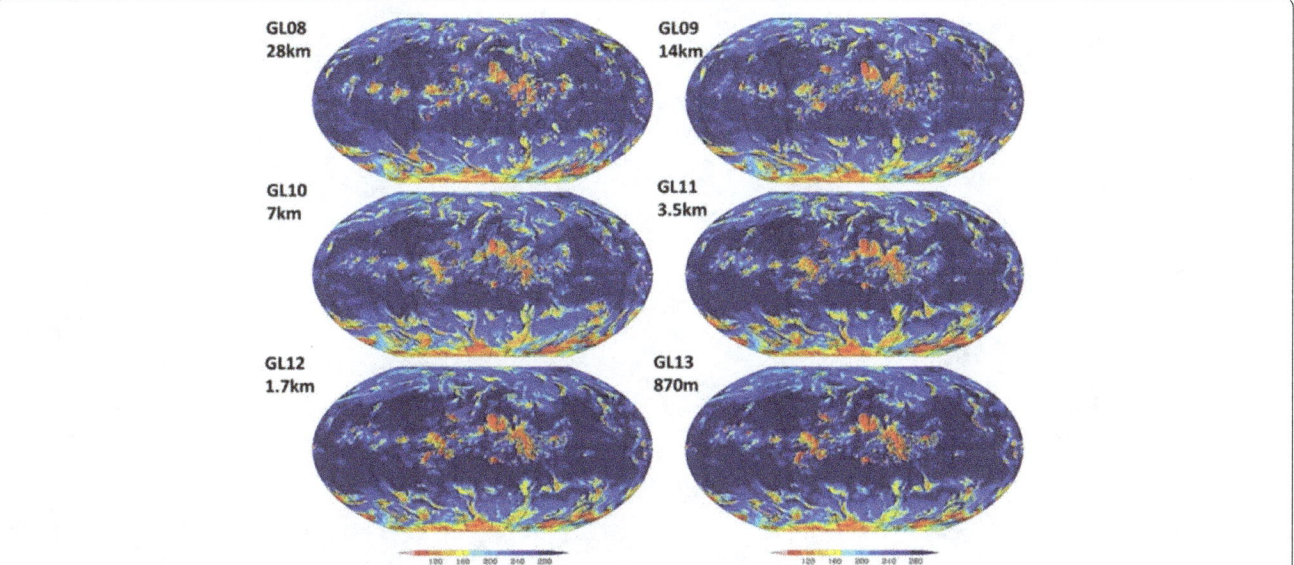

Figure 3 Resolution dependency of cloud structure simulated by NICAM between 28 km and 870 m grid spacing experiments. Cloud distributions simulated by NICAM between 28 km and 870 m grid spacing experiments for 6:00 UTC 25 August 2012, shown by the outgoing longwave radiation at the top of the atmosphere. The unit is W m^{-2}.

variables during computation, so the size of 3D arrays significantly increases with increases in spatial resolution. Large-sized, various, frequent data output is also needed for simulations. Taken together, these factors make it clear that weather/climate models require faster data throughput in all layers of a computer system. NICAM has been tactically designed to achieve efficient throughput at the 1) memory layer, 2) network layer, and 3) file input/output (I/O) layer.

Since numerous simple arithmetic operations are used for large arrays in NICAM algorithms, memory throughput is the most important aspect governing total computational efficiency. For example, when using Earth Simulator, the ratio of memory bandwidth to the floating-point performance (Byte/Flops ratio; hereinafter B/F ratio) is 4, which makes it easy to execute NICAM with a high degree of efficiency. However, in the past decade, the B/F ratio has decreased during the recent evolution of supercomputers. As a result, when NICAM was implemented on the K computer at B/F = 0.5, numerous optimizations had to be applied to improve the efficiency of computer performance. Saving memory transfers is one effective optimization tactic. To accomplish this, we suppressed the memory copies, reduced the intermediate arrays, and avoided unnecessary zero-filling. These changes had to be performed manually and required labor-intensive efforts. However, as a result of these optimizations, we succeeded in reducing the model execution time by 30%.

To improve network throughput, the finite volume method with an explicit scheme was adopted to minimize the global communications, as described in the section 'Dynamics'. When 640 nodes on the K computer are used, the network communication of NICAM accounts for about 10 % of the total elapsed time at the case of g-level 9 ($\bar{\Delta}$ = 14 km). This percentage increases as the number of nodes increases. Kodama et al. (2014b) developed an algorithm for optimally assigning of nodes in a network topology such as a torus/mesh.

For the file I/O throughput, NICAM adopts a distributed file I/O. If the output data are aggregated into a single file at the time of output, it would be easy to handle the file. However, NICAM cannot maximize the I/O throughput using such a single output file. The data of each region on the icosahedral grid of NICAM are output to a file at the node, and the icosahedral grid is converted into a lat-lon grid during the post-process. Generally, we use a simple bilinear interpolation for the post-process, although we have an option of another interpolation to keep conservation of area integration. A recently developed post-process program can run in parallel with the main model using a coupler (see 'Coupler' section). However, an increase in file size due to the higher resolution is a current issue, and we are planning to change the file type to a compressible standard format, such as Network Common Data Form version 4 (NetCDF4) (http://www.unidata.ucar.edu/software/netcdf/). Recently, a benchmark run on the K computer has achieved 10% of the peak performance using five nodes (40 cores), and usage has been verified with 81,920 nodes (655,360 cores). Furthermore, a subkilometer mesh global domain simulation (Miyamoto et al. 2013) was executed with a performance of 230 Tflops in double precision for 68 billion grid cells while using 20,480 nodes of the K computer.

Table 2 Timeline of the development of NICAM

	2000	2001	2002	2003	2004	2005	2006	2007	2008	2009	2010	2011	2012	2013	2014
Grid system		→Modified grid using the spring dynamics						→Grid stretching method							→Diamond NICAM, plane NICAM
Dynamics															
Basic scheme		→SWM		→3D dynamical core											
Advection scheme			→Non-hydrostatic scheme				→Miura scheme		→CWC consideration						
Physics															
Cumulus parameterizations model					→Grabowski						→Tiedke	→Chikira			
Cloud microphysics model								→NSW6			→NDW6				
Land surface model					→Bucket	→MATSIRO									
Ocean surface model					→Fixed SST		→Slab Ocean							→Coupled Ocean	
PBL model					→MY2.5					→MYNN					
Radiation model					→MSTRN-X										
Aerosol/chemistry						→SPRINTARS							→CHASER		
Frameworks										→NICAM-LETKF	→NICAM-TM	→Coupler (Jcup)			
Supercomputers			→ES							→ES2			→K computer		
Milestone experiments					→Aqua-Planet Experiment			→MJO Experiment	→Athena Project					→Subkilometer Experiment	

See the 'Abbreviations' section for the expanded forms of the abbreviations used in this table.

Further information on the NICAM development team and the initial developments can be found at http://nicam.jp. NICAM team members consist of researchers belonging to several research institutions in Japan. The source code is primarily written in Fortran 95 and follows the Meteorological Research Institute/Japan Meteorological Agency (MRI/JMA) standard coding rule (Muroi et al. 2002). The program is parallelized using Message Passing Interface (MPI) process parallelization via automatic compiler thread parallelization. OpenMP is not actively used as the current program. Each component is managed as one or more modules. In order to maintain readability of the source code, C preprocessor macros are not used. The source code has been managed at repositories using the Concurrent Versions System (CVS), and the NICAM development team is currently moving to multi-branch development using the Git distributed revision control and source code management system. Due to its flexible portability, NICAM can run on various computer platforms, such as Earth Simulator, the K computer, IBM AIX, Linux, and Mac OSX. We have also tested several combinations of MPI libraries and Fortran compilers.

NICAM is also used as a benchmark program or a test platform for new architectures. For this purpose, we have released the dynamical core of NICAM as open source software (NICAM-DC), which is distributed under the terms of the Berkeley Software Development (BSD) 2-Clause License and is available for download (http://scale.aics.riken.jp/nicamdc/). The NICAM development team also participates in a working group for the common base library environment of weather and climate models and in cooperative development efforts with other modeling groups in Japan. In the field of computational science, NICAM has made significant contributions as a benchmark program and is one of the main application programs used in Exascale computing research.

Dynamics
Grid configuration and advection scheme
As reviewed in the 'Introduction' section, the use of grid models with finite-difference and finite-volume methods for weather and climate models was reconsidered as massively parallel computer architectures entered widespread use. An icosahedral grid is a globally quasi-uniform and is possibly the most suitable base for the development of an atmospheric model that can be run on a supercomputer. Additionally, the adoption of an icosahedral grid for the horizontal grid of NICAM was partially motivated by the results of Heikes and Randall (1995), in which a two-dimensional (2D) shallow water system model was developed using the Z-grid arrangement of the prognostic variables (Randall 1994; Figure 4d), which followed Masuda and Ohnishi (1986). The Z-grid model, however, had a

disadvantage in that the model needed to diagnose the velocity fields from the vorticity and divergence fields by solving two Poisson equations in each time step. Furthermore, it was not obvious that an efficient Poisson solver would become available for use on the massively parallel computers, at least when the NICAM development began.

Tomita et al. (2001) did not follow any previous study (such as Heikes and Randall 1995; Masuda and Ohnishi 1986) and instead examined a new method by using the Arakawa A-grid (Arakawa and Lamb 1977), which located prognostic variables (fluid depth and velocity) at the same nodes (Figure 4a) and thus did not require Poisson solvers. The accuracy and stability of the shallow water model were improved after applying a grid optimization that is now commonly known as 'spring dynamics' (Tomita et al. 2002), and the physical performance of the model was evaluated using the standard tests provided in Williamson et al. (1992). The methods used to discretize the horizontal gradient, divergence, rotation, and Laplacian operators in NICAM are similar to those described in Tomita et al. (2001) and Tomita and Satoh (2004), except for some minor updates in the passive tracer transport (Miura 2007; Niwa et al. 2011a) and damping operators.

When the model was first developed by Tomita et al. (2005), NICAM was capable of producing a global simulation with horizontal grid spacing of 3.5 km. It was also the first model to produce a global simulation with horizontal grid spacing of less than 1 km (Miyamoto et al. 2013). These characteristics indicate that the A-grid arrangement is a good choice due to its simplicity and the relative ease with which it attains high computational performance on Earth Simulator and the K computer. It should be noted that Tomita et al. (2001) provided the first example of the shallow water model on an icosahedral grid, which uses the finite-volume method and adopts velocity components as prognostic variables. However, it is well known that the A-grid model is unsuitable for simulations of inertia-gravity waves and geostrophic adjustments and that the model suffers from grid-scale noise (Randall 1994). Therefore, other options should be considered for the horizontal variable arrangement, such as those described below.

The disadvantages of the B/ZM-grid (Figure 4b) and C-grid (Figure 4c) arrangements, which were pointed out by Ničkovic et al. (2002), on the hexagonal grid were partially solved by Ringler and Randall (2002) and Thuburn et al. (2009), respectively. Although problems with the computational modes remain due to the mismatch of the degrees of freedom between the fluid depth/pressure and horizontal velocity, the B/ZM-grid and C-grid not only allow a more realistic representation of inertia-gravity waves and geostrophic adjustments than

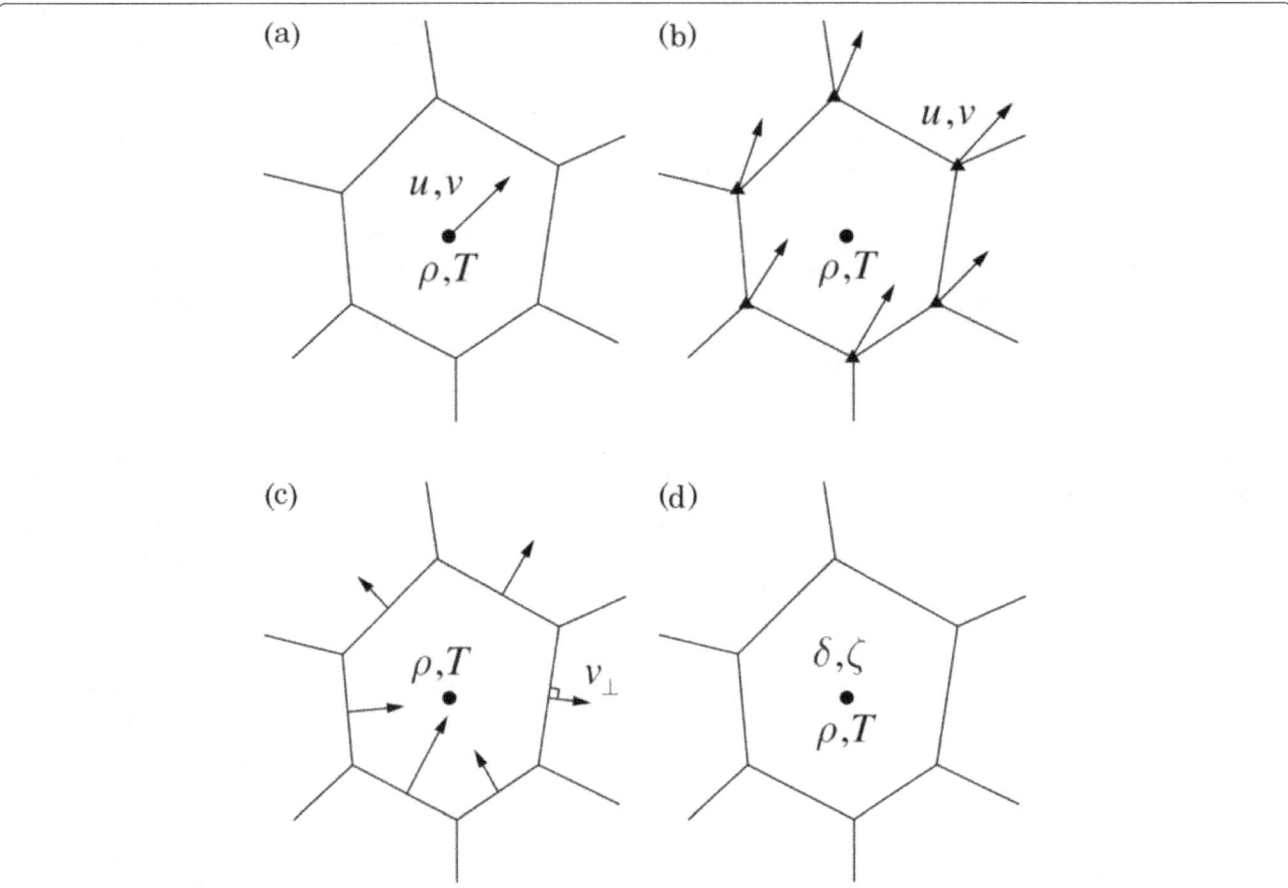

Figure 4 Arrangements of variables on the hexagonal grid. Arrangements of prognostic variables on the hexagonal grid for the **(a)** A-grid, **(b)** B/ZM-grid, **(c)** C-grid, and **(d)** Z-grid staggering. The symbols ρ, T, u, v, $v_\perp$, δ, and ζ refer to density, temperature, zonal components, and meridional components of the flow velocity, respectively, that is defined **(a)** at the center or **(b)** at the vertex of the hexagon, normal component of the flow velocity on each edge, divergence of the horizontal flow, and vertical component of vorticity.

the A-grid but also are free from the checkerboard pattern. Currently, we are investigating the B/ZM-grid as a candidate to replace the A-grid, and the difficulty in managing the computational mode has been largely eliminated by revising the operators of Ringler and Randall (2002). This activity is ongoing, and details of the improvements will be reported in forthcoming papers. The update of the dynamical core will also include the newly developed transport algorithms of Miura and Skamarock (2013) and Miura (2013).

Vertical resolution issues

NICAM adopts the Lorenz grid and z^* (terrain-following) coordinates (Gal-Chen and Somerville 1975) as a vertical coordinate system (Figure 3 of Satoh et al. 2008). Density, horizontal velocity, and internal energy are defined at the full levels, and vertical velocity is defined at the half levels. In NICAM, z^* is generally specified so as

to linearly depend on the geometric height z between the model surface and top (the linear-z^* coordinate). Using the standard vertical levels Z_k, geometric height $z_{X,k}$ is formulated as

$$z_{X,k} = Z_k + \left(1 - \frac{Z_k}{H_{\text{top}}}\right) Z_{X,sfc},$$

where k is an index for the vertical level, X represents the location of the horizontal grid, and H_{top} and $Z_{X,sfc}$ are the heights of the model top and bottom, respectively. In general, $H_{\text{top}} = 40$ km is used with 40 vertical levels. For standard cases, the lowermost full level is located at 81 m. The vertical interval of the half-level ΔZ_k increases with height, from $\Delta Z_k = 170$ m at the bottom, $\Delta Z_k < 1$ km below 10 km, and $\Delta Z_k \sim 3$ km near the model top.

In some situations, the hybrid-z^* coordinate (Simmons and Burridge 1981) is employed to reduce the amount of pressure gradient force numerical errors that arise from topography. It is formulated as

$$z_{X,k} = Z_k + \left(\frac{\sinh \frac{H_{top} - Z_k}{H_{efold}}}{\sinh \frac{H_{top}}{H_{efold}}} \right) Z_{X,sfc},$$

where H_{efold} is a height scale parameter (e.g., H_{efold} = 10 km). In the hybrid-z^* coordinate, intervals of the vertical level approach those of the standard vertical level for $Z_k \gg H_{efold}$, whereas in the linear-z^* coordinate, it is constant with height even near the model top. This hybrid-z^* coordinate advantage is obtained at the expense of packing more vertical levels just above the higher mountain, which may cause the vertical CFL condition to become a serious issue for the hybrid-z^* coordinate. Klemp (2011) also proposed a hybrid approach, in which the coordinate surfaces are smoothed with height in order to remove smaller scale terrain structures from the surfaces.

Recently, we constructed a new configuration for the vertical levels of NICAM in order to perform simulations with finer vertical resolutions in the troposphere, stratosphere, and mesosphere levels. The development of this 'middle atmosphere NICAM' is motivated by the results of recent climate studies in which the tropospheric circulations were found to be affected by the stratosphere. It should be noted that the inclusion of the middle atmosphere into a climate model is a worldwide trend in the climate model community, since it improves the reproducibility of the large-scale tropospheric circulation (e.g., Shindell et al. 1999). Currently, the middle atmosphere NICAM employs the hybrid-z^* coordinate, and its model top height is set to 80 km. The vertical intervals in the lower atmosphere are the same as the standard vertical coordinate. As the vertical interval increases with height and reaches a specific criteria (Δz_{max}), the vertical level is set with the uniform interval of Δz_{max} (except near the model top) in order to appropriately simulate the propagation of atmospheric gravity waves. We have tested Δz_{max} = 2, 1, and 0.5 km with a horizontal resolution of g-level 7 ($\bar{\Delta}$ = 120 km), as shown in Figure 5. As the vertical resolution increases, the reproducibility of the zonal mean temperature and zonal wind tends to improve in the upper troposphere and middle atmosphere. Interestingly, westerly winds around the equatorial lower troposphere in June 2004 area only well simulated when the vertical resolution is set to 1 km or less. NICAM has an option that allows the deep atmosphere configuration (Tomita and Satoh 2004) to be switched. For the middle atmosphere simulations, the choice of the deep atmosphere is considered appropriate.

Determining the proper vertical coordinates is an ongoing issue for high-resolution non-hydrostatic models.

For global non-hydrostatic models in particular, the numerical instability of the terrain-following coordinates becomes more severe because the steep topography over the Andes or the Antarctica is within the simulation domain. For the present simulations, relatively coarser topographic data were used by applying a smoother for the NICAM simulations. For example, for the 3.5 km mesh simulations, 14 km mesh coarsened topographic data were generally used. In addition, a minimum threshold value of the Brunt-Väisälä frequency was specified for the calculations in the PBL schemes, that is and the Mellor-Yamada Nakanishi-Niino (MYNN) scheme, in order to suppress an unusual increase in vertical shears near the surface.

To overcome the problem of terrain-following coordinates, different candidates have been proposed. For example, the possibility of a vertical coordinate that is based on geopotential height, i.e., the z-coordinate, has been investigated. Although this approach is common for ocean models (Adcroft et al. 1997), it must be understood that the implementation of the z-coordinate to non-hydrostatic atmospheric models is not straightforward (Gallus and Klemp 2000). The immersed boundary method is well known for computational fluid dynamics (Mittal and Iaccarino 2003). Various approaches to overcome the problems of the z-coordinate were taken into consideration, such as the cut-cell method or the thin-wall approximation (e.g., Bonaventura 2000; Steppeler et al. 2002, 2006; Yamazaki and Satomura 2008, 2010), have been taken into consideration. We plan to introduce such an option for the z-coordinate of NICAM in the near future.

Physics
Cloud microphysics schemes
Cloud microphysics can be used to solve the growth equations of a single hydrometeor particle. The related processes are mainly categorized into four components: nucleation processes, phase change, collisional processes, and gravitational sedimentation. These processes have not been calculated explicitly and have been parameterized in conventional AGCMs since the characteristic timescale of these processes and spatial variability of cloud ensembles cannot be resolved using the time-step and grid resolution of GCMs. Global non-hydrostatic models operate by calculating these processes via cloud microphysics schemes in order to explicitly evaluate time evolution of cloud growth over the global domain. The implemented schemes are generally developed for regional non-hydrostatic models. This section introduces cloud microphysics models that are applicable to global non-hydrostatic models. For a more fundamental theoretical basis of cloud microphysics processes, refer to Pruppacher and Klett (1997).

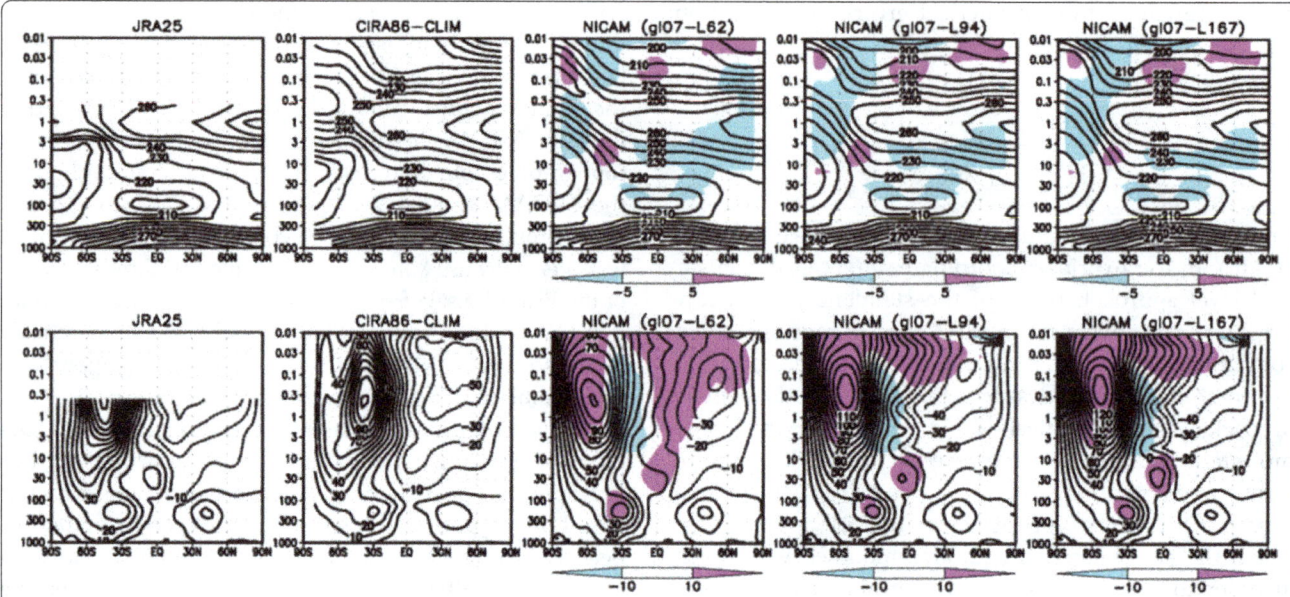

Figure 5 Dependency of the simulated NICAM climatology on the vertical resolution. Dependency on the vertical resolution of the (upper) zonal mean temperature (K), and (lower) zonal mean zonal wind (m s^{-1}). From left to right: JRA25 reanalysis (Onogi et al. 2007), CIRA86 (http://badc. nerc.ac.uk/view/badc.nerc.ac.uk__ATOM__dataent_CIRA), NICAM ($\Delta z_{max} = 2$ km), NICAM ($\Delta z_{max} = 1$ km), and NICAM ($\Delta z_{max} = 0.5$ km). For CIRA86, the climatology for June is shown. All the other panels are the averages for June 2004. Shading denotes deviations from the CIRA86 climatology.

Multi-moment bulk method Hydrometeor particles typically have a radius r from 10 μm to 1 mm and populations from 10^2 to 10^9 m^{-3}. To manage numerous particles efficiently (in terms of computational cost), cloud microphysics modeling generally assumes that particles with the same radius are homogeneously distributed within a grid cell of the model and develop in the same manner while under the same environmental conditions. Growth equations of hydrometeor particles with similar radii are solved together and the particle size distribution (PSD) $f(r)$, which is the number density sorted by the radius of the hydrometeors, is then predicted. Numerous observational studies have shown that liquid hydrometeor particles have three major modes for the PSD (Pruppacher and Klett 1997): particles in the condensational growth mode (which have a mode radius around 10 μm), particles in the collisional growth mode (which have a mode radius larger than 100 μm), and the drizzling mode (which is a transitional mode between the two aforementioned modes).

There are two main methods that can be used to predict the time evolution of the PSD. One is the multi-moment bulk method, which predicts the integrand of each mode by approximating the modes via the gamma distribution (e.g., Milbrandt and Yau 2005b; Seifert and Beheng 2006):

$$f(r) = Kr^\nu \exp(-\lambda r^\mu),$$

where κ, λ, μ, and ν are the diagnosed parameters. The k-th moment of the PSD, $M^{(k)}$, is defined as

$$M^{(k)} \equiv \int_0^\infty f(r)r^k dr,$$

and the time evolution of the prognostic moments is evaluated by integrating the growth rates of a single hydrometeor particle over the entire range of the PSD (e.g., Seifert and Beheng 2006). The other is the spectral bin method, which solves growth equations by discretizing the PSD with tens of bins covering these modes (Khain et al. 2000, 2008). Because of their computational efficiency, the multi-moment bulk methods are widely used for 3D non-hydrostatic simulations.

The multi-moment bulk method is categorized by the number of its prognostic moments: the single-moment bulk method predicts the mass concentration of hydrometeors (Kessler 1969; Lin et al. 1983; Rutledge and Hobbs 1983; Walko et al. 1995; Grabowski 1998; Hong et al. 2004; Thompson et al. 2008; Tomita 2008b), and the double-moment bulk method predicts the number concentration in addition to the mass concentration (Meyers et al. 1997; Feingold et al. 1998; Morrison et al. 2005; Seifert and Beheng 2006; Phillips et al. 2007; Morrison and Gettelman 2008; Lim and Hong 2010).

The four parameters used to represent the PSD (κ, λ, μ, and ν) are diagnosed by prognostic moments. Hence, as the number of the prognostic moments is increased, a variety of PSD shapes can be represented. For example, increases in the mode radius and skewness of the PSD are observed in collisional growth (Berry and Reinhardt 1974). These distortions of the PSD are important for initiation of precipitation (Seifert and Beheng 2001). Furthermore, once precipitation occurs, the gravitational sedimentation of the hydrometeor particles induces further distortion of the PSD due to the gravitational size sorting mechanism and the time evolution of the PSD affects the subsequent chain of particle growth (Wacker and Seifert 2001; Milbrandt and Yau 2005a; Milbrandt and McTaggart-Cowan 2010). In addition, higher order moments are required for the closure of the governing equations in the collisional processes (e.g., Seifert and Beheng 2001). Thus, an increase in the number of prognostic moments is linked to improvements in the time evolution of the PSD and complexity of the cloud microphysics modeling. Additionally, complexity and calculation costs increase as the number of prognostic moments increases.

For climate studies, collisional processes are particularly important to the radiative budget because such processes dominate the persistence of warm-phase clouds due to the rapid growth rate of cloud droplets during the precipitating process, along with the strong cloud albedo perturbations caused by the spatial and temporal variabilities of their growth rates (Albrecht 1989; Lohmann and Feichter 2005). The double-moment bulk method enables us to explicitly evaluate the dependence of the collisional growth rate on the mode radii (Berry and Reinhardt 1974; Feingold et al. 1998; Khairoutdinov and Kogan 2000; Seifert and Beheng 2001). In addition, cloud radiative forcing can be more accurately estimated because the cloud optical properties are calculated using the mass concentration and number concentration (Hansen and Travis 1974). In contrast, the single-moment bulk method assumes that the radii of the hydrometeors are empirical values. Empirical values do not reproduce the temporal and spatial variability of atmospheric states, and their validity has not been confirmed for various types of climate change, such as global warming conditions. The triple-moment bulk method, which was recently developed and used (Milbrandt and Yau 2005a, b; Seifert 2008; Shipway and Hill 2012), predicts or diagnoses the radar reflectivity factor in addition to the aforementioned two moments and is applied to the regional simulations of severe storms (Milbrandt and Yau 2005a, 2006). Of course, the higher order moment bulk methods also contain empirical formulations and parameters that can be used to calculate cloud growth and optical properties. However, previous process studies have shown that the differences of simulated results between double-moment and triple-moment bulk methods are smaller than those between single-moment and double-moment bulk methods

(e.g., Milbrandt and Yau 2005a; Shipway and Hill 2012). In addition, it has also been demonstrated that a well-optimized double-moment method can capture the characteristics of cloud growth simulated by a spectral bin method (Seifert and Beheng 2001; Seifert et al. 2006; Seifert 2008; Milbrandt and Yau 2005a, 2006). These facts indicate that the double-moment bulk method is practical and applicable in terms of cost performance of numerical simulations.

Cloud microphysics schemes in NICAM NICAM initially adopted a single-moment bulk cloud microphysics scheme that was proposed by Grabowski (1998) in order to explore the multi-scale interaction of moist convective systems because the organizations of deep convective systems do not depend on the details of cloud microphysics schemes. Instead, meso-scale circulations are reproduced if the latent heat release in updrafts and the evaporation cooling caused by rain drops are considered by predicting the mass concentration of hydrometeor particles, as introduced in the single-moment bulk scheme. In the latest version, the schemes proposed by Kessler (1969), Lin et al. (1983), Hong et al. (2004), and Tomita (2008b) are implemented. In particular, NICAM single-moment scheme with six water categories proposed by Tomita (2008b) (hereinafter, referred to as NSW6) has been well evaluated by comparisons with satellite observations (Satoh et al. 2010; Kodama et al. 2012; Hashino et al. 2013; Roh and Satoh 2014) and, hence, is used by default. Because of the relatively low calculation cost, the use of single-moment schemes enables us to conduct very high-resolution global non-hydrostatic simulations at g-level 13 ($\bar{\Delta}$ = 870 m; Miyamoto et al. 2013), or for longer integrations up to several decades.

Recently, Seiki and Nakajima (2014) implemented a double-moment bulk cloud microphysics scheme (NICAM double-moment scheme with six water categories; hereinafter, referred to as NDW6) into NICAM. Seiki et al. (2014) demonstrated that the NDW6 scheme well reproduces cloud radiative forcing by calculating cloud ice number concentration with using accurate databases of the cloud optical properties. Figure 6 shows the simulated cloud optical thickness (COT) and outgoing shortwave radiative flux at the top of the atmosphere that were obtained using the NDW6 scheme, along with the observed outgoing shortwave flux at the top of the atmosphere that was obtained from the satellite product of CERES SYN1deg_Ed3A (Wielicki et al. 1996; Kato et al. 2011). The simulation showed that convective clouds with a high COT were scattered over the tropics and formed regional cloud clusters (e.g., over the central Pacific). For example, Typhoon Fengshen remained over the Philippines and high COT clouds were distributed along its vortical structure. Furthermore, synoptic scale disturbances were

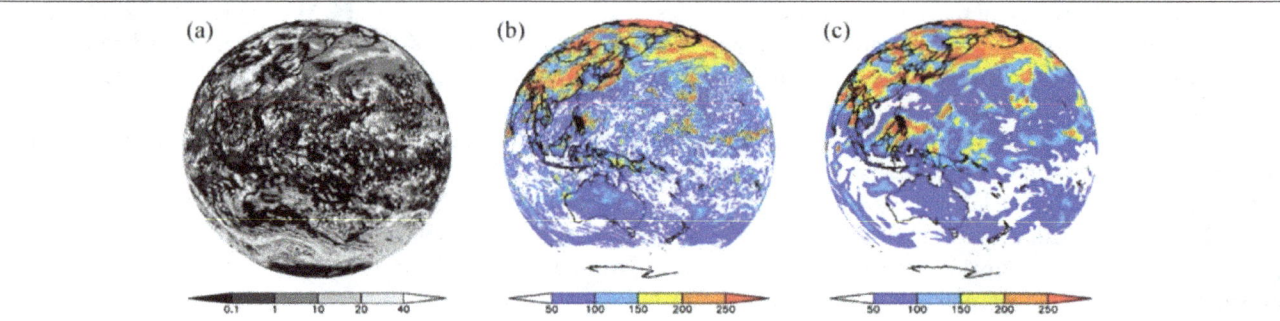

Figure 6 Simulated results for the 14 km mesh NICAM with the double-moment cloud microphysics scheme. Simulated results for the double-moment cloud microphysics scheme for the 14 km mesh NICAM on 20 June 2004: **(a)** simulated cloud optical thickness at a wavelength of 550 nm at 00:00 UTC, **(b)** daily averaged outgoing shortwave flux (W m^{-2}) at the top of the atmosphere, and **(c)** the daily averaged outgoing shortwave flux (W m^{-2}) at the top of atmosphere, as produced by the CERES SYN1deg_Ed3A product.

organized over the mid-latitude regions and optically thick regions were located along the front. A shortwave flux corresponding to these high COT regions was largely reflected by clouds. These features agree well with the satellite observation. As shown by the NDW6 scheme, the impact of the cloud radiative forcing, which results in improvements to the climatological biases in NICAM, should be evaluated in future studies.

Subgrid-scale turbulence of the planetary boundary layer
Boundary layer turbulence plays an important role in driving large-scale circulations of the atmosphere as it transports heat upward from the Earth's surface in the form of sensible and latent heats, which are converted from incoming solar energy at the surface (Trenberth et al. 2009). To simulate atmospheric circulations, a parameterization scheme that computes the effects of such turbulent transport in the subgrid scale of a numerical model needs to be incorporated, unless the resolution is high enough to resolve turbulence.

NICAM adopts the MYNN scheme, which is a modified version of the original Mellor-Yamada (MY) model (Mellor and Yamada 1982; Nakanishi and Niino 2006, 2009), for such turbulent transport models (Noda et al. 2010). There are three major advantages to employing the MY-type model: 1) the MY-type model can consistently and simultaneously compute subgrid-scale condensation and resultant turbulent fluxes by assuming the probability density functions regarding the inhomogeneity of temperature and water on a subgrid scale, 2) the model can consistently evaluate the impacts of the differences of turbulent moments, and 3) the model can run with low computational cost because it omits minor terms in the turbulent transport equation system. Based on the degree of approximation from the original equations, level 1 to level 4 can be included in the MY-type model. The level-2, level-2.5, and level-3 MYNN version models are installed in NICAM, and the level-2 model is used as a standard PBL scheme.

Figure 7 compares the impacts of these different level models on the spatial distributions of simulated clouds. As can be seen in the figure, all turbulent models properly simulate the spatial characteristics of, for example, high and mid-level clouds, even though the magnitude of the high clouds tends to be larger than those seen in satellite observations. For low clouds, the level-2 model properly represents the characteristics of the regional contrast between a clear and cloudy sky, such as those seen over the eastern and western regions of the Pacific Ocean. Conversely, the low clouds that systematically develop along and off the Peruvian coast are not simulated well. Compared with the results of the level-2, level-2.5 and level-3 models produce a much smaller fraction of low clouds. It is speculated that these two models calculate more vertical turbulent heat transports of heat from the boundary layer to the free atmosphere, which results in a less humid boundary-layer state.

Due to the remarkable progress of computational technology, the model resolution of NICAM is steadily being increased, with the highest resolution to date having been achieved when a horizontal mesh size of 870 m was used (Miyamoto et al. 2013). For a global non-hydrostatic model, boundary-layer scale turbulence becomes explicitly and partially calculatable around this resolution, but will not be sufficiently resolved until the mesh size is less than 100 m (Khairoutdinov et al. 2009; Moeng et al. 2009). It is a challenging issue to parameterize turbulent transport processes in such an intermediate resolution (Wyngaard 2004), and continuous efforts are needed to improve the subgrid-scale turbulence parameterization schemes along with shallow cumulus convection.

Radiation
Atmospheric radiative transfer models calculate heating rates in the atmosphere and radiative flux at the Earth's surface using the 3D distribution of clouds/gases/aerosols and land/ocean surface properties with the atmospheric state. NICAM adopts the broadband radiative transfer

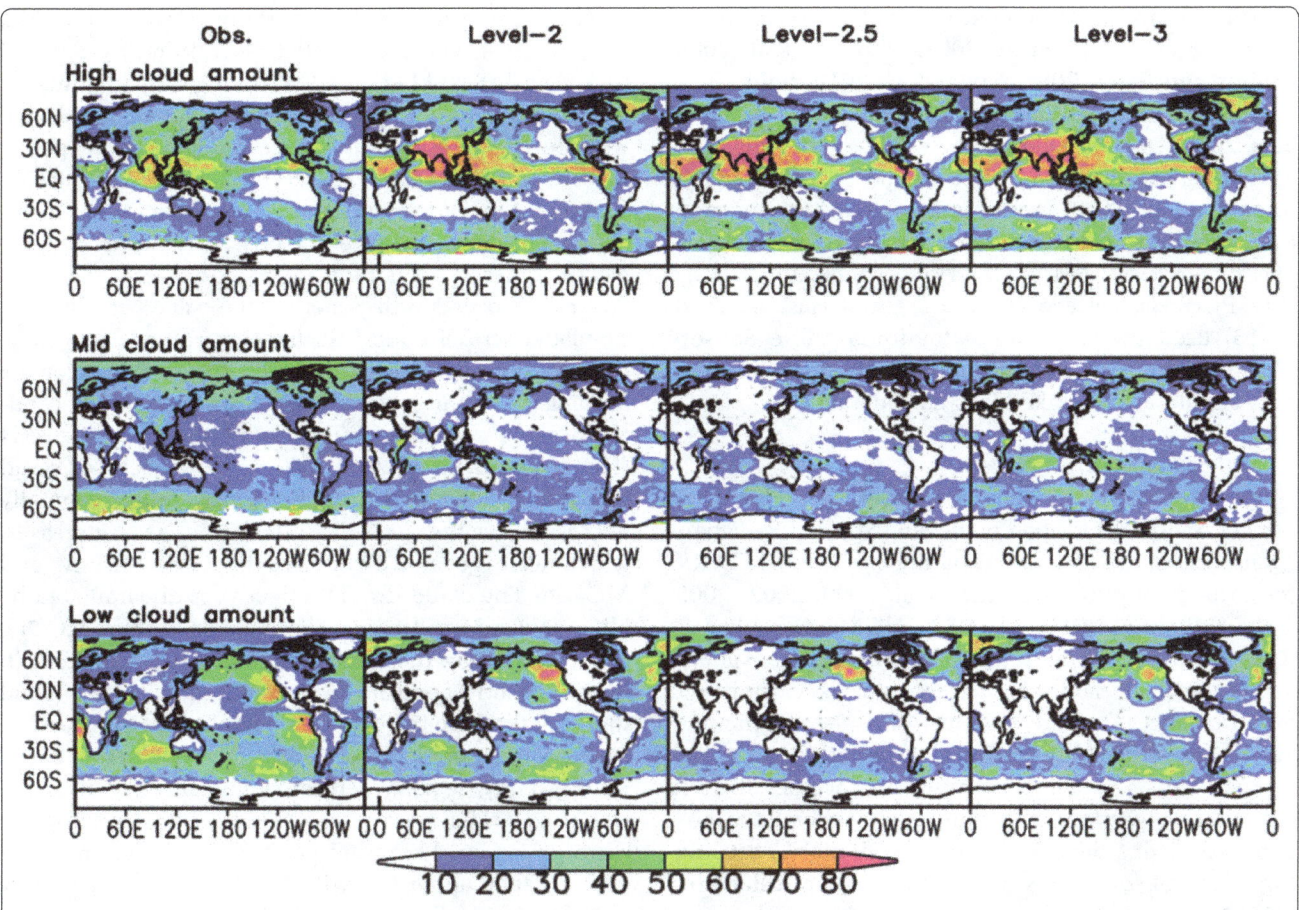

Figure 7 Comparison of the cloud amounts simulated by using different turbulent models to the satellite observation. The comparisons of the cloud amounts from the ISCCP satellite data, level-2, level-2.5, and level-3 MYNN turbulent schemes were obtained by using the 14 km mesh NICAM. The NICAM data shown are averages during 29 July 1982 and 15 August 1982, and that of the satellite observation are the climatology for July and August.

model 'MSTRN', which is based on the discrete ordinate method with a delta two-stream approximation and the correlated k-distribution method. The MSTRN model calculates gas absorption (H_2O, CO_2, O_3, N_2O, CH_4, and O_2) and the scattering/absorption of hydrometeors and aerosols (dusts, sea salts, carbonaceous, and sulfates) over an arbitrary time interval and then outputs short- and long-wave radiation fluxes. A basic description of this model is written in Nakajima et al. (2000). In the latest package, 'MSTRN-X' (Sekiguchi and Nakaima 2008), the gas absorption processes are improved and successfully used to reduce the error of the heating rate. The scheme was introduced into NICAM through the AGCM version implemented for MIROC from the original one-dimensional (1D) model. The calculation of solar insolation (Berger 1978) was added to the AGCM version. The other cloud overlapping method (maximum-random overlapping) was incorporated since this version of the radiation scheme was included in MIROC3.2 (Hasumi and Emori 2004).

The calculation cost of the radiation scheme is very high in the AGCM, because both computing speed and precision are required for the scheme. The latest MSTRN-X scheme improved the optimization method in order to reduce the integration points in the correlated k-distribution approximation without decreasing accuracy. In the NICAM default setting, the spectrum between 0.2 and 200 μm is divided into 29 spectral bands with 111 integration points. Cloud-radiation feedback is one of the most important issues in the study of climate change. The high-resolution experiments performed by NICAM have an advantage in that they reduce the uncertainties of the horizontal/vertical cloud distributions. In the recent studies described above, both the cloud scheme and radiation scheme are improved by considering the detailed size distribution of the hydrometeors and non-sphericity of the ice particles. This cooperative development of the cloud and radiation schemes will facilitate improvements in the future development of aerosol and chemistry modules (see 'Aerosol and chemistry modules' section).

NICAM also introduces the ISCCP simulator (Klein and Jakob 1999; Webb et al. 2001), with which the distribution of every cloud category can be calculated online.

The results from the simulator have been used in numerous studies (Iga et al. 2007a; Suzuki et al. 2008; Collins and Satoh 2009; Noda et al. 2010; Sohn et al. 2010; Satoh et al. 2012b; Miyamoto et al. 2013; Tsushima et al. 2014). In some studies, new satellite simulator packages have been applied offline to the NICAM dataset. Kodama et al. (2012) used a series of satellite simulators that were part of the Cloud Feedback Model Intercomparison Project (CFMIP) Observation Simulator Package (COSP) (Bodas-Salcedo et al. 2011), and Hashino et al. (2013) used the Joint Simulator for Satellite Sensors (J-simulator). Therefore, the online use of these packages should also be taken into consideration.

Aerosol and chemistry modules

Currently, NICAM is implemented with a global 3D aerosol transport-radiation model, the Spectral Radiation-Transport Model for Aerosol Species (SPRINTARS) (described fully in Takemura et al. 2000, 2002, 2005, 2009; Goto et al. 2011a). The SPRINTARS module predicts the mass mixing ratios of the main tropospheric aerosol species, i.e., carbonaceous (black carbon and organic matter), sulfate, soil dust, sea salt, and the precursor gases of sulfate (sulfur dioxide and dimethylsulfide (DMS)). The aerosol transport processes include emission, advection, diffusion, sulfur chemistry, wet deposition, dry deposition, and gravitational settling. Emissions of soil dust, sea salt, and DMS are calculated using the internal state of the model, whereas the emissions for the other primary aerosols and the distributions of oxidants (hydroxyl radical, ozone, and hydrogen peroxide) for the sulfur chemistry are prescribed by externally assumed data. In the module itself, meteorological fields such as clouds and precipitation, which are critical to determining aerosol lifetimes, are provided from the outer module after the cloud microphysics calculations are computed at each time step. In contrast, the calculated mass mixing ratios of the aerosols in each grid cell are calculated by the aerosol module SPRINTARS and then returned back to the outer module. In addition, aerosol variables are forwarded to the radiative transfer module, MSTRN-X (Sekiguchi and Nakaima 2008), in order to represent the direct aerosol effects, and also to cloud schemes such as NSW6, to represent the indirect aerosol effects.

SPRINTARS, which was originally coupled to MIROC (Watanabe et al. 2010) with a low spatial resolution of 100 to 300 km, has compared well with various measurements (Goto et al. 2011a, b, 2012) and other global aerosol models participating in the AeroCom international project (e.g., Kinne et al. 2006). The results of the SPRINTARS module coupled to NICAM were also shown to generally agree with the satellite observations of aerosols and clouds for versions based on low spatial resolution of g-level 5 ($\bar{\Delta}$ = 220 km) (Dai et al. 2014a, b; Goto 2014) and high spatial resolution of g-level 10 ($\bar{\Delta}$ = 7 km) (Suzuki et al. 2008). Using Earth Simulator, Suzuki et al. (2008) conducted a global simulation with the finest resolution to date and showed the global aerosol and cloud distributions for July 2006 (Figure 8). Most of the aerosol plumes were in generally good agreement with the satellite observations, except for North America, Australia, and most remote oceans. Especially, over North America, the values of NICAM-simulated aerosol optical thickness were underestimated, possibly because of insufficient emissions from biomass burning and anthropogenic sources, or the lack of secondary organic aerosols formed from isoprene in forests (Henze and Seinfeld 2006). Through the aerosol-cloud interaction, the cloud droplet effective radius generally decreases, as the aerosol optical thickness increases, which were generally captured by both NICAM and MODIS. The cloud droplet effective radii simulated by NICAM were smaller than those obtained from MODIS, probably because those retrieved from 2.2 μm (shown in Figure 8) correspond to those captured below the cloud top heights where the cloud droplet effective radii are larger than those at cloud top height (Nakajima et al. 2010). Very recently, regional aerosol simulations that used the stretched grid system with a high spatial resolution of approximately 10 km were performed for the Kanto region of Japan under the MEXT/RECCA/SALSA project (SALSA 2014; Goto et al. 2014). These simulations were carried out at relatively low computational cost by using two supercomputers at the University of Tokyo (SR16000 and FX10).

The CHASER atmospheric chemistry module has also been implemented into NICAM (SALSA 2014). This module primary focuses on tropospheric chemistry while considering the chemical cycle of O_x-NO_x-HO_x-CH_4-CO relation to the oxidation of volatile organic compounds (VOCs) and halogen chemistry in order to calculate the concentrations of 92 chemical species with 262 chemical reactions (58 photolytic, 183 kinetic, and 21 heterogeneous reactions) (Sudo et al. 2002a, b; Watanabe et al. 2011). Using the CHASER module, the mixing ratios of atmospheric gases, which mainly include ozone, NO_x, CO, SO_2, VOCs, and sulfate aerosols, can be predicted. The transport processes of the chemical tracers include emission, advection, diffusion, chemical reaction, wet deposition, and dry deposition. The emissions of lighting NO_x and DMS are calculated by using the internal parameters of the model, whereas the emissions of the other gases are prescribed. The photolysis rates are calculated online by using the temperature and radiation fluxes computed in the radiative transfer module. Like the aerosol module, the meteorological field information is input from the outer module after the cloud

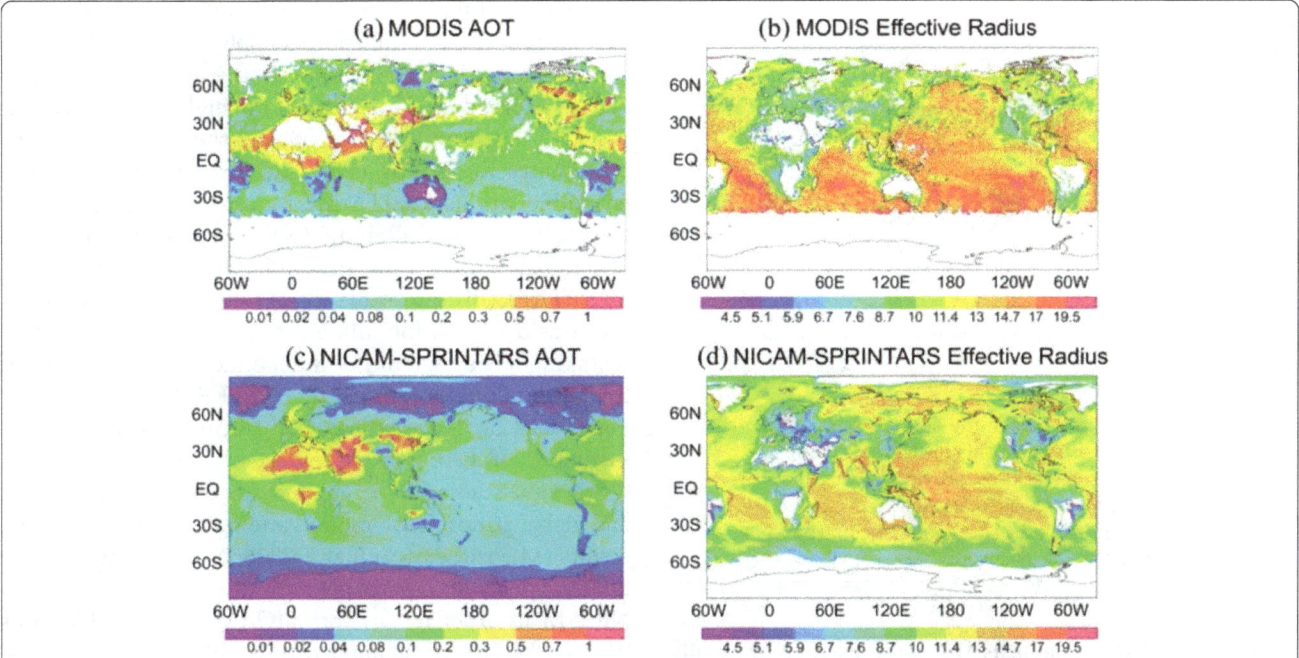

Figure 8 Distributions of aerosol optical thickness and cloud droplet effective radius from the NICAM-SPRINTARS simulations. Global geographical distributions of (a, c) aerosol optical thickness and (b, d) cloud droplet effective radius from (c, d) the NICAM-SPRINTARS simulations in comparison to those obtained from (a, b) the MODIS satellite observations for 1 to 8 July 2006 (cited from Suzuki et al. 2008). The unit of cloud droplet effective radius is micrometers.

microphysics calculations and to the chemistry module at each time step. In contrast, the calculated mass mixing ratios of the tracers in each grid cell within the chemistry module are returned back to the outer module. In addition, ozone, N_2O, and CH_4 are calculated online and used in the radiative transfer module to represent the atmospheric warming effects.

The distributions of the trace gases obtained by CHASER, which is currently coupled to MIROC at a low spatial resolution of approximately 300 km, are generally compared to the observations (Sudo et al. 2002b; Sudo and Akimoto 2007; Nagashima et al. 2010) and to other chemistry models under international model comparison projects such as the Atmospheric Chemistry and Climate Model Intercomparison Project (ACCMIP) (e.g., Lamarque et al. 2013). Under the MEXT/RECCA/SALSA project, the results of CHASER, along with the globally uniform and regionally stretched grid systems, were compared with the observations (SALSA 2014). Furthermore, the chemical module has been currently expanded to a fully coupled aerosol-chemistry module by coupling CHASER with SPRINTARS, as developed in Watanabe et al. (2011) and used by MIROC.

The unified module, consisting of the fully coupled aerosol-chemistry, coupled to NICAM (hereinafter referred to as NICAM-Chem) can explicitly consider various interactions between aerosols and gases, as shown in Figure 9. For example, as an aspect of the aerosol simulation, the

unified module can precisely calculate the sulfate formation with online oxidants such as hydroxyl radicals and ozone, whereas SPRINTARS uses assumed oxidants. In contrast, as an aspect of short-lived gases, calculations of the heterogeneous reactions onto the aerosol surface are expected to provide the greatest advantage when using the unified model. In addition to the sophisticated interactions between aerosols and gases, the chemistry of nitrate will be implemented into NICAM-Chem by using a thermodynamic equilibrium condition and secondary organic aerosols (SOAs). As a result, the new unified chemistry module is expected to provide better distributions of aerosols and gases, and, in the near future, we expect to be able to provide such distributions globally with a high spatial resolution of less than 10 km.

Land model

We implemented two land surface schemes in NICAM in order to calculate the water and energy balance over land. The first scheme is a simple bucket land surface model (Kondo 1993), and the other involves the minimal advanced treatments of surface interaction and runoff (MATSIRO) model (Takata et al. 2003). When used with NICAM, the simple bucket land surface model, which was first used by Miura et al. (2007a), has three vertical layers for predicting soil temperature and one layer for predicting the soil moisture.

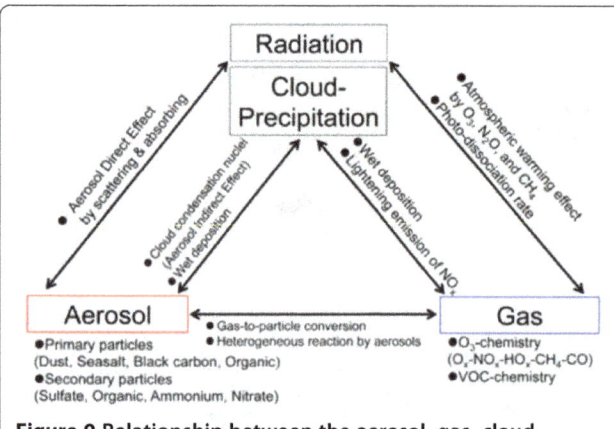

Figure 9 Relationship between the aerosol, gas, cloud, precipitation, and radiation fields for NICAM-Chem.
Relationship between the aerosol, gas, cloud, precipitation, and radiation fields for the unified aerosol-chemistry module (NICAM-Chem). The aerosol and gas modules correspond to SPRINTARS and CHASER, respectively.

The MATSIRO scheme is the same as that originally employed in MIROC (Hasumi and Emori 2004). Kodama et al. (2012) were the first to use MATSIRO with NICAM. This scheme has five vertical layers for predicting soil temperature and soil moisture, as well as a snow scheme with thee vertical layers.

Even in a several-kilometer grid simulation, land use and land cover cannot be regarded as homogeneous in one grid cell if it is to be used to evaluate the land surface heterogeneities smaller than the horizontal grid scale. Accordingly, we are currently implementing a mosaic treatment of the land surface models in order to improve the accuracy of the water and energy fluxes at the land surface. We will also attempt to incorporate an urban canopy model in order to simulate the global urban energy balance.

Assimilation

Local ensemble transform Kalman filter applied to NICAM: NICAM-LETKF

The introduction of a data assimilation system into NICAM is a challenge because it might enable us to produce a high-resolution dataset with O (km) mesh size that can be used for improved better numerical simulations and research analysis. Data assimilation is an analysis technique that provides an accurate estimate of the atmospheric state by using observation data and model forecasts. It plays an important role in improving numerical weather prediction by providing an accurate initial condition. The ensemble Kalman filter (EnKF) is an advanced data assimilation method that uses flow-dependent forecast error statistics that are represented by the ensemble prediction (Evensen 1994). For a linear system, the Kalman filter (KF) gives a minimum mean

square error if the error is a Gaussian distribution (Kalman 1960). However, it is difficult to apply the KF because the computational cost for an atmospheric model with a high degree of freedom (order of 10^7) is high. To resolve this problem, Evensen (1994) proposed the EnKF, which approximates the KF by using the forecast spread of the ensemble prediction. Furthermore, the local ensemble transform Kalman filter (LETKF) (Hunt et al. 2007) was developed based on the ensemble transform Kalman filter (ETKF) (Bishop et al. 2001) and used an algorithm that was designed for suitability with parallel computers by taking advantage of the independent local analyses of the local ensemble Kalman filter (LEKF) (Ott et al. 2004). For details on the mathematical formulation and implementation of these approaches, refer to each reference.

Previous studies have applied the LETKF to regional and global atmospheric models and have shown promising results (Miyoshi and Aranami 2006; Miyoshi and Yamane 2007). Kondo and Tanaka (2009) applied the LETKF to NICAM in order to develop NICAM-LETKF, which was the first LETKF to be applied to a global non-hydrostatic model. They then investigated the feasibility and stability of NICAM-LETKF using a perfect model and determined that the system works appropriately for a realistic non-hydrostatic model (NICAM). However, the horizontal resolution used in their study was not very high (g-level 5 or $\bar{\Delta}$ = 220 km), where the model would behave hydrostatically. Additionally, it is unclear whether NICAM-LETKF will work stably when real observations are used. Therefore, a future task will be to investigate data assimilation with a high-resolution global non-hydrostatic regime using real observations. This would not only contribute to an improvement in weather prediction but also to the research on multi-scale interaction, by using the product of the assimilation. We are now developing a version of NICAM-LETKF that will be based on the LETKF code that was developed and continuously improved by Miyoshi (e.g., Miyoshi and Kunii 2011) and are currently testing assimilation of precipitation data observed by satellites in order to obtain a fine-mesh precipitation database using the method proposed by Lien et al. (2013).

NICAM-based transport model: NICAM-TM

An accurate estimation of the past and current CO_2 budget for the Earth's surface is needed for reliable predictions of carbon cycle changes under global warming conditions. Therefore, an inversion method coupled with a tracer transport model was used to estimate the spatial and temporal variations of CO_2 sources and sinks based on observations of CO_2 concentrations (e.g., Tans et al. 1990). In an inversion analysis, we quantitatively estimated regional CO_2 fluxes by considering the *a priori* estimate and uncertainties of the CO_2 fluxes, which were derived from our present understanding of CO_2

source/sink mechanisms and uncertainties of observations (including model representative errors). This analysis is based on Bayesian theory and is similar to data assimilations for meteorological forecasts (Tarantola 2005). However, in contrast to the observations of meteorological parameters, CO_2 concentrations are not sufficiently observed around the globe. Therefore, estimates of CO_2 sources and sinks are not well constrained by actual observations and are extremely sensitive to model transport errors (e.g., Gurney et al. 2002).

In response to the need for an accurate transport model, Niwa (2010) developed the NICAM-based transport model (NICAM-TM). For transport simulations of CO_2, NICAM has an advantageous property: consistency with continuity (CWC) (Jöckel et al. 2001; Gross et al. 2002). Under the CWC condition, the mass conservation and Lagrangian conservation of a tracer are simultaneously guaranteed; that is, the volume integral of mass is conserved and a constant specific mass of a tracer is maintained along flow trajectories. The ability to achieve the CWC property is attributed to the fact that NICAM uses the finite volume method for its meteorological field integration and performs tracer transport on a common dynamical frame (Niwa et al. 2011a). The tracer mass conservation is strictly required for analyses of the CO_2 budget, which is the main target of inversion studies. Furthermore, the Lagrangian conservation property is imperative for CO_2 because it is an abundantly existing tracer (the global average CO_2 concentration was about 390 ppm for 2013) and the concentration changes that are analyzed are very small (no more than a few tens of ppm).

NICAM-TM consists of online and offline transport models, as well as an adjoint model for the tracer transport. In the online model, tracer transport is calculated concomitantly but independently with the integration of the meteorological field. More specifically, the tracers other than those for water are passive. Compared to the online model, the offline model is computationally inexpensive because it only calculates the tracer transport by using meteorological data calculated by the online NICAM-TM and stored in a disk beforehand. The adjoint model was developed based on the offline model, which calculates the sensitivities to certain tracer variables backward in time. Because long-term and multiple simulations are required for the inverse calculation, NICAM-TM is currently run with a relatively low-resolution for CO_2 transport simulations; for the most part, a resolution of g-level 5 or 6 is used. However, after conducting intercomparison experiments (Law et al. 2008; Patra et al. 2008; Niwa et al. 2011b), we found that the performances of CO_2 transport obtained are comparable to those of other models. Generally, the modeled winds are nudged towards the analyzed data (e.g., JMA Climate Data Assimilation System (JCDAS); Onogi et al.

2007) in order to match the simulated CO_2 concentrations to the observed concentrations.

Niwa et al. (2012) performed an inversion analysis by using the synthesis inversion method (Enting 2002) with NICAM-TM. This was the first inversion study in which aircraft data were extensively used to constrain the global and regional CO_2 budgets. The aircraft data were obtained from the Comprehensive Observation Network for Trace Gases by Airliner project (CONTRAIL) (Machida et al. 2008; Matsueda et al. 2008; Sawa et al. 2012). By adding the CONTRAIL data (Figure 10a) to the conventionally used surface observation network, Niwa et al. (2012) successfully reduced uncertainties of the CO_2 flux estimates for Asia. Moreover, this study revealed the existence of a strong summer uptake by the biosphere in South Asia (Figure 10b), where the CO_2 observations from CONTRAIL in the middle to upper troposphere have a predominant constraint on the flux estimation. Because biosphere models have not shown such a strong summer uptake (Patra et al. 2013), Niwa et al. (2012) suggested the existence of some unknown mechanisms of the CO_2 sources and sinks in this region. To further improve the CO_2 inversion, a four-dimensional (4D) variational method system is now being developed by using the adjoint model of NICAM-TM. Using this system, a large amount of CO_2 concentration data that have been recently available from regular aircraft observations and satellite observations (e.g., Greenhouse Gases Observing Satellite (GOSAT)) will be exploited to further constrain the estimates of CO_2 sources and sinks.

Models related to NICAM
Stretch NICAM and diamond NICAM
As described in 'Grid configuration and advection scheme' section, NICAM was originally designed for use with an icosahedral grid, which covers a sphere of the Earth quasi-uniformly and has a grid system based on 10 large diamonds. As extensions to the original configuration, NICAM can also be applied to a non-uniform resolution distribution grid or a regional grid by changing the positions of the grid points, the total number of diamonds, or the topology of their combination.

NICAM can also be adapted to simulations that target a particular area. Two approaches to performing a partially high-resolution simulation are typically used with NICAM. One includes the stretched icosahedral grid system developed by Tomita (2008a). In this approach, the icosahedral grid is stretched using the Schmidt transformation method and the grid size becomes gradually finer as grid points are concentrated onto the area of interest. In the stretch NICAM, the inhomogeneity of the horizontal grid sizes requires resolution-dependent parameters for physical schemes if the global domain is of interest. However, if a limited region is of interest,

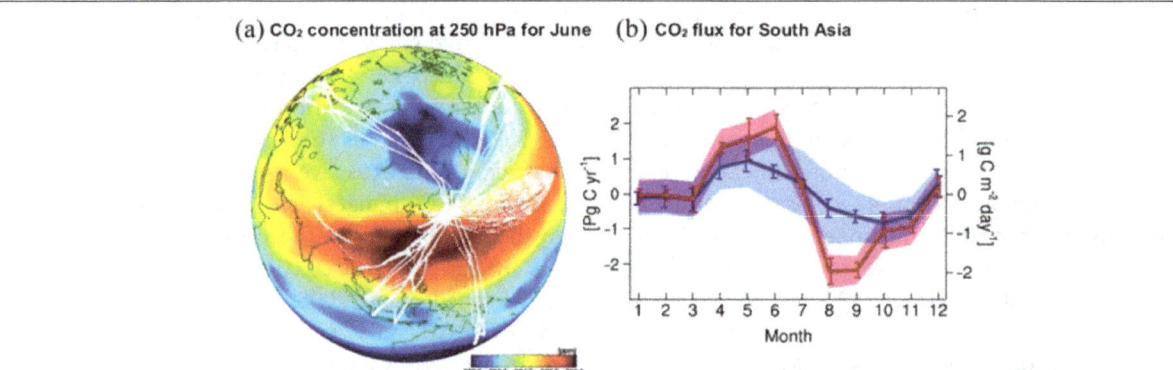

Figure 10 Simulated CO₂ concentration field and monthly variation of the CO₂ flux by NICAM-TM. (a) Mean CO_2 concentration field at 250 hPa for June 2007, as simulated by NICAM-TM, and the locations of the CONTRAIL measurements for 2007 (white dots). **(b)** Monthly variation of the CO_2 flux averaged for 2006 to 2008, estimated by the inversion analysis of Niwa et al. (2012) (this figure is slightly modified from Figure 5d in Niwa et al. 2012). The blue and red lines are the CO_2 fluxes (positive values indicate sources and negative values indicate sinks) estimated by only using surface data and by using both surface and CONTRAIL data, respectively, and the colored shades and error bars denote the range of the estimated flux uncertainty and standard deviation for 2006 to 2008, respectively.

such as within regions with finer mesh sizes, resolution-dependent parameters are not required. The stretch NICAM is widely used for studies on tropical convective systems (Satoh et al. 2010; Satoh and Kitao 2013; Roh and Satoh 2014), tropical cyclones (Yanase et al. 2010a, b; Satoh et al. 2013; Arakane et al. 2014), and MJOs (Nasuno 2013).

The other approach is similar to the use of a regional model. Only 1 of the 10 diamond regions that make up the icosahedral grid is used for the simulation (hereinafter, the diamond NICAM). The simulation domain of the diamond NICAM consists of 1 diamond (2 triangles of an icosahedron), even though the original NICAM global domain consists of 10 diamonds (20 triangles). The diamond NICAM is used as a non-hydrostatic regional model. By using this approach, we can share the same code for both the global domain and limited-area models. This is advantageous for the maintenance of the source code because all components of NICAM are used in common in both models.

To enable the limited-area simulations using the diamond NICAM, we developed a nudging scheme in which the strength of the nudge depends on the distance from a certain point on the globe. Figure 11 shows the simulation domain of a regional climate simulation targeted on Japan. The lateral boundary of the simulation domain (outer edge of the gray circle in Figure 11) is forced by six-hourly reanalysis data. Verification of the diamond NICAM is still under investigation and the results of test simulations performed using the diamond NICAM were compared with the results of the high-resolution experiments using the stretch and global NICAM. The comparison results showed good agreement with those produced by the original global NICAM simulation. The temporal evolution of a convective system, such as the locations of triggering,

development, and decay, was also well simulated during the integration.

Plane NICAM
Although the stretch and diamond NICAM can be used for limited-area simulations, an *f*-plane system is useful when conducting some types of numerical simulations, such as idealized simulations on an *f*-plane and

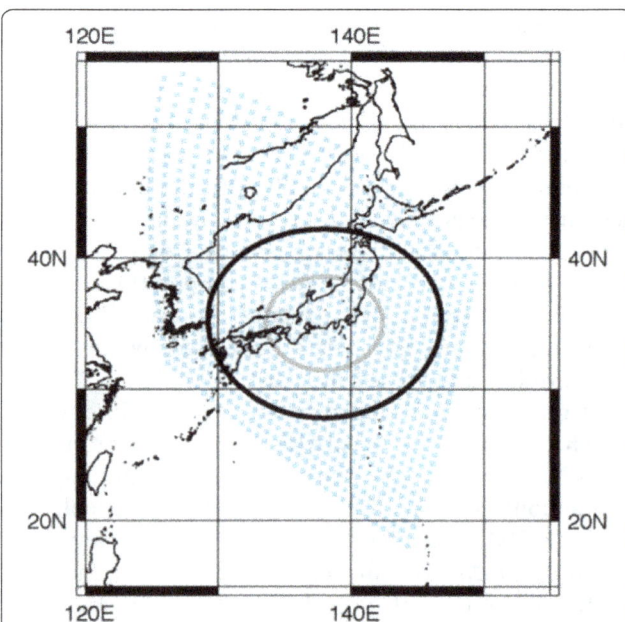

Figure 11 Simulation domain of the diamond NICAM targeted on Japan. Light blue dots indicate the grid points. Reanalysis nudging was applied to the outer side of the gray circle. This grid is based on g-level 5. Stretching parameter (*β* in Equation 8, Tomita et al. 2008) of grid transformation is 10. The minimum grid interval is about 59 km.

simulations with a narrow region in which the curvature of the Earth is not important. For this reason, an f-plane model (hereinafter, the plane NICAM) whose dynamical core is the same as the original NICAM was developed. To save maintenance costs, the plane NICAM used the same code as the global NICAM, which also makes knowledge gained by the plane NICAM immediately and directly applicable to the global NICAM.

In the plane NICAM, the shape of the calculation domain is a diamond composed of two regular triangles with a double periodic lateral boundary condition. Figure 12 shows the configuration of the calculation domain for the plane NICAM, as indicated by black bold lines. The outward fluxes across side AB are equivalent to the inward fluxes across side DC. The red, green, blue, and yellow areas inside the diamond area surrounded by the black bold lines correspond to the hatched areas of the same respective color. Therefore, if the double periodic boundary condition is applied to the diamond-shaped domain with side length L, the calculation domain is equivalent to the regular hexagonal domain with side length $l = L/\sqrt{3}$, as represented by the hatched area in Figure 12. Regular hexagonal geometry is better suited than square or regular triangle geometry in terms of isotropy.

As an example of the use of the plane NICAM, we briefly show a result of an idealized tropical cyclone simulation on an f-plane (Ohno and Satoh 2014). The model was initialized with an axisymmetric cyclonic vortex that had a maximum azimuthal wind of 12 m s^{-1} as formulated by Rotunno and Emanuel (1987). The initial thermodynamic structure of the unperturbed model atmosphere is defined by moist adiabatic lapse rate up to 17 km and a constant temperature above this height.

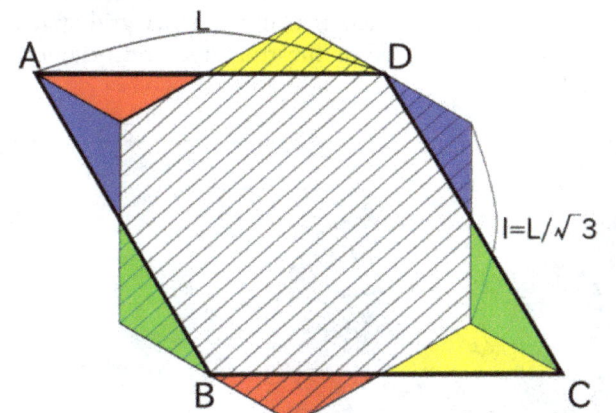

Figure 12 A schematic image of the calculation domain for the plane NICAM. A schematic image showing the relationship between the calculation domain and a unit cell for the plane NICAM. The diamond-shaped area (ABCD) surrounded by black bold lines indicates the calculation domain, and the hatched regular hexagonal area indicates the unit cell.

The length of the sides of the numerical domain is L = 3,000 km, and a horizontal resolution of 2.7 km is used for the entire domain. We used 50 vertical levels up to 45 km where the intervals of the vertical levels become smaller in the boundary layer. The SST is fixed at 31°C and an f-plane with a constant Coriolis parameter at 18 N is assumed. After an initial shock, the minimum sea level pressure continued to drop for 140 h and a quasi-steady stage was achieved. The cyclone reached a minimum central surface pressure of 931 hPa at about 160 h. A plan view of the distribution of the hourly precipitation amounts after 128 h of the time integration is shown in Figure 13, which indicates that the tropical cyclone has an eye, an eyewall, inner core rainbands, and distant rainbands. The array of rainbands and the eyewall are typical features of a tropical cyclone (Houze 2010).

Further modified grids

In addition to the original icosahedral grid, alternative spherical grids denoted by extended triangular meshes (XTMS) (Iga 2014a) are also implemented on NICAM. The mesh topologies of these grids are different from the topology of an icosahedral grid and are generated by three processes: grid relaxation via spring dynamics, transformation via an analytical function, and the Schmidt transformation. By changing mesh topologies and transformation functions, it is possible to obtain a grid with non-uniform distributions of resolution on a sphere, which is applicable to various situations. For example, the resolution can be increased at a particular region or along a tropical zone as shown by Figure 14a,b. The grid structure shown by Figure 14a looks similar to that of the stretch NICAM, but the variability of the horizontal mesh spacing is smaller and the variability of the resolution is smoother than that of the stretch NICAM. The resolution is approximately proportional to the combination of map factors of two polar stereographic projections. As for the grid structure shown by Figure 14b, resolution is enhanced along the equatorial belt and is approximately proportional to the combination of map factors of two Lambert conformal conic projections and one Mercator projection (hereinafter, LML grid). The LML grid is more useful for studying the multi-scale structure of convective systems in the tropics. The performance of the LML grid was examined by aqua planet experiments (Iga 2014b).

Coupler

Currently, high-resolution NICAM is mainly used for short-term simulations such as for tropical cyclones and intra-seasonal variability. Since deep sea circulations are not important for these simulations, a simple mixed-layer ocean model is implemented in NICAM. However, in order to cover wider areas of research, such as

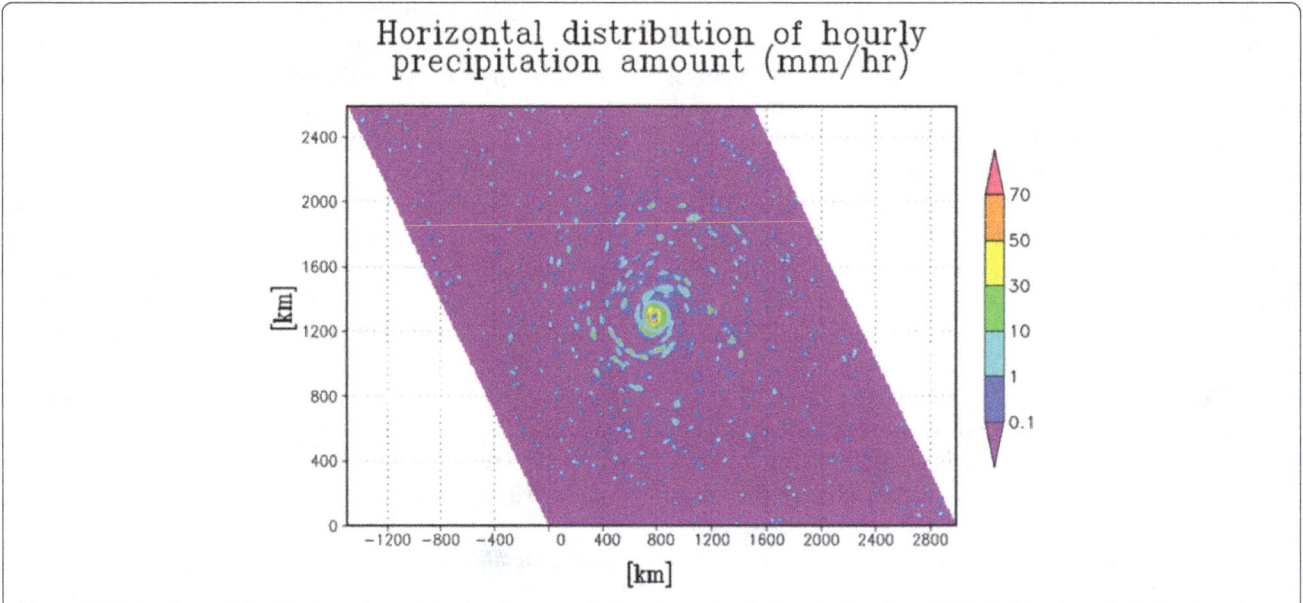

Figure 13 A horizontal distribution of precipitation for a tropical cyclone simulation by the plane NICAM. A horizontal distribution of hourly precipitation amount by the plane NICAM after a time integration of 128 h for a 2.7 km mesh tropical cyclone simulation.

climate projection, coupling with an ocean model is necessary when NICAM is used as an atmospheric component of an Earth system model. To couple an atmosphere model and an ocean model, we used a coupler called Jcup (Figure 15). Because Jcup connects models with different grid systems, it can be used as a grid transformation tool. Thus, in addition to ocean coupling, Jcup is used as an I/O tool of NICAM by converting the output data from the icosahedral grid to the lat-lon grid which is generally used for analysis. In this section, the Jcup coupling library that was used as a core of the coupled model system will be introduced first, after which the ocean and I/O coupling will be described.

The Jcup coupling library is a collaborative work of JMA/MRI, JAMSTEC, and the Research Organization for Information Science and Technology (RIST) (Arakawa and Yoshimura 2009; Arakawa et al. 2011), that is freely available as an open source program at https://sites.google.com/a/rist.jp/jcup/. Jcup inherited the development experience and design concept of Scup, which was developed by Yoshimura and Yukimoto (2008). The most remarkable feature of Jcup is its wide applicability. Generally speaking, the applicability range of a coupler is restricted by the specific grid structures and interpolation schemes it supports. For example, OASIS version 4 supported logically rectangular (i.e., 2D structured) or reduced Gaussian grids, along with bilinear, trilinear, bicubic, nearest-neighbor, and 2D conservative interpolations (Redler et al. 2009). This indicates that the models that utilize other grid systems, such as non-structured or substructured grids, cannot be coupled.

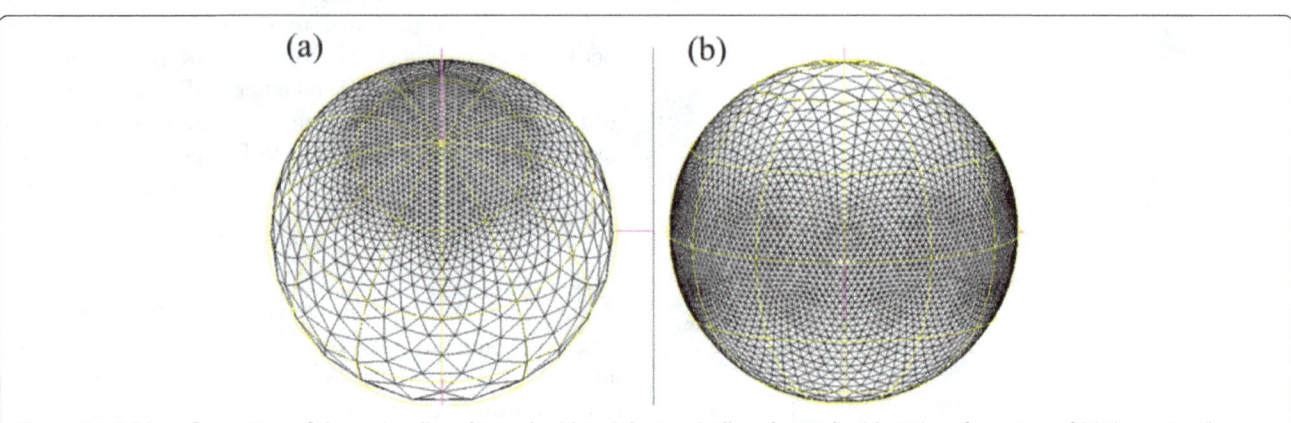

Figure 14 Grid configurations of the regionally enhanced grid and the tropically enhanced grid. Grid configurations of **(a)** the regionally enhanced grid and **(b)** the tropically enhanced grid (after Iga 2014a).

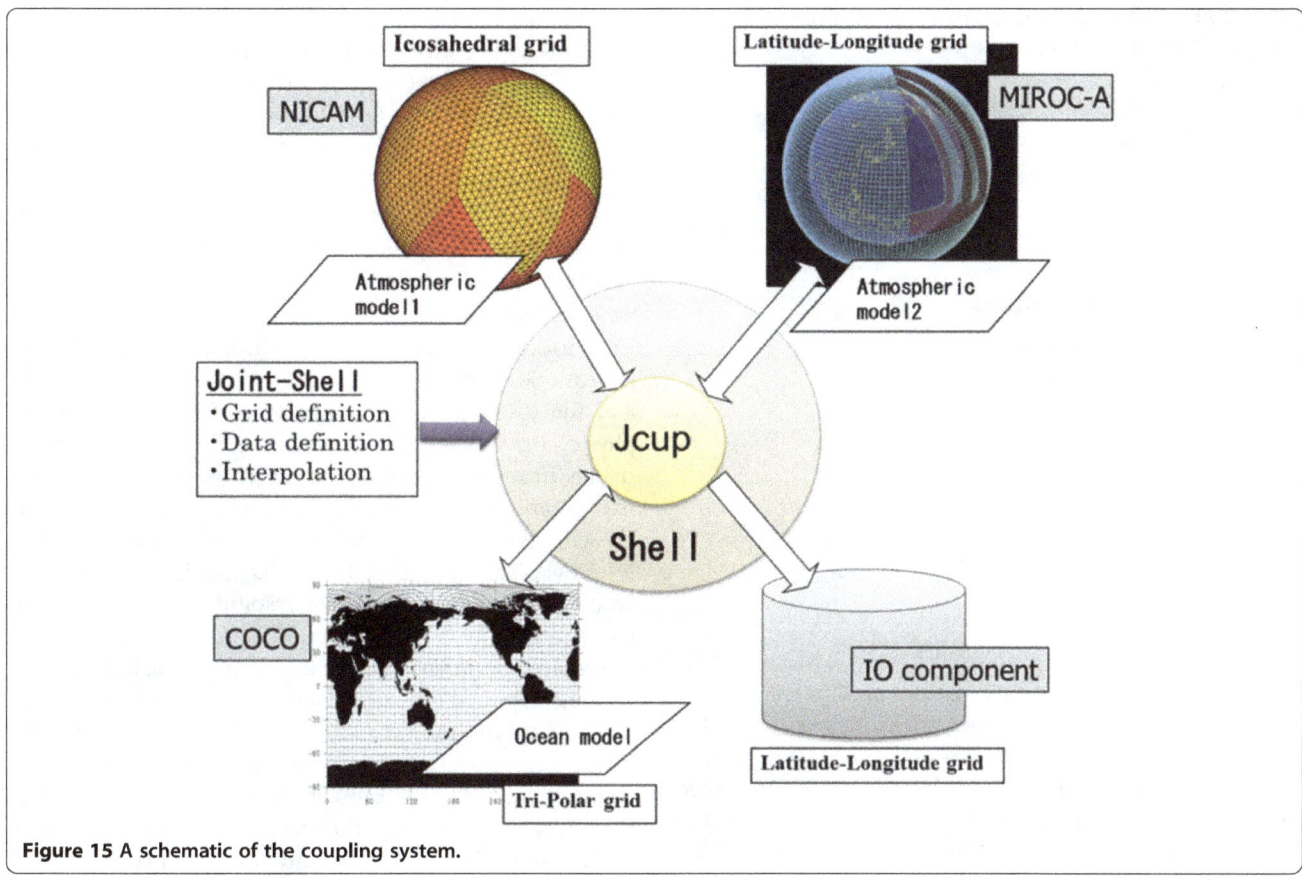

Figure 15 A schematic of the coupling system.

The spectral method has been widely used for global atmospheric models. However, this method is not suitable for massively parallel processor (MPP) supercomputers because all-to-all communication is required. This is one reason why substructured grid models such as NICAM were developed energetically by the research organizations of numerous countries. Therefore, it is also necessary to develop a coupler that can handle such models. Jcup is specifically designed to couple different grid systems, so users can (and must) implement their own interpolation code and prepare a 'mapping table' in advance. See Arakawa et al. (2011) for details.

Next, we will describe NICAM and ocean model coupling. Currently, the Center for Climate System Research (CCSR) Ocean Component Model (COCO) (Hasumi 2000, 2006) is used for the ocean model coupled with NICAM. In the coupler of the NICAM-COCO coupled model, Jcup is used as a core library and is implemented via a package called Joint-shell, which contains definition of the grid structures, exchange data, and the interpolation code (Figure 15). The interpolation code is based on the first-order conservative remapping scheme in Jones (1999). The 'mapping table' that must be set up as external files for beforehand defines grid-to-grid relations and Jcup's weight for grid systems of coupled models.

Furthermore, we also implement a utility program to calculate such table files as a part of the Joint-shell. Note that COCO adopts a tri-polar grid, in which the northern polar region grid points do not follow latitude-longitude lines. As a result, coupling NICAM and COCO requires coupling two substructured grid models, which is why Jcup is used as a core of the coupler.

To couple NICAM, a new module has been implemented to replace the mixed-layer ocean module that has already implemented. Most of the other parts, such as the dynamical core, cloud microphysics, and land models, are untouched. The physical quantities exchanged between NICAM and COCO are listed in Table 3. The most significant issue was the inconsistency of ocean definition points between the two models. More specifically, some points that are recognized as ocean points by NICAM are recognized as land points by COCO. At such grid points, no data are provided to NICAM. Inconsistencies such as this occur if the land-ocean distribution has been set independently between the two models. In the future, we plan to improve pre-process utilities in order to create a NICAM land-ocean map that will maintain consistency with the COCO map. However, to avoid this problem in the interim, we applied the mixed-layer ocean module, which is pre-implemented in NICAM to such grid points.

Table 3 Physical quantities exchanged between NICAM and COCO through the coupler

NICAM to COCO	COCO to NICAM
Wind stress to ocean (eastward)	Sea surface temperature
Wind stress to ocean (northward)	Sea ice thickness
Wind stress to sea ice (eastward)	Ice covered ratio
Wind stress to sea ice (northward)	Snow amount on sea ice
Energy flux from atmosphere to ocean	Temperature of sea ice
Energy flux from atmosphere to sea ice	
Energy flux from sea ice to ocean	
Radiation flux (short wave)	
Sublimation from sea ice	
Evaporation	
Precipitation	
Snowfall	
Runoff	

The icosahedral grid employed by NICAM cannot be used to analyze the results. For example, the calculation of the zonal mean values is not straightforward and the visualization tools assume a lat-lon grid in many cases. Because of this, we generally use a conversion tool called 'ico2ll' to move data from the icosahedral grid to the lat-lon grid for post-processing of simulations. However, in many cases, running ico2ll as a post-process results in a bottle neck because it requires numerous I/O transfers and cannot parallelize itself since it is necessary to create a single file for the lat-lon grid of each variable. Jcup allows NICAM-I/O coupling, which leads to an efficient I/O conversion, to be successfully achieved. More specifically, the I/O module converts the icosahedral grid to the lat-lon grid and is executed in parallel with NICAM. The implemented conversion schemes include a bi-linear interpolation, a control volume weighted average, and the nearest-neighbor method. A bi-linear interpolation is originally implemented method in the stand-alone version of ico2ll', and this method has first-order accuracy. The control volume weighted average scheme is the same with the one used in NICAM-COCO coupling and also has first-order accuracy, but the global averaged fluxes are conserved. The nearest-neighbor method has less accuracy and conservation properties than the other two methods. This method was implemented due to the demands from NICAM users and is used to provide a quick look at the raw values by using the general latitude-longitude graphic tools, such as the Grid Analysis and Display System (GrADS), without an averaging process. These schemes can be set for each output data type and can be switched by altering the configuration file.

Figure 15 is a schematic of the coupling system described above. The coupling system is designed so that NICAM

automatically detects the coupling pattern at runtime without taking any other configuration. For example, when COCO is executed in parallel with NICAM, subroutines for NICAM-COCO coupling are used, and if not, subroutines of the mixed-layer ocean model are called. I/O is also the same as the case of COCO, in which NICAM automatically sends output data to I/O only when the I/O component is executed.

Conclusions

This paper reviewed recent activities of the global non-hydrostatic model, NICAM, and described its development and the details of each component of the dynamics and physics, the strategy of assimilation, and the related models. In the 'Introduction' section, a review of other global non-hydrostatic models, scientific overviews of NICAM, and the current design/structure and a future timeframe of NICAM were described. In the 'Dynamics' section, horizontal grid issues and vertical resolution issues were discussed. In the 'Physics' section, components of cloud microphysics, subgrid-scale turbulence, radiation, aerosol and chemistry, and land surface were described. As for assimilation, NICAM-LETKF was presented, and NICAM-TM was introduced to describe the assimilation of the transport model. The 'Models related to NICAM' section showed a variety of ways that NICAM can be used, specifically, stretch NICAM, diamond NICAM, plane NICAM, the tropically enhanced grid model, and coupling with the ocean model or the I/O module.

A number of comprehensive NICAM-related review articles have been published (Satoh et al. 2008; Satoh 2013) since the 3.5 km mesh global non-hydrostatic experiments were conducted by Tomita et al. (2005) and Miura et al. (2007b). However, these were primarily limited to descriptions of the governing equations, and because many of the model components continue to evolve and develop as the entire package expands, we were requested to provide a comprehensive overview of NICAM activities and future projections. NICAM has been a pioneering global non-hydrostatic model for almost a decade, and now that other global non-hydrostatic models are emerging, we believe that our experiences and considerations will be useful for other high-resolution modeling groups. Mutually beneficial information exchanges between different modeling groups will be required for future fruitful studies during the forthcoming era of global non-hydrostatic modeling.

Abbreviations
AICS: RIKEN Advanced Institute for Computational Science;
AORI: Atmosphere and Ocean Research Institute, The University of Tokyo;
AGCM: atmospheric general circulation model; AMIP: Atmospheric Model Intercomparison Project; CALIPSO: Cloud-Aerosol Lidar and Infrared Pathfinder Satellite Observations; CCSR: Center for Climate System Research, The University of Tokyo; CFL condition: Courant-Friedrichs-Lewy condition; CFMIP: Cloud Feedback Model Intercomparison Project; CHASER: Chemical Atmospheric General Circulation Model for Study of

Atmospheric Environment and Radiative Forcing; CINDY2011: Cooperative Indian Ocean Experiment on Intraseasonal Variability in the Year 2011; CONTRAIL: Comprehensive Observation Network for Trace Gases by Airliner Project; COCO: CCSR Ocean Component Model; COSP: CFMIP Observation Simulator Package; CWC: consistency with continuity; DYNAMO: Dynamics of the MJO; EnKF: ensemble Kalman filter; ETKF: ensemble transform Kalman filter; ES: Earth Simulator; ES2: Earth Simulator 2; GrADS: Grid Analysis and Display System; GOSAT: Greenhouse Gases Observing Satellite; g-level: grid division level; ISCCP: International Satellite Cloud Climatology Project; JAMSTEC: Japan Agency for Marine-Earth Science and Technology; JCDAS: JMA Climate Data Assimilation System; JMA: Japan Meteorological Agency; J-simulator: Joint Simulator for Satellite Sensors; LETKF: local ensemble transform Kalman filter; MATSIRO: Minimal Advanced Treatments of Surface Interaction and Runoff; MEXT/RECCA/SALSA: Development of Seamless Chemical Assimilation System and its Application for Atmospheric Environmental Material (SALSA) Project of the Research Program on Climate Change Adaptation (RECCA) in Ministry of Education, Culture, Sports, Science and Technology, Japan (MEXT); MIROC: Model for Interdisciplinary Research on Climate; MIROC ESM: MIROC Earth System Model; MISMO: Mirai Indian Ocean cruise for the study of the MJO-convection onset; MRI: Meteorological Research Institute; MSTRN-X: the latest version of the broadband radiative transfer model MSTRN based on the discrete ordinate method with a delta two-stream approximation; MTSAT-1R: Multi-functional Transport Satellite; MYNN: Mellor-Yamada Nakanishi-Niino scheme; MY: Mellor-Yamada Scheme; NDW6: NICAM Double Moment Scheme with six water categories; NSW6: NICAM Single Moment Scheme with six water categories; NICAM: non-hydrostatic icosahedral atmospheric model; NICAM-Chem: Unified Aerosol-Chemistry Model coupled to NICAM; NIES: National Institute for Environmental Studies; RIST: Research Organization for Information Science and Technology; SPRINTARS: Spectral Radiation-Transport Model for Aerosol Species; SWM: Shallow Water Model; TRMM: Tropical Rainfall Measuring Mission; VOC: volatile organic compound; XTMS: extended triangular meshes.

Competing interests

The authors declare that they have no competing interests. For more information, visit the website, 'http://nicam.jp/'.

Authors' contributions

MS and HT coordinated the structure of the paper and wrote the 'Introduction' and 'Conclusions' sections. HY wrote the 'Design, structure, development, and timeline' and 'Radiation' sections. In the 'Dynamics' section, HM wrote the 'Grid configuration and advection scheme' section, while CK and MS wrote the 'Vertical resolution issues' section. In the 'Physics' section, TS wrote the 'Cloud microphysics schemes' section, while AN wrote the 'Subgrid-scale turbulence of the planetary boundary layer' section using experimental data simulated by YY. DG wrote the 'Aerosol and chemistry modules' section, and MH wrote the 'Land model' section. In the 'Assimilation' section, MSw and TM wrote the 'Local ensemble transform Kalman filter applied to NICAM: NICAM-LETKF' section, while YN wrote the 'NICAM-based transport model: NICAM-TM' section. In the 'Models related to NICAM' section, MH wrote the 'Stretch NICAM and diamond NICAM' section, while TO wrote the 'Plane NICAM' section. The 'Further modified grids' section was written by SI, and the 'Coupler' section was written by TA, TI, and HK. All authors read and approved the final manuscript.

Acknowledgements

This research used the computational resources of the High Performance Computing Infrastructure (HPCI) system provided by the Information Technology Center of The University of Tokyo, through the HPCI System Research Project (Project ID:hp120190). The K computer at AICS was used under the supported by Strategic Programs for Innovative Research (SPIRE) Field 3 (Projection of Planet Earth Variations for Mitigating Natural Disasters), which is funded by MEXT (ID: hp120279 and hp130010). Earth Simulator at JAMSTEC and FX10 at The University of Tokyo were also used. The development of coupler is supported by Japan Science and Technology Agency/Core Research for Evolutional Science and Technology (JST/CREST), 'ppOpen-HPC: Open Source Infrastructure for Development and Execution of Large-Scale Scientific Applications on Post-Peta-Scale Supercomputers with Automatic Tuning (AT)'. The development of the diamond NICAM and NICAM-Chem is supported by MEXT/RECCA/SALSA.

Author details

[1]Atmosphere and Ocean Research Institute, The University of Tokyo, 5-1-5 Kashiwanoha, Kashiwa, Chiba 277-85648, Japan. [2]Japan Agency for Marine-Earth Science and Technology, 3173-15, Showa-machi, Kanazawa-ku, Yokohama, Kanagawa 236-0001, Japan. [3]RIKEN Advanced Institute for Computational Science, 7-1-26, Minatojima-minami-machi, Chuo-ku, Kobe, Hyogo 650-0047, Japan. [4]Department of Earth and Planetary Science, The University of Tokyo, 7-3-1 Hongo, Bunkyo-ku, Tokyo 113-0033, Japan. [5]National Institute for Environmental Studies, 16-2 Onogawa, Tsukuba, Ibaraki 305-8568, Japan. [6]Meteorological Research Institute, 1-1 Nagamine, Tsukuba, Ibaraki 305-0052, Japan. [7]Research Organization for Information Science and Technology, 2-32-3, Kitashinagawa, Shinagawa-ku, Tokyo 140-0001, Japan.

References

Adcroft A, Hill C, Marshall J (1997) Representation of topography by shaved cells in a height coordinate ocean model. Mon Wea Rev 125:2293–2315

Albrecht BA (1989) Aerosols, cloud microphysics, and fractional cloudiness. Science 245:1227–1230

Arakane T, Satoh M, Yanase W (2014) The excitation of the deep convection to the north of tropical storm Bebinca (2006). J Meteor Soc Japan 92:141–161

Arakawa A (2004) The cumulus parameterization problem: past, present, and future. J Clim 17:2493–2525

Arakawa A, Lamb VR (1977) Computational design of the basic dynamical processes of the UCLA general circulation model. Methods Comput Phys 17:173–265

Arakawa T, Yoshimura H (2009) Performance evaluation of a coupling software for climate modeling (in Japanese). J Inform Process Japan 2:95_110

Arakawa T, Yoshimra H, Saito F, Ogochi K (2011) Data exchange algorithm and software design of KAKUSHIN coupler Jcup. Procedia Comp Sci 4:1516–1525

Berger AL (1978) Long-term variations of daily insolation and quaternary climatic changes. J Atmos Sci 35:2362–2367

Berry EX, Reinhardt RL (1974) An analysis of cloud drop growth by collection: part II. Single initial distribution. J Atmos Sci 31:1825–1831

Bishop CH, Etherton B, Majumdar SJ (2001) Adaptive sampling with the ensemble transform Kalman filter. Part I: theoretical aspects. Mon Wea Rev 129:420–436

Blackburn M, Williamson DL, Nakajima K, Ohfuchi W, Takahashi YO, Hayashi Y-Y, Nakamura H, Ishiwatari M, McGregor J, Borth H, Wirth V, Frank H, Bechtold P, Wedi NP, Tomita H, Satoh M, Zhao M, Held IM, Suarez MJ, Lee M-I, Watanabe M, Kimoto M, Liu Y, Wang Z, Molod A, Rajendran K, Kitoh A, Stratton R (2013) The Aqua-Planet Experiment (APE): Control SST simulation. J Meteor Soc Japan 91A:17–56

Bodas-Salcedo A, Webb MJ, Bony S, Chepfer H, Dufresne JL, Klein SA, Marchand R, Haynes JN, Pincus R, John VO (2011) COSP: satellite simulation software for model assessment. Bull Am Meteorol Soc 91:1023–1043

Bonaventura L (2000) A semi-implicit semi-Lagrangian scheme using the height coordinate for a nonhydrostatic and fully elastic model of atmospheric flows. J Comp Phys 158:186–213

Bryan GH, Wyngaard JC, Fritsch JM (2003) Resolution requirements for the simulation of deep moist convection. Mon Wea Rev 131:2394–2416

Cheong H-B (2006) A dynamical core with double Fourier series: comparison with the spherical harmonics method. Mon Wea Rev 134:1299–1315

Collins WD, Satoh M (2009) Simulating global clouds, past, present, and future. In: Heintzenberg J, Charlson RJ (eds) Clouds in the perturbed climate system: their relationship to energy balance, atmospheric dynamics, and precipitation, Struengmann Forum Report, vol. 2. The MIT Press, Cambridge, pp 469–486

Côté J, Gravel S, Méthot A, Patoine A (1998) The operational CMC-MRB global environmental multiscale (GEM) model. Part I: design considerations and formulation. Mon Wea Rev 126:1373_1395

Cullen MJP, Davies T, Mawson MH, James JA, Coutler SC, Malcolm A (1997) An Overview of Numerical Methods for the Next Generation U. K. NWP and Climate Model. In: Lin CA et al (eds) Numerical Methods in Atmospheric and Oceanic Modelling. The Andrew J. Robert Memorial Volume, NRC Research Press, Ottawa, pp 425–444

Dai T, Goto D, Schutgens NAJ, Dong X, Shi G, Nakajima T (2014a) Simulated aerosol key optical properties over global scale using an aerosol transport coupled with a new type of dynamic core. Atmos Environ 82:71–82

Dai T, Schutgens NAJ, Goto D, Shi G, Nakajima T (2014b) Improvement of aerosol optical properties modeling over Eastern Asia with MODIS AOD assimilation in a global non-hydrostatic icosahedral aerosol transport model. Environ Pollut doi:10.1016/j.envpol.2014.06.021

Davies T, Cullen MJP, Malcolm AJ, Mawson MH, Staniforth A, White AA, Wood N (2005) A new dynamical core for the Met Office's global and regional modelling of the atmosphere. Quart J Roy Meteor Soc 131:1759–1782

Dirmeyer PA, Cash BA, Kinter JL III, Jung T, Marx L, Satoh M, Stan C, Tomita H, Towers P, Wedi N, Achuthavarier D, Adams JM, Altshuler EL, Huang B, Jin EK, Manganello J (2012) Simulating the diurnal cycle of rainfall in global climate models: resolution versus parameterization. Clim Dyn 39:399–418

Emanuel K, Oouchi K, Satoh M, Tomita H, Yamada Y (2010) Comparison of explicitly simulated and downscaled tropical cyclone activity in a high-resolution global climate model. J Adv Model Earth Syst 2:9, doi:10.3894/JAMES.2010.2.9

Enting IG (2002) Inverse problems in atmospheric constituent transport. Cambridge University Press, Cambridge, p 392, doi:10.1017/CBO9780511535741

Evensen G (1994) Sequential data assimilation with a nonlinear quasi-geostrophic model using Monte-Carlo methods to forecast error statistics. J Geophys Res 99:10143–10162

Feingold G, Walko RL, Stevens B, Cotton WR (1998) Simulations of marine stratocumulus using a new microphysical parameterization scheme. Atmos Res 47–48:505–528

Fu X, Wang B (2004) Differences of boreal summer intraseasonal oscillations simulated in an atmosphere–ocean coupled model and an atmosphere-only model. J Clim 17:1263–1271

Fudeyasu H, Wang Y, Satoh M, Nasuno T, Miura H, Yanase W (2008) The global cloud-system-resolving model NICAM successfully simulated the lifecycles of two real tropical cyclones. Geophys Res Lett 35:L22808, doi:10.1029/2008GL0360033

Fudeyasu H, Wang Y, Satoh M, Nasuno T, Miura H, Yanase W (2010a) Multiscale interactions in the lifecycle of a tropical cyclone simulated in a global cloud-system-resolving model. Part I: large scale aspects. Mon Wea Rev 138:4285–4304

Fudeyasu H, Wang Y, Satoh M, Nasuno T, Miura H, Yanase W (2010b) Multiscale interactions in the lifecycle of a tropical cyclone simulated in a global cloud-system-resolving model. Part II: mesoscale and storm-scale processes. Mon Wea Rev 137:3254–3268

Fujita M, Yoneyama K, Mori S, Nasuno T, Satoh M (2011) Diurnal convection peaks over the eastern Indian Ocean off Sumatra during different MJO phases. J Meteor Soc Japan 89A:317–330

Gal-Chen T, Somerville CJ (1975) On the use of a coordinate transformation for the solution of the Navier-Stokes equations. J Comp Phys 17:209–228

Gallus WA Jr, Klemp JB (2000) Behaviour of flow over step orography. Mon Wea Rev 128:1153–1164

Goto D (2014) Modeling of black carbon in Asia using a global-to-regional seamless aerosol-transport model. Environ Pollut doi:10.1016/j.envpol.2014.06.006

Goto D, Nakajima T, Takemura T, Sudo K (2011a) A study of uncertainties in the sulfate distribution and its radiative forcing associated with sulfur chemistry in a global aerosol model. Atmos Chem Phys 11:10889–10910

Goto D, Takemura T, Nakajima T, Badarinath KVS (2011b) Global aerosol model-derived black carbon concentration and single scattering albedo over Indian region and its comparison with ground observations. Atmos Environ 45:3277–3285

Goto D, Kanazawa S, Nakajima T, Takemura T (2012) Evaluation of a relationship between aerosols and surface downward shortwave flux through an integrative analysis of modeling and observation. Atmos Environ 49:294–301

Goto D, Dai T, Satoh M, Tomita H, Uchida J, Misawa S, Inoue T, Tsuruta H, Ueda K, Ng CFS, Takami A, Sugimoto N, Shimizu A, Ohara T, Nakajima T (2014) Application of a global nonhydrostatic model with a stretched-grid system to regional aerosol simulations around Japan. Geosci Model Dev Discuss 7:131–179

Grabowski WW (1998) Toward cloud resolving modeling of large-scale tropical circulations: a simple cloud microphysics parameterization. J Atmos Sci 55:3283–3298

Gross ES, Bonaventura L, Rosatti G (2002) Consistency with continuity in conservative advection schemes for free-surface models. Int J Numer Meth Fluids 38:307–327

Gurney KR, Law RM, Denning AS, Rayner PJ, Baker D, Bousquet P, Bruhwiler L, Chen YH, Ciais P, Fan S, Fung IY, Gloor M, Heimann M, Higuchi K, John J, Maki T, Maksyutov S, Masarie L, Peylin P, Prather M, Pak BC, Randerson J,

Sarmiento J, Taguchi S, Takahashi T, Yuen CW (2002) Towards robust estimates of CO_2 sources and sinks using atmospheric transport models. Nature 415:626–630

Ham S-H, Sohn B-J, Kato S, Satoh M (2013) Vertical inhomogeneity of ice cloud layers from CloudSat and CALIPSO measurements and comparison to NICAM simulations. J Geophys Res 118:9930–9947

Hansen JE, Travis LD (1974) Light scattering in planetary atmosphere. Space Sci Rev 16:527–610

Hashino T, Satoh M, Hagihara Y, Kubota T, Matsui T, Nasuno T, Okamoto H (2013) Evaluating global cloud distribution and microphysics from the NICAM against CloudSat and CALIPSO. J Geophys Res 118:7273–7292

Hasumi H (2000) CCSR Ocean Component Model (COCO). CCSR Rep 13. The University of Tokyo, Chiba, Japan

Hasumi H (2006) CCSR Ocean Component Model (COCO) version 4.0. CCSR Rep 25. The University of Tokyo, Chiba, Japan

Hasumi H, Emori S (2004) K-1 coupled model (MIROC) description. K-1 Tech Rep. p 34, http://ccsr.aori.u-tokyo.ac.jp/~hasumi/miroc_description.pdf. Accessed at 29 Sep 2014

Hayashi YY, Sumi A (1986) The 30–40 day oscillations simulated in an aqua-planet model. J Meteorol Soc Japan 64:451–467

Heikes R, Randall DA (1995) Numerical integration of the shallow-water equations on a twisted icosahedral grid. Part I: basic design and results of tests. Mon Wea Rev 123:1862–1880

Heikes R, Randall DA, Konor CS (2013) Optimized icosahedral grids: performance of finite-difference operators and multigrid solver. Mon Wea Rev 141:445–4469

Henze DK, Seinfeld JH (2006) Global secondary organic aerosol from isoprene oxidation. Geophys Res Lett 33:L09812, doi:10.1029/2006GL025976

Holloway CE, Woolnough SJ, Lister GMS (2013) The effects of explicit versus parameterized convection on the MJO in a large-domain high-resolution tropical case study. Part I: characterization of large-scale organization and propagation. J Atmos Sci 70:1342–1369

Hong SY, Dudhia J, Chen SH (2004) A revised approach to ice microphysical processes for the bulk parameterization of clouds and precipitation. Mon Wea Rev 132:103–120

Houze RA (2010) Clouds in tropical cyclones. Mon Wea Rev 138:293–344

Hunt BR, Kostelich EJ, Szunyogh I (2007) Efficient data assimilation for spatiotemporal chaos: a local ensemble transform Kalman filter. Physica D 230:112–126

Iga S (2014a) Smooth, seamless and structured grid generation with flexibility in resolution distribution on a sphere based on conformal mapping and spring dynamics method. J Compt Phys submitted

Iga S (2014b) Aqua-planet experiment on an AGCM with tropics-enhanced grid. SOLA, submitted

Iga S, Tomita H (2014) Improved smoothness and homogeneity of icosahedral grids using the spring dynamics method. J Comp Phys 258:208–226

Iga S, Tomita H, Tsushima Y, Satoh M (2007a) Climatology of a nonhydrostatic global model with explicit cloud processes. Geophys Res Lett 34:L22814, doi:10.1029/2007GL031048

Iga S, Tomita H, Satoh M, Goto K (2007b) Mountain-wave-like spurious waves due to inconsistency of horizontal and vertical resolution associated with cold fronts. Mon Wea Rev 135:2629–2641

Iga S, Tomita H, Tsushima Y, Satoh M (2011) Sensitivity of upper tropospheric ice clouds and their impacts on the Hadley circulation using a global cloud-system resolving model. J Climate 24:2666–2679

Inoue T, Satoh M, Miura H, Mapes B (2008) Characteristics of cloud size of deep convection simulated by a global cloud resolving model. J Meteor Soc Japan 86A:1–15

Inoue T, Satoh M, Hagihara Y, Miura H, Schmetz J (2010) Comparison of high-level clouds represented in a global cloud system-resolving model with CALIPSO/CloudSat and geostationary satellite observations. J Geophys Res 115:D00H22, doi:10.1029/2009JD012371

Iwasaki T, Yamada S, Tada K (1989) A parameterization scheme of orographic gravity wave drag with two different vertical partitionings. Part I: impacts on medium-range forecasts. J Meteor Soc Japan 67:11–27

Jöckel P, von Kuhlmann R, Lawrence M, Steil B, Brenninkmeijer C, Crutzen P, Rasch P, Eaton B (2001) On a fundamental problem in implementing flux-form advection schemes for tracer transport in 3-dimensional general circulation and chemistry transport models. Quart J Roy Meteor Soc 127:1035–1052

Jones PW (1999) First- and second-order conservative remapping schemes for grids in spherical coordinates. Mon Wea Rev 127:2204–2210

Jung T, Miller MJ, Palmer TN, Towers P, Wedi N, Achuthavarier D, Adams JM, Altshuler EL, Cash BA, Kinter JL III (2012) High-resolution global climate simulations with the ECMWF model in Project Athena: experimental design, model climate, and seasonal forecast skill. J Climate 25:3155–3172

Kalman RE (1960) A new approach to linear filtering and prediction problems. J Basic Eng Trans ASME 82:35–45

Kato S, Rose FG, SunMack S, Miller WF, Chen Y, Rutan DA, Stephens GL, Loeb NG, Minnis P, Wielicki BA, Winker DA, Charlock TP, Stackhouse PW Jr, Xu KM, Collins WD (2011) Improvements of top-of-atmosphere and surface irradiance computations with CALIPSO-, CloudSat-, and MODIS-derived cloud and aerosol properties. J Geophys Res 116:D19209, doi:10.1029/2011JD016050

Kessler E (1969) On the distribution and continuity of water substance in atmospheric circulations. Meteorologial Monograph. Amer Meteor Soc 32:1–84

Khain A, Ovtchinnikov M, Pinsky M, Pokrovsky A, Krugliak H (2000) Notes on the state-of-the-art numerical modeling of cloud microphysics. Atmos Res 55:159–224

Khain A, BenMoshe N, Pokrovsky A (2008) Factors determining the impact of aerosols on surface precipitation from clouds: an attempt at classification. J Atmos Sci 65:1721–1748

Khairoutdinov M, Kogan Y (2000) A new cloud physics parameterization in a large-eddy simulation model of marine stratocumulus. Mon Wea Rev 128:229–243

Khairoutdinov M, Krueger SK, Moeng CH, Bogenschutz PA, Randall DA (2009) Large-eddy simulation of maritime deep tropical convection. J Adv Model Earth Systems 1:15, doi:10.3894/JAMES.2009.1.15

Kikuchi K, Wang B (2010) Formation of tropical cyclones in the northern Indian Ocean associated with two types of tropical intraseasonal oscillation modes. J Meteor Soc Japan 88:475–496

Kinne S, Schulz M, Textor C, Guibert S, Balkanski Y, Bauer SE, Berntsen T, Berglen TF, Boucher O, Chin M, Collins W, Dentener F, Diehl T, Easter R, Feichter J, Fillmore D, Ghan S, Ginoux P, Gong S, Grini A, Hendricks J, Herzog M, Horowitz L, Isaksen I, Iversen T, Kirkevag A, Kloster S, Koch D, Kristjansson JE, Krol M et al (2006) An AeroCom initial assessment - optical properties in aerosol component modules of global models. Atmos Chem Phys 6:1815–1834

Kinter JL III, Cash B, Achuthavarier D, Adams J, Altshuler E, Dirmeyer P, Doty B, Huang B, Marx L, Manganello J, Stan C, Wakefield T, Jin E, Palmer T, Hamrud M, Jung T, Miller M, Towers P, Wedi N, Satoh M, Tomita H, Kodama C, Nasuno T, Oouchi K, Yamada Y, Taniguchi H, Andrews P, Baer T, Ezell M, Halloy C et al (2013) Revolutionizing climate modeling - Project Athena: a multi-institutional, international collaboration. Bull Am Meteor Soc 94:231–245

Klein SA, Jakob C (1999) Validation and sensitivities of frontal clouds simulated by the ECMWF model. Mon Wea Rev 127:2514–2531

Klemp JB (2011) A terrain-following coordinate with smoothed coordinate surfaces. Mon Wea Rev 139:2163–2169

Klemp JB, Wilhemson RB (1978) The simulation of three-dimensional convective storm dynamics. J Atmos Sci 35:1070–1096

Kodama C, Noda AT, Satoh M (2012) An assessment of the cloud signals simulated by NICAM using ISCCP, CALIPSO, and CloudSat satellite simulators. J Geophys Res 117:D12210, doi:10.1029/2011JD017317

Kodama C, Iga S, Satoh M (2014a) Impact of the sea surface temperature rise on storm-track clouds in global non-hydrostatic aqua-planet simulations. Geophys Res Lett 41:3545–3552, doi:10.1002/2014GL059972

Kodama C, Terai M, Noda AT, Yamada Y, Satoh M, Seiki T, Iga S, Yashiro H, Tomita H, Minami K (2014b) Scalable rank-mapping algorithm for an icosahedral grid system on the massive parallel computer with a 3-D torus network. Parallel Comput 40:362–373

Kondo J (1993) A new bucket model for predicting water content in the surface model. J Japan Soc Hydrol and Water Resour 6:344–349 (in Japanese with English abstract)

Kondo K, Tanaka HL (2009) Applying the local ensemble transform Kalman filter to the Nonhydrostatic Icosahedral Atmospheric Model (NICAM). SOLA 5:121–124

Kubokawa H, Fujiwara M, Nasuno T, Satoh M (2010) Analysis of the tropical tropopause layer using the Nonhydrostatic Icosahedral Atmospheric Model (NICAM): aqua planet experiments. J Geophys Res 115:D08102, doi:10.1029/2009JD012686

Kubokawa H, Fujiwara M, Nasuno T, Miura H, Yamamoto M, Satoh M (2012) Analysis of the tropical tropopause layer using the Nonhydrostatic Icosahedral Atmospheric Model (NICAM): 2. An experiment under the atmospheric conditions of December 2006 to January 2007. J Geophys Res 117:D17114

Lamarque JF, Shindell DT, Josse B, Young PJ, Cionni I, Eyring V, Bergmann D, Cameron-Smith P, Collins WJ, Doherty R, Dalsoren S, Faluvegi G, Folberth G, Ghan SJ, Horowitz LW, Lee YH, MacKenzie IA, Nagashima T, Naik V, Plummer D, Righi M, Rumbold ST, Schulz M, Skeie RB, Stevenson DS, Strode S, Sudo K, Szopa S, Voulgarakis A, Zeng G (2013) The Atmospheric Chemistry and Climate Model Intercomparison Project (ACCMIP): overview and description of models, simulations and climate diagnostics. Geosci Model Dev 6:179–206

Laprise R (2008) Regional climate modelling. J Comp Phys 227:641–3666

Law RM, Peters W, Rödenbeck C, Aulagnier C, Baker I, Bergmann DJ, Bousquet P, Brandt J, Bruhwiler L, Cameron-Smith PJ, Christensen JH, Delage F, Denning AS, Fan S, Geels C, Houweling S, Imasu R, Karstens U, Kawa SR, Kleist J, Krol MC, Lin SJ, Lokupitiya R, Maki T, Maksyutov S, Niwa Y, Onishi R, Parazoo N, Patra PK, Pieterse G et al (2008) TransCom model simulations of hourly atmospheric CO_2: experimental overview and diurnal cycle results for 2002. Global Biogeochem Cycles 22:GB3009, doi:10.1029/2007GB003050

Lien G-Y, Kalnay E, Miyoshi T (2013) Effective assimilation of global precipitation: simulation experiments. Tellus A 65:19915, http://dx.doi.org/10.3402/tellusa.v65i0.19915

Lim K-SS, Hong S-Y (2010) Development of an effective double-moment cloud microphysics scheme with prognostic cloud condensation nuclei (CCN) for weather and climate models. Mon Wea Rev 138:1587–1612

Lin SJ (2004) A "vertically Lagrangian" finite-volume dynamical core for global models. Mon Wea Rev 132:2293–2307

Lin Y-L, Farley RD, Orville HD (1983) Bulk parameterization of the snow field in a cloud model. J Climate Appl Meteor 22:1065–1092

Liu P, Satoh M, Wang B, Fudeyasu H, Nasuno T, Li T, Miura H, Taniguchi H, Masunaga H, Fu X, Annamalai H (2009) A MJO simulated by the NICAM at 14-km and 7-km resolutions. Mon Wea Rev 137:3254–3268

Lohmann U, Feichter J (2005) Global indirect aerosol effects: a review. Atmos Chem Phys 5:715–737

Machida T, Matsueda H, Sawa Y, Nakagawa Y, Hirotani K, Kondo N, Goto K, Nakazawa T, Ishikawa K, Ogawa T (2008) Worldwide measurements of atmospheric CO_2 and other trace gas species using commercial airlines. J Atmos Oceanic Technol 25:1744–1754

Madden RA, Julian PR (1971) Description of a 40–50 day oscillation in the tropics. J Atmos Sci 28:702–708

Madden RA, Julian PR (1972) Description of global-scale circulation cells in the tropics with a 40–50 day period. J Atmos Sci 29:1109–1123

Majewski D, Liermann D, Prohl P, Ritter B, Buchhold M, Hanisch T, Paul G, Wergen W (2002) The operational global icosahedral-hexagonal gridpoint model GME: description and high-resolution tests. Mon Wea Rev 130:319–338

Mapes B, Tulich S, Nasuno T, Satoh M (2008) Predictability aspects of global aqua-planet simulations with explicit convection. J Meteor Soc Japan 86A:175–185

Masuda Y, Ohnishi H (1986) An integration scheme of the primitive equation model with an icosahedral-hexagonal grid system and its application to the shallow water equations. Short- and Medium-Range Numerical Weather Prediction, Collection of Papers Presented at the WMO/IUGG NWP Symposium. Japan Meteorological Society, Tokyo, pp 317–326

Masunaga H, Satoh M, Miura H (2008) A joint satellite and global cloud-resolving model analysis of a Madden-Julian Oscillation event: model diagnosis. J Geophys Res 113:D17210, doi:10.1029/2008JD009986

Matsueda H, Machida T, Sawa Y, Nakagawa Y, Hirotani K, Ikeda H, Kondo N, Goto K (2008) Evaluation of atmospheric CO_2 measurements from new flask air sampling of JAL airliner observations. Pap Meteorol Geophys 59:1–17

McGregor JL (1996) Semi-Lagrangian advection on conformal-cubic grids. Mon Wea Rev 124:1311–1322

Mellor GL, Yamada T (1982) Development of a turbulence closure model for geophysical fluid problems. Rev Geophys Space Phys 20:851–875

Meyers MP, Walko RL, Harrington JY, Cotton WR (1997) New RAMS cloud microphysics parameterization. Part II: the two-moment scheme. Atmos Res 45:3–39

Milbrandt JA, McTaggart-Cowan M (2010) Sedimentation-induced errors in bulk microphysics schemes. J Atmos Sci 67:3931–3948

Milbrandt JA, Yau MK (2005a) A multimoment bulk microphysics parameterization. Part I: analysis of the role of the spectral shape parameter. J Atmos Sci 62:3051–3064

Milbrandt JA, Yau MK (2005b) A multimoment bulk microphysics parameterization. Part II: a proposed three-moment closure and scheme description. J Atmos Sci 62:3065–3081

Milbrandt JA, Yau MK (2006) A multimoment bulk microphysics parameterization. Part IV: sensitivity experiments. J Atmos Sci 63:3137–3159

Mittal R, Iaccarino G (2003) Immersed boundary methods. Annu Rev Fluid Mech 37:239–261

Miura H (2007) An upwind-biased conservative advection scheme for spherical hexagonal-pentagonal grids. Mon Wea Rev 135:4038–4044

Miura H (2013) An upwind-biased conservative transport scheme for multi-stage temporal integrations on spherical icosahedral grids. Mon Wea Rev 141:4049–4068

Miura H, Skamarock WC (2013) An upwind-biased transport scheme using a quadratic reconstruction on spherical icosahedral grids. Mon Wea Rev 141:832–847

Miura H, Tomita H, Nasuno T, Iga S, Satoh M, Matsuno T (2005) A climate sensitivity test using a global cloud resolving model under an aqua planet condition. Geophys Res Lett 32:L19717, doi:1029/2005GL023672

Miura H, Satoh M, Tomita H, Nasuno T, Iga S, Noda AT (2007a) A short-duration global cloud-resolving simulation with a realistic land and sea distribution. Geophys Res Lett 34:L02804, doi:10.1029/2006GL027448

Miura H, Satoh M, Nasuno T, Noda AT, Oouchi K (2007b) A Madden-Julian Oscillation event realistically simulated by a global cloud-resolving model. Science 318:1763–1765

Miura H, Satoh M, Katsumata M (2009) Spontaneous onset of a Madden-Julian oscillation event in a cloud-system-resolving simulation. Geophys Res Lett 36:L13802, doi:10.1029/2009GL039056

Miyakawa T, Takayabu YN, Nasuno T, Miura H, Satoh M, Moncrieff MW (2012) Convective momentum transport by rainbands within a Madden-Julian oscillation in a global nonhydrostatic model with explicit deep convective processes. Part I: methodology and general results. J Atmos Sci 69:1317–1338

Miyakawa T, Satoh M, Miura H, Tomita H, Yashiro H, Noda AT, Yamada Y, Kodama C, Kimoto M, Yoneyama K (2014) Madden-Julian oscillation prediction skill of a new-generation global model demonstrated using a supercomputer. Nat Commun 5:3769, doi:10.1038/ncomms4769

Miyamoto Y, Kajikawa Y, Yoshida R, Yamaura T, Yashiro H, Tomita H (2013) Deep moist atmospheric convection in a sub-kilometer global simulation. Geophys Res Lett 40:4922–4926

Miyoshi T, Aranami K (2006) Applying a four-dimensional local ensemble transform Kalman filter (4D-LETKF) to the JMA Nonhydrostatic Model (NHM). SOLA 2:128–131

Miyoshi T, Kunii M (2011) The local ensemble transform Kalman filter with the weather research and forecasting model: experiments with real observations. Pure Appl Geophys 169:321–333

Miyoshi T, Yamane S (2007) Local ensemble transform Kalman filtering with an AGCM at a T159/L48 resolution. Mon Wea Rev 135:3841–3861

Mizuta R, Oouchi K, Yoshimura H, Noda A, Katayama K, Yukimoto S, Hosaka M, Kusunoki S, Kawai H, Nakagawa M (2005) 20-km-mesh global climate simulations using JMA-GSM model - mean climate states. J Meteor Soc Japan 84:165–185

Moeng CH, LeMone MA, Khairoutdinov M, Krueger SK, Bogenschutz PA, Randall DA (2009) The tropical marine boundary layer under a deep convection system: a large-eddy simulation study. J Model Earth Systems 1:16, doi:10.3894/JAMES.2009.1.16

Morrison H, Gettelman A (2008) A new two-moment bulk stratiform cloud microphysics scheme in the Community Atmospheric Model, version 3 (CAM3). Part I: description and numerical tests. J Climate 21:3642–3659

Morrison H, Curry JA, Khvorostyanov VI (2005) A new double-moment microphysics parameterization for application in cloud and climate models. Part I: description. J Atmos Sci 62:1665–1677

Muroi C, Toyoda E, Yoshimura H, Hosaka M, Sugi M (2002) Standard coding rule. Tenki 49:91–95 (in Japanese)

Nagashima T, Ohara T, Sudo K, Akimoto H (2010) The relative importance of various source regions on East Asian surface ozone. Atmos Chem Phys 10:11305–11322

Nakajima T, Tsukamoto M, Tsushima Y, Numaguti A, Kimura T (2000) Modeling of the radiative process in an atmospheric general circulation model. Appl Opt 39:4869–4878

Nakajima TY, Suzuki K, Stephens GL (2010) Droplet growth in warm water clouds observed by the A-Tran Part 1: sensitivity analysis of the MODIS-derived cloud droplet sizes. J Atmos Sci 67:1884–1896

Nakanishi M, Niino H (2006) An improved Mellor-Yamada level-3 model: its numerical stability and application to a regional prediction of advection fog. Boundary-Layer Meteor 119:397–407

Nakanishi M, Niino H (2009) Development of an improved turbulence closure model for the atmospheric boundary layer. J Meteor Soc Japan 87:895–912

Nakazawa T (1988) Tropical super clusters within intraseasonal variations over the western Pacific. J Meteor Soc Japan 66:823–839

Nasuno T (2008) Equatorial mean zonal wind in a global nonhydrostatic aquaplanet experiment. J Meteor Soc Japan 86A:219–236

Nasuno T (2013) Forecast skill of Madden-Julian Oscillation events in a global nonhydrostatic model during the CINDY2011/DYNAMO observation period. SOLA 9:69–73

Nasuno T, Satoh M (2011a) Properties of precipitation and in-cloud vertical motion in a global nonhydrostatic aquaplanet experiment. J Meteor Soc Japan 89:413–439

Nasuno T, Satoh M (2011b) Statistical relationship between maximum vertical velocity and surface precipitation of Tropical convective clouds in global nonhydrostatic aquaplanet experiment. J Meteor Soc Japan 89:553–561

Nasuno T, Tomita H, Iga S, Miura H, Satoh M (2007) Multi-scale organization of convection simulated with explicit cloud processes on an aqua planet. J Atmos Sci 64:1902–1921

Nasuno T, Tomita H, Iga S, Miura H, Satoh M (2008) Convectively coupled equatorial waves simulated by a global nonhydrostatic experiment on an aqua planet. J Atmos Sci 65:1246–1265

Nasuno T, Miura H, Satoh M, Noda AT, Oouchi K (2009) Multi-scale organization of convection in a global numerical simulation of the December 2006 MJO event using explicit moist processes. J Meteor Soc Japan 87:335–345

Neale RB, Hoskins BJ (2001) A standard test for AGCMs including their physical parameterizations: I: the proposal. Atmos Sci Lett 1:153–155, doi:10.1006/asle.2000.0019

Ničkovic S, Gavrilov MB, Tosic IA (2002) Geostrophic adjustment on hexagonal grids. Mon Wea Rev 130:668–683

Niwa Y (2010) Numerical Study on Atmospheric Transport and Surface Source/Sink of Carbon Dioxide, Dissertation. The University of Tokyo, Tokyo

Niwa Y, Tomita H, Satoh M, Imasu R (2011a) A three-dimensional icosahedral grid advection scheme preserving monotonicity and consistency with continuity for atmospheric tracer transport. J Meteorol Soc Jpn 89:255–268

Niwa Y, Patra PK, Sawa Y, Machida T, Matsueda H, Belikov D, Maki T, Ikegami M, Imasu R, Maksyutov S, Oda T, Satoh M, Takigawa M (2011b) Three-dimensional variations of atmospheric CO_2: aircraft measurements and multi-transport model simulations. Atmos Chem Phys 11:3359–13375

Niwa Y, Machida T, Sawa Y, Matsueda H, Schuck TJ, Brenninkmeijer CAM, Imasu R, Satoh M (2012) Imposing strong constraints on tropical terrestrial CO_2 fluxes using passenger aircraft based measurements. J Geophys Res 117:D11303, doi:10.1029/2012JD017474

Noda AT, Oouchi K, Satoh M, Tomita H, Iga S, Tsushima Y (2010) Importance of the subgrid-scale turbulent moist process: Cloud distribution in global cloud-resolving simulatioins. Atmos Res 96:208–217

Noda AT, Oouchi K, Satoh M, Tomita H (2012) Quantitative assessment of diurnal variation of tropical convection simulated by a global nonhydrostatic model without cumulus parameterization. J Climate 25:5119–5134

Ohfuchi W, Nakamura H, Yoshioka MK, Enomoto T, Takaya K, Peng X, Yamane S, Nishimura T, Kurihara Y, Ninomiya K (2004) 10-km mesh meso-scale resolving simulations of the global atmosphere on the Earth Simulator: Preliminary outcomes of AFES (AGCM for the Earth Simulator). J Earth Simulator 1:8–34

Ohno T, Satoh M (2014) On the warm core of the tropical cyclone formed near the tropopause. J Atmos Sci in press

Onogi K, Tsutsui H, Koide H, Sakamoto M, Kobayashi S, Hatsushika H, Matsumoto T, Yamazaki N, Kamahori H, Takahashi K, Kadokura S, Wada K, Kato K, Oyama R, Ose T, Mannoji N, Taira T (2007) The JRA-25 reanalysis. J Meteorol Soc Jpn 85:369–432

Oouchi K, Noda AT, Satoh M, Miura H, Tomita H, Nasuno T, Iga S (2009a) A simulated preconditioning of typhoon genesis controlled by a boreal summer Madden-Julian Oscillation event in a global cloud-system-resolving model. SOLA 5:65–68

Oouchi K, Noda AT, Satoh M, Wang B, Xie S-P, Takahashi HG, Yasunari T (2009b) Asian summer monsoon simulated by a global cloud-system-resolving model: Diurnal to intra-seasonal variability. Geophys Res Lett 36:L11815, doi:10.1029/2009GL038271

Oouchi K, Taniguchi H, Nasuno T, Satoh M, Tomita H, Yamada Y, Ikeda M, Shirooka R, Yamada H, Yoneyama K (2012) A Prototype Quasi Real-Time Intra-Seasonal Forecasting of Tropical Convection Over the Warm Pool Region: A new Challenge of Global Cloud-System-Resolving Model for a Field Campaign. In: Oouchi K, Fudeyasu H (eds) Cyclones: Formation, Triggers and Control. Nova Science Publishers Inc, pp 233–248

Oouchi K, Satoh M, Yamada Y, Tomita H, Sugi M (2014) A hypothesis and a case-study projection of an influence of MJO modulation on boreal-summer tropical cyclogenesis in a warmer climate with a global non-hydrostatic model: a transition toward the central Pacific? Front Earth Sci 2:1, doi:10.3389/feart.2014.00001 (accepted)

Ott E, Hunt BR, Szunyogh I, Zimin AV, Kostelich EJ, Corazza M, Kalnay E, Patil DJ, Yorke JA (2004) A local ensemble Kalman filter for atmospheric data assimilation. Tellus 56A:415–428

Patra PK, Law RM, Peters W, Rödenbeck C, Takigawa M, Aulagnier C, Baker I, Bergmann DJ, Bousquet P, Brandt J, Bruhwiler L, Cameron-Smith PJ, Christensen JH, Delage F, Denning AS, Fan S, Geels C, Houweling S, Imasu R, Karstens U, Kawa SR, Kleist J, Krol MC, Lin SJ, Lokupitiya R, Maki T, Maksyutov S, Niwa Y, Onishi R, Parazoo N et al (2008) TransCom model simulations of hourly atmospheric CO_2: analysis of synoptic scale variations for the period 2002–2003. Global Biogeochem Cycles 22:GB4013, doi:10.1029/2007GB003081

Patra PK, Canadell JG, Houghton RA, Piao SL, Oh N-H, Ciais P, Manjunath KR, Chhabra A, Wang T, Bhattacharya T, Bousquet P, Hartman J, Ito A, Mayorga E, Niwa Y, Raymond P, Sarma VSS, Lasco R (2013) The carbon budget of South Asia. Biogeoscience 10:513–527

Phillips VTJ, Donner LJ, Garner ST (2007) Nucleation processes in deep convection simulated by a cloud-system-resolving model with double-moment bulk cloud microphysics. J Atmos Sci 64:738–761

Pruppacher HR, Klett JD (1997) Microphysics of Clouds and Precipitation. Kluwer Academic Publisher, Heidelberg, p 954

Putman WM, Suarez M (2011) Cloud-system resolving simulations with the NASA Goddard Earth Observing System global atmospheric model (GEOS-5). Geophys Res Lett 38:L16809, doi:10.1029/2011GL048438

Qian J-H, Semazzi FHM, Scroggs JS (1998) A global nonhydrostatic semi-Lagrangian atmospheric model with orography. Mon Wea Rev 126:747–771

Randall DA (1994) Geostrophic adjustment and the finite-difference shallow-water equations. Mon Wea Rev 122:1371–1377

Randall DA, Heikes R, Ringler T (2000) Global Atmospheric Modeling Using a Geodesic Grid With an Isentropic Vertical Coordinate. In: General Circulation Model Development, Chapter 17. Academic Press, California London, pp 509–538

Randall DA, Khairoutdinov M, Arakawa A, Grabowski WW (2003) Breaking the cloud-parameterization deadlock. Bull Amer Meteor Soc 84:1547–1564

Redler R, Valcke S, Ritzdorf H (2009) OASIS-4 – a coupling software for next generation earth system modeling. Geoscientific Model Development Discussions 2:797–843

Ringler TD, Randall DA (2002) A potential enstrophy and energy conserving numerical scheme for solution of the shallow-water equations on a geodesic grid. Mon Wea Rev 130:1397–1410

Ringler TD, Heikes RH, Randall DA (2000) Modeling the atmospheric general circulation using a spherical geodesic grid: a new class of dynamical cores. Mon Wea Rev 128:2471–2490

Roh W, Satoh M (2014) Evaluation of precipitating hydrometeor parameterizations in a single-moment bulk microphysics scheme for deep convective systems over the tropical open ocean. J Atmos Sci 71:2654–2673

Rossow WB, Schiffe RA (1999) Advances in understanding clouds from ISCCP. Bull Am Meteorol Soc 80:2261–2287

Rotunno R, Emanuel KA (1987) An air-sea interaction theory for tropical cyclones. Part II: evolutionary study using a nonhydrostatic axisymmetric numerical model. J Atmos Sci 44:542–561

Rutledge SA, Hobbs P (1983) The mesoscale and microscale structure and organization of clouds and precipitation in midlatitude cyclones. VIII: a model for the "seeder-feeder" process in warm-frontal rainbands. J Atmos Sci 40:1185–1206

SALSA (2014) Annual report of Development of Seamless Chemical AssimiLation System and its Application for Atmospheric Environmental Materials (SALSA). In: Project of the Research Program on Climate Change Adaptation (RECCA) in Ministry of Education and Sports in Japan (MEXT). Aavailable at http://157.82.240.167/~salsa/Program/SALSA_Annual_Report_FY2013.pdf. Accessed at 29 Sep 2014

Sato T, Miura H, Satoh M (2007) Spring diurnal cycle of clouds over Tibetan Plateau: global cloud-resolving simulations and satellite observations. Geophys Res Lett 34:L18816, doi:10.1029/2007GL030782

Sato T, Yoshikane T, Satoh M, Miura H, Fujinami H (2008) Resolution dependency of the diurnal cycle of convective clouds over the Tibetan Plateau in a mesoscale model. J Meteor Soc Japan 86A:17–31

Sato T, Miura H, Satoh M, Takayabu YN, Wang Y (2009) Diurnal cycle of precipitation in the tropics simulated in a global cloud-resolving model. J Climate 22:4809–4826

Satoh M (2002) Conservative scheme for the compressible non-hydrostatic models with the horizontally explicit and vertically implicit time integration scheme. Mon Wea Rev 130:1227–1245

Satoh M (2003) Conservative scheme for a compressible nonhydrostatic model with moist processes. Mon Wea Rev 131:1033–1050

Satoh M (2013) Atmospheric Circulation Dynamics and General Circulation Models, 2nd edn. Springer-PRAXIS, Heidelberg, p 730

Satoh M, Kitao Y (2013) Numerical examination of the diurnal variation of summer precipitation over southern China. SOLA 9:129–133

Satoh M, Matsuda Y (2009) Statistics of high-cloud areas and its sensitivity to cloud microphysics with single cloud experiments. J Atmos Sci 66:2659–2677

Satoh M, Tomita H, Miura H, Iga S, Nasuno T (2005) Development of a global cloud resolving model - a multi-scale structure of tropical convections. J Earth Simulator 3:11–19

Satoh M, Matsuno T, Tomita H, Miura H, Nasuno T, Iga S (2008) Nonhydrostatic icosahedral atmospheric model (NICAM) for global cloud resolving simulations. J Comput Phys 227:3486–3514

Satoh M, Inoue T, Miura H (2010) Evaluations of cloud properties of global and local cloud system resolving models using CALIPSO and CloudSat simulators. J Geophys Res 115:D00H14, doi:10.1029/2009JD012247

Satoh M, Oouchi K, Nasuno T, Taniguchi H, Yamada Y, Tomita H, Kodama C, Kinter J III, Achuthavarier D, Manganello J, Cash B, Jung T, Palmer T, Wedi N (2012a) The Intra-Seasonal Oscillation and its control of tropical cyclones simulated by high-resolution global atmospheric models. Clim Dyn 39:2185–2206

Satoh M, Iga S, Tomita H, Tsushima Y, Noda AT (2012b) Response of upper clouds due to global warming tested by a global atmospheric model with explicit cloud processes. J Climate 25:2178–2191

Satoh M, Nihonmatsu R, Kubokawa H (2013) Environmental conditions for tropical cyclogenesis associated with African easterly waves. SOLA 9:120–124

Sawa Y, Machida T, Matsueda H (2012) Aircraft observation of the seasonal variation in the transport of CO_2 in the upper atmosphere. J Geophys Res 117:D05305, doi:10.1029/2011JD016933

Seifert A (2008) On the parameterization of evaporation of raindrops as simulated by a one-dimensional rainshaft model. J Atmos Sci 65:3608–3619

Seifert A, Beheng KD (2001) A double-moment parameterization for simulating autoconversion, accretion and selfcollection. Atmos Res 59–60:265–281

Seifert A, Beheng KD (2006) A two-moment cloud microphysics parameterization for mixed-phase clouds. Part I: model description. Meteorol Atmos Phys 92:45–66

Seifert A, Khain A, Pokrovsky A, Beheng KD (2006) A comparison of spectral bin and two-moment bulk mixed-phase cloud microphysics. Atmos Res 80:46–66

Seiki T, Nakajima T (2014) Aerosol effects of the condensation process on a convective cloud simulation. J Atmos Sci 71:833–853

Seiki T, Satoh M, Tomita H, Nakajima T (2014) Simultaneous evaluation of ice cloud microphysics and non-sphericity of the cloud optical properties using hydrometeor video sonde and radiometer satellite in-situ observations. J Geophys Res 119:6681–6701, doi:10.1002/2013JD021086

Sekiguchi M, Nakaima T (2008) A k-distribution-based radiation code and its computational optimization for an atmospheric general circulation model. J Quant Spectrosc Radiat Transfer 109:2779–2793

Semazzi FHM, Qian J-H, Scroggs JS (1995) A global nonhydrostatic semi-Lagrangian atmospheric model without orography. Mon Wea Rev 123:2534–2550

Sherwood SC, Ingram W, Tsushima Y, Satoh M, Roberts M (2010) Relative humidity changes in a warmer climate. J Geophys Res 115:D09104, doi:10.1029/2009JD012585

Shindell DT, Miller RL, Schmidt GA, Pandolfo L (1999) Simulation of recent northern winter climate trends by greenhouse-gas forcing. Nature 399:452–455

Shipway BJ, Hill AA (2012) Diagnosis of systematic differences between multiple parametrizations of warm rain microphysics using a kinematic framework. Quart J Roy Meteor Soc 138:2196–2211

Simmons AJ, Burridge DM (1981) An energy and angular-momentum conserving vertical finite-difference scheme and hybrid vertical-coordinates. Mon Wea Rev 109:758–766

Skamarock WC, Klemp JB, Duda MG, Fowler LD, Park S-H (2012) A multi-scale nonhydrostatic atmospheric model using centroid Vornoi tesselations and C-grid staggering. Mon Wea Rev 140:3090–3105

Sohn BJ, Nakajima T, Satoh M, Jang H-S (2010) Impact of different definitions of clear-sky flux on the determination of longwave cloud radiative forcing: NICAM simulation results. Atmos Chem Phys 10:11641–11646

Staniforth A, Wood N (2008) Aspects of the dynamical core of a nonhydrostatic, deep-atmosphere, unified weather and climate-prediction model. J Comput Phys 227:3445–3464

Steppeler J, Bitzer HW, Minotte M, Bonaventura L (2002) Nonhydrostatic atmospheric modeling using a z-coordinate representation. Mon Wea Rev 130:2143–2149

Steppeler J, Bitzer HW, Janjic Z, Schättler U, Prohl P, Gjertsen U, Torrisi L, Parfinievicz J, Avgoustoglou E, Damrath U (2006) Prediction of clouds and rain using a z-coordinate nonhydrostatic model. Mon Wea Rev 134:3625–3643

Stuhne GR, Peltier WR (1996) Vortex erosion and amalgamation in a new model of large scale flow on the sphere. J Comput Phys 128:58–81

Sudo K, Akimoto H (2007) Global source attribution of tropospheric ozone: long-range transport from various source regions. J Geophys Res 112:D12302, doi:10.1029/2006JD007992

Sudo K, Takahashi M, Kurokawa J, Akimoto H (2002a) CHASER: A global chemical model of the troposphere: 1. Model description. J Geophy Res 107:4339, doi:10.1029/2001JD001113

Sudo K, Takahashi M, Akimoto H (2002b) CHASER: a global chemical model of the troposphere 2. Model results and evaluation. J Geophys Res 107:4586

Suzuki K, Nakajima T, Satoh M, Tomita H, Takemura T, Nakajima TY, Stephens GL (2008) Global cloud-system-resolving simulation of aerosol effect on warm clouds. Geophys Res Lett 35:L19817, doi:10.1029/2008GL035449

Takata K, Emori S, Watanabe T (2003) Development of the minimal advanced treatments of surface interaction and runoff. Global and Planetary Change 38:209–222

Takayabu YN, Iguchi T, Kachi M, Shibata A, Kanzawa H (1999) Abrupt termination of the 1997–98 El Nino in response to a Madden-Julian oscillation. Nature 402:279–282

Takemura T, Okamoto H, Maruyama Y, Numaguti A, Higurashi A, Nakajima T (2000) Global three-dimensional simulation of aerosol optical thickness distribution of various origins. J Geophys Res 105:17853–17873

Takemura T, Nakajima T, Dubovik O, Holben BN, Kinne S (2002) Single scattering albedo and radiative forcing of various aerosol species with a global three-dimensional model. J Climate 15:333–352

Takemura T, Nozawa T, Emori S, Nakajima TY, Nakajima T (2005) Simulation of climate response to aerosol direct and indirect effects with aerosol transport-radiation model. J Geophys Res 110:D02202, doi:10.1029/2004JD005029

Takemura T, Egashira M, Matsuzawa L, Ichijo H, O'ishi R, Abe-Ouchi A (2009) A simulation of the global distribution and radiative forcing of soil dust aerosols at the Last Glacial Maximum. Atmos Chem Phys 9:3061–3073

Taniguchi H, Yanase W, Satoh M (2010) Ensemble simulation of cyclone Nargis by a global cloud-system-resolving model–modulation of cyclogenesis by the Madden-Julian Oscillation. J Meteor Soc Japan 88:571–591

Tans PP, Fung IY, Takahashi T (1990) Observational constrains on the global atmospheric CO_2 budget. Science 274:1431–1438

Tarantola A (2005) Inverse problem theory and methods for model parameter estimation. Soc Ind Appl Math, Philadelphia, p 342, doi:10.1137/1.9780898717921

Taylor M, Tribbia J, Iskandarani M (1997) The spectral element method for the shallow water equations on the sphere. J Comp Phys 130:92–108

Terasaki K, Tanaka HL, Satoh M (2009) Characteristics of the kinetic energy spectrum of NICAM. SOLA 5:180–183

Thompson G, Field PR, Rasmussen RM, Hall WD (2008) Explicit forecasts of winter precipitation using an improved bulk microphysics scheme. Part II: implementation of a new snow parameterization. Mon Wea Rev 136:5095–5115

Thuburn J, Ringler T, Skamarock WC, Klemp JB (2009) Numerical representation of geostrophic modes on arbitrarily structured C-grids. J Comput Phys 228:8321–8335

Tomita H (2008a) A stretched icosahedral grid by a new grid transformation. J Meteor Soc Japan 86A:107–119

Tomita H (2008b) New microphysical schemes with five and six categories by diagnostic generation of cloud ice. J Meteor Soc Japan 86:121–142

Tomita H, Satoh M (2004) A new dynamical framework of nonhydrostatic global model using the icosahedral grid. Fluid Dyn Res 34:357–400

Tomita H, Tsugawa M, Satoh M, Goto K (2001) Shallow water model on a modified icosahedral geodesic grid by using spring dynamics. J Comp Phys 174:579–613

Tomita H, Satoh M, Goto K (2002) An optimization of icosahedral grid modified by spring dynamics. J Comp Phys 183:307–331

Tomita H, Miura H, Iga S, Nasuno T, Satoh M (2005) A global cloud-resolving simulation: preliminary results from an aqua planet experiment. Geophys Res Lett 32:L08805, doi:10.1029/2005GL022459

Tomita H, Goto K, Satoh M (2008) A new approach of atmospheric general circulation model: Global cloud resolving model NICAM and its computational performance. SIAM J Sci Comp 30:2755–2776

Trenberth K, Fasullo FT, Kiehl J (2009) Earth's global energy budget. Bull Amer Meteor Soc 90:311–323

Tsuchiya C, Sato K, Nasuno T, Noda AT, Satoh M (2011) Universal frequency spectra of surface meteorological fluctuations. J Climate 24:4718–4732

Tsushima Y, Iga S, Tomita H, Satoh M, Noda AT, Webb M (2014) High cloud increase in a perturbed SST experiment with a global nonhydrostatic model including explicit convective processes. J Adv Model Earth Syst 06:, doi:10.1002/2013MS000301

Wacker U, Seifert A (2001) Evolution of rain water profiles resulting from pure sedimentation: spectral vs. parameterized description. Atmos Res 58:19–39

Walko RL, Cotton WR, Meyers MP, Harrington JY (1995) New RAMS cloud microphysics parameterization. Part I: the single-moment scheme. Atmos Res 38:29–62

Watanabe M, Emori S, Satoh M, Miura H (2009) A PDF-based hybrid prognostic cloud scheme for general circulation models. Clim Dyn 33:795–816

Watanabe M, Suzuki T, O'ishi R, Komuro Y, Watanabe S, Emori S, Takemura T, Chikira M, Ogura T, Sekiguchi M, Takata K, Yamazaki D, Yokohata T, Nozawa T, Hasumi H, Tatebe H, Kimoto M (2010) Improved climate simulation by MIROC 5: mean states, variability, and climate sensitivity. J Climate 23:6312–6335

Watanabe S, Hajima T, Sudo K, Nagashima T, Takemura T, Okajima H, Nozawa T, Kawase H, Abe M, Yokohata T, Ise T, Sato H, Kato E, Takata K, Emori S, Kawamiya M (2011) MIROC-ESM 2010: model description and basic results of CMIP5-20c3m experiments. Geosci Model Dev 4:845–872

Webb M, Senior C, Bony S, Morcrette JJ (2001) Combining ERBE and ISCCP data to assess clouds in the Hadley Centre, ECMWF and LMD atmospheric climate models. Clim Dyn 17:905–922

Wedi NP (2014) Increasing horizontal resolution in numerical weather prediction and climate simulations: illusion or panacea? Philos Trans R A 372:20130289

Wedi NP, Smolarkiewicz PK (2009) A framework for testing global non-hydrostatic models. Q J R Meteorol Soc 135:469–484

Wielicki BA, Barkstrom BR, Harrison EF, Lee RB, Smith GL, Cooper JE (1996) Clouds and the Earth's Radiant Energy System (CERES): an earth observing system experiment. Bull Amer Meteor Soc 77:853–868

Williamson DL, Drake JB, Hack JJ, Jakob R, Swarztrauber PN (1992) A standard test set for numerical approximations to the shallow water equations in spherical geometry. J Comp Phys 102:211–224

Wood N, Staniforth A, White A, Allen T, Diamantakis M, Gross M, Melvin T, Smith C, Vosper S, Zerroukat M, Thuburn J (2013) An inherently mass-conserving semi-implicit semi-Lagrangian discretization of the deep-atmosphere global non-hydrostatic equations. Q J R Meteorol Soc 140:1505–1520, doi:10.1002/qj.2235

Wyngaard JC (2004) Toward numerical modeling in the "Terra Incognita". J Atmos Sci 61:1816–1826

Xiao F, Okazaki T, Satoh M (2003) An accurate semi-Lagrangian scheme for rain drop sedimentation. Mon Wea Rev 131:974–983

Yamada Y, Satoh M (2013) Response of ice and liquid water paths of tropical cyclones to global warming simulated by a global nonhydrostatic model with explicit cloud microphysics. J Climate 26:9931–9945

Yamada Y, Oouchi K, Satoh M, Tomita H, Yanase W (2010) Projection of changes in tropical cyclone activity and cloud height due to greenhouse warming: global cloud-system-resolving approach. Geophys Res Lett 37:L07709, doi:10.1029/2010GL042518

Yamada Y, Oouchi K, Satoh M, Noda AT, Tomita H (2012) Sensitivity of Tropical Cyclones to Large-Scale Environment in a Global non-Hydrostatic Model With Explicit Cloud Microphysics. In: Oouchi K, Fudeyasu H (eds) Cyclones: formation, Triggers and Control. Nova Science, pp 145–159

Yamaura T, Kajikawa Y, Tomita H, Satoh M (2013) Possible impact of a tropical cyclone on the northward migration of the Baiu frontal zone. SOLA 9:89–93

Yamazaki H, Satomura T (2008) Vertically combined shaved cell method in z-coordinate for non-hydrostatic atmospheric model. Atmos Sci Lett 9:171–175

Yamazaki H, Satomura T (2010) Nonhydrostatic atmospheric modeling using a combined cartesian grid. Mon Wea Rev 138:3932–3945

Yanase W, Taniguchi H, Satoh M (2010a) Environmental modulation and numerical predictability associated with the genesis of tropical cyclone Nargis (2008). J Meteor Soc Japan 88:497–519

Yanase W, Satoh M, Yamada H, Yasunaga K, Moteki Q (2010b) Continual influences of tropical waves on the genesis and rapid intensification of typhoon Durian (2006). Geophys Res Lett 37:L08809, doi:10.1029/2010GL042516

Yanase W, Satoh M, Taniguchi H, Fujinami H (2012a) Seasonal and intraseasonal modulation of tropical cyclogenesis environment over the Bay of Bengal during the extended summer monsoon. J Clim 25:2914–2930

Yanase W, Satoh M, Iga S, Chan JCL, Fudeyasu H, Wang Y, Oouchi K (2012b) Multi-Scale Dynamics of Tropical Cyclone Formations in an Equilibrium Simulation Using a Global Cloud-System Resolving Model. In: Oouchi K, Fudeyasu H (eds) Cyclones: formation, triggers and control. Nova Science, pp 221–231

Yasunaga K, Nasuno T, Miura H, Takayabu YN, Yoshizaki M (2013) Afternoon precipitation peak simulated in an aqua-planet global non-hydrostatic model (aqua-planet-NICAM). J Meteor Soc Japan 91A:217–229

Yeh K-S, Côté J, Gravel S, Méthot A, Patoine A, Roch M, Staniforth A (2002) The CMC-MRB global environmental multiscale (GEM) model. Part III: nonhydrostatic formulation. Mon Wea Rev 120:329–356

Yoneyama K, Katsumata M, Mizuno K, Yoshizaki M, Shirooka R, Yasunaga K, Yamada H, Sato N, Ushiyama T, Moteki Q, Seiki A, Fujita M, Ando K, Hase H, Ueki I, Horii T, Masumoto Y, Kuroda Y, Takayabu YN, Shareef A, Fujiyoshi Y, McPhaden MJ, Murty VSN, Yokoyama C, Miyakawa T (2008) MISMO field experiment in the equatorial Indian Ocean. Bull Am Meteorol Soc 89:1889–1903

Yoneyama K, Zhang C, Long CN (2013) Tracking pulses of the Madden-Julian Oscillation. Bull Amer Meteor Soc 94:1871–1891

Yoshimura H, Yukimoto S (2008) Development of a simple coupler (Scup) for earth system modeling. Pap Met Geophys 59:19–29

Yoshizaki M, Iga S, Satoh M (2012a) Eastward propagating property of large-scale precipitation systems simulated in the coarse-resolution NICAM and an explanation of its formation. SOLA 8:21–24

Yoshizaki M, Yasunaga K, Iga S, Satoh M, Nasuno T, Noda AT, Tomita H (2012b) Why do super clusters and Madden Julian Oscillation exist over the equatorial region? SOLA 8:33–36

Zängl G, Tomita H, Satoh M, Ludwig T, Linardakis L, Thuburn J, Dubos T (2011) ICOMEX: ICOsahedral-grid Models for EXascale Earth system simulations. IS-ENES Workshop, Lecce

Zängl G, Reinert D, Rípodas P, Baldauf M (2014) The ICON (ICOsahedral Non-hydrostatic) modelling framework of DWD and MPI-M: Description of the non-hydrostatic dynamical core. Quart J Roy Meteor Soc, doi:10.1002/qj.2378

12

Deep learning approach for detecting tropical cyclones and their precursors in the simulation by a cloud-resolving global non-hydrostatic atmospheric model

Daisuke Matsuoka[1,2]* (iD), Masuo Nakano[3], Daisuke Sugiyama[1] and Seiichi Uchida[4]

Abstract

We propose a deep learning approach for identifying tropical cyclones (TCs) and their precursors. Twenty year simulated outgoing longwave radiation (OLR) calculated using a cloud-resolving global atmospheric simulation is used for training two-dimensional deep convolutional neural networks (CNNs). The CNNs are trained with 50,000 TCs and their precursors and 500,000 non-TC data for binary classification. Ensemble CNN classifiers are applied to 10 year independent global OLR data for detecting precursors and TCs. The performance of the CNNs is investigated for various basins, seasons, and lead times. The CNN model successfully detects TCs and their precursors in the western North Pacific in the period from July to November with a probability of detection (POD) of 79.9–89.1% and a false alarm ratio (FAR) of 32.8–53.4%. Detection results include 91.2%, 77.8%, and 74.8% of precursors 2, 5, and 7 days before their formation, respectively, in the western North Pacific. Furthermore, although the detection performance is correlated with the amount of training data and TC lifetimes, it is possible to achieve high detectability with a POD exceeding 70% and a FAR below 50% during TC season for several ocean basins, such as the North Atlantic, with a limited sample size and short lifetime.

Keywords: Tropical cyclogenesis, Cloud resolving atmospheric model, Deep learning, Convolutional neural network

Introduction

Tropical cyclones (TCs), also referred to as typhoons, cyclones, and hurricanes, cause significant damage to human life, agriculture, forestry, fisheries, and infrastructure. For example, Typhoon Lionrock in 2016 caused record-breaking heavy rainfall, which resulted in severe floods and the loss of 23 lives in Japan. Moreover, TCs occasionally form very close to and approach countries at low latitudes (e.g., the Philippines) with rapid intensification. Therefore, accurate prediction of TC track and intensity is necessary. Early prediction of TC formation is important not only from an academic but also from a disaster mitigation perspective.

* Correspondence: daisuke@jamstec.go.jp
[1]Center for Earth Information Science and Technology (CEIST), Japan Agency for Marine-Earth Science and Technology (JAMSTEC), 3173-25 Showa-machi, Kanazawa-ku, Yokohama, Kanagawa 236-0001, Japan
[2]PRESTO, Japan Science and Technology Agency (JST), 4-1-8 Honcho, Kawaguchi, Saitama 332-0012, Japan
Full list of author information is available at the end of the article

TCs form from convective disturbances in the tropics (Riehl 1954). Dynamic environmental conditions (e.g., small vertical wind shear, low-level cyclonic vorticity, and non-zero planetary vorticity) and thermodynamically favorable environmental conditions (e.g., sea surface temperature $> 26\,^\circ$C, existence of convective instability, and mid-tropospheric moisture) necessary for TC formation were proposed in the pioneering work of Gray (1968, 1975). However, because only a small percentage of convective disturbances in the tropics develop into TCs under favorable environmental conditions (Emanuel 1989), accurate and early prediction of TC formation is still a developing area of research. The Japan Meteorological Agency (JMA) extended the Dvorak method (Dvorak 1975; Dvorak 1984), which estimates TC intensity based on satellite infrared imagery (IR), to tropical depressions (maximum sustained surface wind speed $< 17.5\,\mathrm{m\,s^{-1}}$). This extension, known as early-stage Dvorak analysis (EDA), has been utilized in operational forecasts since 2001 (Tsuchiya et al. 2001), and the JMA issues early warnings on typhoon occurrence 1 day before its formation based on EDA. The National Hurricane Center (NHC) and Central Pacific Hurricane Center (CPHC) also use the advanced Dvorak method for predicting TC genesis with lead time and accuracies of 48 h and 15–57%, respectively (Cossuth et al. 2013). Yamaguchi and Koide (2017) demonstrated that that the predictability of TC genesis could be improved to 35–79% by combining the Dvorak method and multi-model ensemble forecasts. On the other hand, with recent advancements in high-performance computing and numerical weather prediction, TC formation could be simulated 2 weeks in advance in case studies of eight typhoons in August 2004 (Nakano et al. 2015), Hurricane Sandy in 2012, and Super Typhoon Haiyan in 2013 (Xiang et al. 2015).

In recent years, deep learning, a machine learning method based on neural networks, has been receiving increasing attention and is being applied to various pattern recognition tasks (Krizhevsky et al. 2012; Simonyan and Zisserman 2015). In meteorology, several studies have proposed applying deep neural networks to existing TC detection (Liu et al. 2016; Kim et al. 2017), tornado prediction (Trafalis et al. 2014), hurricane pathway prediction (Kordmahalleh et al. 2015), and extreme rain fall prediction (Gope et al. 2016). Although several studies have used deep learning approaches for TCs after their formation, no research has considered this approach for detecting TCs before their formation.

In general, there are two approaches to detecting extreme events such as TCs: the model-driven approach (deductive approach), including numerical simulation, and the data-driven approach (inductive approach), including machine learning. The model-driven approach has the limitation that the prediction error increases with lead time because numerical models are inherently dependent on initial values. On the other hand, machine learning, as a data-driven approach, requires a large amount of high-quality training data. Most related works use reanalysis data and/or satellite observational data and labeled data as TCs or precursors based on the best track data provided by meteorological agencies. However, best track data for a TC's occurrence well ahead of its formation is limited because the best track data is basically generated using the EDA technique and are limited in accuracy and elapsed time. For example, the best track data from the Regional Specialized Meteorological Center (RSMC), Tokyo, captures precursors 60 h before TC formation, on average, whereas simulation data from cloud-resolving global atmospheric models (Kodama et al. 2015) and TC tracking algorithms capture TC formation 107 h ahead, on average.

In this work, we adopted the deep learning approach to detect precursors of TCs before their formation using only two-dimensional (2D) Outgoing Longwave Radiation (OLR) data, which is equivalent to IR and is a good proxy of atmospheric deep convection and cloud cover. In our 2D deep convolutional neural networks (CNNs), we use 30 year cloud-resolving global atmospheric simulation data (20 year data for training and 10 year data for verification) and a TC tracking algorithm for automatic labeling. Although the basic concept, simulation data, and TC tracking algorithm of this work are the same as those in our previous conference paper (Matsuoka et al. 2017), the present study improves the deep learning architecture and investigates predictive ability for various basins, seasons, and elapsed times.

The manuscript is organized as follows. The "Data" section presents the climate simulation data and TC tracking algorithm. The "Method" section explains the training data preparation, deep convolutional neural networks, and evaluation metrics of prediction results. The "Results and discussion" section examines the detection results, including detectability for each ocean basin, spatial detectability, seasonal detectability, and long-term detectability. The "Conclusions" section provides a summary of the main conclusions of the present work.

Data
Climate simulation data
Thirty year atmospheric simulation data were produced by the Nonhydrostatic Icosahedral Atmospheric Model (NICAM) with a 14 km horizontal resolution (Kodama et al. 2015). This model employs fully compressible nonhydrostatic equations and guarantees the conservation of mass and energy. Equations were discretized by the finite volume method. One characteristic feature of this model is that it explicitly calculates deep convective circulations without using any cumulus parameterizations. Moist processes are calculated using a single-moment bulk cloud microphysics scheme (NSW6) (Tomita 2008). HadISST (Rayner et al. 2003) is used for lower boundary condition. The seasonal march of TC genesis, TC track, and TC intensity in each basin is well simulated, as described in Kodama et al. (2015). The dataset includes simulated OLR, precipitation, wind velocity, pressure, temperature, water vapor, and cloud (liquid, ice, rain, snow, and graupel) for 30 years since January 1979. OLR and precipitation are output every hour, and other physical quantities are output every 6 h. An example of simulation results of OLR is depicted in Fig. 1a. Three TCs and three precursors are reproduced at 0:00:00 08/17/2008 UTC.

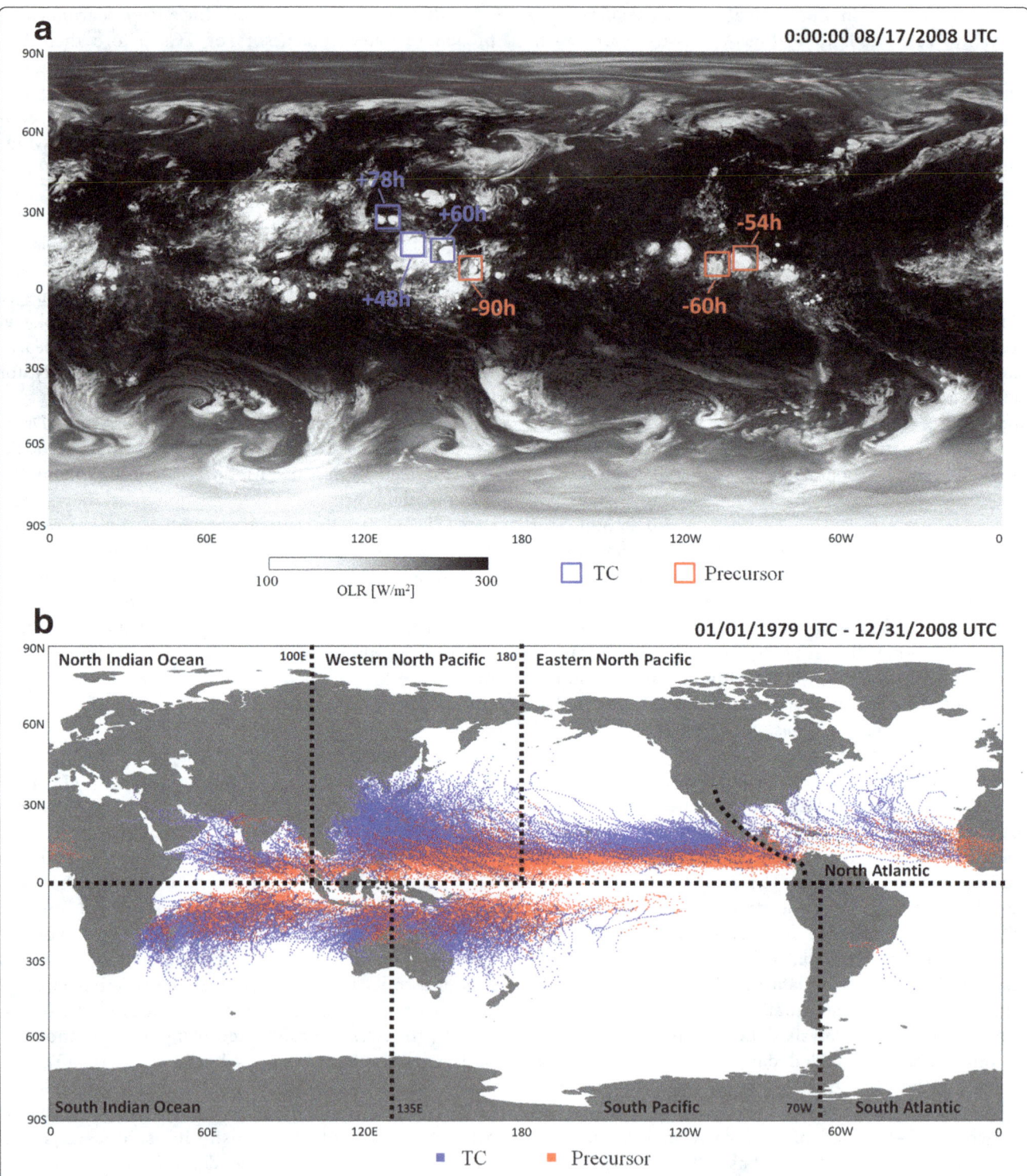

Fig. 1 a An example of the simulated OLR. Blue and red squares show TCs and their precursors, respectively. Numeric values next to the squares represent elapsed time (positive numbers: TCs, negative numbers: precursors). **b** Detection results of TCs (blue dots) and their precursors (red dots) during 1979–2008 from NICAM. The definition of basins is based on Fudeyasu et al. (2014)

This model is suitable for reproduction of tropical phenomena such as TCs (Nakano et al. 2015; Nakano et al. 2017a; Nakano et al. 2017b) and the Madden–Julian oscillation (MJO) (Miura et al. 2007). For additional details on this model, please see the original and review papers (Tomita and Satoh 2004; Satoh et al. 2014).

TC tracking

To detect TCs and precursors, we employed a TC tracking algorithm for six-hourly outputs of the horizontal components of wind, temperature, and sea level pressure (SLP). This algorithm was originally proposed by Sugi et al. (2002) and optimized for NICAM data by Nakano et al. (2015) and Yamada et al. (2017). In the first step, grid points at which the SLP was 0.5 hPa less than the mean of its surrounding area (eight-neighbor grids) were selected as candidate TC centers. In this step, grid points that met the following criteria were considered as "TCs": (i) the maximum wind speed at 10 m is greater than 17.5 m/s, (ii) the maximum relative vorticity at 850 hPa is greater than $1.0 \times 10^{-3}\,\mathrm{s}^{-1}$, (iii) the sum of temperature deviations at 300, 500, and 700 hPa is greater than 2 K, (iv) the wind speed at 850 hPa is greater than that at 300 hPa, (v) the duration of each detected storm is greater than 36 h, and (vi) the TC is formed within a limited range of latitudes (30° S–30° N). In the second step, these grid points were connected with nearest neighbors in time, and tracks of "precursors" (before becoming TCs), "TCs," and "extratropical cyclones" were subsequently obtained.

Figure 1b shows the result of applying the above algorithm to the 30 year NICAM data. TCs and precursors are represented by blue and red dots. In 30 years, 2532 TCs were detected (72–103 TCs per year). The numbers of TC tracks, detected samples, positive samples in training data, and average lifetime in each ocean basin (North Indian Ocean, western North Pacific, eastern North Pacific, North Atlantic, South Indian Ocean, South Pacific, and South Atlantic) are listed in Table 1. The definition of basins is taken from Fudeyasu et al. (2014) and illustrated in Fig. 1b. Detected TCs and precursors were used not only for labeling data, but also for ground truth. Ground truth was provided beforehand as the center point of TCs and precursors to evaluate identification results.

Methods/Experimental
Training data preparation

In this work, we performed a CNN-based binary classification that categorizes 2D cloud data (OLR) into "developing TCs and their precursors" or "non-developing depressions." Binary classification is the task of categorizing (or classifying) objects into two groups (positive examples and negative examples) on the basis of classification rules. Before classification, it is necessary to prepare a labeled training data set, which is a series of sample data with the labels (positive or negative).

In the present work, at first glance, it appears natural to categorize the data into the following three classes: TCs, precursors, and non-developing depressions. However, since they are defined by the threshold value of maximum wind speed, it is difficult to identify them from cloud images. Figure 2a indicates a 2D histogram of cloud cover and elapsed time of all detected TCs including precursors by using the TC tracking algorithm. Here, we define cloud cover as $(\mathrm{OLR}_{max} - \mathrm{OLR}_{mean})/(\mathrm{OLR}_{max} - \mathrm{OLR}_{min})$, where OLR_{mean} is the mean value of OLR in 64×64 grids, and $\mathrm{OLR}_{max} = 300.0\,\mathrm{W/m^2}$ and $\mathrm{OLR}_{min} = 100.0\,\mathrm{W/m^2}$. Figure 2a shows the developing phase and dissipation phase in which the cloud cover increases and decreases over time, respectively. All precursors (elapsed time < 0) were in the developing phase; therefore, we could identify both precursors and TCs in the developing phase (inside yellow dotted line) under the same category "developing TCs and their precursors" (referred to here as "TCs") for labeling supervised data. The range was 30.0–95.0% for cloud cover and was from 10 days before to 7 days after formation, which could cover 97.0% of all precursors of TCs (in Matsuoka et al. 2017, the cloud cover range was 30.0–90.0% and covered 92.0% of all precursors of TCs). The other category, "non-developing depressions" (hereinafter referred as "non-TCs"), are low pressure clouds that were candidates for TCs but did not satisfy criteria (i)–(vi). For the binary classification,

Table 1 The numbers of TC tracks, detected samples, positive samples in training data, and average lifetime in each basin

	Number of TC tracks	Average lifetime [day]		Number of detected samples (number of positive samples in training data)	
		TCs	Pre-TCs	TCs	Pre-TCs
North Indian Ocean	169	4.6	2.9	3011 (1422)	2162 (1184)
Western North Pacific	754	6.8	3.1	21,546 (9549)	13,514 (8023)
Eastern North Pacific	589	4.8	7.8	11,880 (4478)	14,392 (7976)
North Atlantic	125	4.4	4.1	2582 (788)	1767 (758)
South Indian Ocean	525	5.5	3.7	7989 (4757)	7989 (4148)
South Pacific	367	4.2	4.0	6649 (3503)	5346 (3193)
South Atlantic	3	1.6	2.3	22 (15)	27 (26)
Total	2532	5.4	4.5	53,679 (24,512)	25,488 (25,308)

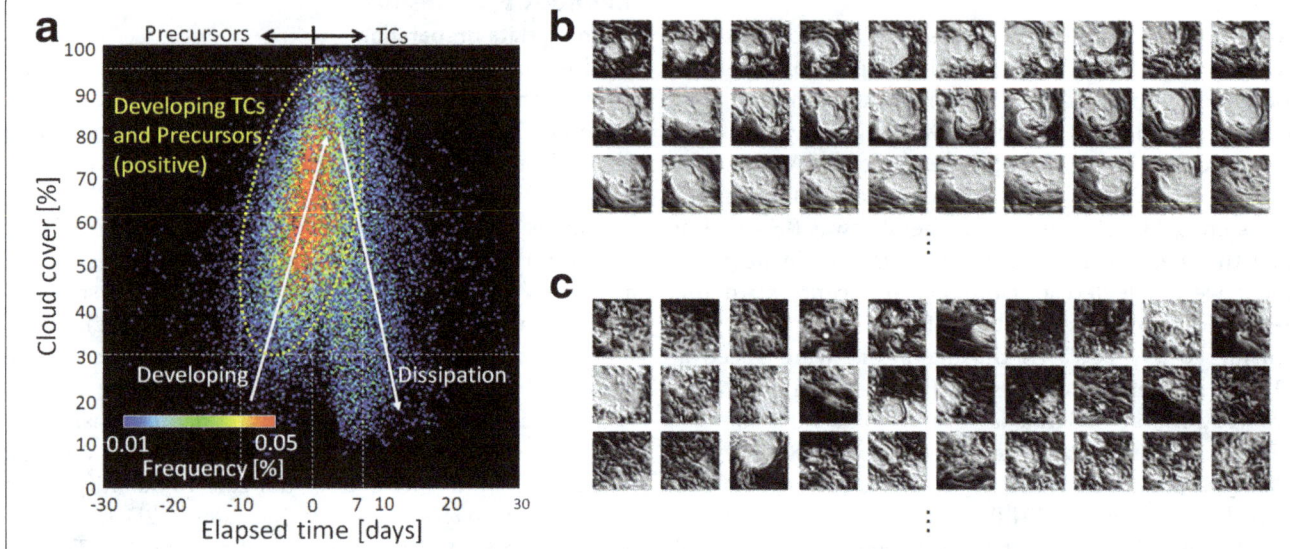

Fig. 2 a Histogram of cloud cover and elapsed time for developing TCs and their precursors. Positive and negative values of elapsed time indicate before formation (precursors) and after formation (TCs), respectively. **b, c** Examples of training data for binary classification of **b** developing TCs and precursors (positive examples) and **c** non-TCs (negative examples)

we labeled "TCs" and "non-TCs" data as "positive" and "negative" examples, respectively.

Examples of "TCs" and "non-TCs" are shown in Fig. 2b, c. Although these figures are visualized images of OLR, the actual training and test data are single-precision floating point numbers. Their horizontal sizes were approximately 1000 km × 1000 km (64 × 64 grids). For training, 20 year data (1979–1998) were used, and the remaining 10 year data (1999–2008) were used for prediction tests. The numbers of positive data (TCs and precursors) and negative data (non-TCs) for training were approximately 50,000 and 1000,000, respectively (the numbers of positive data in training data in each basin are listed in Table 1). Generally, the numbers of positive and negative data are often set to same number in binary classification. In this work, in order to train the CNN with a vast number of negative data, ten training data sets including the same 50,000 positive data and 50,000 randomly chosen negative data were generated for ten deep CNNs. By combining multiple CNNs, the influence of initial value dependence becomes smaller than when only single CNN is used (Freund and Schapire, 1997; Kearns and Valiant 1989; Breiman 1996; Breiman 2001).

Training and prediction using deep convolutional neural networks

We used a 2D deep CNN for binary classification (Table 2). CNNs are algorithms of neural networks used for image recognition and classification and for directly learning visual patterns from images. CNNs usually consist of convolutional layers, pooling layers, and fully connected layers (LeCun et al. 1999; Krizhevsky et al.

2012). Convolutional layers extract local features (feature maps) of input images, pooling layers allow spatial invariance by reducing the resolution of the image, and fully connected layers determine which features most correlate to a particular class. Dropout is a regularization technique where randomly selected neurons are ignored during training for preventing overfitting in a neural network.

Our CNN architecture comprises four convolutional layers, three pooling layers, and three fully connected layers. Input data were 64 × 64 grids of OLR data

Table 2 The architecture of our deep CNN. The parameters of the input layer, convolutional layers, pooling layers, and fully connected layers are denoted as [input data size] (e.g., 64 × 64), [filter size]@[number of feature maps] (e.g., 3 × 3@32), [pooling window size] (e.g., 2 × 2), and [number of units] (e.g., 2048), respectively

Layer	Shape
Input	64 × 64
Convolution 1	3 × 3@32
Convolution 2	3 × 3@64
Pooling	2 × 2
Convolution 3	3 × 3@64
Pooling	2 × 2
Convolution 4	7 × 7@128
Pooling	2 × 2
Fully-connected	2048
Fully-connected	2048
Fully-connected	2

consisting of single-precision real numbers, and output was generated in two classes (1 or positive: TCs; 0 or negative: non-TCs). Hyper parameters were optimized on the basis of the cross-validation test, which was conducted to evaluate the performance of the CNN using a random part of the training data. We examined the validation accuracy of 216 combinations of architecture settings: the number of convolutional (1–5) and pooling layers (1–5), the number of parameters in the fully connected layer (128, 256, 512, 1024, 2048), drop out ratio (0.2, 0.3, 0.4, 0.5), size of the convolutional filter (3×3, 5×5, 7×7), and number of feature maps (16, 32, 64, 128). Accordingly, the architecture with the highest level of performance was adopted, as shown in Table 2. The Adam optimizer (Kingma and Ba 2015) was applied to the CNN to update the network parameter to minimize the loss function called binary cross entropy. Batch normalization (Ioffe and Szegedy, 2015) was also applied to the CNN to minimize the initial-value dependence of the parameters.

The source code for deep learning was implemented in Python 3.6.3 with Keras (TensorFlow backend) (Chollet 2015), running on an NVIDIA Tesla P100 (1 node). Training 100,000 data over one epoch consumed approximately 3 min.

The accuracy of ten CNN classifiers using 100,000 data (50,000 for each of the two classes) for training and 5000 data for cross-validation is shown in Table 3. The maximum, minimum, and average values were 90.99%, 89.58%, and 90.30%, respectively (the number of epochs ranged over 19–46). The metric "Accuracy" is defined as follows:

$$\text{Accuracy} = \frac{\text{TP} + \text{TN}}{\text{TP} + \text{TN} + \text{FP} + \text{FN}} \quad (1)$$

Here, TP (true positive), TN (true negative), FP (false positive), and FN (false negative) correspond to "correctly

predicted positive example as positive," "correctly predicted negative examples as negative," "incorrectly predicted negative example as positive," and "incorrectly predicted positive examples as negative," respectively. Compared with the average accuracy of ten classifiers in Matsuoka et al. (2017), which was 86.60%, the average accuracy increased by 3.7 percentage point. Note that, although simulation data and the TC tracking algorithm were the same as in Matsuoka et al. 2017, in this study, the target range of cloud cover was expanded from 30.0–80.0% to 30.0–95.0%.

In the present study, in order to effectively train imbalanced data (positive 50,000, negative 1000,000), ten classifiers (*Classifier* 1, 2, ..., 10) were generated by training ten sets of 100,000 data on the same neural network, as shown in Fig. 3a. Each classifier was trained with the same 50,000 positive data and randomly selected 50,000 negative data. In this manner, our CNNs could train 50,000 positive examples and 500,000 negative examples simultaneously.

To verify the model's performance, classifiers trained using the 20 year data were applied to the test data (untrained 10 year simulation data). Candidate regions in the test data to be predicted by applying trained classifiers were clipped with a sliding window, which is widely used for object detection (Kumar 2013). We slid a rectangular area (approximately 1000×1000 km) with a 125 km (eight-grid) stride and continued sliding the window over the whole data within latitudes of 30° N to 30° S because three pooling layers of our CNN assumed eight grids of horizontal shift. Furthermore, in order to reduce the number of candidate regions, we set a limit to the cloud cover in the range of 30.0 to 95.0% and 50% or more over sea areas. In this manner, 97.0% of precursors of TCs in the simulation data could be covered.

Our ensemble CNNs output the ensemble average using the weight value of each trained classifier, as shown in Fig. 3b. The weight value given by the accuracy of the ten classifiers is listed in Table 3. The final probability p for detecting the presence of developing TCs and precursors in an arbitrary region is defined as follows:

$$p = \frac{1}{10} \sum_{i=1}^{10} \frac{w_i x_i}{w_i} \quad (2)$$

where w_i is the weight value of classifier i, and x_i is the output value obtained by *Classifier i* (0: non-TCs, 1: TC). When the threshold value p_{th} is given, arbitrary candidate areas that satisfy $p \geq p_{th}$ are regarded as positive. Although we adopt binary classification to facilitate the evaluation of prediction results, we can also output detection results as probabilistic information using p.

Table 3 Accuracy values (also used as weights) of the ten classifiers

Model number: i	Accuracy (weight: w_i)
Classifier 1	0.9099
Classifier 2	0.9050
Classifier 3	0.9013
Classifier 4	0.9014
Classifier 5	0.8958
Classifier 6	0.9085
Classifier 7	0.8987
Classifier 8	0.9065
Classifier 9	0.9025
Classifier 10	0.9007

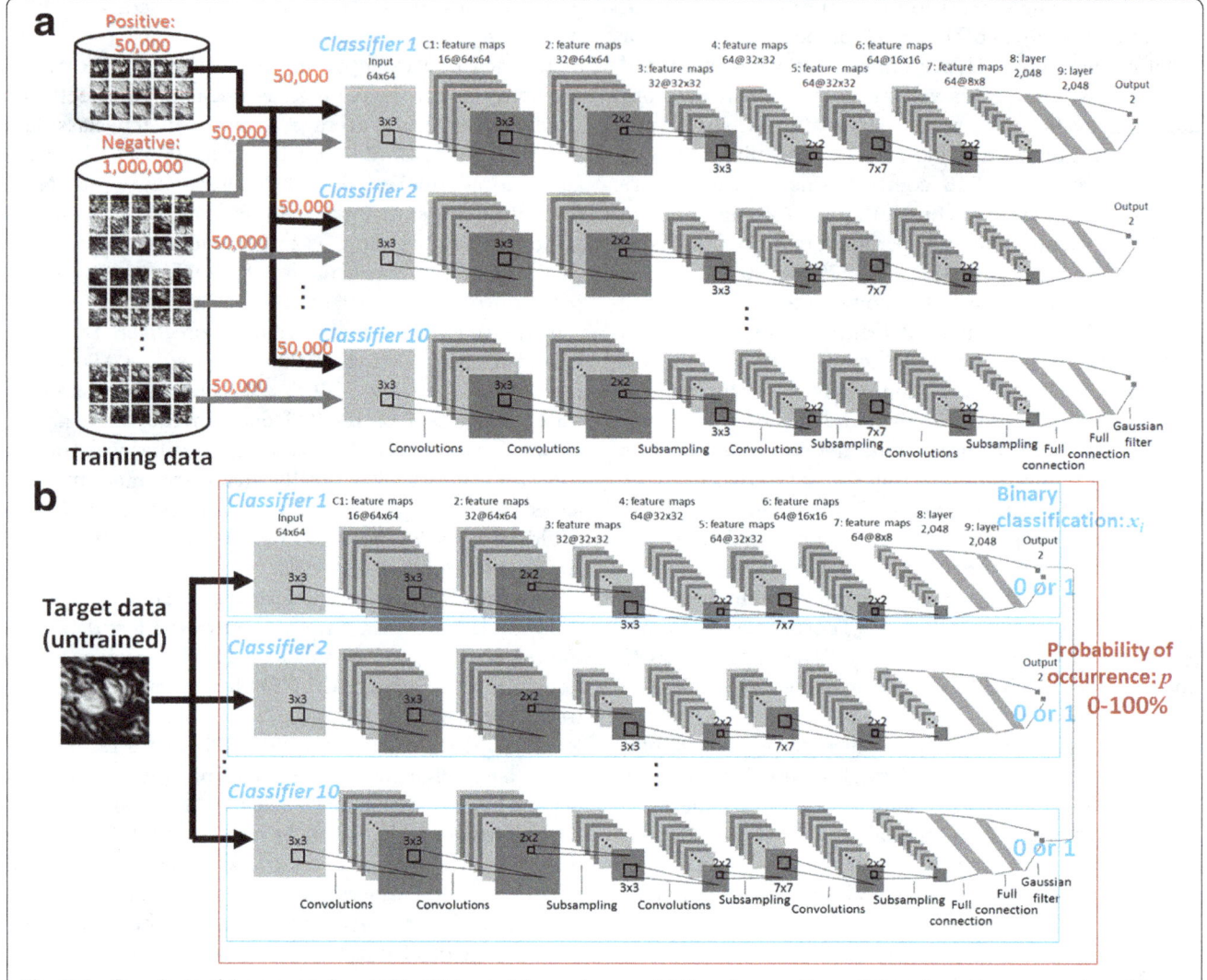

Fig. 3 The flow charts of the proposed ensemble CNNs: **a** training and cross-validation phase and **b** prediction phase. Hyper parameters of the CNN architecture are listed in Table 2

Evaluation metrics of prediction results

The false alarm ratio (*FAR*) and the probability of detection (*POD*) are often used as evaluation metrics of prediction results in weather forecasting (Jolliffe and Stephenson, 2003; Wilks 2006; Barnes et al. 2007). *FAR*, incorrectness of positive prediction, is defined using "correctly predicted positive examples as positive (*TP*, true positive)" and "incorrectly predicted negative examples as positive (*FP* false positive)" as follows:

$$FAR = \frac{FP}{TP + FP} \qquad (3)$$

As shown in Fig. 4a, when a positive predicted area captures the ground truth, the area is *TP*. Similarly, when a positive predicted area does not capture the ground truth, the area is *FP*. In the example of Fig. 4a,

TP is 3, *FP* is 1, and *FAR* is 25.0%. Although *FAR* is closely related to *Precision* = *TP*/(*TP* + *FP*), which is widely used in computer vision and pattern recognition (Forsyth 2011), *Precision* is one minus *FAR* and means correctness of positive prediction.

POD is another important metric in prediction; it indicates the amount of ground truth that can be correctly predicted. *POD* is conceptually the same as *Recall* = *TP*/(*TP* + *FN*) used in computer vision except for cases in which multiple positive predicted areas overlap, as shown in Fig. 4b. This is because the denominator of *POD* is the value of the ground truth, whereas the denominator of *Recall* is the number of predicted areas corresponding to the ground truth. Therefore, *TP* is given to TCs instead of prediction area, and we define the *POD* at multiple areas with the same detected ground truth as follows:

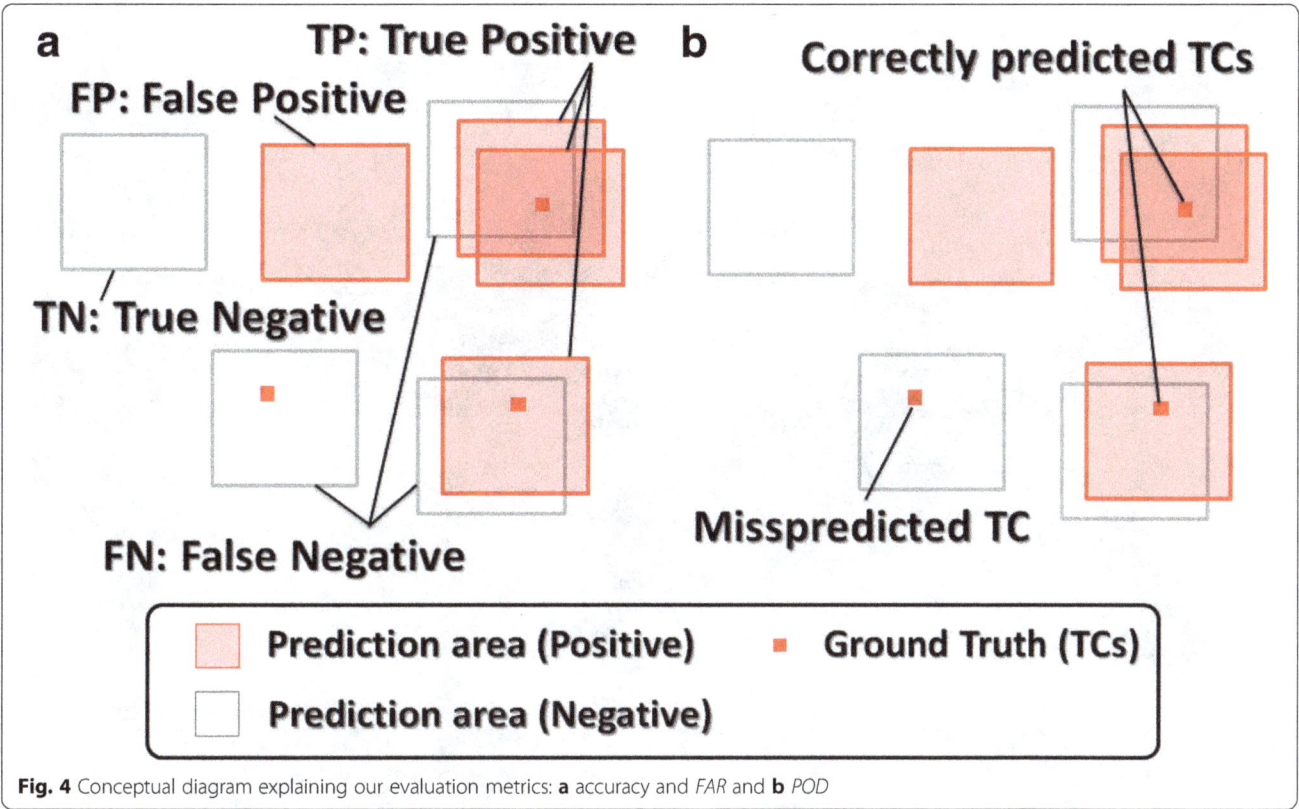

Fig. 4 Conceptual diagram explaining our evaluation metrics: **a** accuracy and *FAR* and **b** *POD*

$$POD = \frac{\text{number of correctly predicted TCs}}{\text{number of TCs}} \quad (4)$$

Here, the numbers of both TCs and correctly predicted TCs include precursors as mentioned above.

Results and discussion
Detection results

This section first introduces one of the best cases of detection results under the condition that the number of TCs and precursors is larger than eight and *POD* is larger than 80.0%. Similarly, one of the worst cases under the condition that the number of TCs and precursors is larger than five is also introduced.

Figure 5 shows the best case of detection results during the 10 year period (October 21, 2003, 18:00:00 UTC) for (a) $p_{th} = 1.0$ and (b) $p_{th} = 0.6$. Red and white boxes represent positive predicted areas (TCs) and negative predicted areas (non-TCs), respectively. Furthermore, blue and red dots represent the central points of actual TCs and precursors (as ground truth) calculated by the TC tracking algorithm, respectively. In Fig. 5a, five developing TCs and three precursors of nine ground truths can be correctly predicted; hence *POD* is 88.9% (= 8/9). Furthermore, 74 of 82 positive prediction areas could be correctly predicted; hence, *FAR* is 9.8% (= 1−74/82). Figure 5b shows the prediction results

after decreasing the threshold p_{th} to 0.6. In this case, the correctly predicted area (true positive) increases (*POD* is 100.0%) because the positive predicted area is expanded. However, the false alarm rate also increases (*FAR* is 34.1%).

Representing the worst case, prediction results of August 17, 2006 18:00:00 UTC are shown in Fig. 6, in which many TCs and precursors with less cloud cover were missed and there are numerous false alarms. While the *POD*s range from 20 to 60%, the *FAR*s were high (72.7 to 78.3%).

Detectability for each basin

The performance for various p_{th} was evaluated on each basin. Figure 8 shows the relationship between *POD* and *FAR* when p_{th} was varied from 10 to 100% in 10% increments. It represents the average value over 10 years (1999–2008) for each basin. The South Atlantic was not considered in this study because the number of TCs that occur in that area is extremely small. Although there are differences in values of detection performance depending on the basin, *POD* and *FAR* exhibited a trade-off relationship for all cases. When p_{th} is increased, the positive prediction area is narrowed down, and both *FAR* and *POD* decrease. In contrast, if p_{th} is decreased, the positive predicted area becomes wider, and both *FAR* and *POD* increase because dropout is reduced.

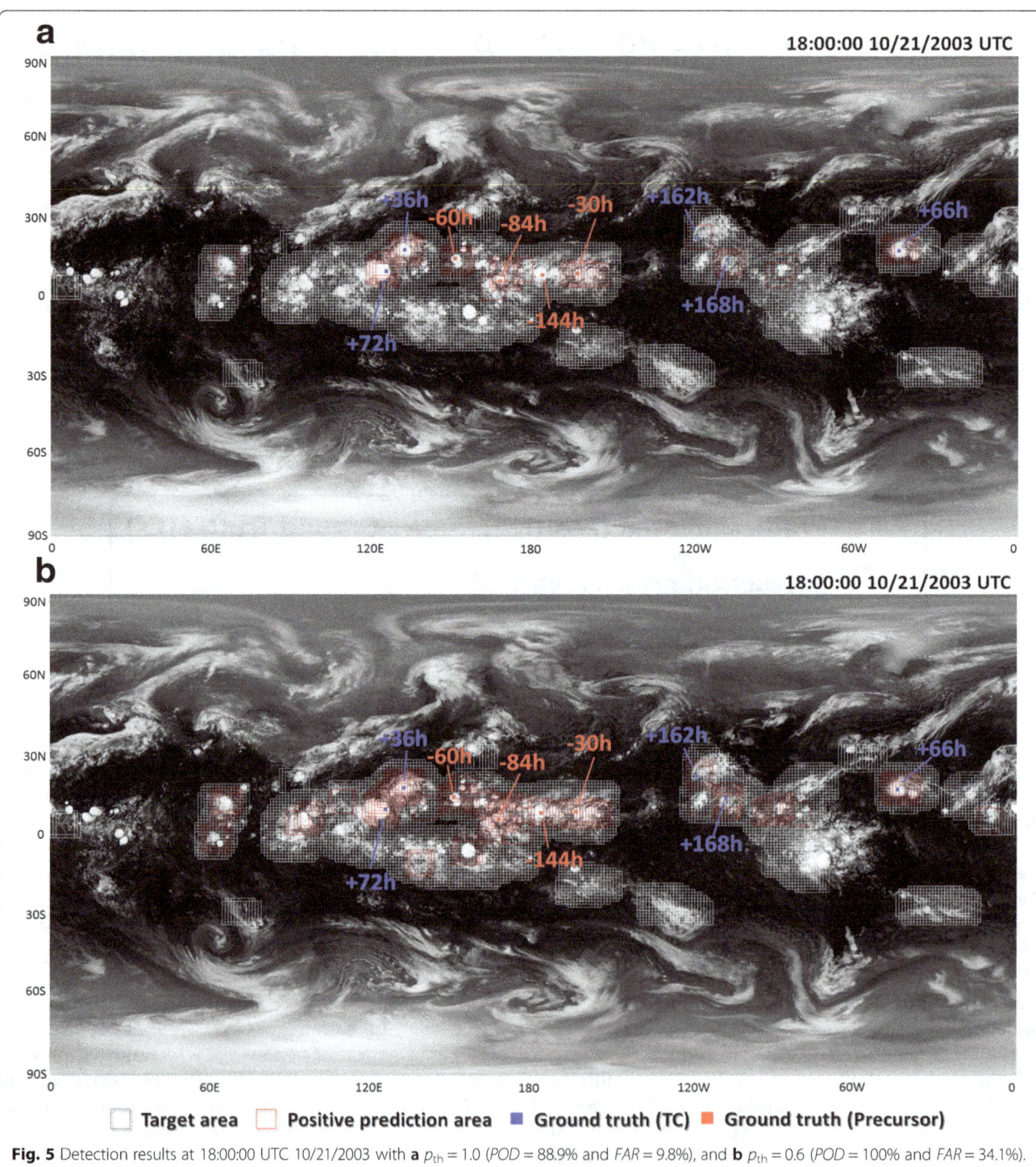

Fig. 5 Detection results at 18:00:00 UTC 10/21/2003 with **a** $p_{th} = 1.0$ (*POD* = 88.9% and *FAR* = 9.8%), and **b** $p_{th} = 0.6$ (*POD* = 100% and *FAR* = 34.1%). Numbers next to the blue or red letters indicate the elapsed time of the ground truth of TC or pre-TC, respectively

There are several reasons for the variation in detection performance for different basins. First, it is known that the pattern of TC genesis is different for each basin (Holland 2008), and the detectability for each pattern may differ (as will be described in the "Conclusion" section, investigation of the detectability for each generation pattern will be undertaken in future studies).

Second, in general, the performance of supervised machine learning depends on the number of training data for each pattern. In our results, the correlation coefficient between *POD* and the number of training data is 0.749, and that between the *FAR* and the number of training data is − 0.756. Lastly, since the cloud pattern of TCs is broken over time in the dissipation phase,

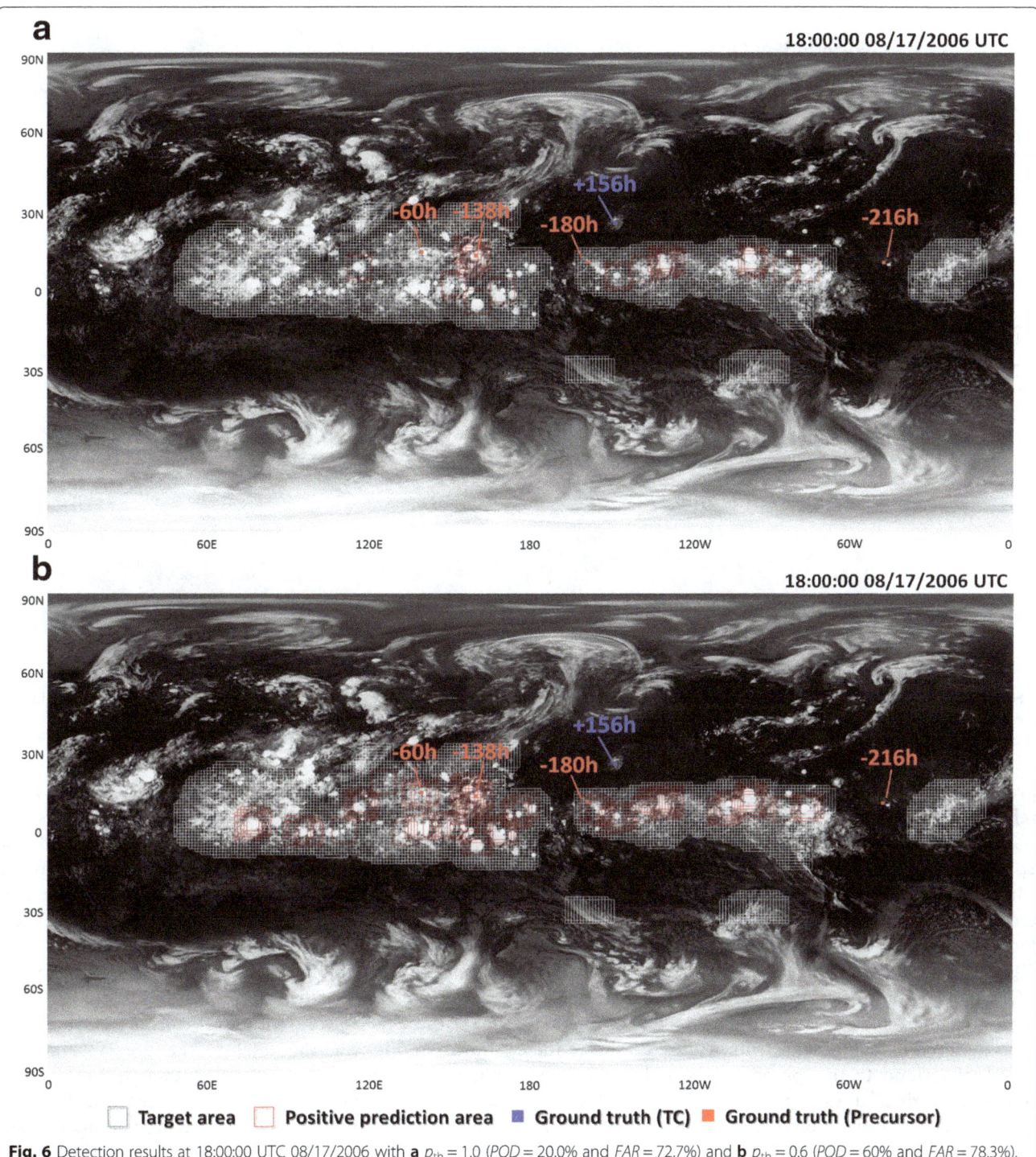

Fig. 6 Detection results at 18:00:00 UTC 08/17/2006 with **a** $p_{th} = 1.0$ ($POD = 20.0\%$ and $FAR = 72.7\%$) and **b** $p_{th} = 0.6$ ($POD = 60\%$ and $FAR = 78.3\%$). Numbers next to the blue or red letters indicate the elapsed time of the ground truth of the TC or pre-TC, respectively

their detectability should decrease. In other words, the detectability should be high for TCs with long lifetimes. The correlation coefficient between *POD* and average TC lifetime in each basin is 0.821 and that between *FAR* and average lifetime is − 0.802.

For example, as seen in Fig. 7, it was found that the basin with the best detection performance was the western North Pacific and that with the worst detection performance was the North Indian Ocean. In the western North Pacific, the number of training data was the largest (TCs 9549, pre-TCs 8023) and the average lifetime was also the longest (6.8 days). On the other hand, in the North Indian Ocean, the number of training data was relatively small (TCs 1422,

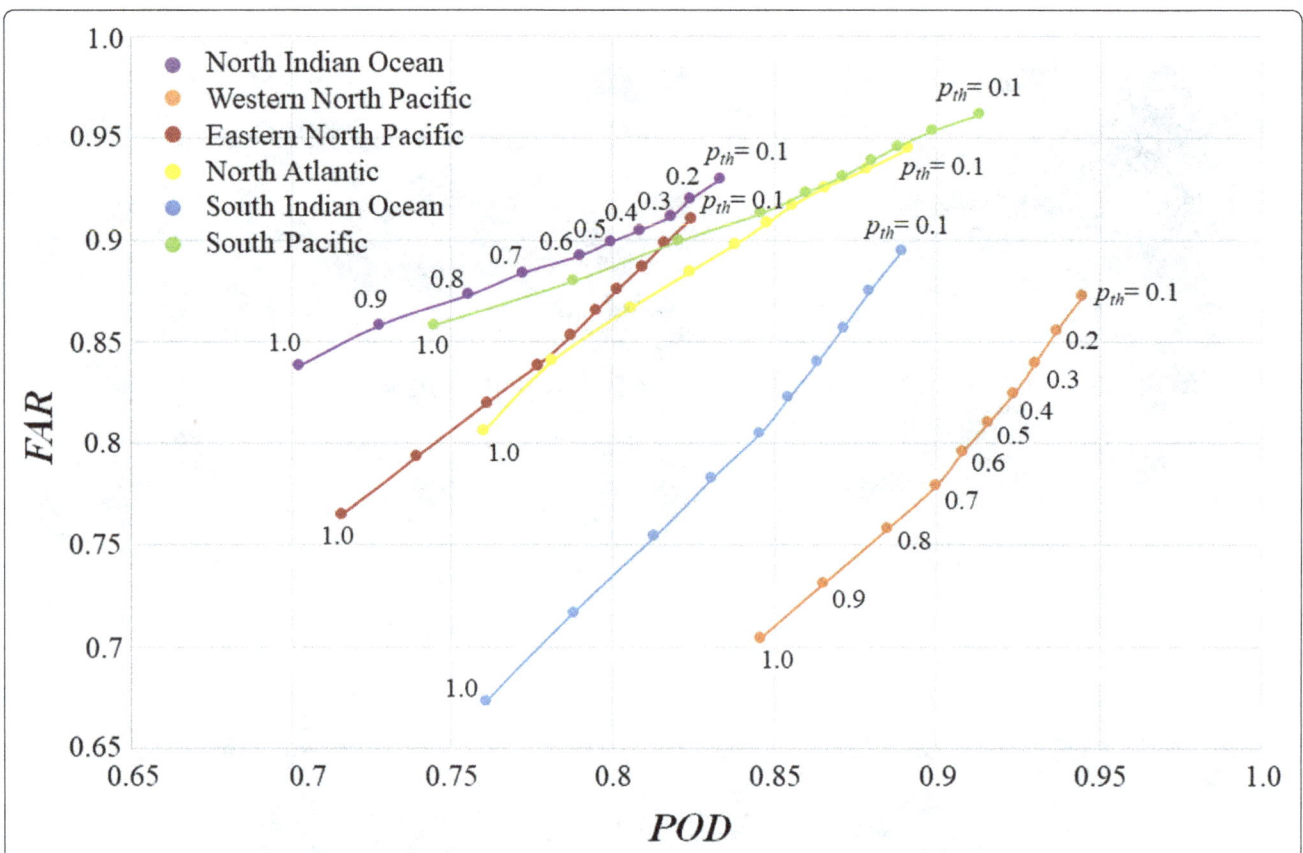

Fig. 7 Relationship of the *POD* and *FAR* with various p_{th} for different basins. The *POD* and *FAR* are 10 year average values from 1999 to 2008 for each basin

pre-TCs 1184) and the average lifetime was also relatively short (4.6 days).

Spatial detectability

This section first shows the spatial distribution of detection performance for a threshold value of $p_{th} = 1.0$. Figure 8a, b shows the spatial distributions of *POD* and *FAR*, respectively. Here, the spatial distributions of *POD* and *FAR* are calculated as their 10 year means at each grid point and are defined in a 64 × 64 rectangular area centered on the ground truth. Figure 8c shows the count of ground truths for both TCs and pre-TCs in the training data at each grid point. Figure 8d shows the count of positive prediction areas at each grid point. Areas with many TCs and pre-TCs are represented by a gray and white dotted line in Fig. 8a, b, d.

In most basins, both *POD* and *FAR* appeared to be roughly associated with the ground truth count in the training data. Especially in the Indian Ocean and the Pacific, *POD* was higher and *FAR* was lower in regions that had higher ground truth counts. As an exception, *POD* also exceeds 80% and *FAR* falls to 60% in a part of the North Atlantic, despite limited training data. Previous studies reported that there is a pattern of TC

genesis unique to the basin (Ritchie and Holland 1999; Yoshida and Ishikawa 2013; Fudeyasu and Yoshida, 2018). For example, Russell et al. (2017) reported that 72% of TCs in the North Atlantic are caused by African Easterly Waves (AEW). Accordingly, our results may indicate that TCs and pre-TCs caused by the AEW are easy to detect using CNNs.

Next, the detection performance of the TC area and pre-TC area are compared. In each basin, pre-TC areas are located at lower latitudes than the TC area. As shown in Fig. 8b, the *FAR* of the pre-TC area was higher than that of the TC area. Although there was no significant difference in *POD* between the TC and pre-TC areas, the count of positive predictions was larger in the pre-TC area than in the TC area. That is, the count of misdetection is larger in the pre-TC area than in the TC area. Intuitively, the pattern of developed TCs is simpler than that before formation, and therefore, it is reasonable that the detectability of TCs is higher than that of pre-TCs.

In the western South Pacific and South Atlantic, although the *FAR* was close to 100%, there were few positive prediction areas. In other words, the number of misdetections (false positives) was small. In contrast,

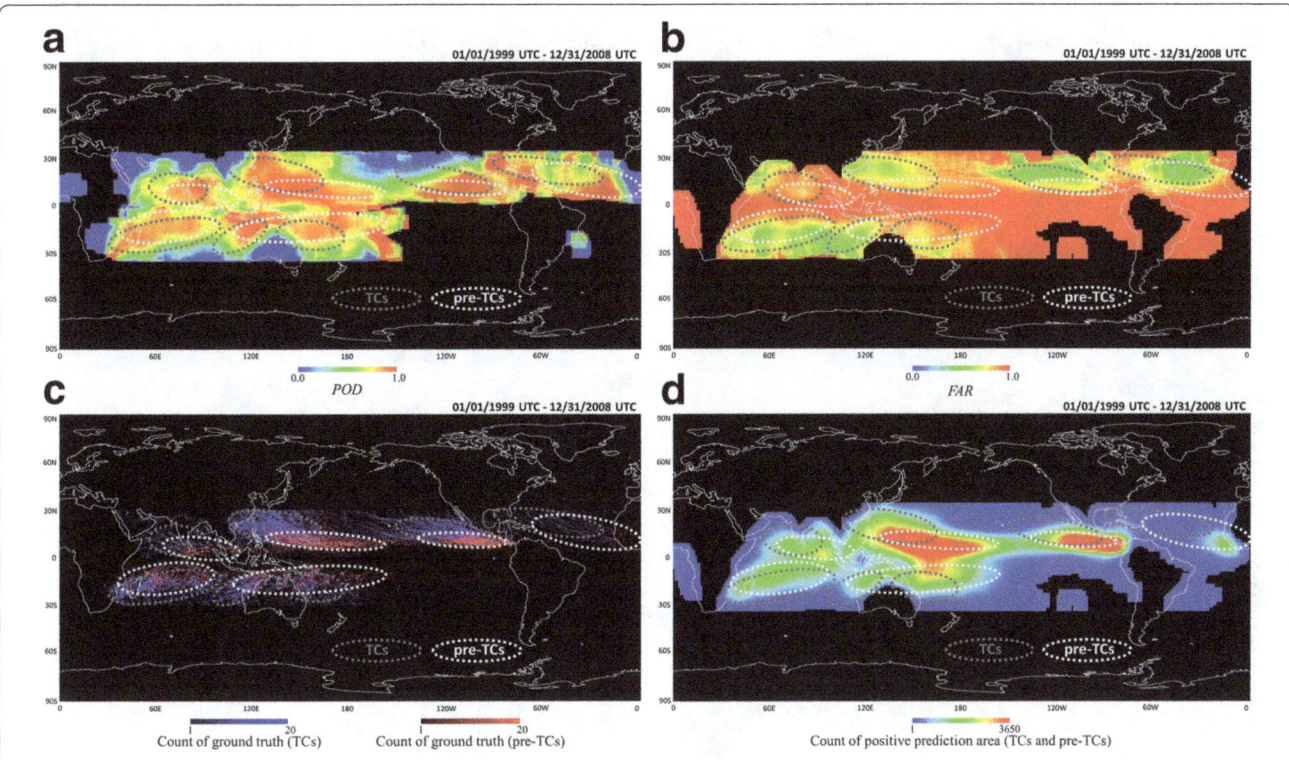

Fig. 8 Ten year means of **a** *POD* and **b** *FAR* with $p_{th} = 1.0$, respectively. **c** Spatial distribution of the count of ground truth in the training data of TCs and precursors. **d** Spatial distribution of the count of positive prediction areas

near the equator of the Indian Ocean and the eastern Pacific, although the *FAR* was also close to 100%, the number of positive prediction areas was large. That is, the number of misdetections was large in these areas.

Seasonal detectability

Seasonal detectability, monthly mean of *POD*, and *FAR* from 1999 to 2008 in each basin are shown in Fig. 9. Monthly variability of the number of training data (positive) and ground truth in each month and each basin are also shown in the same figure. In each basin, monthly changes in *POD* and *FAR* almost correspond to the number of training data. In other words, seasonal TCs can be detected without generating numerous false alarms. In particular, seasonal TCs in the western North Pacific (from July to November) could be detected with a *POD* of 79.0–89.1% and a *FAR* of 32.8–53.4% ($p_{th} = 1.0$). Similarly, seasonal TCs in the South Indian Ocean (from December to March) could be detected with a *POD* of 76.7–78.0% and a *FAR* of 31.1–40.3%. Furthermore, seasonal TCs in the North Atlantic (from August to November) could be detected with a *POD* of 75.0–78.2% and a *FAR* of 36.7–51.0%.

In contrast, numerous false alarms were generated during seasons with a low frequency of TCs. In the western North Pacific, although *POD* was not very low (65.7–83.0%) during January to May, the *FAR* was

remarkably high (75.4–95.3%). It is noteworthy that *POD* in the North Pacific is unlikely to decrease even during seasons with a low frequency of TCs, except for months with extremely small numbers of positive ground truth. In the western North Pacific in December, the *POD* was 78.7% and the *FAR* was 65.9% ($p_{th} = 1.0$). Similarly, in the eastern North Pacific in November, the *POD* was 72.6% and the *FAR* was 68.4% ($p_{th} = 1.0$).

Long-term detectability

Detectability (*POD*), the number of training data (positive), and ground truth of each elapsed time frame in different basins are shown in Fig. 10. In each basin, the day with the highest *POD* is the day TCs formed (elapsed time = 0–2 days). One of the reasons for this is that this period is the transition period from the developing phase to the dissipation phase for most TCs, as shown in Fig. 2a. Obviously, fully developed TCs are easy to recognize. Another reason is that the training of the CNNs tends to focus on large samples near this period to reduce errors.

The *POD* decreases as lead time is increased because it is difficult to detect the precursors many days in advance. For example, 91.2%, 77.8%, and 74.8% of precursors 2, 5, and 7 days before their formation can be detected in the western North Pacific, respectively (Fig. 10b). Similarly, 91.7%, 76.2%, and 70.1% of

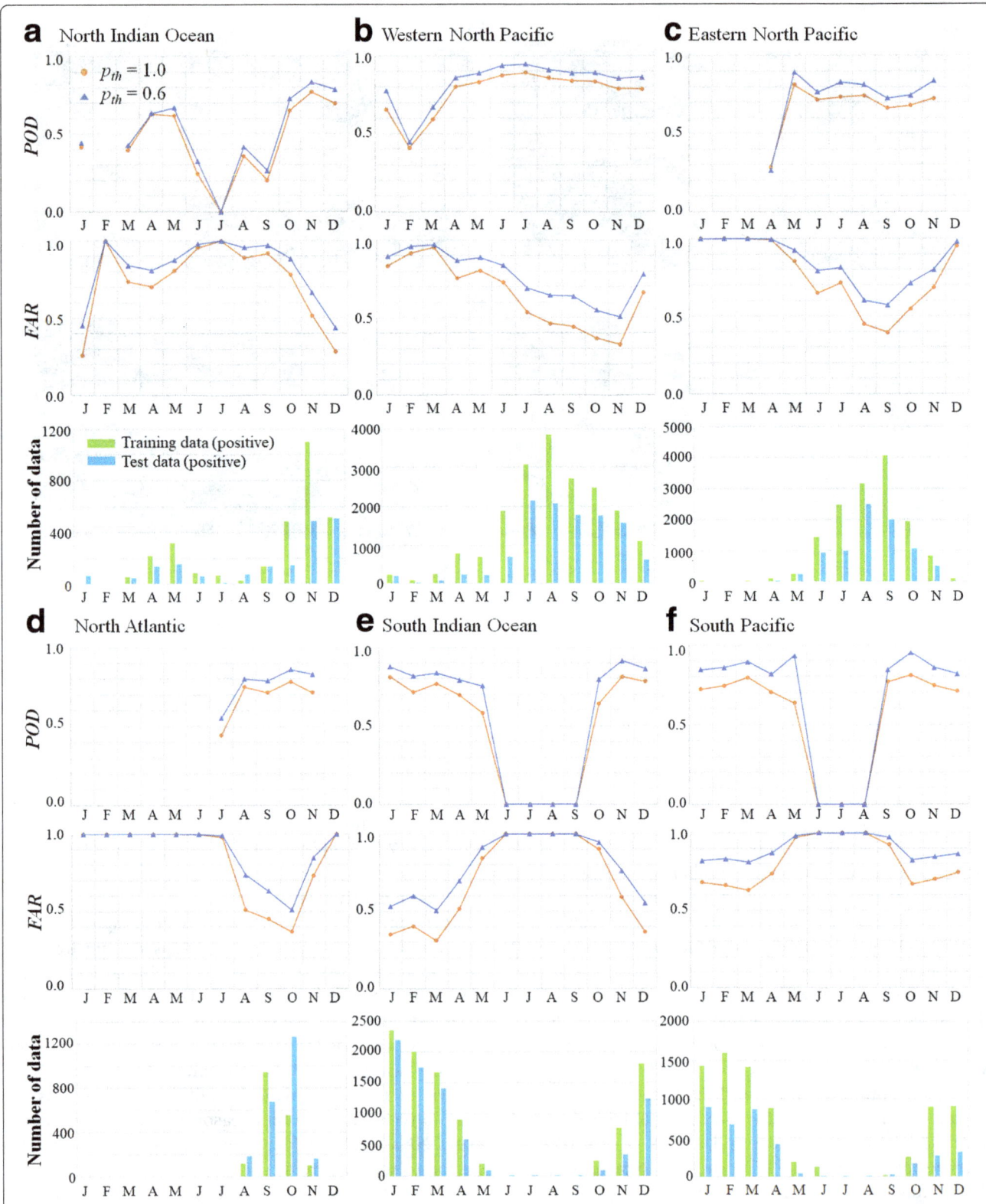

Fig. 9 *POD* and *FAR* ($p_{th} = 1.0$ and $p_{th} = 0.6$) and the number of data (training and test data) for each month in each basin. Note that the *POD* cannot be defined when the number of TCs is zero according to Eq. (2). In the same manner, the *FAR* cannot be calculated when the positive prediction area is zero, according to Eq. (3)

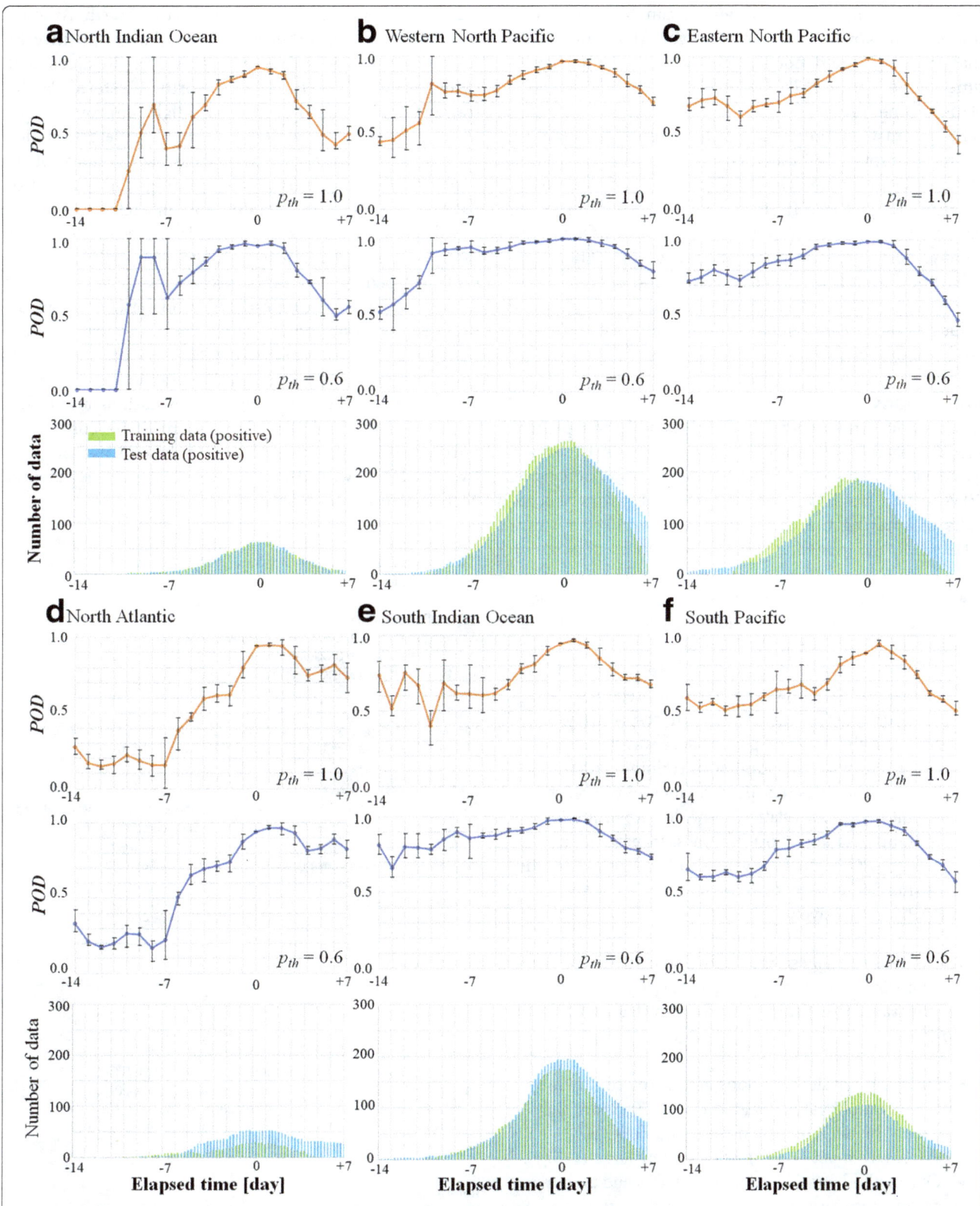

Fig. 10 Average-max-min charts of *POD* (p_{th} = 1.0 and p_{th} = 0.6) and bar graphs of the number of data (training and test data) for each elapsed time frame in each basin

precursors 2, 5, and 7 days ahead can be detected in the eastern North Pacific, respectively (Fig. 10c). Note that the *FAR* could not be calculated in each elapsed time frame because *TP* could not be defined in each elapsed time frame.

In contrast to these basins, long-term detectability decreases rapidly in basins with less training data, such as the North Indian Ocean (Fig. 10a) and the North Atlantic (Fig. 10d). In most basins, the number of ground truth matches of test data is small from 7 to 14 days before the formation of precursors. Therefore, this time frame involves large errors and decreased reliability of *POD*. Note that these results are reference values because the *POD* and *FAR* could not be evaluated simultaneously.

Conclusions

In this work, the detectability of TCs and their precursors for each basin, season, and lead time was investigated based on a deep neural network approach using 20 year simulated OLR by the NICAM. From the results of applying the CNNs to untrained 10 year simulated OLR, the following conclusions can be drawn:

- Particularly in the western North Pacific, we could successfully detect TCs and their precursors during July to November with *POD* values of 79.0–89.1% and *FAR* values of 32.8–53.4%. Detection results include 91.2%, 77.8%, and 74.8% of precursors 2, 5, and 7 days before their formation, respectively.
- Although the detection performance was approximately consistent with the number of training data and TC lifetime, the detection performance in the North Atlantic was not relatively low despite limited training data and short lifetimes. In particular, the average *POD* and *FAR* values in the North Atlantic during September to October were 74.8% and 40.9%, respectively.

These results suggest the high potential of the data-driven approach for studying tropical cyclogenesis.

In contrast, the limitations of our framework are as follows:

- Since the candidate regions are narrowed down by the threshold value of cloud cover (30–95%), they cannot be detected when the cloud cover is extremely small (< 30%) or extremely large (> 95%).
- Our method considers developing TCs and precursors as one category. To evaluate the detection performance of only pre-TCs, it is necessary to classify them by improving the CNN model.
- Our CNN classifiers may have model-specific biases arising from training using only NICAM data.

In areas with less data such as the North Atlantic, training data from other basins may have a positive influence on prediction results. However, in areas with sufficient data, such as the North Western Pacific, training according to each basin might improve detectability. In order to verify and improve detection performance, it is necessary to analyze the influence of the data in different basins by training the generation patterns and environmental factors in each basin.

This paper describes the preliminary results of detecting precursors of tropical cyclones using only simulated OLR; we plan to use other proxies of convection such as rain rate and mixing ratio of solid water for improving the detection performance. Furthermore, we also plan to apply our ensemble CNNs to reanalysis data and satellite observation data for practical use. For this purpose, data of different spatial resolutions and different variables or satellite channels must be considered. Furthermore, time-sequence data as well as comparative analyses with the results of the Early Dvorak Method are also required.

Abbreviations
CNN: Convolutional neural network; NICAM: Nonhydrostatic ICosahedral Atmospheric Model

Acknowledgements
We are grateful to Dr. C. Kodama for the production of simulation data and Dr. Y. Yamada for the production of best-track data on tropical cyclones. We also thank Editage for the English language review.

Funding
This work was supported by JST, PRESTO Grant Number JPMJPR1777, JSPS KAKENHI Grant Number 16 K13885, 26700010, and 17 K13010.

Authors' contributions
DM proposed the topic and conceived and designed the study. MN analyzed the data and helped in their interpretation. DS conducted the experimental study. SU collaborated with the corresponding author in the preparation of the manuscript. All authors have read and approved the final manuscript.

Competing interests
The authors declare that they have no competing interest.

Author details
[1]Center for Earth Information Science and Technology (CEIST), Japan Agency for Marine-Earth Science and Technology (JAMSTEC), 3173-25 Showa-machi, Kanazawa-ku, Yokohama, Kanagawa 236-0001, Japan. [2]PRESTO, Japan Science and Technology Agency (JST), 4-1-8 Honcho, Kawaguchi, Saitama 332-0012, Japan. [3]Department of Seamless Environmental Prediction Research, Japan Agency for Marine-Earth Science and Technology (JAMSTEC), 3173-25 Showa-machi, Kanazawa-ku, Yokohama, Kanagawa 236-0001, Japan. [4]Graduate School and Faculty of Information Science and Electrical Engineering, Kyushu University, 744 Motooka, Nishi-ku, Fukuoka 819-0395, Japan.

References
Barnes LR, Gruntfest EC, Hayden MH, Schultz DM, Benight C (2007) False alarms and close calls: a conceptual model of warning accuracy. Wea. Forecasting 22:1140–1147
Breiman L (1996) Bagging predictors. Mach Learn 24:123–140
Breiman L (2001) Random forests. Mach Learn 45(1):53–52
Chollet F (2015) Keras, GitHub

Cossuth J, Knabb TD, Brown DP, Hart RE (2013) Tropical cyclone formation guidance using pregenesis Dvorak climatology. Part I: operational forecasting and predictive potential. Wea. Forecasting 28:100–118

Dvorak VG (1975) Tropical cyclone intensity analysis and forecasting from satellite imagery. Mon. Wea. Rev. 103:420–430

Dvorak VG (1984) Tropical cyclone intensity analysis using satellite data. NOAA Technical Report NESDIS 11:1–47

Emanuel KA (1989) The finite-amplitude nature of tropical cyclogenesis. J Atmos Sci 46:3431–3456

Forsyth DA (2011) Computer Vision: A Modern Approach, 2nd edn. Pearson India, Delhi.

Freund Y, Schapire RE (1997) A decision-theoretic generalization of on-line learning and an application to boosting. J Comput Syst Sci 55:119–139

Fudeyasu H, Hirose S, Yoshioka H, Kumazawa R, Yamasaki S (2014) A global view of the landfall characteristics of tropical cyclones. Tropical Cyclone Research Review 3:178–192

Fudeyasu H, Yoshida R (2018) Western north pacific tropical cyclone characteristics stratified by genesis environment. Mon. Wea. Rev. 146:435–446

Gope S, Sarkar S, Mitra P (2016) In: Banerjee A, Ding W, Dy J, Lyubchich V, Rhines A (eds) Prediction of extreme rainfall using hybrid convolutional-long short term memory networks. Proceedings of the 6th International Workshop on Climate Informatics, Boulder 2016

Gray WM (1968) Global view of the origin of tropical disturbances and storms. Mon. Wea. Rev. 96:669–700

Gray WM (1975) Tropical cyclone genesis. Atmospheric science Paper 234. Colorado State University, Fort Collins

Holland GJ (2008) Tropical cyclones. In: Introduction to Tropical Meteorology, 1st edn. The COMET program, Boulder

Ioffe S, Szegedy C (2015) Batch normalization: accelerating deep network training by reducing internal covariate shift. Proceedings of the 32nd international conference on machine learning, Lille Grand Palais, Lille 6–11 July 2015

Jolliffe IT, Stephenson DB (2003) Forecast verification: a practitioner's guide in atmospheric science. Wiley, Hoboken

Kearns M, Valiant L (1989) Cryptographic limitations on learning Boolean formulae and finite automata. Proceedings of the 21st annual ACM Symposium on Theory of Computing, Seattle 14–17 May 1989

Kim SK, Ames S, Lee J, Zhang C, Wilson AC, Williams D (2017) In: Ebert-Uphoff I, Monteleoni C, Nychka D (eds) Massive scale deep learning for detecting extreme climate events. Proceedings of the 7th International Workshop on Climate Informatics, Boulder 2017

Kingma DP, Ba J (2015) Adam: a method for stochastic optimization. Proceedings of the 3rd International Conference on Learning Representations (ICLR2015), Hilton San Diego Resort & Spa, San Diego 7–9 May 2015

Kodama C, Yamada Y, Noda AT, Kikuchi K, Kajikawa Y, Nasuno T, Tomita T, Yamaura T, Takahashi HG, Hara M, Kawatani Y (2015) A 20-year climatology of a NICAM AMIP-type simulation. J Meteorol Soc Jpn 93(4):393–424

Kordmahalleh MM, Sefidmazgi MG, Homaifar A, Liess S (2015) In: Dy JG, Emile-Geay J, Lakshmanan V, Liu Y (eds) Hurricane trajectory prediction via a sparse recurrent neural network. Proceedings of the 5th International Workshop on Climate Informatics, Boulder 2015

Krizhevsky A, Sutskever I, Hinton GE (2012) ImageNet classification with deep convolutional neural networks. Paper presented at neural information processing systems (NIPS) 2012, Harrahs and Harveys, Lake Tahoe 3–8 December 2012

Kumar DS (2013) Context and subcategories for sliding window object recognition. LAP LAMBERT Academic Publishing, Saarbrücken

LeCun Y, Haffner P, Bottou L, Bengio Y (1999) Object recognition with gradient-based learning. In: Forsyth DA, Mundy JL, Vd G, Cipolla R (eds) Shape, contour and grouping in computer vision. Lecture Notes in Computer Science, vol 1681. Springer, Berlin, Heidelberg, pp 319–345

Liu Y, Racah E, Prabhat, Correa J, Khosrowshahi A, Lavers D, Kunkel K, Wehner M, Collins W (2016) Application of deep convolutional neural networks for detecting extreme weather in climate datasets. arXiv reprint arXiv:1605.01156

Matsuoka D, Nakano M, Sugiyama D, Uchida S (2017) In: Ebert-Uphoff I, Monteleoni C, Nychka D (eds) Detecting precursors of tropical cyclone using deep neural networks. Proceedings of the 7th International Workshop on Climate Informatics, Boulder 2017

Miura H, Satoh M, Nasuno T, Noda AT, Oouchi K (2007) A Madden-Julien oscillation event realistically simulated by a global cloud-resolving model. Science 318:1763–1765

Nakano M, Kubota H, Miyakawa T, Nasuno T, Satoh M (2017b) Genesis of super cyclone pam (2015): modulation of low-frequency large-scale circulations

and the madden-Julian oscillation by sea surface temperature anomalies. Mon. Wea. Rev. 145:3143–3159

Nakano M, Sawada M, Nasuno T, Satoh M (2015) Intraseasonal variability and tropical cyclogenesis in the Western North Pacific simulated by a global nonhydrostatic atmospheric model. Geophys. Res. Lett. 42(2):565–571

Nakano M, Wada A, Sawada M, Yoshimura H, Onishi R, Kawahara S, Sasaki W, Nasuno T, Yamaguchi M, Iriguchi T, Sugi M, Takeuchi Y (2017a) Global 7 km mesh nonhydrostatic Model Intercomparison Project for improving TYphoon forecast (TYMIP-G7): experimental design and preliminary results. Geosci Model Dev 10:1363–1381

Rayner NA, Parker DE, Horton EB, Folland CK, Alexander LV, Rowell DP, Kent EC, Kaplan A (2003) Global analyses of sea surface temperature, sea ice, and night marine air temperature since the late nineteenth century. J Geophys Res 108:4407. https://doi.org/10.1029/2002JD002670

Riehl H (1954) Tropical meteorology. McGraw-Hill, New York

Ritchie EA, Holland GJ (1999) Large-scale patterns associated with tropical cyclogenesis in the Western Pacific. Mon. Wea. Rev. 127:2027–2043

Russell JO, Aiyyer A, White JD, Hannah W (2017) Revisiting the connection between African easterly waves and Atlantic tropical cyclogenesis. Geophys Res Lett 44(1):587–595

Satoh M, Tomita H, Yashiro H, Miura H, Kodama C, Seiki T, Noda AT, Yamada Y, Goto D, Sawada M, Miyoshi T, Niwa Y, Hara M, Ohno T, Iga S, Arakawa T, Inoue T, Kubokawa H (2014) The non-hydrostatic icosahedral atmospheric model: description and development. Prog Earth Planet Sci 1:1–32

Simonyan K, Zisserman Z (2015) Very deep convolutional networks for large-scale image recognition. Paper presented at International Conference on Learning Representation (ICLR) 2015, The Hilton San Diego Resort & Spa, San Diego 7–9 May 2015

Sugi M, Noda A, Sato N (2002) Influence of the global warming on tropical cyclone climatology: an experiment with the JMA global model. J Meteorol Soc Jpn 80(2):249–272

Tomita H (2008) New microphysical schemes with five and six categories by diagnostic generation of cloud ice. J Meteor. Soc. Jpn 86A:121–142

Tomita H, Satoh M (2004) A new dynamical framework of nonhydrostatic global modeling using the icosahedral grid. Fluid Dyn. 1(8):357–400

Trafalis T, Adrianto I, Richman M, Lakshmivarahan S (2014) Machine-learning classifiers for imbalanced tornado data. Comput Manag Sci 11:403–418

Tsuchiya A, Mikawa T, Kikuchi A (2001) Method of distinguishing between early stage cloud systems that develop into tropical storms and ones that do not. Geophys Mag 1-4:49–59

Wilks DS (2006) Statistical methods in the atmospheric sciences, 2nd edn. Academic Press/Elsevier, New York

Xiang B, Lin S-J, Zhao M, Zhang S, Vecchi G, Li T, Jiang X, Harris L, Chen J-H (2015) Beyond weather time-scale prediction for hurricane Sandy and super typhoon Haiyan in a global climate model. Mon. Wea. Rev. 143:524–535

Yamada Y, Satoh M, Sugi M, Kodama C, Noda AT, Nakano M, Nasuno T (2017) Response of tropical cyclone activity and structure to global warming in a high-resolution global nonhydrostatic model. J Clim 30:9703–9724

Yamaguchi M, Koide N (2017) Tropical cyclone genesis guidance using the early stage Dvorak analysis and global ensembles. Wea. Forecasting 32:2133–2141

Yoshida R, Ishikawa H (2013) Environmental factors contributing to tropical cyclone genesis over the Western North Pacific. Mon. Wea. Rev. 141:451–467

13

Ensemble experiments using a nested LETKF system to reproduce intense vortices associated with tornadoes of 6 May 2012 in Japan

Hiromu Seko[1,2*], Masaru Kunii[1], Sho Yokota[1], Tadashi Tsuyuki[4,1] and Takemasa Miyoshi[3]

Abstract

Experiments simulating intense vortices associated with tornadoes that occurred on 6 May 2012 on the Kanto Plain, Japan, were performed with a nested local ensemble transform Kalman filter (LETKF) system. Intense vortices were reproduced by downscale experiments with a 12-member ensemble in which the initial conditions were obtained from the nested LETKF system analyses. The downscale experiments successfully generated intense vortices in three regions similar to the observed vortices, whereas only one tornado was reproduced by a deterministic forecast. The intense vorticity of the strongest tornado, which was observed in the southernmost region, was successfully reproduced by 10 of the 12 ensemble members. An examination of the results of the ensemble downscale experiments showed that the duration of intense vorticities tended to be longer when the vertical shear of the horizontal wind was larger and the lower airflow was more humid. Overall, the study results show that ensemble forecasts have the following merits: (1) probabilistic forecasts of the outbreak of intense vortices associated with tornadoes are possible; (2) the miss rate of outbreaks should decrease; and (3) environmental factors favoring outbreaks can be obtained by comparing the multiple possible scenarios of the ensemble forecasts.

Keywords: Ensemble forecast, Tornadoes, LETKF

Background

Local heavy rainfalls and tornadoes cause severe damage due to flash floods, landslides, and strong winds. Accurate forecasts of these phenomena, for example, by numerical models, could reduce the damages they cause. However, numerical models are not perfect, and the initial conditions of numerical forecasts include errors. In particular, forecasts of convection cells generated in areas of weak convergence, which frequently occur in urban areas such as the Tokyo Metropolitan Area in summer, are affected by small errors in the initial conditions. This means that forecasts of the position and timing at which a thunderstorm will be generated from a weak convergence are

sensitive to errors in the initial conditions. Because of the resulting uncertainty in forecasts of convection cell generation, ensemble forecast techniques must be employed to produce probabilistic forecasts. In addition to probabilistic forecasts, ensemble forecasts produce multiple possible scenarios. Thus, ensemble forecasts have the following merits: (1) analyzed fields of ensemble forecasts (i.e., the ensemble average of multiple scenarios) are statistically more accurate than those of deterministic forecasts and (2) multiple possible scenarios provide information regarding uncertainty (i.e., when there is large scatter among the scenarios, the uncertainty is larger).

The ensemble forecasts operationally performed by the Japan Meteorological Agency (JMA) with the highest horizontal resolution are typhoon forecasts and 1 week forecasts. However, these ensemble forecasts cannot represent convection cells of thunderstorms that cause local heavy rainfall and tornadoes because the horizontal grid interval (40 km) is too large to resolve the convection

* Correspondence: hseko@mri-jma.go.jp
[1]Meteorological Research Institute, Japan Meteorological Agency, 1-1 Nagamine, Tsukuba, Ibaraki 305-0052, Japan
[2]Japan Agency for Marine-Earth Science and Technology, 3173-25, Showa-machi, Kanazawa-ku, Yokohama, Kanagawa 236-0001, Japan
Full list of author information is available at the end of the article

cells. Recently, several ensemble forecast experiments in which heavy rainfall or local heavy rainfall was reproduced have been performed with a local ensemble transform Kalman filter (LETKF; Hunt et al. 2007; Miyoshi and Aranami 2006). For instance, Kunii (2013) reproduced a heavy rainfall in northern Kyushu by assimilating JMA's operational observation data with a LETKF system with a grid interval of 5 km. Probability distributions for a 3 h rainfall amount exceeding 50 mm were obtained by changing the initial time of the ensemble forecasts. Heavy rainfall of the northern Kyushu heavy rainfall event could be forecast with high probability even when the initial time of the ensemble forecasts was 12 h before the event.

One method of representing convection cells is to perform downscale forecasts with the model with a grid interval of a few kilometers and to obtain the initial conditions by spatial and temporal interpolation of the analyzed or predicted fields of a data assimilation system. For example, a local heavy rainfall, which caused a flash flood in the Toga River in Kobe that swept away six people, was reproduced by downscale experiments with JMA's non-hydrostatic model (JMANHM), the initial fields of which were obtained with a LETKF system (Seko et al. 2011). Though the linear band of intense rainfall responsible for the local heavy rainfall was represented as a weak rainfall region by the NHM-LETKF system with a grid interval of 20 km, downscale simulations with a grid interval of 1.6 km were successful in reproducing the linear rainfall band. This result is one example of a local heavy rainfall being reproduced by modifying the environment around the rainfall system through the assimilation of observed data.

A local heavy rainfall that occurred on the Osaka Plain on 5 September 2008 was reproduced by a nested LETKF system (Seko et al. 2013). In this experiment, high-resolution data (Doppler radar and GPS data) were assimilated to reproduce the local heavy rainfall. The number of ensemble members that reproduced the local heavy rainfall was increased by the assimilation of high-resolution data, compared with the number when the analyzed fields were obtained with assimilating only low-resolution conventional data. The results of these ensemble experiments show that the use of a LETKF system is a promising approach to the analysis and prediction of local heavy rainfall events.

In this study, we applied a LETKF system to the prediction of intense vortices associated with tornadoes. To date, no numerical experiments have been conducted using ensemble Kalman filters with a horizontal resolution of a few kilometers to reproduce intense vortices associated with tornadoes occurring in Japan, although a number of studies have used ensemble Kalman filters to reproduce or to investigate tornadoes that occurred in

the USA. For instance, Snook et al. (2011) showed that it was possible to obtain probabilistic forecasts by using an ensemble Kalman filter analysis system that assimilated Doppler radar data from the Collaborative Adaptive Sensing of the Atmosphere and Weather Surveillance Radar-1988 systems. Yussouf et al. (2013) conducted simulations with a mesoscale and convective scale ensemble data assimilation system consisting of the Weather Research and Forecasting model and an ensemble adjustment Kalman filter that used a double-moment microphysics scheme, and they successfully reproduced a supercell storm. Clark et al. (2013) investigated the relation between the path lengths of tornadoes and simulated updraft helicity by using an ensemble Kalman filter system with a horizontal grid interval of 1.25 km. Recently, Snook et al. (2015) demonstrated that use of an ensemble Kalman filter system could provide skillful analyses and ensemble-based probabilistic forecasts of tornadic mesoscale convective systems in the USA.

Tornadoes in Japan are different from those in the USA; Japanese tornadoes have an intensity equal to or less than F3 on the Fujita scale, and the low-level atmosphere in Japan is much more humid. Ensemble experiments to reproduce Japanese tornadoes are desirable for improving the forecast accuracy of tornadoes in Japan and for understanding the generation mechanisms of Japanese tornadoes.

We applied a nested LETKF system to a tornado event on 6 May 2012, during which three tornadoes were generated on the Kanto Plain. To express the vortices associated with these tornadoes, we performed downscaling experiments using the JMANHM model (Saito et al. 2006) with a horizontal grid interval of 350 m (NHM-350) and the analyzed fields produced by the LETKF system. We expected the multiple possible scenarios obtained from these ensemble forecast experiments to be useful for inferring environmental factors favoring the generation of intense vortices.

The rest of this paper is organized as follows: In the "Methods" section, the tornadoes of 6 May 2012 are first briefly described, and then, the LETKF system and downscale forecast experiments are explained. In the "Results and discussion" section, we present the results of the ensemble downscale experiments and discuss the factors that favor the generation of the intense vortices associated with tornadoes. In the "Conclusions" section, we present our conclusions.

Methods
Observed features of the tornadoes of 6 May 2012
On 6 May 2012, three tornadoes occurred on the northern Kanto Plain in Japan. The surface weather map for 09:00 Japan Standard Time (JST = UTC + 9) on 6 May (Fig. 1a) shows that there was a low-pressure system

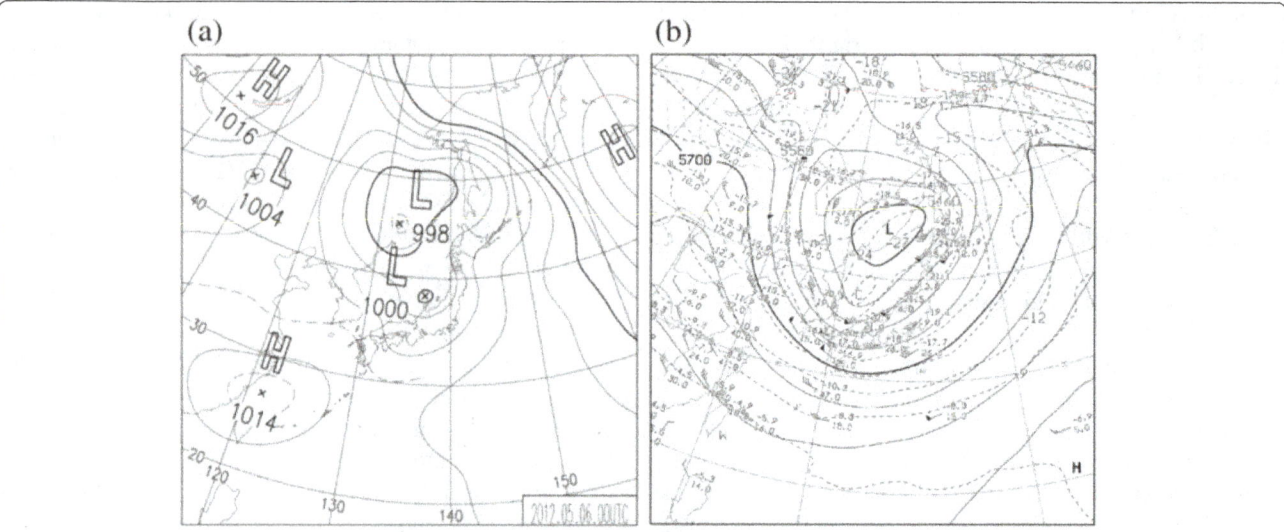

Fig. 1 a Surface weather map and **b** weather map at the height of 500 hPa at 09:00 JST on 6 May 2012. **b** Solid and *broken contours* indicate height (m) and temperature (°C), respectively. Intervals of the solid and broken contours are 3 °C and 60 m, respectively

over eastern Russia and the Japan Sea. The presence of the low-pressure system over the Japan Sea caused the isobar contours over central Japan, including the Kanto Plain, to be relatively densely spaced. As a result, strong southerly winds were expected over the plain. At the same time, a low-pressure system existed at the 500 hPa height that was centered in eastern Russia, at the same location as the center of the surface low-pressure system (Fig. 1b). Over the Japan Sea, there was a cold air mass with temperature lower than −18 °C, which was associated with this high-level low-pressure system. This cold

air mass was moving southeast and approaching the Kanto Plain.

At 12:00 JST, a cloud region, visible in a satellite image, extended from southwest to northeast over central Japan (Fig. 2a). Because this cloud region was located along the eastern edge of the cold air mass at the height of 500 hPa, we deduced that it had been generated partly by the upper cold air mass as it approached the Kanto Plain, though surface temperature also contributed to the cloud generation. On the Kanto Plain, a brighter cloud area (A in Fig. 2a) indicated the presence of intense convection cells.

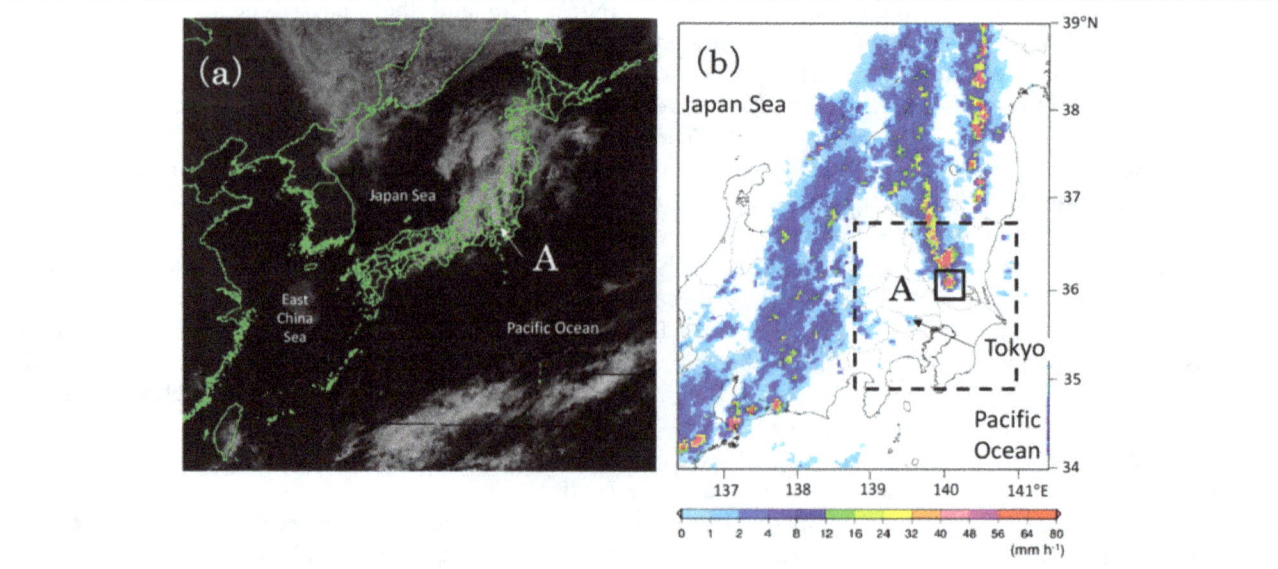

Fig. 2 a Visible image acquired at 12:00 JST on 6 May 2012 by a geostationary meteorological satellite of the Japan Meteorological Agency (JMA). **b** Rainfall intensity at 12:40 JST on 6 May 2012 observed by operational radars of JMA. The regions shown in Figs. 3 and 4 are indicated by the *broken line* and *solid line rectangles*, respectively

At 12:40 JST, the operational radars of JMA showed that several rainfall regions (Fig. 2b), which corresponded to the cloud region over central Japan observed by satellite, extended northeastward from central Japan. In addition, another rainfall region (region A) that extended northward from the Kanto Plain corresponded to the bright cloud area seen over the northern Kanto Plain. Rainfall intensity in the southern part of region A in some places exceeded 80 mm h^{-1}. These intense rainfall areas in region A were moving northeastward.

JMA determined the paths of the tornadoes that were generated on 6 May from the damage left in their wake (Japan Meteorological Agency, Meteorological Research Institute, Tokyo District Meteorological Observatory and Sendai District Meteorological Observatory 2012). Three tornadoes were generated at around 12:30 JST over the northern Kanto Plain. The intensity of the southernmost tornado was F3 on the Fujita scale, the highest intensity of tornadoes observed in Japan, and this tornado damaged about 800 houses, killed one person, and injured 37 people. The two northern tornadoes had intensities of F2 and F1.

Temperature and surface horizontal winds observed by JMA's Automated Meteorological Data Acquisition System (AMeDAS) showed that a warm southerly and a cold northerly airflow converged over the northern Kanto Plain on 6 May (Fig. 3). Rainfall regions in which rainfall intensity exceeded 56 mm h^{-1} were located in an area with a steep temperature gradient produced by the convergence that extended from southwest to northeast over the plain. This distribution indicates that these

intense rainfall regions were produced by the convergence of the warm southerly flow and cold northerly flow. Just after the outbreak of the tornadoes (12:30 JST), two distinct regions of intense rainfall were observed (I and II in Fig. 3b). Wind speeds of the southerly flow east of these intense rainfall regions were markedly higher than wind speeds in the surrounding area. The tornado tracks (red lines in Fig. 3b) extended northeastward from near the southern edges of intense rainfall regions I and II.

Because the southernmost tornado passed near Narita city, the reflectivity and radial winds within the rainfall region in which the southernmost tornado was generated were observed by the Doppler radar at Narita International Airport (red dot in Fig. 3). A pairing of intense winds of opposite direction, indicating the mesocyclone of the southernmost tornado, was observed in a filament-like rainfall band (arrows in Fig. 4) at the southern end of the intense rainfall region. This configuration of a rainfall region with a tornado located near the filament-like rainfall band at its southern tip, resembling configurations that have been observed in the USA (figure 8.10 in Houze 1994), suggested that the rainfall system that generated the southernmost tornado had a classic supercell structure.

The Meteorological Research Institute (MRI) has analyzed the environmental conditions when these tornadoes were generated by using JMA's operational mesoscale analysis data (Meteorological Research Institute 2012). The water vapor distribution at 12:00 JST on 6 May showed that a moist airflow supplied the intense rainfall regions

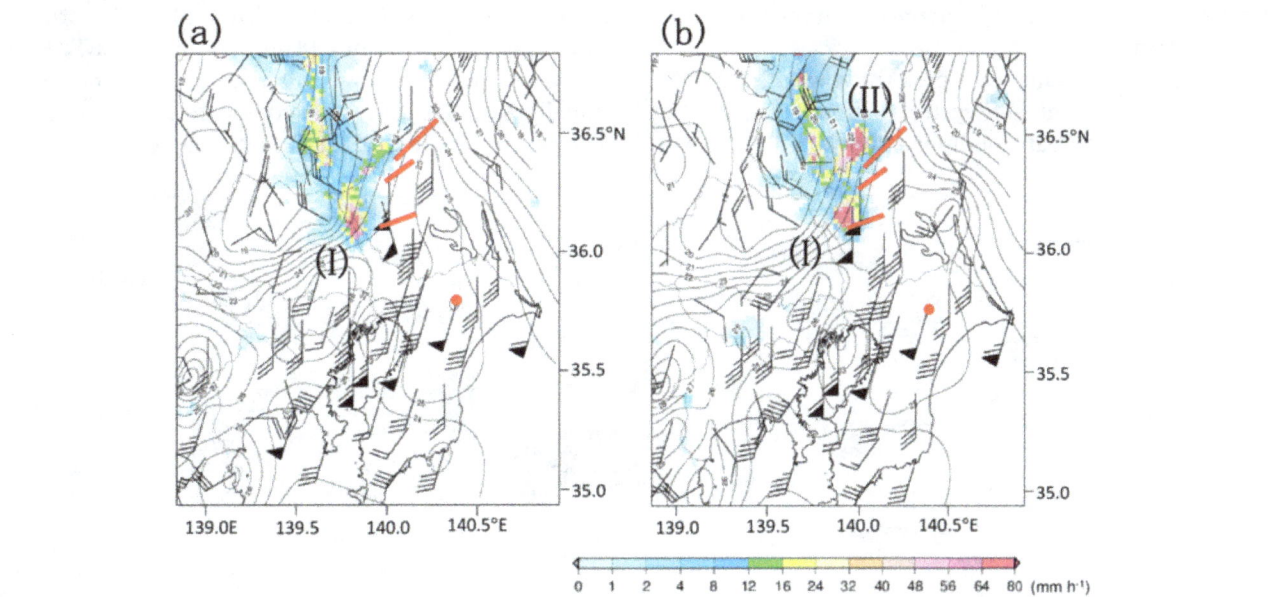

Fig. 3 Horizontal distributions of horizontal wind and temperature observed by AMeDAS at **a** 12:20 JST and **b** 12:30 JST on 6 May 2012 in the area enclosed by the *broken line rectangle* in Fig. 2. *Colors* indicate rainfall intensity observed by the operational radars of JMA. The *red lines* indicate the tornado tracks, as revealed by the damage they caused. The *red dot* in each panel indicates the position of the Narita radar

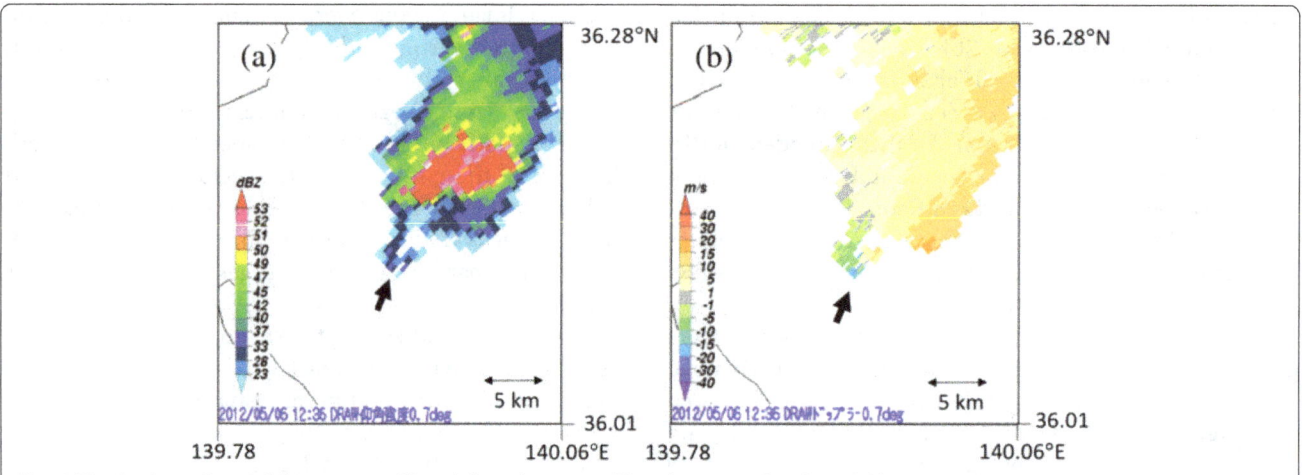

Fig. 4 Distributions of **a** rainfall intensity and **b** radial wind at 12:36 JST in the area within the *solid line rectangle* in Fig. 2, observed by the Narita airport radar (elevation angle 0.7°). The *arrows* indicate a filament-like rainfall band

where the tornadoes were generated. In particular, a moist airflow with a water vapor mixing ratio that exceeded 12 g kg^{-1} contributed to the generation of these tornadoes (Meteorological Research Institute 2012). To investigate the detailed structure of the convection cell that produced the southernmost tornado, MRI conducted a deterministic experiment using a model with a horizontal grid interval of 250 m. Although this experiment reproduced only the southernmost tornado, it showed that the rainfall system that generated the tornadoes had a supercell structure. In the present study, we performed ensemble experiments to further investigate these three tornadoes generated on the northern Kanto Plain.

Outline of the experiments performed with the nested LETKF system

Downscale experiments were performed with NHM-350 to reproduce the intense vortices associated with the tornadoes. The initial conditions for NHM-350 were

obtained from outputs of a nested LETKF system (Seko et al. 2013). The nested LETKF system (Fig. 5) was composed of two LETKF systems, each with 12 ensemble members; the grid interval of the outer LETKF was 15 km and that of the inner LETKF was about 1.875 km. The outer LETKF was used to reproduce the large-scale convergence by assimilation of part of JMA's conventional data (horizontal wind, temperature, and relative humidity data from upper air soundings and aircraft and wind profiler observations and surface pressure data recorded by meteorological observatories). The outer LETKF domain consisted of 80 × 80 horizontal grids covering Japan, except for Hokkaido and Kyushu, and the domain of the inner LETKF, which was within the outer LETKF domain, consisted of 160 × 160 horizontal grids covering the Kanto Plain. The inner LETKF grid interval of 1.875 km was adopted so that the intense rainfall regions that generated the tornadoes could be represented. In both LETKFs, the number of vertical

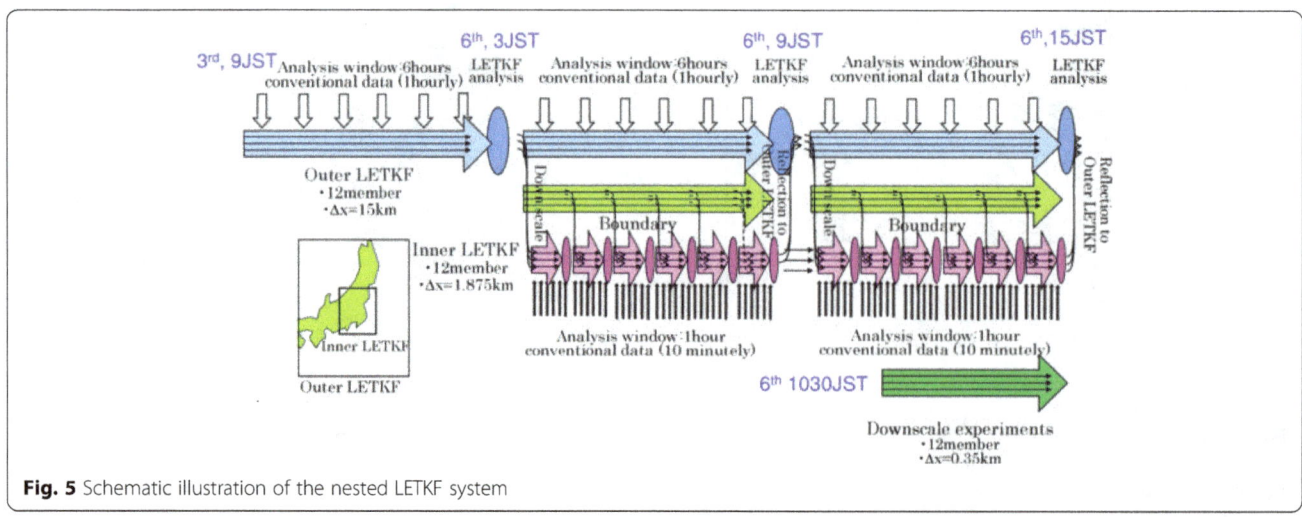

Fig. 5 Schematic illustration of the nested LETKF system

layers was 50, and the layer depth stretched from 20 m at the bottom to 890 m at the top of the domain. The altitude of the top was 22.6 km. The assimilation window of the outer LETKF was 6 h, and the data assimilation slot interval was 1 h. To reduce the convergence time required to get the analyzed fields of the outer LETKF, initial seeds for the outer LETKF were obtained at several prior points in time from the initial time of the outer LETKF from analyzed fields of JMA's mesoscale model. Because analyzed fields were used to produce the boundary conditions of the outer LETKF, the results of this study are, strictly speaking, not forecasts but simulations of the tornadoes.

The assimilation window of the inner LETKF was 1 h, and the slot interval was 10 min. In this study, conventional data (described above) from JMA were assimilated in both LETKFs. Although higher resolution observation data (e.g., radial wind from Doppler radars and GPS precipitable water vapor) are expected to be better able to reproduce smaller scale distributions of the initial conditions than conventional data, we did not use such high-resolution data in this study because our main purpose was to explore the merits of ensemble forecasts. Yokota et al. (2015) have shown that the assimilation of high-resolution data (e.g., radial winds from Doppler radars and dense surface observation data of AMeDAS and the Environmental Sensor Network of NTT DOCOMO, Inc.) could improve the positions of the reproduced vortices of the tornadoes of 6 May 2012. The analyzed fields of the inner LETKF were reflected to those of the outer LETKF every 6 h, at which time the analyzed fields of both the inner and outer LETKFs were available. The data assimilation cycles of the outer and inner LETKFs began at 09:00 JST on 3 September and 03:00 JST on 6 September, respectively.

NHM-350, which performed the downscale experiments to represent the intense vortices associated with the tornadoes, had 600×600 horizontal and 70 vertical grids, and its initial time was set to 10:30 JST on 6 September. The initial and boundary conditions for NHM-350 were produced from the output of the inner LETKF. NHM-350 used a double-moment microphysics parameterization scheme and the Deardorff (1980) planetary boundary scheme.

Results and discussion

Intense vortices produced by the inner LETKF and NHM-350

NHM-350 successfully reproduced the rainfall regions from 11:00 JST to 12:00 JST, which includes the period of the intense vortices associated with the tornadoes (Fig. 6). Because the rainfall regions were moving, the hourly rainfall region distribution extended northeastward over the northern Kanto Plain. An intense southerly flow from the south and a relatively weak flow under the relatively

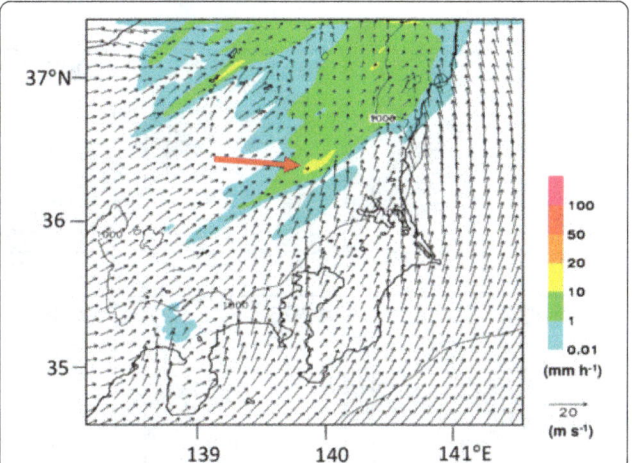

Fig. 6 The ensemble mean distributions of 1 h rainfall and horizontal wind at the height of 20 m at 12:00 JST, analyzed by the inner LETKF. A *red arrow* indicates the relatively intense rainfall region near which the relatively weak flow was generated

intense rainfall region (indicated by a red arrow in Fig. 6) were seen to converge along the southern and eastern edges of the rainfall regions. These distributions indicate that the rainfall regions were generated by this convergence. In general, this relationship between the rainfall regions and horizontal winds was the same as that observed, except that the northerly component of the airflow reproduced on the north side of the convergence line was much weaker than the observed airflow.

All 12 ensemble members of NHM-350 reproduced a few rainfall regions at 12:00 JST that extended to the north-northeast (Fig. 7a). As mentioned above, these rainfall regions were moving northeastward. The positions of the rainfall regions and the rainfall intensities varied among the ensemble members. Intense vortices, where vertical vorticity exceeded $0.1 \ s^{-1}$ at 12:00 JST (red dots in Fig. 7), were located near the southern tip of the rainfall regions generated by ensemble members #000, #002, #003, #004, and #008; this position corresponds to that of the southernmost tornado generated in intense rainfall region I (Fig. 3). In #006, an intense vortex was generated near the southern tip of the northern rainfall region; this position corresponds to one of the northern tornadoes generated in the observed intense rainfall region II (Fig. 3).

Ensemble members #004 and #007 generated the strongest and weakest vorticities among the ensemble members, respectively, at the height of 20 m. An intense vortex was generated near the southern tip of the intense convection by #004 at 11:45 JST (Fig. 7b), 45 min earlier than the observed southernmost tornado was generated, but only weak vorticity was generated by #007 (Fig. 7c). The position of the southernmost intense vortex relative to the rainfall region in #004 was

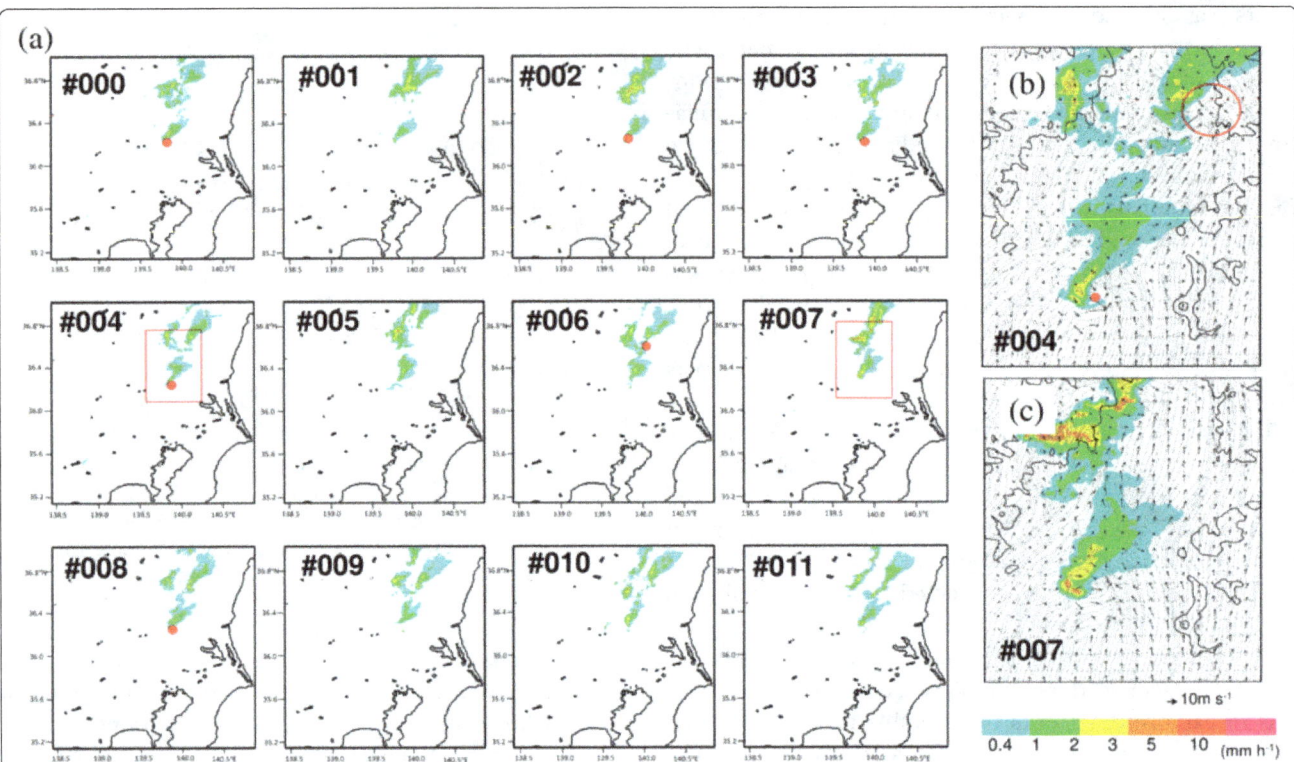

Fig. 7 a Rainfall regions at the height of 20 m at 12:00 JST reproduced by each of the 12 ensemble members of NHM-350. *Enlarged views* of the distributions of rainfall and horizontal wind by **b** #004 and **c** #007. The *red rectangles* in **a** indicate the regions shown in **b** and **c**. The *red dots* in **a** and **b** show the positions of vorticity exceeding 0.1 s^{-1} at 12:00 JST, and the *red circle* in **b** shows where convergence is occurring near the northern rainfall region. The thin contours are the stream lines of horizontal wind

the same as that of the observed vortex relative to the observed rainfall region. This result indicates that #004 reproduced the observed rainfall region that generated the tornado. An intense vorticity was not reproduced by #007, even though #007 produced more intense rainfall than #004 did. The relationship between vorticity and rainfall intensity in #007 suggests that rainfall intensity might not be a factor controlling the generation of intense vortices. The convergence reproduced by #004 at the southern edge of the northern rainfall region (red circle in Fig. 7b) corresponds to the northern intense vortex that was reproduced by #006 (Fig. 7a), though the vorticity generated by #004 was weaker than 0.1 s^{-1}. The stream lines, which show the small-scale distributions of horizontal wind, indicate that in #004 (Fig. 7b) the airflow that converged at the position of the northern vortex passed near the southern rainfall region. If the southern rainfall region modified the airflows that were supplied to the northern rainfall region, it would prevent the development of the northern rainfall region.

We investigated the airflows that entered the northern rainfall region by performing a back trajectory analysis (Fig. 8). The end points of the tracers were positioned around the southern tip of the northern rainfall region at 12:00 JST and traced backward in time for 90 min.

The tracers were assumed to move with the speed of the horizontal wind at the height of 0.8 km, so that the paths of the air inflows before their arrival at the northern rainfall region could be used to check the influence of the southern rainfall region on the northern vortices.

We chose ensemble members #002 and #006 for the trajectory analysis because intense vorticity was generated near the northern rainfall region by #006 but not by #002 (Fig. 7).

Both members showed most airflows approaching the southern tip of the northern rainfall region from the south. In #002, the airflows passed the southern rainfall region at 11:30 JST, and their potential temperature decreased before their arrival at the northern rainfall region. In #006, in contrast, some airflows passed far to the east of the southern rainfall region, and their temperature did not decrease. After the passage of these airflows from the east of the southern rainfall region, their direction of movement was changed by the presence of low pressure near the rainfall region, causing them to approach the southern tip of the northern rainfall region. That is, the southern rainfall region in #006 is deduced to have a smaller effect on the airflows that were supplied to the northern rainfall region. This difference of the paths, as a result, might be one reason why #006 was

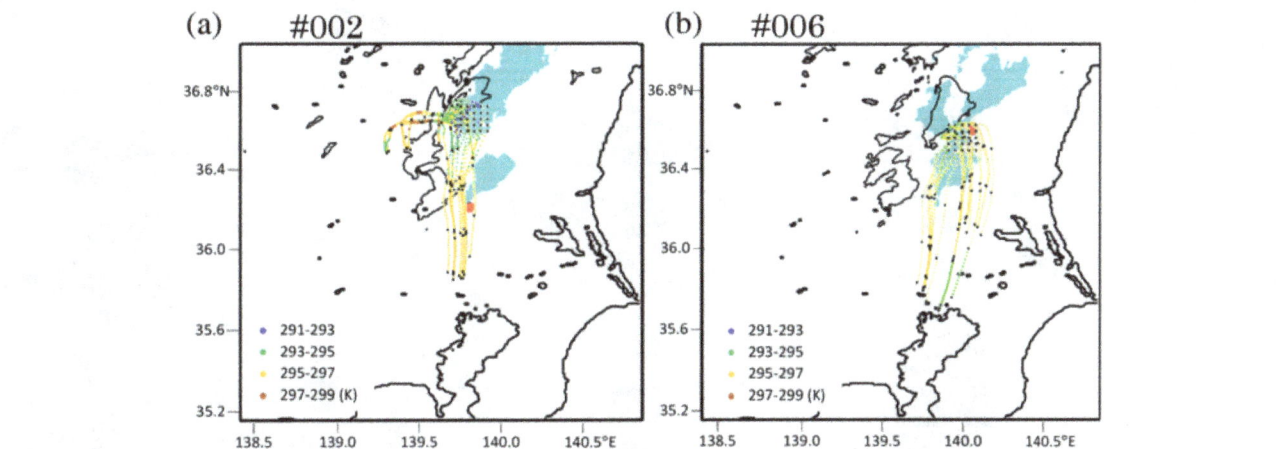

Fig. 8 Rainfall regions at the height of 20 m at 11:30 JST (*contours*) and 12:00 JST (*blue regions*) reproduced by two NHM-350 ensemble members: **a** #002 and **b** #006. The *red dots* in each panel show where vorticity exceeded 0.1 s^{-1} at 12:00 JST. The *small dots* show the trajectories traced backward in time from positions around the southern edge of the northern rainfall region. The *colors of the dots* indicate the potential temperature at the tracer positions. The *black dots* on the tracer paths show the tracer positions, from south to north, at 10:30, 11:00, 11:30, and 12:00 JST

able to generate intense vorticity near the northern rainfall region. This relationship between the southern rainfall and the intense vortex near the northern rainfall region was obtained with the NHM-350 system. To confirm the existence of a corresponding observed relationship, observation data of airflows within the lower atmosphere over the Kanto Plain would be needed.

The intense vortices, with vertical vorticity larger than 0.1 s^{-1}, reproduced by the 12 ensemble members from 11:00 JST to 13:00 JST in the downscale experiments (colored dots in Fig. 9) occurred in the same three regions as the observed tornadoes (Fig. 3), though the precise positions of the observed tornadoes were not reproduced. The southernmost intense vorticity, which was also reproduced by the deterministic forecast mentioned above, was generated by most of the ensemble members. In fact, an intense vortex associated with the southernmost tornado was reproduced by 10 of the 12 ensemble members (thus, the occurrence probability of an intense vortex was 83 %). The ensemble experiment also reproduced the two northern intense vortices, though their generation probability was low. This result indicates that probabilistic forecasts of intense vortices associated with tornadoes are possible and that ensemble forecasts can decrease the number of tornado outbreaks that are missed. Among the ensemble members, the paths of the southernmost intense vortex were clustered, whereas those of the northern vortices were scattered. This result indicates that the uncertainty of the paths was larger in the case of the northern vortices than it was in the case of the southernmost vortex. The ability to obtain information on uncertainty is one of the merits of ensemble forecasts mentioned in the "Background" section.

Next, we compared the outputs of NHM-350 around the mature vortices among the ensemble members to investigate the environmental factors favoring the generation of the intense vortices. We focused on the relationship between the durations of intense vortices and the values of certain environmental factors, because a tornado with longer duration causes damage over a wider area (Fig. 10). Here, for each member, duration was defined as the period during which vorticities exceeded 0.1 s^{-1}, and the environmental values (water vapor at 0.8 km height, vertical wind shear between the heights

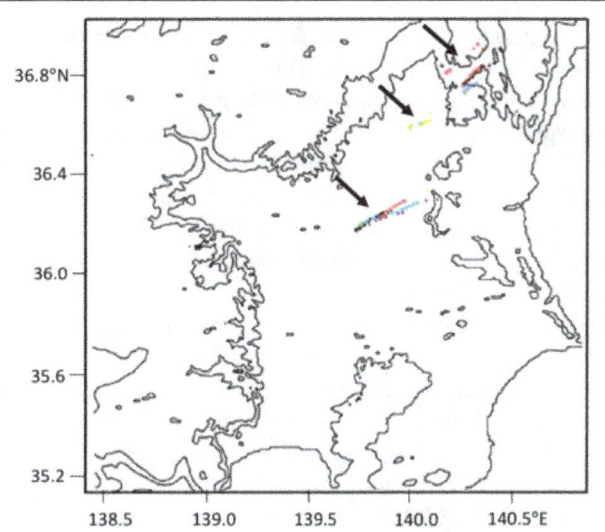

Fig. 9 Positions of intense vortices reproduced by NHM-350. The *colored dots* indicate points where ensemble members reproduced vorticities exceeding 0.1 s^{-1}. *Contours* indicate coastlines and land elevations of 300 and 600 m, respectively

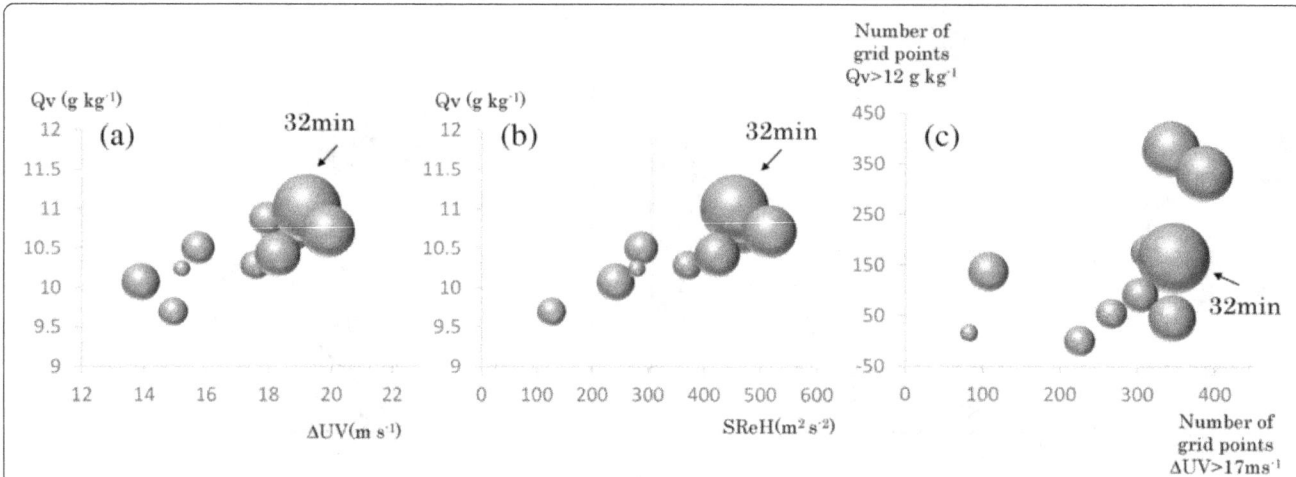

Fig. 10 Relationships between the duration of vortices (*larger circles* indicate longer durations) and environmental factors, obtained by NHM-350. **a** The vertical shear of horizontal wind (ΔUV) between the heights of 0.8 and 2.5 km (*horizontal axis*) and low-level water vapor (Qv) at the height of 0.8 km (*vertical axis*). **b** Storm relative helicity (SReH; *horizontal axis*) and low-level water vapor at the height of 0.8 km (*vertical axis*). **c** The number of grids in the averaged region where vertical shear 17 m s^{-1} (*horizontal axis*) and water vapor exceeded 12 g kg^{-1} (*vertical axis*). The averaged region was a rectangle (6.3 km east–west by 8.75 km north–south) on the southeastern side of the southern rainfall region and moved with the rainfall region

of 0.8 and 2.5 km, and storm relative helicity (SReH)) are the spatial averages within a rectangle (6.3 km east–west × 8.75 km north–south) on the southeastern side of the southern rainfall region at the time the vorticities reached 0.1 s^{-1}. This rectangle moved with the moving speed of the southern rainfall region. SReH is defined by the following equation:

$$\text{SReH} = \int_0^{3\text{km}} \left[\frac{\partial \boldsymbol{V}(z)}{\partial z} \times (\boldsymbol{V}(z) - \boldsymbol{C}) \right] \cdot \boldsymbol{k} \mathrm{d}z \qquad (1)$$

where $\boldsymbol{V}(z)$ is the horizontal wind vector at height z, $\boldsymbol{k}$ is the vertical unit vector, and $\boldsymbol{C}$ is the moving vector of the storm. To estimate SReH, the moving speed $\boldsymbol{C}$ of the southern rainfall region is needed. In this study, the moving speeds of the reproduced southern rainfall regions were used. A scatter diagram of the durations of intense vortices and the values of certain environmental factors shows that the duration became longer as the low-level layer became more humid and the vertical wind shear became larger (Fig. 10a). The duration also became longer when the SReH in the averaged region was larger.

In addition, comparison of duration with the number of grids in which water vapor and vertical wind shear in the averaged area exceeded 12 g kg^{-1} and 17 m s^{-1}, respectively (Fig. 10c), showed that the durations of the intense vortices became longer as the number of grids exceeding these thresholds became larger. This result was expected because the number of grids in which values exceeded these thresholds was correlated positively with the averaged values shown in Fig. 10a. A larger number of grids with values exceeding these

thresholds indicated that the favorable region for intense vortex generation within the averaged area, that is, within the rectangle on the inflow side of the southern rainfall region, was larger. Therefore, these results indicate that ensemble members reproducing large areas with a favorable environment reproduce vortices with a long duration. Thus, favorable environmental factors can be extracted by comparing the outputs of the ensemble members.

Environmental factors affecting intense vortex generation
The temporal variations of intense vortices and the horizontal distributions of environmental factors (e.g., low-level water vapor and vertical shear of horizontal winds) are examined in this section to show how environmental factors affect intense vortex generation. Figure 11 shows the temporal variations of maximum vorticity obtained by NHM-350 within the whole domain of NHM-350 from forecast time (FT) = 0 min to 180 min. Because the vorticities of ensemble members #004 and #007 were the strongest and weakest, respectively, we expected them to provide information about the key factors that affect the outbreak of intense vortices. To show the impacts of environmental factors on the intensity of vertical vorticities more clearly, we performed a series of experiments in which the initial conditions were produced by combining the initial conditions of #004 and #007 with different weights as follows:

$$I_{\text{combined}} = w \times I_{004} + (1-w) \times I_{007} \qquad (2)$$

where w is the weight of ensemble member #004.

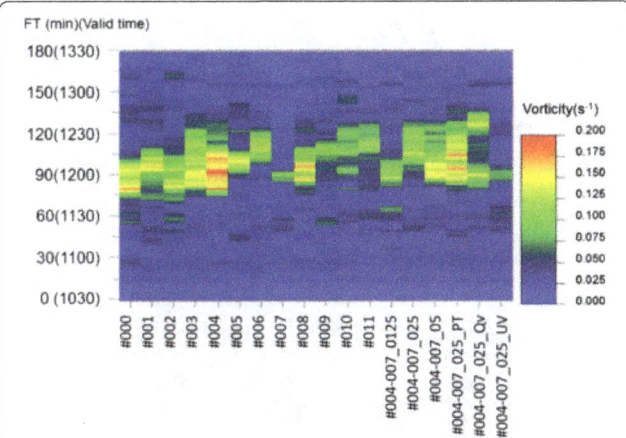

Fig. 11 Temporal variations of maximum vorticity from FT = 0 min (10:30 JST) to 180 min (13:30 JST) reproduced by each ensemble member of NHM-350. The *last six columns* show the results of a series of experiments in which the initial conditions were produced by combining the initial conditions of #004 and #007 with different weights and by replacing water vapor, horizontal wind, or potential temperature in the initial conditions of #004-007_025 with the corresponding values of #007

As *w* became smaller, the maximum vorticity generated during FT = 0 min to 180 min became weaker (Fig. 11). This result is expected because the impact of #004 is reduced when *w* is smaller; the timing of the maximum vorticity, however, did not change linearly from that of #007 as *w* decreased. When the weight was changed from 0.25 (#004-007_025) to 0.125 (#004-007_0125), the vorticity decreased greatly. This large change of vertical vorticity suggests that the #004-007_025 experiment was particularly sensitive to environmental factors; therefore, we expected this experiment to show the impact of environmental factors more clearly than the others. Thus, we investigated the impact of each environmental factor by replacing water vapor, horizontal wind, or potential temperature in the initial conditions of #004-007_025 with those of #007.

When the horizontal wind or water vapor was replaced with those of #007 (#004-007_025_UV, #004-007_025_QV), the vorticities became weaker. In contrast, when the potential temperature was replaced with that of #007 (#004-007_025_PT), vorticity became more intense. These results indicate that the horizontal wind and water vapor in the initial conditions of #004 were favorable for the generation of an intense vortex but that the temperature distribution of #004 suppressed the generation of an intense vortex.

We therefore examined the horizontal distributions of the vertical shear of horizontal winds and of the low-level water vapor at 11:40 JST (5 min before the outbreak of the reproduced southernmost intense vortex in #004) reproduced by #004 and #007 to see where the differences existed between them (Fig. 12). On the

southeastern side of the rainfall regions, that is, on the inflow side of the southernmost intense vortex, both #004 and #007 simulated an intense vertical shear region where wind shear between the heights of 0.8 and 2.5 km was larger than 12 m s^{-1} (circles in Fig. 12a, b). As this region of intense vertical shear moved northward and the rainfall regions approached it from the west, the intensity of the vertical shear became larger. This region in which vertical shear exceeded 12 m s^{-1}, however, was larger in #004 than in #007 (Fig. 12a, b). This result suggests that the intensity of vertical shear affects the outbreak of intense vortices, with the likelihood of an outbreak increasing with the intensity of vertical shear. The distributions of low-level water vapor show that a humid region also existed on the eastern side of the rainfall regions in both members #004 and #007 and that more water vapor was supplied to the rainfall regions in #004 than in #007 (Fig. 12c, d). This difference in low-level water vapor between #004 and #007 might be one reason why intense vorticity was not generated in #007.

The temporal variations of water vapor and vertical shear of horizontal wind on the southeastern side of the southernmost intense vortex in #004 are compared with those of vorticity in Fig. 13. The estimation region of the temporal variations was the same as that used in producing Fig. 10. The water vapor supplied to the rainfall region gradually increased up until the outbreak of the southernmost intense vortex (outbreak time in #004, 11:45 JST), though there was a relatively small water vapor period from 11:30 JST to 11:40 JST. The vorticity became rapidly larger just after an abrupt increase in the vertical shear of the horizontal winds. These temporal variations in water vapor and vertical shear show that, in the case of this event, the intense vertical shear determined the timing of the outbreak of intense vorticity.

Conclusions

The intense vortices associated with tornadoes occurring on 6 May 2012 were reproduced by a nested LETKF system that simulated the environments and convection cells simultaneously. The results of this study demonstrated three merits of ensemble forecasts based on the outputs of the nested LETKF system. The occurrence probability of an intense vortex with a vertical vorticity exceeding 0.1 s^{-1} during this tornado event was 83 %. This result shows the first merit: probabilistic forecasts of the outbreak of intense vortices associated with tornadoes are possible. In addition, the multiple possible scenarios revealed two other merits of ensemble forecasts. Although a deterministic forecast had reproduced only the southernmost of the three tornadoes, several ensemble members reproduced the intense vortices associated with the two northern tornadoes. Thus, the second merit is that ensemble forecasts can be expected to decrease

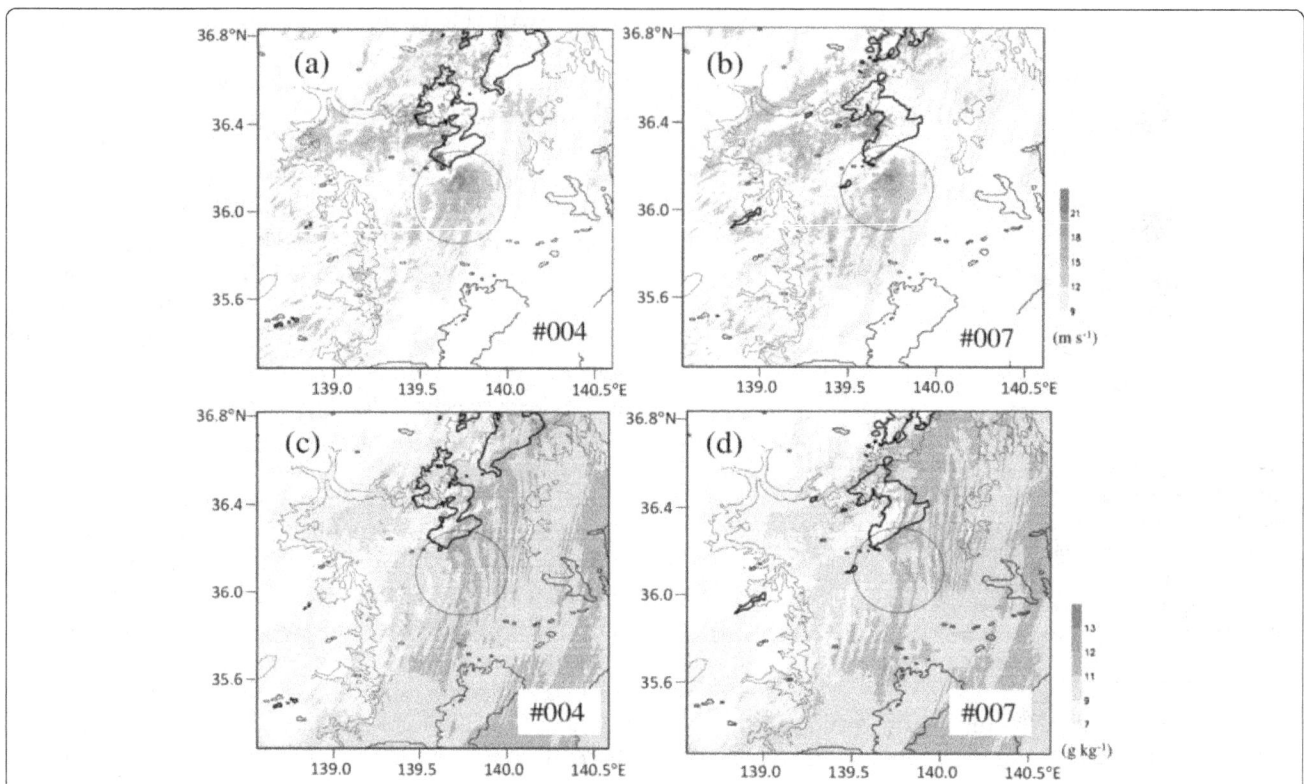

Fig. 12 Vertical shear of horizontal wind between the heights of 2.5 and 0.8 km at FT = 70 min (11:40 JST) in **a** #004 and **b** #007 and the horizontal distribution of the water vapor mixing ratio at the height of 0.8 km at FT = 70 min (11:40 JST) in **c** #004 and **d** #007. The *thick black contours* indicate a rain mixing ratio of 1.0 g kg⁻¹. The *circles* indicate the area in which airflows were supplied to the rainfall regions

the miss rate of outbreaks because they can reproduce low-probability phenomena. Comparison of the multiple possible scenarios showed that the duration of the southernmost tornado was related to the low-level water vapor and the vertical wind shear between heights of 0.8 and 2.5 km. Thus, the third merit is that the multiple possible scenarios of the ensemble forecast make it possible to determine the environmental factors favorable for the formation of severe weather phenomena such as intense vortices associated with tornadoes.

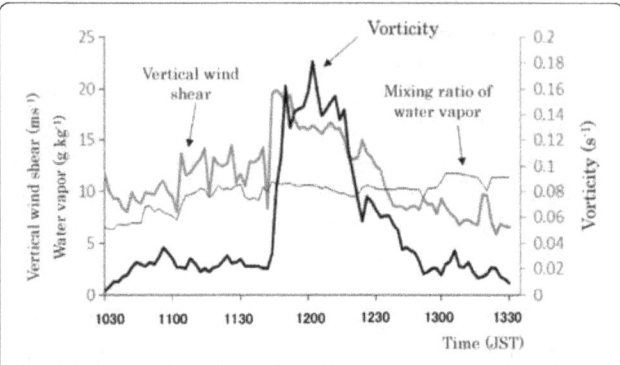

Fig. 13 Temporal variations of vorticity, water vapor mixing ratio, and vertical wind shear reproduced by #004 of NHM-350

Although we demonstrated the merits of ensemble forecasts in this paper, there are several areas where improvements are needed. First, the initial seeds used might influence the probability of outbreaks of intense vortices. In this study, the initial seeds of the outer LETKF were provided by JMA's mesoscale model analysis fields from before the occurrence of the observed tornadoes, not by adding random perturbations. In this experiment, differences among the initial seeds might have been too small to cause dispersion of intense vortices to be large.

The second area is the number of ensemble members. The merits of the ensemble forecasts of convective scale phenomena, such as the possible distribution of intense vortices, can be demonstrated with as few as 12 ensemble members. For quantitative discussion, however, more ensemble members would be preferable. In this study, we investigated the relationships between environmental parameters and intense vortices in each ensemble member, but we did not discuss the ensemble mean distributions in depth. As shown in Fig. 3, a region of intense southerly winds existed east of the rainfall regions. Because of the small ensemble size, the ensemble mean distribution, if it was produced by simple averaging of the outputs of the ensemble members, would show several separate regions of intense southerly winds;

as a result, the relationships between the southerly wind and the rainfall regions would be obscured. Therefore, the relationships in this study were extracted from the output of each ensemble member. It would also be possible to extract these relationships from composite distributions relative to the positions and timings of the occurrence of intense vortices; this approach was discussed by Yokota et al. (2014).

The third area is the assimilation data of the inner LETKF. High-resolution and high-frequency data are needed to reproduce convection cells in rainfall regions. Because such data were not used in this study, the intense vortices were not reproduced at the positions where they were actually observed. For more realistic reproduction of rainfall regions and tornadoes, high-resolution and high-frequency data should be used.

Abbreviations
AMeDAS: Automated Meteorological Data Acquisition System; JMA: Japan Meteorological Agency; JMANHM: Japan Meteorological Agency non-hydrostatic model; JST: Japan Standard Time; LETKF: Local Ensemble Transform Kalman Filter; MRI: Meteorological Research Institute; NHM-350: Non-Hydrostatic Model with a grid interval of 350 m.

Competing interests
The authors declare that they have no competing interests.

Authors' contributions
HS, TT, MK, and TM developed the nested LETKF system. HS performed the analyses obtained from the nested LETKF and downscale experiments. HS and SY developed the interpretations. All authors read and approved the final manuscript.

Acknowledgements
The authors express their gratitude to Drs. Kazuo Saito, Teruyuki Kato, and Wataru Mashiko of MRI, to Mr. Hiroshi Yamauchi of JMA, and to two anonymous reviewers for valuable comments. Mesoscale analysis data and conventional observation data from JMA were used to produce the initial seeds and boundary conditions of the outer LETKF and in the data assimilation. "Kasaneru 3D" and "Multi-screen monitoring tool" were used to produce Figs. 2, 4, and 5. The authors extend their gratitude to the Tokyo District Meteorological Observatory and the Observation Department and Numerical Prediction Division of JMA. This study was supported in part by "Strategic Programs for Innovative Research (SPIRE), Field 3 (ID: hp120282, hp130012, hp140220)" and "Tokyo Metropolitan Area Convection Study for Extreme Weather Resilient Cities (TOMACS)." Science section to which this research article belongs: 2) Atmospheric and hydrospheric sciences

Author details
[1]Meteorological Research Institute, Japan Meteorological Agency, 1-1 Nagamine, Tsukuba, Ibaraki 305-0052, Japan. [2]Japan Agency for Marine-Earth Science and Technology, 3173-25, Showa-machi, Kanazawa-ku, Yokohama, Kanagawa 236-0001, Japan. [3]RIKEN Advanced Institute for Computational Science, 7-1-26, Minatojima-minami-machi, Chuo-ku, Kobe, Hyogo 650-0047, Japan. [4]Meteorological College, 7-4-81 Asahicho, Kashiwa, Chiba 277-0852, Japan.

References
Clark AJ, Gao J, Marth PT, Smith T, Kain JS, Correia Jr J, et al. Tornado pathlength forecasts from 2010 to 2011 using ensemble updraft helicity. Monthly Weather Review. 2013;28:387–407.

Deardorff JW. Stratocumulus-capped mixed layers derived from a three-dimensional model. Bound-Layer Meteor. 1980;18:495–527.

Houze Jr RA. Cloud dynamics. San Diego: Academic Press; 1994.

Hunt BR, Kostelich EJ, Szunyogh I. Efficient data assimilation for spatiotemporal chaos: a local ensemble transform Kalman filter. Physica D: Nonlinear Phenomena. 2007. 230:112–126.

Japan Meteorological Agency, Meteorological Research Institute, Tokyo District Meteorological Observatory and Sendai District Meteorological Observatory. Tornadoes generated on 6th May 2012. 2012. http://www.jma.go.jp/jma/menu/tatsumaki-portal/tyousa-houkoku.pdf, Accessed 21 July 2014 (in Japanese).

Kunii M. Mesoscale data assimilation for a local severe rainfall event with the NHM-LETKF system. Weather and Forecasting. 2013;29:1093–105.

Meteorological Research Institute. Tornadoes generated near Tsukuba, Ibaraki on 6th May 2012. 2012. http://www.jma.go.jp/jma/press/1205/11c/120511tsukuba_tornado.pdf, Accessed 21 July 2014 (in Japanese).

Miyoshi T, Aranami K. Applying a four-dimensional local ensemble transform Kalman filter (4D-LETKF) to the JMA nonhydrostatic model (NHM). SOLA. 2006;2:128–31.

Saito K, Fujita T, Yamada Y, Ishida J, Kumagai Y, Aranami K, et al. The operational JMA nonhydrostatic mesoscale model. Monthly Weather Review. 2006;134: 1266–98.

Seko H, Miyoshi T, Shoji Y, Saito K. Data assimilation experiments of precipitable water vapour using the LETKF system: intense rainfall event over Japan 28 July 2008. Tellus A. 2011;63:402–14.

Seko H, Tsuyuki T, Saito K, Miyoshi T. Development of a two-way nested-LETKF system for cloud-resolving model. In: Park SK, Xu L, editors. Data assimilation for atmospheric, oceanic and hydrological applications, vol. 2. Springer: Heidelberg; 2013. p. 489–507.

Snook N, Xue M, Jung Y. Analysis of a tornadic mesoscale convective vortex based on ensemble Kalman filter assimilation of CASA X-band and WSR-88D Radar data. Monthly Weather Review. 2011;139:3446–68.

Snook N, Xue M, Jung Y. Multiscale EnKF assimilation of radar and conventional observations and ensemble forecasting for a tornadic mesoscale convective system. Monthly Weather Review. 2015;143:1035–57.

Yokota S, Kunii M, Seko H. Doppler radar radial wind assimilation for the tornado outbreak on May 6, 2012. CAS/JSC WGNE Research Activities in Atmospheric and Oceanic Modelling. 2014;1:29–1.30.

Yokota S, Kunii M, Seko H, Yamauchi H. Assimilation of rainwater estimated by the polarimetric radar for tornado outbreaks on 6 May 2012. CAS/JSC WGNE Research Activities in Atmospheric and Oceanic Modelling. 2015;1:27–1.28.

Yussouf N, Mansell ER, Wicker LJ, Wheatley DM, Stensrud DJ. The ensemble Kalman filter analyses and forecasts of the 8 May 2003 Oklahoma City tornadic supercell storm using single- and double-moment microphysics schemes. Monthly Weather Review. 2013;141:3388–412.

Permissions

List of Contributors

Toshitaka Tsuda
Research Institute for Sustainable Humanosphere (RISH), Kyoto University, Gokasho, Uji 611-0011, Japan

Eugenio Realini
Research Institute for Sustainable Humanosphere (RISH), Kyoto University, Gokasho, Uji 611-0011, Japan
Geomatics Research & Development (GReD) srl, Como, Italy

Kazutoshi Sato
Research Institute for Sustainable Humanosphere (RISH), Kyoto University, Gokasho, Uji 611-0011, Japan
Satellite Navigation Office, Satellite Applications Mission Directorate I, Japan Aerospace Exploration Agency (JAXA), Tsukuba, Japan

Susilo
Indonesian Geospatial Information Agency (BIG), Cibinong 16911, Indonesia

Timbul Manik
National Institute of Aeronautics and Space (LAPAN), Bandung 40173, Indonesia

Masaki Satoh and Tomoki Miyakawa
The University of Tokyo, 5-1-5 Kashiwanoha, Kashiwa, Chiba 277-8568, Japan

Hirofumi Tomita, Hisashi Yashiro and Tsuyoshi Yamaura
RIKEN Advanced Institute for Computational Science, 7-1-26, Minatojima-minami-machi, Chuo-ku, Kobe, Hyogo 650-0047, Japan

Yoshiyuki Kajikawa
RIKEN Advanced Institute for Computational Science, 7-1-26, Minatojima-minami-machi, Chuo-ku, Kobe, Hyogo 650-0047, Japan
Research Center for Urban Safety and Security, Kobe University, 1-1, Rokko-dai, Nada-ku, Kobe 657-8501, Japan

Yoshiaki Miyamoto
Rosenstiel School of Marine and Atmospheric Science, University of Miami, 4600 Rickenbacker Causeway, Miami, FL 33149, USA
RIKEN Advanced Institute for Computational Science, 7-1-26, Minatojima-minami-machi, Chuo-ku, Kobe, Hyogo 650-0047, Japan

Masuo Nakano, Chihiro Kodama, Akira T. Noda, Tomoe Nasuno and Yohei Yamada
Japan Agency for Marine-Earth Science and Technology, 3173-15, Showa-machi, Kanazawa-ku, Yokohama, Kanagawa 236-0001, Japan

Yoshiki Fukutomi
Institute for Space-Earth Environmental Research, Nagoya University, Furo-cho, Chikusa-ku, Nagoya 464-8601, Japan

Yasuhito Igarashi, Mizuo Kajino, Yuji Zaizen and Kouji Adachi
Meteorological Research Institute, 1-1 Nagamine, Tsukuba, Ibaraki 305-0052, Japan

Masao Mikami
Meteorological Research Institute, 1-1 Nagamine, Tsukuba, Ibaraki 305-0052, Japan
Japan Meteorological Business Support Center, Chiyoda-ku, Tokyo 101-0054, Japan

Yoshiyuki Kajikawa, Yoshiaki Miyamoto, Ryuji Yoshida, Tsuyoshi Yamaura and Hisashi Yashiro
RIKEN Advanced Institute for Computational Science, 7-1-26 Minatojima-minami-machi, Chuo-ku, Kobe, Hyogo 650-0047, Japan

Hirofumi Tomita
RIKEN Advanced Institute for Computational Science, 7-1-26 Minatojima-minami-machi, Chuo-ku, Kobe, Hyogo 650-0047, Japan
Japan Agency for Marine-Earth Science and Technology, 2-15, Natsushima-cho, Yokosuka, Kanagawa 237-0061, Japan

Yousuke Sato, Seiya Nishizawa, Hisashi Yashiro, Yoshiaki Miyamoto, Yoshiyuki Kajikawa and Hirofumi Tomita
RIKEN Advanced Institute for Computational Science, 7-1-26 Minatojima-Minami-machi, Chuo-ku, Kobe, Hyogo 650-0047, Japan

Shigenori Otsuka
RIKEN Advanced Institute for Computational Science, 7-1-26 Minatojima-minami-machi, Chuo-ku, 650-0047 Kobe, Japan

Megumi Takeshita
Weather Caster Network, 1-14-21 Ueno-Sakuragi, Taito-ku, 110-0002 Tokyo, Japan

Shigeo Yoden
Graduate School of Science, Kyoto University, Kitashirakawa-Oiwake-cho, Sakyo-ku, 606-8502 Kyoto, Japan

Takeshi Kinase
Graduate School of Science and Engineering, Ibaraki University, 2-1-1 Bunkyo, Mito, Ibaraki 310-8512, Japan
Hitachi Power Solutions Co., Ltd, 2-2-3 Saiwaicho, Hitachi, Ibaraki 317-0073, Japan

Kazuyuki Kita
College of Science, Ibaraki University, 2-1-1 Bunkyo, Mito, Ibaraki 310-8512, Japan

Yasuhito Igarashi and Kouji Adachi
Atmospheric Environment and Applied Meteorology Research Department, Meteorological Research Institute, 1-1 Nagamine, Tsukuba, Ibaraki 305-0052, Japan

Kazuhiko Ninomiya andAtsushi Shinohara
Graduate School of Science, Osaka University, 1-1, Machikaneyama, Toyonaka, Osaka 560-0043, Japan

Hiroshi Okochi and Hiroko Ogata
Faculty of Science and Engineering, Waseda University, 3-4-1 Okubo, Shinjuku, Tokyo 169-8555, Japan

Masahide Ishizuka
Faculty of Engineering, Kagawa University, 2217-20 Hayashi-cho, Takamatsu, Kagawa 761-0396, Japan

Sakae Toyoda and Keita Yamada
Department of Chemical Science and Engineering, School of Materials and Chemical Technology, Tokyo Institute of Technology, 4259 Nagatsuta, Yokohama, Kanagawa 226-8502, Japan

Naohiro Yoshida
Department of Chemical Science and Engineering, School of Materials and Chemical Technology, Tokyo Institute of Technology, 4259 Nagatsuta, Yokohama, Kanagawa 226-8502, Japan
Earth-Life Science Institute, Tokyo Institute of Technology, 2-12-1 Ookayama, Meguro, Tokyo152-8551, Japan

Yuji Zaizen
Forecast Research Department, Meteorological Research Institute, Japan Meteorological Agency, 1-1 Nagamine, Tsukuba, Ibaraki 305-0052, Japan

Masao Mikami
Japan Meteorological Business Support Center, 3-17 Nishikicho, Kanda, Chiyoda, Tokyo 101-0054, Japan

Hiroyuki Demizu
College of Engineering, Ibaraki University, 4-12-1 Narusawa, Hitachi, Ibaraki 316-8511, Japan

Yuichi Onda
Center for Research in Isotopes and Environmental Dynamics, University of Tsukuba, 1-1-1 Tennodai, Tsukuba, Ibaraki 305-8577, Japan

Hironobu Iwabuchi, Masanori Saito, Yuka Tokoro and Nurfiena Sagita Putri
Center for Atmospheric and Oceanic Studies, Graduate School of Science, Tohoku University, 6-3 Aoba, Aramakiaza, Aoba-ku, Sendai, Miyagi 980-8578, Japan

Miho Sekiguchi
The Graduate School of Marine Science and Technology, Tokyo University of Marine Science and Technology, 2-1-6 Etchujima, Koto-ku, Tokyo 135-8533, Japan

Shuichi Mori, Miki Hattori, Pei-Ming Wu and Masaki Katsumata
Japan Agency for Marine-Earth Science and Technology (JAMSTEC), 2-15 Natsushima-cho, Yokosuka 237-006, Japan

Jun-Ichi Hamada
Japan Agency for Marine-Earth Science and Technology (JAMSTEC), 2-15 Natsushima-cho, Yokosuka 237-006, Japan
Research Center for Climatology and Department of Geography, Tokyo Metropolitan University, 1-1 Minami-Osawa, Hachioji 192-0397, Japan

Nobuhiko Endo
National Institute for Agro-Environmental Sciences (NIAES), National Agriculture and Food Research Organization (NARO), 3-1-3 Kan-non-dai, Tsukuba 305-8604, Japan
Japan Agency for Marine-Earth Science and Technology (JAMSTEC), 2-15 Natsushima-cho, Yokosuka 237-006, Japan

Kimpei Ichiyanagi
Japan Agency for Marine-Earth Science and Technology (JAMSTEC), 2-15 Natsushima-cho, Yokosuka 237-006, Japan
Graduate School of Advanced Science and Technology, Kumamoto University, 2-39-1 Kurokami, Chuo-ku, Kumamoto 860-8555, Japan

Hiroyuki Hashiguchi
Research Institute for Sustainable Humanosphere (RISH), Kyoto University, Gokasho, Uji 611-0011, Japan

Reni Sulistyowati and Fadli Syamsudin
Center for Regional Resources Development Technology (PTPSW), Agency for the Assessment and Application of Technology (BPPT), Jl. Raya Puspiptek, South Tangerang 15314, Indonesia

Ardhi A. Arbain
Center for Regional Resources Development Technology (PTPSW), Agency for the Assessment and Application of Technology (BPPT), Jl. Raya Puspiptek, South Tangerang 15314, Indonesia
Atmosphere and Ocean Research Institute (AORI), The University of Tokyo, 5-1-5, Kashiwanoha, Kashiwa-shi, Chiba 277-8564, Japan

Sopia Lestari
Center for Regional Resources Development Technology (PTPSW), Agency for the Assessment and Application of Technology (BPPT), Jl. Raya Puspiptek, South Tangerang 15314, Indonesia
School of Earth Sciences, The University of Melbourne, Victoria 3010, Australia

Timbul Manik
Space Science Center, National Institute of Aeronautics and Space (LAPAN), Jl. Dr. Junjunan, Bandung 40173, Indonesia

Manabu D. Yamanaka
Research Institute for Humanity and Nature (RIHN), Kamigamo-motoyama, Kita-ku, Kyoto 603-8047, Japan
Kobe University, Rokkodai-cho, Nada-ku, Kobe 657-8501, Japan

Takamichi Iguchi
Earth System Science Interdisciplinary Center, University of Maryland, College Park, MD, USA
Code 612 NASA Goddard Space Flight Center, Greenbelt, MD, USA

In-Jin Choi
Korea Institute of Atmospheric Prediction Systems, Seoul, Republic of Korea

Yousuke Sato
RIKEN Advanced Institute for Computational Science, Kobe, Japan

Kentaroh Suzuki
Atmosphere and Ocean Research Institute, The University of Tokyo, Kashiwa, Japan

Teruyuki Nakajima
Earth Observing Research Center, Japan Aerospace Exploration Agency, Tsukuba, Japan

Masahiro Sawada, Tomoki Ohno and Hiroyasu Kubokawa
Atmosphere and Ocean Research Institute, The University of Tokyo, 5-1-5 Kashiwanoha, Kashiwa, Chiba 277-85648, Japan

Masaki Satoh
Atmosphere and Ocean Research Institute, The University of Tokyo, 5-1-5 Kashiwanoha, Kashiwa, Chiba 277-85648, Japan
Japan Agency for Marine-Earth Science and Technology, 3173-15, Showa-machi, Kanazawa-ku, Yokohama, Kanagawa 236-0001, Japan

Masayuki Hara, Chihiro Kodama, Tatsuya Seiki and Akira T Noda
Japan Agency for Marine-Earth Science and Technology, 3173-15, Showa-machi, Kanazawa-ku, Yokohama, Kanagawa 236-0001, Japan

Yohei Yamada
Japan Agency for Marine-Earth Science and Technology, 3173-15, Showa-machi, Kanazawa-ku, Yokohama, Kanagawa 236-0001, Japan
Atmosphere and Ocean Research Institute, The University of Tokyo, 5-1-5 Kashiwanoha, Kashiwa, Chiba 277-85648, Japan

Takemasa Miyoshi, Shin-ichi Iga and Hisashi Yashiro
RIKEN Advanced Institute for Computational Science, 7-1-26, Minatojima-minami-machi, Chuo-ku, Kobe, Hyogo 650-0047, Japan

Hirofumi Tomita
RIKEN Advanced Institute for Computational Science, 7-1-26, Minatojima-minami-machi, Chuo-ku, Kobe, Hyogo 650-0047, Japan
Japan Agency for Marine-Earth Science and Technology, 3173-15, Showa-machi, Kanazawa-ku, Yokohama, Kanagawa 236-0001, Japan

Hiroaki Miura
Department of Earth and Planetary Science, The University of Tokyo, 7-3-1 Hongo, Bunkyo-ku, Tokyo 113-0033, Japan
Japan Agency for Marine-Earth Science and Technology, 3173-15, Showa-machi, Kanazawa-ku, Yokohama, Kanagawa 236-0001, Japan
RIKEN Advanced Institute for Computational Science, 7-1-26, Minatojima-minami-machi, Chuo-ku, Kobe, Hyogo 650-0047, Japan

Daisuke Goto
National Institute for Environmental Studies, 16-2 Onogawa, Tsukuba, Ibaraki 305-8568, Japan

Yosuke Niwa
Meteorological Research Institute, 1-1 Nagamine, Tsukuba, Ibaraki 305-0052, Japan

Takashi Arakawa and Takahiro Inoue
Research Organization for Information Science and Technology, 2-32-3, Kitashinagawa, Shinagawku, Tokyo 140-0001, Japan
Japan Agency for Marine-Earth Science and Technology, 3173-15, Showa-machi, Kanazawa-ku, Yokohama, Kanagawa 236-0001, Japan

Daisuke Sugiyama
Center for Earth Information Science and Technology (CEIST), Japan Agency for Marine-Earth Science and Technology (JAMSTEC), 3173-25 Showa-machi, Kanazawa-ku, Yokohama, Kanagawa 236-0001, Japan

Daisuke Matsuoka
Center for Earth Information Science and Technology (CEIST), Japan Agency for Marine-Earth Science and Technology (JAMSTEC), 3173-25 Showa-machi, Kanazawa-ku, Yokohama, Kanagawa 236-0001, Japan
PRESTO, Japan Science and Technology Agency (JST), 4-1-8 Honcho, Kawaguchi, Saitama 332-0012, Japan

Masuo Nakano
Department of Seamless Environmental Prediction Research, Japan Agency for Marine-Earth Science and Technology (JAMSTEC), 3173-25 Showa-machi, Kanazawa-ku, Yokohama, Kanagawa 236-0001, Japan

Seiichi Uchida
Graduate School and Faculty of Information Science and Electrical Engineering, Kyushu University, 744 Motooka, Nishi-ku, Fukuoka 819-0395, Japan

Masaru Kunii1 and Sho Yokota
Meteorological Research Institute, Japan Meteorological Agency, 1-1 Nagamine, Tsukuba, Ibaraki 305-0052, Japan

Hiromu Seko
Meteorological Research Institute, Japan Meteorological Agency, 1-1 Nagamine, Tsukuba, Ibaraki 305-0052, Japan
Japan Agency for Marine-Earth Science and Technology, 3173-25, Showa-machi, Kanazawa-ku, Yokohama, Kanagawa 236-0001, Japan

Tadashi Tsuyuki
Meteorological College, 7-4-81 Asahicho, Kashiwa, Chiba 277-0852,Japan
Meteorological Research Institute, Japan Meteorological Agency, 1-1 Nagamine, Tsukuba, Ibaraki 305-0052, Japan

Index

CPSIA information can be obtained
at www.ICGtesting.com
Printed in the USA
BVHW062122290822
645775BV00003B/104